AF388660

Energy Transition and Climate Change in Decision-making Processes

Energy Transition and Climate Change in Decision-making Processes

Editors

Georgios Tsantopoulos
Evangelia Karasmanaki

MDPI • Basel • Beijing • Wuhan • Barcelona • Belgrade • Manchester • Tokyo • Cluj • Tianjin

Editors

Georgios Tsantopoulos
Department of Forestry and
Management of the Environment
and Natural Resources
Democritus University of Thrace
Orestiada
Greece

Evangelia Karasmanaki
Department of Forestry and
Management of the Environment
and Natural Resources
Democritus University of Thrace
Orestiada
Greece

Editorial Office
MDPI
St. Alban-Anlage 66
4052 Basel, Switzerland

This is a reprint of articles from the Special Issue published online in the open access journal *Sustainability* (ISSN 2071-1050) (available at: www.mdpi.com/journal/sustainability/special_issues/ Energy_Transition_Climate_Change_Decision_making_Processes).

For citation purposes, cite each article independently as indicated on the article page online and as indicated below:

LastName, A.A.; LastName, B.B.; LastName, C.C. Article Title. *Journal Name* **Year**, *Volume Number*, Page Range.

ISBN 978-3-0365-2707-9 (Hbk)
ISBN 978-3-0365-2706-2 (PDF)

Contents

About the Editors

Georgios Tsantopoulos

Georgios Tsantopoulos is a Professor at the Department of Forestry and Management of the Environment and Natural Resources, Democritus University of Thrace, Orestiada, Greece. He achieved his Ph.D. at the Department of Forestry and Natural Environment of the Aristotle University of Thessaloniki in 2000. His research interests are forest extension, environmental communication, and public relations. He has written more than 160 papers in international journals, chapters in books, proceedings of international and national conferences, etc. He has also participated in 23 research programs financed by the European Union and National Funds.

Evangelia Karasmanaki

Evangelia Karasmanaki is a Ph.D. Candidate in Environmental Communication and Policy at the Department of Forestry and Management of the Environment and Natural Resources, School of Agricultural and Forestry Sciences, Democritus University of Thrace, Orestiada, Greece. She has published papers in peer-reviewed journals and conference proceedings. Moreover, she has reviewed papers in various international scientific journals. Some of her research interests are environmental education, environmental communication, environmental policy, and environmental awareness.

sustainability

MDPI

Editorial

Energy Transition and Climate Change in Decision-Making Processes

Georgios Tsantopoulos and Evangelia Karasmanaki *

Department of Forestry and Management of the Environment and Natural Resources, Democritus University of Thrace, Pantazidou 193, 68200 Orestiada, Greece; tsantopo@fmenr.duth.gr
* Correspondence: evagkara2@fmenr.duth.gr

Citation: Tsantopoulos, G.; Karasmanaki, E. Energy Transition and Climate Change in Decision-Making Processes. *Sustainability* **2021**, *13*, 13404. https://doi.org/10.3390/su132313404

Received: 30 November 2021
Accepted: 2 December 2021
Published: 3 December 2021

Publisher's Note: MDPI stays neutral with regard to jurisdictional claims in published maps and institutional affiliations.

Humans have been using fossil fuels for centuries, and the development of fossil fuel technology reshaped society in lasting ways. From an economic perspective, countries with significant quantities of fossil fuel deposits are regarded as privileged. They have been able not only to avoid expensive fuel imports but also to develop cost-effective electricity sectors, which, in turn, brought economic development to rural areas lacking other avenues for economic development [1].

The problem, however, is that the combustion of fossil fuels in the energy, residential, and transport sectors is a major source of carbon dioxide emissions which trigger climate change, the most dangerous environmental problem that threatens the survival of all living beings on the planet [2,3]. As concerns about the environmental impact of fossil fuels are growing, the idea of producing clean, inexhaustible, and sustainable energy from alternative energy sources such as wind and sun is gaining attention around the world. Renewable energies generate comparably lower emissions, and even when estimating the emission rates of a renewable facility at all stages (construction, installation, operation, decommission), the emissions of renewables in comparison to fossil fuels are still notably lower. Moreover, the deployment of renewables can contribute to the diversification of energy supply, establish locally-produced power, help countries decrease their dependence on expensive fuel imports, create new job positions, etc.

The risk of climate change and the potential of renewable energy to mitigate emissions are reflected in the policy agenda. Over the past years, strong policies aiming at lowering the emissions of the fossil fuel-fired electricity system have been established. Indicatively, tight regulations are forcing businesses across sectors to reduce their environmental impact while many countries provide incentives to encourage investments in renewable energy sources [4].

That being said, the technology of renewable energy sources is not as developed as that of fossil fuels, with the latter having the momentum of two centuries of development. This means that it is much easier and more affordable to establish new fossil fuel projects rather than renewable ones [5]. Beside this obstacle, the deployment of renewable energy requires public support and acceptance because the public can influence actions aimed at realizing this transition. To avoid conflicts and bolster the efforts to deploy renewables, public attitudes towards climate change, energy transition, and energy must be examined [6,7].

The aim of this Special Issue is to publish research and review papers that will offer insights new into various aspects of the new energy landscape. Such insights may help policymakers reach decisions that will facilitate the shift to a low-carbon economy. The Special Issue includes papers focusing on various topics, including the effectiveness of energy policies, the technological performance of renewable systems, informatics tools and software, as well as public attitudes towards energy topics.

The Special Issue includes the following reviewed works:

- Efthimios Zervas, Leonidas Vatikiotis, Zoe Gareiou, Stella Manika, Ruth Herrero-Martin: 'Assessment of the Greek National Plan of Energy and Climate Change—Critical Remarks'
- Cyril Anak John, Lian See Tan, Jully Tan, Peck Loo Kiew, Azmi Mohd Shariff, and Hairul Nazirah Abdul Halim: 'Selection of Renewable Energy in Rural Area Via Life Cycle Assessment-Analytical Hierarchy Process (LCA-AHP): A Case Study of Tatau, Sarawak'
- Ali Rafiei Sefiddashti, Reza Shirmohammadi, and Fontina Petrakopoulou: 'Efficiency Enhancement of Gas Turbine Systems with Air Injection Driven by Natural Gas Turboexpanders'
- Hyung-Seok Jeong, Ju-Hee Kim, and Seung-Hoon Yoo: 'South Korean Public Acceptance of the Fuel Transition from Coal to Natural Gas in Power Generation'
- John Moodie, Carlos Tapia, Linnea Löfving, Nora Sánchez Gassen, and Elin Cedergren: 'Towards a Territorially Just Climate Transition—Assessing the Swedish EU Territorial Just Transition Plan Development Process'
- Konstantinos Ioannou and Dimitrios Myronidis: 'Automatic Detection of Photovoltaic Farms Using Satellite Imagery and Convolutional Neural Networks'
- Christina Diakaki and Evangelos Grigoroudis: 'Improving Energy Efficiency in Buildings Using an Interactive Mathematical Programming Approach'
- Dmitry A. Ruban, Natalia N. Yashalova, and Vladimir A. Ermolaev: 'Is Environment a Strategic Priority of the Leading Energy Companies? Evidence from Mission Statements'
- Piotr Bugajski, Agnieszka Operacz, Dariusz Młyński, Andrzej Wałęga, and Karolina Kurek: 'Optimizing Treatment of Cesspool Wastewater at an Activated Sludge Plant'
- Yongrok Choi, Hyoungsuk Lee, and Jahira Debbarma: 'Are Global Companies Better in Environmental Efficiency in India? Based on Metafrontier Malmquist CO_2 Performance'
- Ulf Liebe and Geesche M. Dobers: 'Measurement of Fairness Perceptions in Energy Transition Research: A Factorial Survey Approach'
- Dmitry A. Ruban, Natalia N. Yashalova, Olga A. Cherednichenko, and Natalya A. Dovgot'ko: 'Climate Change, Agriculture, and Energy Transition: What Do the Thirty Most-Cited Articles Tell Us?'
- Veronika Andrea, Stilianos Tampakis, Paraskevi Karanikola, and Maria Georgopoulou: 'The Citizens' Views on Adaptation to Bioclimatic Housing Design: Case Study from Greece'
- Christiana Koliouska and Zacharoula Andreopoulou: 'A Multicriteria Approach for Assessing the Impact of ICT on EU Sustainable Regional Policy'
- Katja Witte: 'Social Acceptance of Carbon Capture and Storage (CCS) from Industrial Applications'
- Nazanin Nasrollahi, Amir Ghosouri, Jamal Khodakarami, and Mohammad Taleghani: 'Heat-Mitigation Strategies to Improve Pedestrian Thermal Comfort in Urban Environments: A Review'.

Finally, we are grateful to many people for helping us complete this Special Issue successfully. It would be no exaggeration to say that nothing would have been possible without their contribution. First, we would like to thank all authors who responded to our invitation and submitted their works to our Special Issue. We would like to thank Julie Suo, our Special Issue's Managing Editor, for her continuous support, attentiveness, and kindness. Her role in the successful completion of this Special Issue has been critical. The support and conscientiousness of the Editorial Board of *Sustainability* must also be acknowledged. We would like to thank the academic editors responsible for each submission as well as the reviewers who have generously dedicated a part of their valuable time to reviewing papers for our Special Issue. The success of the journal relies on their meticulousness and competence. Having served as Guest Editors of *Sustainability*, we are certain that *Sustainability* will continue to publish high-quality research and review

Sustainability **2021**, *13*, 13404

papers that provide state-of-the-art knowledge about topics related to the environment, energy, and decision-making. We would also like to express the hope that this Special Issue will make a notable contribution to energy transition and will be used by policymakers in decision-making processes.

Conflicts of Interest: The authors declare no conflict of interest.

References

1. Grigoroudis, E.; Kouikoglou, V.S.; Phillis, Y.A.; Kanellos, F.D. Energy sustainability: A definition and assessment model. *Oper. Res.* **2021**, *21*, 1845–1885. [CrossRef]
2. Kaldellis, J.K.; Kapsali, M.; Katsanou, E. Renewable energy applications in Greece-What is the public attitude? *Energy Policy* **2012**, *42*, 37–48. [CrossRef]
3. Karasmanaki, E.; Tsantopoulos, G. Exploring future scientists' awareness about and attitudes towards renewable energy sources. *Energy Policy* **2019**, *131*, 111–119. [CrossRef]
4. Karasmanaki, E.; Ioannou, K.; Katsaounis, K.; Tsantopoulos, G. The attitude of the local community towards investments in lignite before transitioning to the post-lignite era: The case of Western Macedonia, Greece. *Resour. Policy* **2020**, *68*, 101781. [CrossRef]
5. Eleftheriadis, I.M.; Anagnostopoulou, E.G. Identifying barriers in the diffusion of renewable energy sources. *Energy Policy* **2015**, *80*, 153–164. [CrossRef]
6. Karytsas, S.; Theodoropoulou, H. Socioeconomic and demographic factors that influence publics' awareness on the different forms of renewable energy sources. *Renew. Energy* **2014**, *71*, 480–485. [CrossRef]
7. Dimitropoulos, A.; Kontoleon, A. Assessing the determinants of local acceptability of wind-farm investment: A choice experiment in the Greek Aegean Islands. *Energy Policy* **2009**, *37*, 1842–1854. [CrossRef]

 *sustainability*

Article

Improving Energy Efficiency in Buildings Using an Interactive Mathematical Programming Approach

Christina Diakaki [1,2,*] and **Evangelos Grigoroudis** [1]

1   School of Production Engineering and Management, Technical University of Crete, 73100 Chania, Greece; vangelis@ergasya.tuc.gr
2   School of Social Sciences, Hellenic Open University, 26335 Patra, Greece
*   Correspondence: cdiakaki@isc.tuc.gr

Abstract: Improving energy efficiency in buildings is a major priority and challenge worldwide. The employed measures vary in nature, and the decision analyst, who is typically the architect, the engineer, or the building expert that has undertaken the task to suggest energy efficient solutions, faces a complex decision problem comprising numerous decision variables and multiple, usually competitive objectives. The solution of such multi-objective problems typically involves some sort of objectives aggregation, which reflects the preferences of the involved final decision maker that is the building's user, occupant, and/or owner. The preferences elicitation, however, is a difficult task, and this paper aims to provide an interactive framework that will allow their consideration in a relatively easy manner. More specifically, a mathematical programming approach is proposed herein, which allows the elicitation and incorporation of the decision maker's preferences in the decision model via the assessment of his/her utility function with the assistance of the multicriteria decision aid method UTASTAR. To study the feasibility and efficiency of the proposed approach, the case of a simple building is examined as an application example. The study results suggest that the proposed approach is capable of helping the decision analyst to suggest energy measures that satisfy, as much as possible, the decision maker's preferences, without having to precisely prescribe them beforehand.

Keywords: buildings; energy efficiency; energy efficiency improvement; multi-objective optimization; preference disaggregation; preference elicitation; value system; utility function

Citation: Diakaki, C.; Grigoroudis, E. Improving Energy Efficiency in Buildings Using an Interactive Mathematical Programming Approach. *Sustainability* **2021**, *13*, 4436. https://doi.org/10.3390/su13084436

Academic Editor: Georgios Tsantopoulos

Received: 13 March 2021
Accepted: 14 April 2021
Published: 15 April 2021

Publisher's Note: MDPI stays neutral with regard to jurisdictional claims in published maps and institutional affiliations.

1. Introduction

Despite the long-lasting research and development in the particular field, the problem of improving energy efficiency in buildings still remains under investigation, according to recent reviews [1,2], due to its inherent complexity. The complexity of the problem stems from the involvement of several, typically competitive objectives (e.g., cost versus energy consumption) and the availability of numerous alternative measures (e.g., addition of insulation, change of color, use of cool coatings and renewables, etc.) [3], based on which, a final choice has to be made.

In practice, the specific measures to be adopted are typically suggested by the architect, the engineer, or the building expert, who undertakes the task to study the problem, thus playing the role of the decision analyst (DA). However, for any suggestions to be accepted by the final decision maker (DM), who may be the building's user, occupant, and/or owner, they have to satisfy his/her specific requirements and preferences. This further increases the complexity of the problem, and calls for solution approaches that allow the realistic comparative evaluation of all the available alternatives [4]. Such an approach has been proposed by Diakaki et al. [5], who developed a relevant multi-objective decision model based on the principles of mathematical programming.

The aforementioned model considers as objectives to minimize the primary energy consumption of a building and the released CO_2 emissions during operation, as well as the initial investment cost. The particular formulation lends itself for solution via mathematical optimization techniques [5], as well as evolutionary methods like genetic algorithms [6], should the problem complexity become such that a solution via analytic techniques is no longer feasible. Despite the reduced precision compared to the simulation models typically employed for the evaluation of alternative measures [2], the mathematical programming-based approach has been proved to allow for the realistic comparative evaluation of all the available, alternative measures [7], it has thus been adopted by several researchers in the field (see, e.g., [8–13]).

Irrespective of the particular technique that one may employ for the solution of a multi-objective mathematical programming problem, to reach a single, final solution, which will be satisfactory, thus acceptable by the corresponding DM, weights need to be assigned to the different objectives [2,14]. These weights reflect the relevant importance of each considered objective to the DM, and/or the trade-off that exists among them, due to their competitive nature. The determination of such weights is a difficult task, as it is very unlikely for a DM to be able to explicitly state his/her preferences and satisfaction levels for each and every considered objective. Thus, rather than trying to determine the criteria weights [14], the implicit elicitation and learning of the preferences and value system of the DM, and their incorporation and use in the decision making process, seems more convenient. The development of such an approach for the multi-objective decision problem of improving energy efficiency in buildings is the purpose of the work presented herein.

Specifically, it is the aim of this paper to present an approach, whereby the DA will manage to reach a single, final solution of maximum utility to the DM, as an outcome of an interactive process of individual inter-alternative preference modelling. To this end, the main principles and rationale of a two-phase, iterative procedure proposed by Siskos and Despotis [15] for similar decision problem settings have been adopted. The procedure starts with identifying an initial compromise solution for the energy efficiency improvement problem established in Diakaki et al. [5] (first phase), and then runs iteratively (second phase) as many times as necessary to extract the DM's aspiration levels for each objective, and estimate a respective utility function, which is used in order to reach a single, final solution, which is as close as possible to the DM's actual preferences and value system. Throughout the iterative procedure, interaction is offered at two levels: (i) interactive modification of the DM's satisfaction levels on the different pursued objectives; and (ii) interactive assessment of the DM's utility function via the development and use of the UTASTAR multicriteria decision aid model [16]. UTASTAR is a preference disaggregation approach, which aims at inferring the value or utility function(s) of a DM, given his/her expressed preferences over a reference set of alternatives.

Through the aforementioned interactive procedure, the proposed approach allows the DA to (a) develop the DM's overall utility function for the considered problem; (b) solve the problem by optimizing the developed utility function, rather than aggregating the individual objective functions of the considered problem via potentially arbitrary weights, like in the original multi-objective problem formulation in [5]; and (c) reach a single, final solution of maximum utility to the DM.

To study the feasibility and efficiency of the proposed approach, the case of a simple building is examined as an application example. The study results suggest that the proposed approach is capable of helping the DA to select and suggest energy measures that satisfy, as much as possible, the DM's spectrum of desires, without having to precisely prescribe them beforehand.

The rest of the paper is structured in three more sections. Section 2 introduces the proposed approach, while Section 3 presents the application example. Section 4 discusses the results and findings, and Section 5, finally, summarizes the conclusions of the study and highlights future research directions.

2. Materials and Methods

2.1. Overview

The approach proposed herein builds upon the mixed-integer, non-linear, multi-objective optimization problem developed by Diakaki et al. [5], which may be generally defined as follows:

$$\min[g_1(\mathbf{x}), g_2(\mathbf{x}), \ldots, g_n(\mathbf{x})]$$
$$\text{subject to } \mathbf{x} \in X,$$

(1)

where $\mathbf{x} = (x_1, x_2, \ldots, x_m)$ is the vector of m binary or continuous decision variables reflecting alternative choices (e.g., doors and windows types that can be used in the building, structure of multi-layer components such as walls, ceilings, and floors, materials to be used for their construction, and systems that can be used for heating, cooling and hot water supply), $X \subseteq R^m$ is the feasible region or decision space of the problem under study, which is implicitly dictated by a set of constraints concerning the decision variables and their intermediary relations; and $g_1(\mathbf{x})$, $g_2(\mathbf{x})$, $\ldots$ $g_n(\mathbf{x})$ are the values of n considered objectives. In the problem defined in [5], $n = 3$, as the considered objectives are the total annual primary energy consumption (MJ/year), the CO_2 emissions (kg CO_2/year) released to the environment by the operation of the heating, cooling, and/or hot water supply systems, and the investment cost for the construction or retrofit of the building envelope and the acquisition and installation of the aforementioned systems, respectively.

The decision model (1) is used herein in the following two-phase procedure:

1. In the first phase, each individual objective of (1) is first minimized and then maximized over the set of the feasible solutions, thus providing lower and upper bounds for the objectives. Given that $g_i(\mathbf{x})$ are minimized in (1), the lower bounds represent the ideal values of the objectives, and remain the same throughout the whole process, while the upper bounds represent the anti-ideal ones, and are refined during the second phase of the procedure. In addition, an initial efficient solution, i.e., a solution, which is not dominated by any other acceptable solution in the decision space is estimated that is closest to the ideal one with respect to the weighted Tchebycheff norm [15].

2. In the second phase, an iterative process is followed, which comprises three successive steps. The first step can be viewed as a learning process of the trade-offs among the objectives for the DM. Through questions and answers, this step refines the upper bounds, thus gradually reducing the feasible region of the decision problem. The second step can be viewed as a learning process of the DM's preferences. During this step, the DM is asked to rank, according to his/her preferences, a reference set of fictitious non-dominated decision profiles. This subjective ranking is then used by a UTASTAR model to generate the DM's utility function u, over the intervals created by the lower and upper limits of the objectives' values, and use them in transforming the decision problem (1) in the following:

$$\max u[\mathbf{g}(\mathbf{x})]$$
$$\text{subject to } \mathbf{x} \in X,$$

(2)

where, $\mathbf{g}(\mathbf{x}) = (g_1(\mathbf{x}), g_2(\mathbf{x}), \ldots, g_n(\mathbf{x}))$ is the vector of the values of the objectives of the initial Problem (1). The decision Problem (2) is solved in the third step of the process, the solution is presented to the DM, and the iterations restart until a solution is reached that will be sufficiently satisfactory for the DM, so that he/she will not wish to further improve it.

Figure 1 presents the flowchart of the aforementioned procedure, while the following subsection provides the details of its different phases.

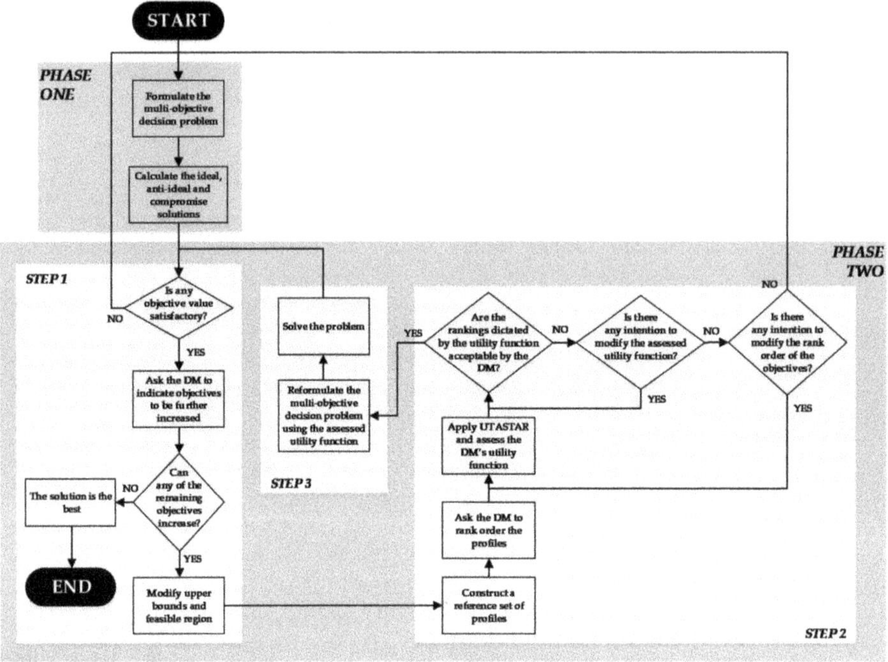

Figure 1. The flowchart of the interactive multi-objective mathematical programming approach.

2.2. The Interactive Mathematical Programming Approach

2.2.1. Phase One

As mentioned earlier, within the first phase of the proposed interactive mathematical programming approach, the individual objectives of decision Problem (1) are minimized and maximized to establish the initial lower and upper bounds of the objectives. More specifically, the lower bound l_i, which is the ideal solution for each objective i, with $i = 1, 2, \ldots, n$, is calculated as follows:

$$l_i = \min[g_i(\mathbf{x})]$$
$$\text{subject to } \mathbf{x} \in X, \tag{3}$$

while for the upper bound h_i, which is the anti-ideal solution, the following problem is solved:

$$h_i = \max[g_i(\mathbf{x})]$$
$$\text{subject to } \mathbf{x} \in X. \tag{4}$$

Then, an initial compromise solution is obtained via the solution of the following problem:

$$\min z$$
$$\text{subject to } \mathbf{x} \in X$$
$$z > m_i(g_i(\mathbf{x}) - l_i), i = 1, 2, \ldots, n \tag{5}$$
$$z > 0$$

where

$$m_i = d_i / \sum_i d_i, i = 1, 2, \ldots, n, \tag{6}$$

and

$$d_i = (h_i - l_i)/h_i, i = 1, 2, \ldots, n. \tag{7}$$

The solution of Problem (5) is the closest one to the ideal values of the objectives calculated via (3) in the sense of the weighted Tchebycheff norm.

2.2.2. Phase Two

The second phase of the proposed interactive mathematical programming approach is the iterative one, so let q be the number of iteration. Let also X^q be the feasible region, h_i^q the upper bound of objective i, and $\mathbf{g}^q$ the vector of the optimal values of the objectives reached in iteration q.

When entering for the first time in phase two, for the upper bounds and the objectives values, the following hold:

- The upper bound values h_i^0 are equal to the solutions of the corresponding problems (4), obtained in phase one;
- The optimal values of the objectives $\mathbf{g}^0$ are equal to the values obtained via the solution of the multi-objective Problem (5) in phase one.

In addition, $X^0 = X$ holds.

Given the above initial values, as well as the lower bounds l_i, i.e., the ideal solutions of the objectives, the three steps described below are successively executed.

Step 1

At the first step of phase two, interaction takes place in order to learn the trade-offs among the objectives for the DM. More specifically, the DM is asked to express his/her satisfaction with respect to the values of the objectives that have been reached so far, i.e., for the values in $\mathbf{g}^{q-1}$.

If the DM does not find any objective value satisfactory, the multi-objective decision problem has no satisfactory solution. In such case, the problem should be reviewed and revised, and the procedure should restart from phase one. However, if some values in $\mathbf{g}^{q-1}$ are satisfactory, the DM is asked to suggest the objectives, which he/she insists to further decrease, and the whole set of objectives G is split in the following two complementary sets:

- G_D: the subset of G, which comprises the objectives that the DM insists to decrease;
- $\overline{G}_D$: the complement of G_D in G.

Given the split of G in the two subsets, the DM is asked again to suggest, if there are any objectives in $\overline{G}_D$, which could be increased to make room for the desired further decrease of the objectives in G_D. If the response to this question is no, there is no room for further improvement, the procedure stops, and the solution reached so far is the best compromise solution to the examined problem. If, however, the response of the DM is yes, the upper bounds of the objectives are updated as follows:

$$h_i^q = \begin{cases} g_i^{q-1} & \text{if } g_i \in G_D \\ h_i^{q-1} & \text{if } g_i \in \overline{G}_D \end{cases} \tag{8}$$

For each $g_j \in G_D$, the following problem is solved:

$$\begin{aligned} &\min g_j(\mathbf{x}) \\ &\text{subject to } \mathbf{x} \in X \\ &\qquad g_i(\mathbf{x}) \leq h_i^q \; i = 1, 2, \ldots, n \text{ and } i \neq j, \end{aligned} \tag{9}$$

and the feasible region is finally reduced as shown below:

$$X^q = X^{q-1} \cap \left\{ \mathbf{x} \in R^m / g_i(x) \leq h_i^q, i = 1, 2, \ldots, n \right\}. \tag{10}$$

Step 2

The second step of phase two is also a learning process aiming at the DM's preferences elicitation. To this end, for an arbitrary chosen integer s, $s + 1$ reference alternative profiles a_k, with $k = 0, 1, \ldots, s$, are generated. Each profile comprises a coordinate a_{ik} for each objective i, which is calculated as follows:

$$a_{ik} = l_i + (k/s)\left(h_i^q - l_i\right) \tag{11}$$

Apparently, any other number of alternative profiles, as well as profiles generation procedure, can be adopted, as long as the generated profiles are representative of the trade-off among the objectives, and do not dominate each other. As their purpose is not to be offered to the DM as possible problem solutions, the generated profiles do not need to be efficient or feasible. They are just presented to the DM, who is asked to rank order them. The ranked set of alternative profiles is then used in the UTASTAR method [16] to assess the DM's utility function $u[\mathbf{g}(\mathbf{x})]$, as described in Appendix A.

Step 3

The utility function assessed in Step 2 is maximized in this last step over the feasible region. In other words, the Problem (2) is modified as follows, to take into account the reduction of the decision space according to (10):

$$\begin{aligned} &\max u[\mathbf{g}(\mathbf{x})] \\ &\text{subject to } \mathbf{x} \in X^q, \end{aligned} \tag{12}$$

and then solved.

The solution of Problem (12) marks the end of the current iteration, and the procedure restarts from step 1 with the new solution, feasible region, and upper objective bounds.

3. Application Example

3.1. Overview of the Decision Problem

To assess the feasibility and efficiency of the proposed approach in suggesting measures that satisfy the competitive objectives of the energy efficiency improvement problem in a way that is compatible with the preferences and value system of the DM, the case of a simple building is studied. The building, taken from the study of Diakaki et al. [5], assumes an envelope, which comprises a floor and ceiling area of 100 m^2, 2 walls of area 24 m^2, 2 walls of area 30 m^2, and a door and window area both of 6 m^2.

The decisions regarding the considered building concern appropriate choices for:

- The type of the building's door and window among the alternatives of Tables 1 and 2, respectively;
- the structure of the building's walls, ceiling, and floor among the alternatives of Tables 3–5, respectively;
- the addition or not, in the building's walls, ceiling, and floor, of an insulation layer of maximum permissible thickness 0.10 m and material chosen among the alternatives of Table 6;
- the space heating system among the alternatives of Tables 7–9;
- the space cooling system among the alternatives of Table 8;
- the hot water supply system(s) among the alternatives of Tables 9 and 10; and
- the addition or not of a solar collector system among the alternatives of Table 11.

Table 1. Alternative door types.

Type	Thermal Transmittance (W/m^2 °C)	Cost (€/m^2)
1. Hollow-core flush door	2.7	800
2. Solid-core flush door with single glazing (17% glass)	2.1	1000

Table 2. Alternative window types.

Type	Subtype	Thermal Transmittance (W/m^2 °C)	Effective Total Solar Energy Transmittance (%)	Cost (€/m^2)
1. Single glazing	1. Typical glazing	5.0	80	40
2. Double glazing	1. 4-20-4, uncoated, air filled	2.6	72	55
	2. 4-12-4, coated, argon filled	1.6	76	65

Table 3. Alternative wall structures.

Structure	Layer	Material	Thickness (m)	Thermal Conductivity (W/m °C)	Cost (€/m^3)
1	1	Plaster	0.025	0.87	10
	2	Brick (complex)	0.150	0.72	23
	3	Plaster	0.025	0.87	10
2	1	Plaster	0.025	0.87	10
	2	Brick (complex)	0.060	0.72	6.2
	3	Brick (complex)	0.060	0.72	6.2
	4	Plaster	0.025	0.87	10

Table 4. Alternative ceiling structures.

Structure	Layer	Material	Thickness (m)	Thermal Conductivity (W/m °C)	Cost (€/m^3)
1	1	Tiles	0.02	1.00	55
	2	Concrete	0.15	0.72	55
2	1	Tiles	0.02	1.00	55
	2	Wood	0.03	0.17	70

Table 5. Alternative floor structures.

Structure	Layer	Material	Thickness (m)	Thermal Conductivity (W/m °C)	Cost (€/m^3)
1	1	Tiles	0.01	1.00	55
	2	Concrete	0.15	0.72	55
2	1	Wood	0.02	0.17	85
	2	Concrete	0.15	0.72	55

Table 6. Alternative insulation materials.

Material	Thermal Conductivity (W/m °C)	Cost (€/m^3)
1. Polystyrene	0.036	200
2. Mineral fiber	0.042	180
3. Plastic fiber	0.020	300

Table 7. Alternative heating-only systems.

Category	Type	Generation Efficiency (%)	Cost (€)
Electrical systems			
1. Resistance-based	1. Dry core storage boiler type 1	100	5000
	2. Dry core storage boiler type 2	85	4200
Non-electrical systems			
1. Oil-based	1. Condensing	83	5300
	2. Standard oil boiler	62	4700
2. Natural-gas based	1. Condensing	85	5800
	2. Floor mounted boiler	55	4500

Table 8. Alternative heating–cooling systems.

Category	Type	Generation Efficiency (%)	Cost (€)
1. Water cooled electric	1. <12,000 BTU	200	500
	2. <18,000 BTU	230	800
	3. <24,000 BTU	250	1200

Table 9. Alternative heating–hot water supply systems.

Category	Type	Generation Efficiency (%)	Cost (€)
Electrical systems			
1. Resistance-based	1. Electric CPSU	100	7200
	2. Water storage boiler	85	5800
Non-electrical systems			
1. Oil-based	1. Condensing combi	81	6200
	2. Combi	70	5800
2. Natural-gas based	1. Condensing combi	84	7200
	2. Combi	65	5700

Table 10. Alternative hot water supply-only systems.

Category	Type	Generation Efficiency (%)	Cost (€)
Electrical systems			
1. Resistance-based	1. Electric immersion	100	1200
	2. Electric instantaneous at point of use	85	1000
Non-electrical systems			
1. Oil-based	1. Oil boiler/circulator	80	1000
	2. Oil single burner	60	800
2. Natural-gas based	1. Circulator built into a gas warm air system type 1	73	850
	2. Circulator built into a gas warm air system type 2	60	650

Table 11. Alternative solar collector systems.

Category	Type	Generation Efficiency (%)	Cost (€)
1. Flat collector	1. Type 1	90	900
	2. Type 2	80	600
2. Vacuum hear pipe CPC collector	1. Type 1	72	780
	2. Type 2	67	500

The values of the thermal and solar transmittance, and the thermal conductivity of construction materials and components in Tables 1–6 have been taken from the ASHRAE database [17], while the cost values in all the aforementioned tables were obtained through a short, unofficial market survey that took place for the needs of the study described in [5].

The application of the multi-objective decision modelling approach to the particular decision problem leads to a mathematical model of the form (1), which includes 18 continuous and 57 binary variables. The model, which is summarized in Appendix B, aims at determining measures that minimize the following three objectives:

- The primary energy consumption $g_1(\mathbf{x})$;
- the release of CO_2 emissions $g_2(\mathbf{x})$; and
- the initial investment cost $g_3(\mathbf{x})$.

These objectives are competitive, since, typically, the cost-efficient solutions are less environmentally friendly and vice versa. Thus, the search for a globally optimal solution is infeasible, and the DA has to search for a feasible solution, which will comply as much as possible with the DM's preferences and value system. To assist the DA in this search, the multi-phase iterative procedure described in Section 2 is applied.

3.2. Application of the Interactive Mathematical Programming Approach
3.2.1. Phase One

In the first phase of the proposed approach, the individual objectives of the examined decision problem are minimized and maximized, according to (3) and (4), respectively, in order to establish the ideal and anti-ideal solutions of the problem. In addition, an initial compromise solution is identified via the solution of Problem (5).

Table 12 summarizes the outcomes of this phase. The outcomes clearly demonstrate that the choices made depend on the pursued objective(s). For example, when the objective is solely to minimize the primary energy consumption, the most energy efficient choices are made in contrast to the choices made when aiming solely at the reduction of the initial investment cost. In this latter case, the cheapest choices are made, which are the worst from the energy efficiency perspective. These two objectives are clearly competitive to each other, but also to the emissions objective. The release of CO_2 emissions does not depend solely on the generation efficiency of the heating, cooling, and hot water supply systems, but also on the utilized fuel. Thus, some energy efficient choices are no longer efficient when emissions come into the picture.

Table 13 summarizes and highlights the basic information about the problem at hand, which has been generated by the proposed approach in phase one. More specifically, the table comprises the ideal and anti-ideal objective values, the initial upper bound for each objective, the initial compromise solution, as well as the rate of closeness of the objectives to their ideal values, being calculated as follows:

$$\text{Rate of closeness to the ideal solution} = 100\frac{g_i^q - l_i}{h_i - l_i} \tag{13}$$

with q the number of iteration; for phase one, $q = 0$ holds. Apparently, the lower the value of the rate, the better.

Table 13 makes clear that the initial compromise solution comprises choices that lead the objectives of primary energy consumption and release of CO_2 emissions very close to their ideal solutions (rates of closeness are 0.85% and 1.59%, respectively). The initial investment cost, on the other hand, is not similarly close to its ideal value (rate of closeness is 38.80%), and this may cause dissatisfaction to the DM. For this reason, the second phase of the proposed approach is activated, to examine the satisfaction level of the DM and refine, if necessary, the problem solution.

Table 12. Summary of phase one outcomes.

Decisions and Criteria	Type of Solution						
	Minimize			Maximize			Compromise
	$g_1(x)$	$g_2(x)$	$g_3(x)$	$g_1(x)$	$g_2(x)$	$g_3(x)$	
Door type	2	2	1	1	1	2	1
Window type/subtype	2/2	2/2	1/1	1/1	1/1	2/2	2/2
Wall structure	1	1	2	2	2	1	2
Wall insulation thickness (m)	0.10	0.10	0.00	0.00	0.00	0.10	0.07
Wall insulation material	3	3	-	-	-	3	3
Ceiling structure	1	1	2	2	2	1	2
Ceiling insulation thickness (m)	0.10	0.10	0.00	0.00	0.00	0.10	0.07
Ceiling insulation material	3	3	-	-	-	3	3
Floor structure	2	2	1	1	1	2	1
Floor insulation thickness (m)	0.10	0.10	0.00	0.00	0.00	0.10	0.07
Floor insulation material	3	3	-	-	-	3	3
Heating system type	EHC	NEH	EHC	EHW	EHW	EHW	EHC
Heating system category/type	1/3	2/1	1/1	1/2	1/2	1/1	1/3
Cooling system type	-	EC	-	EC	EC	EC	-
Cooling system category/type	-	1/3	-	1/1	1/1	1/3	-
Hot water supply system type	NEW	NEW	NEW	-	-	-	NEW
Hot water supply system category/type	1/1	2/1	2/2	-	1/2	-	2/2
Solar collector category/type	1/1	1/1	-	-	-	1/1	2/2
g_1: Primary energy consumption (MJ/year)	13,593	13,970	321,276	722,123	722,123	32,475	19,582
g_2: Release of CO_2 emissions (kg CO_2/year)	1400	810	32,758	74,559	74,559	3353	1986
g_3: Initial investment cost (€)	21,987	27,637	7524	12,674	12,674	28,187	15,540

EHC: electrical system that will be used for both heating and cooling (see Table 8); EHW: electrical system that will be used for both heating and hot water supply (see Table 9); EC: electrical system that will be used only for cooling (see Table 8); NEH: electrical system that will be used only for heating (see Table 7); NEW: non-electrical system that will be used only for hot water supply (see Table 10).

Table 13. Basic information generated in phase one (iteration $q = 0$).

Information	Primary Energy Consumption (MJ/Year) $i = 1$	Release of CO_2 Emissions (kg CO_2/Year $i = 2$	Initial Investment Cost (€) $i = 3$
Ideal solution (lower bound) l_i	13,593	810	7524
Initial compromise solution g_i^0	19,582	1986	15,540
Anti-ideal solution h_i	722,123	74,559	28,187
Initial upper bound h_i^0	722,123	74,559	28,187
Rate of closeness to the ideal solution	0.85%	1.59%	38.80%

3.2.2. Phase Two-Iteration 1-Step 1

Entering in phase two, the basic information of Table 13 is presented to the DM. Assuming that he/she is satisfied by the performance on objectives 1 and 2, but asks for an improvement on objective 3, i.e., a further cost reduction, even at the expense of the other two objectives, the following sets are formed:

- $G = \{\text{objective 1, objective 2, objective 3}\}$;
- $G_R = \{\text{objective 3}\}$;
- $\overline{G}_R = \{\text{objective 1, objective 2}\}$;

and the upper bound of objective 3 is updated as follows:

$$h_3^1 = h_3^0 = 15540. \tag{14}$$

Being members of set $\overline{G}_R$, the upper bounds of the other two objectives remain equal to their initial values, i.e.:

$$h_1^1 = h_1^0 = 722123,$$
$$h_2^1 = h_2^0 = 74559. \tag{15}$$

Then, the Problem (9) is solved for the third objective, which is the only member of set G_R:

$$\min g_3(\mathbf{x})$$
$$\text{subject to } \mathbf{x} \in X$$
$$g_1(\mathbf{x}) \leq h_1^1$$
$$g_2(\mathbf{x}) \leq h_2^1 \tag{16}$$

and the feasible region of the decision problem is reduced as follows:

$$X^1 = X^0 \cap \left\{ \mathbf{x} \in R^{75} / g_i(x) \leq h_i^1, i = 1, 2, 3 \right\}, \tag{17}$$

with X^0 being the decision space X of the initial problem.

3.2.3. Phase Two-Iteration 1-Step 2

On the basis of information from step 1 and assuming $s = 9$, 10 alternative profiles a_k, $k = 0, 1, \ldots, 9$, are generated, according to Equation (11), and presented to the DM in order to rank order them. Table 14 presents the profiles of these alternatives for each objective, along with their assumed ranking r, $r = 1, 2, \ldots, 10$.

Table 14. Reference set of alternatives and DM's ranking.

Profile	Primary Energy Consumption (MJ/Year)	Release of CO$_2$ Emissions (kg CO$_2$/Year)	Initial Investment Cost (€)	DM's Ranking
a_0	13,593	810	15,540	3
a_1	92,319	9004	14,650	2
a_2	171,044	17,199	13,759	1
a_3	249,770	25,393	12,868	4
a_4	328,495	33,587	11,978	5
a_5	407,221	41,782	11,087	6
a_6	485,947	49,976	10,196	7
a_7	564,672	58,171	9306	8
a_8	643,398	66,365	8415	9
a_9	722,123	74,559	7524	10

The information of Table 14 is then used in UTASTAR, leading to the marginal utility functions graphically displayed in Figure 2, which define the global utility of the DM via the following additive function:

$$u[\mathbf{g}(\mathbf{x})] = u_1(g_1(\mathbf{x})) + u_2(g_2(\mathbf{x})) + u_3(g_3(\mathbf{x})), \tag{18}$$

or the equivalent:

$$u[\mathbf{g}(\mathbf{x})] = 0.300 u'_1(g_1(\mathbf{x})) + 0.525 u'_2(g_2(\mathbf{x})) + 0.175 u'_3(g_3(\mathbf{x})), \tag{19}$$

where u'_i, $i = 1, 2, 3$, are the normalized, in the range $[0, 1]$, values of the marginal utilities u_i, graphically displayed in Figure 3.

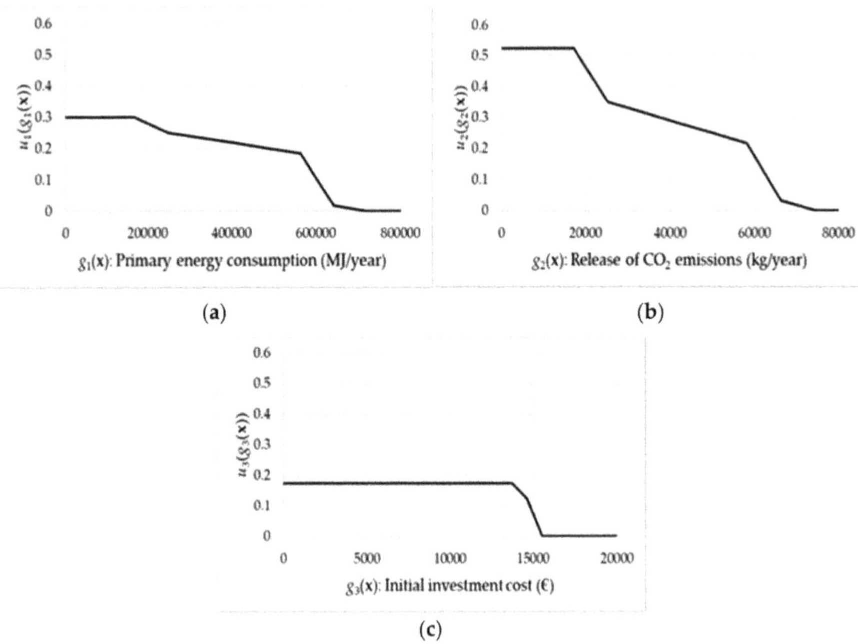

Figure 2. Marginal utility functions of: (**a**) Primary energy consumption; (**b**) release of CO_2 emissions; (**c**) initial investment cost.

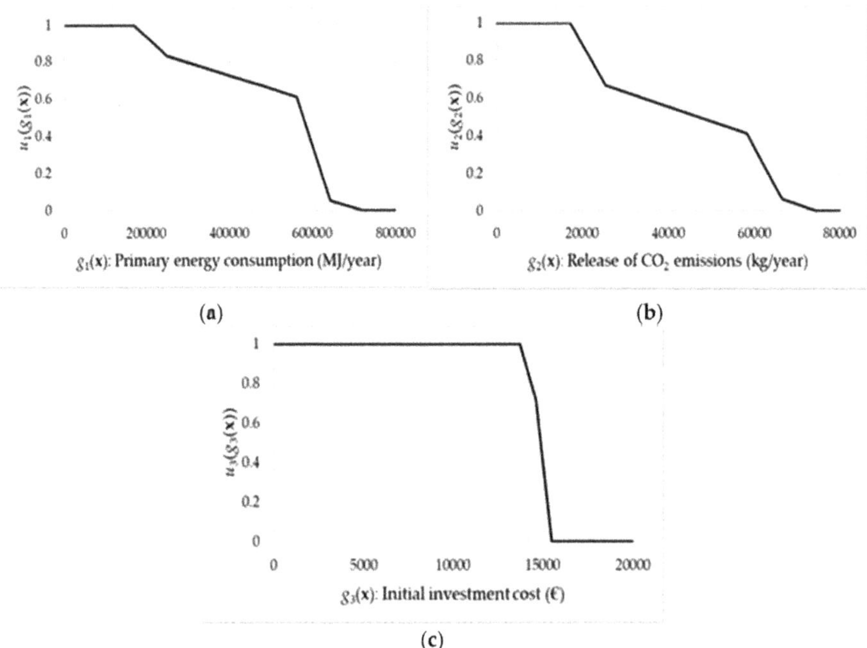

Figure 3. Normalized marginal utility functions of: (**a**) Primary energy consumption; (**b**) release of CO_2 emissions; (**c**) initial investment cost.

3.2.4. Phase Two-Iteration 1-Step 3

In this last step of phase two, the utility Function (18) or (19) is maximized over the decreased feasible solution space X^1, defined in (17). More specifically, the following problem:

$$\max u[\mathbf{g}(\mathbf{x})] = u_1(g_1(\mathbf{x})) + u_2(g_2(\mathbf{x})) + u_3(g_3(\mathbf{x}))$$
$$\text{subject to } \mathbf{x} \in X^1 \tag{20}$$

or its equivalent:

$$\max u[\mathbf{g}(\mathbf{x})] = 0.300u'_1(g_1(\mathbf{x})) + 0.525u'_2(g_2(\mathbf{x})) + 0.175u'_3(g_3(\mathbf{x}))$$
$$\text{subject to } \mathbf{x} \in X^1 \tag{21}$$

is solved.

The solution of any of the aforementioned problems generates the new compromise solution, displayed in Table 15, the current iteration is terminated, and a new iteration starts from step 1.

Table 15. Basic information generated in iteration $q = 1$ of phase two.

Information	Primary Energy Consumption (MJ/Year) $i = 1$	Release of CO$_2$ Emissions (kg CO$_2$/Year) $i = 2$	Initial Investment Cost (€) $i = 3$
Ideal solution (lower limit) l_i	13,593	810	7524
New compromise solution g_i^1	166,640	17,199	11,147
Anti-ideal solution h_i	722,123	74,559	28187
New upper bound h_i^1	722,123	74,559	15,540
Rate of closeness to the ideal solution	21.60%	22.22%	17.53%

3.2.5. Phase Two-Iteration 2-Step 1

The second iteration of phase two starts with the results of Table 15 being presented to the DM. Apparently, the cost objective has been reduced as desired, coming closer to its ideal value; a rate of closeness 17.53% has been achieved, which is also reduced compared to its previous value (38.80%). This improvement, however, has come at the expense of the other two objectives, the values of which, as well as their corresponding rates of closeness, present an increase.

If the consequences of the obtained solution are not satisfactory, the interaction with the DM should continue, like in the previous iteration, until reaching a satisfactory solution. Otherwise, the procedure stops here and the final choices made through this multi-phase procedure (see Table 16) are presented to the DM.

Table 16. Initial and final compromise solutions.

Decisions and Criteria	Initial Compromise Solution	Final Compromise Solution
Door type	1	1
Window type/subtype	2/2	2/1
Wall structure	2	1
Wall insulation thickness (m)	0.07	0
Wall insulation material	3	-
Ceiling structure	2	2
Ceiling insulation thickness (m)	0.07	0.01
Ceiling insulation material	3	1
Floor structure	1	1
Floor insulation thickness (m)	0.07	0.07
Floor insulation material	3	1
Heating system type	EHC	EHC
Heating system category/type	1/3	1/1

Table 16. *Cont.*

Decisions and Criteria	Initial Compromise Solution	Final Compromise Solution
Cooling system type	-	-
Cooling system category/type	-	-
Hot water supply system type	NEW	NEW
Hot water supply system category/type	2/2	2/2
Solar collector category/type	2/2	1/1
g_1: Primary energy consumption (MJ/year)	19,582	166,640
g_2: Release of CO_2 emissions (kg CO_2/year)	1986	17,199
g_3: Initial investment cost (€)	15,540	11,147

EHC: electrical system that will be used for both heating and cooling (see Table 8); NEW: non-electrical system that will be used only for hot water supply (see Table 10).

4. Discussion

The previous two sections presented an interactive mathematical programming approach to the problem of improving energy efficiency in buildings, and demonstrated its use via an example case study. The problem is difficult to solve as it involves multiple, competitive objectives, and a large number of decision variables, given the large number of available, alternative measures, which can be adopted in this respect. In addition, the solution of the problem requires the DM to express his/her preferences to the considered objectives, a fact that further increases the problem's complexity.

The approach proposed herein exploits the mathematical programming model proposed by Diakaki et al. [5] and the UTASTAR value elicitation method proposed by Siskos and Yannacopoulos [16] under an interactive decision framework, which has been developed following the rationale and principles of the decision-oriented method for multi-objective linear programming problems proposed by Siskos and Despotis [15]. The proposed framework assists the decision making procedure so that decisions are made, which comply with the value system of the DM, without the need to prescribe it beforehand.

The proposed approach can be also adopted in other decision settings within, but also beyond, the field of energy and environment. A similar approach, for example, lies on the basis of ADELAIS, an interactive computer program developed to support the search for a satisfactory solution in multi-objective linear programming problems, which has been used as a tool for the selection of stock portfolios [18]. In contrast, however, to both the initial conception in [15] and the ADELAIS program, the decision framework developed herein concerns a mixed-integer nonlinear mathematical programming problem, which aims at minimizing rather than maximizing the considered multiple objectives. This means that the overall framework can be adopted to any possible decision settings, should adequate care be taken to consider any potential particularities; e.g., in a case where objectives with a positive preference direction (e.g., comfort) should also be considered, to incorporate them, preserving at the same time the required cohesiveness of all considered objectives, their preference direction should be reversed by changing the sign of their corresponding functions. In addition, the mathematical programming formulation is quite flexible, allowing the incorporation of additional DM's objectives and preferences.

5. Conclusions

The study presented herein demonstrated the feasibility as well as the strengths of applying an interactive mathematical modelling approach to the problem of energy efficiency improvement. The application of such a systematic approach allows for the simultaneous consideration of all available combinations of alternative actions, the consideration of any logical, physical, technical, or other constraints that may apply, and the incorporation of the preferences and value system of the DM without having to explicitly prescribe them beforehand. In addition, the application of the proposed approach ensures that a single, final solution will be reached, which will be satisfactory, and thus acceptable by the corresponding DM.

The proposed approach addresses the problem of improving energy efficiency in buildings in a systematic way. Thus, it can provide the basis for the development of a corresponding decision support system (DSS), which could assist the respective DAs in their difficult task of identifying, among the large volume of available measures, those that will satisfy the needs, requirements, and preferences of the DMs. According to Li et al. [19], there is still plenty of room for the enhancement of the existing relevant toolkits and the development of new ones, and the proposed approach provides the ground in this direction.

Author Contributions: Conceptualization, C.D. and E.G.; investigation, C.D. and E.G.; writing—original draft, C.D. and E.G.; writing—review and editing, C.D. and E.G. All authors have read and agreed to the published version of the manuscript.

Funding: This research received no external funding.

Institutional Review Board Statement: Not applicable.

Informed Consent Statement: Not applicable.

Data Availability Statement: Not applicable.

Conflicts of Interest: The authors declare no conflict of interest.

Appendix A. The UTASTAR Method

The UTASTAR method proposed by Siskos and Yannacopoulos [16] is a variation of the UTA method, which aims at inferring a set of additive value functions from a given ranking on a reference set A_R of alternative actions $a \in A_R$.

UTASTAR assumes an unweighted additive value function of the form:

$$u(\mathbf{g}) = \sum_{i=1}^{n} u_i(g_i) \tag{A1}$$

under the normalization constraints:

$$\begin{cases} \sum_{i=1}^{n} u_i(g_i^*) = 1 \\ u_i(g_{i*}) = 0 \end{cases} \forall i = 1, 2, \ldots, n \tag{A2}$$

where n is the number of criteria; $\{g_1, g_2, \ldots, g_n\}$ is the set of criteria; $[g_i^*, g_{i*}]$ is the evaluation scale of criterion i, with $i = 1, 2, \ldots, n$ and g_{i*}, g_i^* the worst and best level of criterion i, respectively; u_i is the marginal value function of criterion i.

On the basis of the additive Model (A1) and (A2), the value of each alternative $a \in A_R$ may be expressed as:

$$u[\mathbf{g}(a)] = \sum_{i=1}^{n} u_i[g_i(a)] - \sigma^+(a) - \sigma^-(a), \tag{A3}$$

where σ^+, σ^- are the overestimation and underestimation errors, respectively.

In addition, linear interpolation is used in order to estimate the corresponding marginal value functions in a piecewise linear form. More specifically, for each criterion i, the interval $[g_i^*, g_{i*}]$ is first cut into $(\alpha_i - 1)$ equal intervals, where the points g_i^j are given by the following formula:

$$g_i^j = g_{i*} + \frac{j-1}{\alpha_i - 1}(g_i^* - g_{i*}) \ \forall j = 1, 2, \ldots, \alpha_i \tag{A4}$$

Then, the marginal value of each action $a \in A_R$, for which $g_i(a) \in \left[g_i^j, g_i^{j+1}\right]$ is approximated by the following linear interpolation:

$$u_i[g_i(a)] = u\left(g_i^j\right) + \frac{g_i(a) - g_i^j}{g_i^{j+1} - g_i^j}\left[u\left(g_i^{j+1}\right) - u\left(g_i^j\right)\right] \tag{A5}$$

Furthermore, the set of reference actions $A_R = \{a_1, a_2, \ldots, a_m\}$ is rearranged so that a_1 is the action with the best ranking, a_m is the action with the worst ranking, and for each pair of consecutive actions (a_k, a_{k+1}), either $a_k \succ a_{k+1}$ (preference) or $a_k \sim a_{k+1}$ (indifference) holds, thus if

$$\Delta(a_k, a_{k+1}) = u[\mathbf{g}(a_k)] - u[\mathbf{g}(a_{k+1})], \tag{A6}$$

one of the following holds:

$$\begin{cases} \Delta(a_k, a_{k+1}) \geq \delta \text{ if } a_k \succ a_{k+1} \\ \Delta(a_k, a_{k+1}) = 0 \text{ if } a_k \sim a_{k+1} \end{cases}, \tag{A7}$$

where δ is a small positive number, which, however, allows the equivalence discrimination of two actions, which are successive in the ranking.

A final important modification of the UTASTAR method concerns the monotonicity constraints of the criteria that are taken into account through the following transformations:

$$w_{ij} = u_i\left(g_i^{j+1}\right) - u_i\left(g_i^j\right) \geq 0 \forall i = 1, 2, \ldots, n \text{ and } j = 1, 2, \ldots, \alpha_i - 1, \tag{A8}$$

which allow the replacement of the monotonicity conditions for u_i with non-negative constraints for the variables w_{ij}.

Based on the above, given the ranking over a reference set A_R of alternative actions $a \in A_R$, the UTASTAR method can be implemented via the following four steps:

1. The global value of all reference actions $u[\mathbf{g}(a_k)]$, $k = 1, 2, \ldots, m$, is first expressed in terms of the marginal values $u_i(g_i)$, and then in terms of the variables w_{ij}, according to (A8), through the following relationships:

$$\begin{cases} u_i\left(g_i^1\right) = 0 \ \forall i = 1, 2, \ldots, n \\ u_i\left(g_i^j\right) = \sum\limits_{t=1}^{j-1} w_{ij} \ \forall i = 1, 2, \ldots, n \text{ and } \forall j = 2, 3, \ldots, \alpha_i - 1 \end{cases} \tag{A9}$$

2. For each pair of actions, which are consecutive in the given ranking, error terms are introduced using the following relationship:

$$\Delta(a_k, a_{k+1}) = u[\mathbf{g}(a_k)] - \sigma^+(a_k) + \sigma^-(a_k) - u[\mathbf{g}(a_{k+1})] + \sigma^+(a_{k+1}) - \sigma^-(a_{k+1}) \tag{A10}$$

3. The following linear programming problem is solved:

$$\min z = \sum\limits_{k=1}^{m} [\sigma^+(a_k) + \sigma^-(a_k)]$$

$$\text{subject to } \left. \begin{array}{l} \Delta(a_k, a_{k+1}) \geq \delta \text{ if } a_k \succ a_{k+1} \\ \Delta(a_k, a_{k+1}) = 0 \text{ if } a_k \sim a_{k+1} \end{array} \right\} \forall r$$

$$\sum\limits_{i=1}^{n} \sum\limits_{j=1}^{\alpha_i - 1} w_{ij} = 1$$

$$w_{ij} \geq 0, \sigma^+(a_k) \geq 0, \sigma^-(a_k) \geq 0 \ \forall i, j, k \tag{A11}$$

4. The existence of multiple or near optimal solutions of the Problem (A11) is examined (stability analysis), and the mean additive value function of those (near) optimal solutions is found, which maximize the objective functions:

$$u_i(g_i^*) = \sum_{j=1}^{\alpha_i - 1} w_{it} \ \forall i = 1, 2, \ldots, n \tag{A12}$$

on the polyhedron of the constraints of the Problem (A11), bounded by the following additional constraint:

$$\sum_{k=0}^{m} \left[\sigma^+(a_k) + \sigma^-(a_k) \right] \leq z * + \varepsilon, \tag{A13}$$

where $z*$ is the optimal value of Problem (A11) and ε is a very small positive number.

Appendix B. The Multi-Objective Decision Model of the Application Example

This Appendix provides an overview of the mathematical model of the considered multi-objective problem. The details of the model can be found in [5].

Appendix B.1. Parameters and Decision Variables

Table A1. Doors-related parameters and data.

Parameters	Description
DR	Number of building's doors; here $DR = 1$
dr	Index to DR; $dr = 1, \ldots, DR$
A_{dr}^{DOR}	Area of door dr (m^2); here $A_{dr}^{DOR} = 6$
b_{dr}^{DOR}	Temperature correction factor of construction part dr; here $b_{dr}^{DOR} = 1$
V	Number of available door types
v	Index to V; $v = 1, \ldots, V$
U_v^{DOR}	Thermal transmittance of door type (W/m^2 K)
C_v^{DOR}	Cost of door type v (€/m^2)

Table A2. Windows-related parameters and data.

Parameters	Description
WN	Number of building's windows; here $WN = 1$
wn	Index to WN; $wn = 1, \ldots, WN$
A_{wn}^{WIN}	Area of window wn (m^2); here $A_{wn}^{WIN} = 6$
b_{wn}^{WIN}	Temperature correction factor of construction part wn; here $b_{wn}^{WIN} = 1$
$F_{F,wn}$	Frame factor of window wn (%); here $F_{F,wn} = 0.7$
$F_{S,wn}$	Correction factor for shading of window wn (%); here $F_{S,wn} = 1$
$F_{CM,wn}$	Correction factor for movable devices of window wn (%); here $F_{CM,wn} = 1$
S	Number of available window types
s	Index to S; $s = 1, \ldots, S$
T_s	Number of available sub-types of window type s
t	Index to T_s; $t = 1, \ldots, T_s, \ \forall s = 1, \ldots, S$
g_{st}^{WIN}	Effective total solar energy transmittance of window sub-type t (%)
U_{st}^{WIN}	Thermal transmittance of window sub-type t (W/m^{2}K)
C_{st}^{WIN}	Cost of window sub-type t (€/m^2)

Table A3. Walls-related parameters and data.

Parameters	Description
WL	Number of walls; here $WL = 4$
wl	Index to WL; $wl = 1, \ldots, WL$
A_{wl}^{WAL}	Area of wall wl (m^2); here $A_1^{WAL} = A_2^{WAL} = 30$ and $A_3^{WAL} = A_4^{WAL} = 24$
b_{wl}^{WAL}	Temperature correction factor of construction part wl; here $b_{wl}^{WAL} = 1$
W	Number of available wall structures
w	Index to W; $w = 1, \ldots, W$
KWL_w	Number of known layers of structure w, regarding material and thickness
kwl	Index to KWL_w; $kwl = 1, \ldots, KWL_w, \ \forall w = 1, \ldots, W$
$d_{w,kwl}^{dWAL}$	Thickness of known layer kwl of wall structure w (m)
$kk_{w,kwl}^{mWAL}$	Thermal conductivity of material of known layer kwl of wall structure w (W/mK)
$CK_{w,kwl}^{mWAL}$	Cost of material of known layer kwl of wall structure w (€/m^3)
Y_w	Number of unknown layers of structure w; here $Y_w = 1$ (insulation layer)
y	Index to Y_w; $y = 1, \ldots, Y_w, \ \forall w = 1, \ldots, W$
$d_{\max,wy}^{WAL}$	Maximum permissible thickness of layer y of structure w (m); here $d_{\max,wy}^{WAL} = 0.1$
C_{wy}	Number of available materials for layer y of structure w
c	Index to C_{wy}; $c = 1, \ldots, C_{yw}, \ \forall (y = 1, \ldots, Y_w, \ \forall w = 1, \ldots, W)$
k_{wyc}^{mWAL}	Thermal conductivity of material c of unknown layer y of structure w (W/mK)
C_{wyc}^{mWAL}	Cost of material c of unknown layer y of structure w (€/m^3)

Table A4. Ceilings-related parameters and data.

Parameters	Description
CE	Number of ceilings; here $CE = 1$
ce	Index to CE; $ce = 1, \ldots, CE$
A_{ce}^{CEIL}	Area of ceiling ce (m^2); here $A_{ce}^{CEIL} = 100$
b_{ce}^{CEIL}	Temperature correction factor of construction part ce; here $b_{ce}^{CEIL} = 1$
D	Number of available ceiling structures
d	Index to D; $d = 1, \ldots, D$
KCL_d	Number of known layers of structure d, regarding material and thickness
kcl	Index to KCL_d; $kcl = 1, \ldots, KCL_d, \ \forall d = 1, \ldots, D$
$d_{d,kcl}^{dCEIL}$	Thickness of known layer kcl of structure d (m)
$kk_{d,kcl}^{mCEIL}$	Thermal conductivity of material of known layer kcl of structure d (W/mK)
$CK_{d,kcl}^{mCEIL}$	Cost of material of known layer kcl of structure d (€/m^3)
F_d	Number of unknown layers of structure d; here $F_d = 1$ (insulation layer)
f	Index to F_d; $f = 1, \ldots, F_d, \ \forall d = 1, \ldots, D$
$d_{\max,df}^{CEIL}$	Maximum permissible thickness of layer f of structure d (m); here $d_{\max,df}^{CEIL} = 0.1$
A_{df}	Number of available materials for layer f of structure d
a	Index to A_{df}; $a = 1, \ldots, A_{df}, \ \forall (f = 1, \ldots, F_d \ \forall d = 1, \ldots, D)$
k_{dfa}^{mCEIL}	Thermal conductivity of material a of unknown layer f of structure d (W/mK)
C_{dfa}^{mCEIL}	Cost of material a of unknown layer f of structure d (€/m^3)

Table A5. Floors-related parameters and data.

Parameters	Description
FL	Number of floors; here $fl = 1$
fl	Index to FL; $fl = 1, \ldots, FL$
A_{fl}^{FLO}	Area of floor fl (m^2); here $A_{fl}^{FLO} = 100$
b_{fl}^{FLO}	Temperature correction factor of construction part fl; here $b_{fl}^{FLO} = 1$
H	Number of available floor structures
h	Index to H; $h = 1, \ldots, H$
KFL_h	Number of known layers of structure h, regarding material and thickness
kfl	Index to KFL_h; $kfl = 1, \ldots, KFL_h, \ \forall h = 1, \ldots, H$

Table A5. *Cont.*

Parameters	Description
$d_{h,kfl}^{dFLO}$	Thickness of known layer *kfl* of structure *h* (m)
$kk_{h,kfl}^{mFLO}$	Thermal conductivity of material of known layer *kfl* of structure *h* (W/mK)
$CK_{h,kfl}^{mFLO}$	Cost of material of known layer *kfl* of structure *h* (€/m^3)
E_h	Number of unknown layers of structure *h*; here $E_h = 1$ (insulation layer)
e	Index to E_h; $e = 1, \ldots, E_h,\ \forall h = 1, \ldots, H$
$d_{\max,he}^{FLO}$	Maximum permissible thickness of layer *e* of structure *h* (m); here $d_{\max,he}^{FLO} = 0.1$
G_{he}	Number of available materials for layer *e* of structure *h*
g	Index to G_{he}; $g = 1, \ldots, G_{he},\ \forall(e = 1, \ldots, E_h\ \forall h = 1, \ldots, H)$
k_{heg}^{mFLO}	Thermal conductivity of material *g* of unknown layer *e* of structure *h* (W/mK)
C_{heg}^{mFLO}	Cost of material *g* of unknown layer *e* of structure *h* (€/m^3)

Table A6. Heating-only systems' parameters and data.

Parameters	Description
EHI	Number of available electrical heating systems' categories
ehi	Index to EHI; $ehi = 1, \ldots, EHI$
EHJ_{ehi}	Number of available systems of category *ehi*
ehj	Index to EHJ_{ehi}; $ehj = 1, \ldots, EHJ_{ehi},\ \forall ehi = 1, \ldots, EHI$
$e_{ehi,ehj}^{EH}$	Generation efficiency of system *ehj* of category *ehi* (%)
$CST_{ehi,ehj}^{EH}$	Installation cost of system *ehj* of category *ehi* (€)
$NEHI$	Number of available non-electrical heating systems' categories
$nehi$	Index to $NEHI$; $nehi = 1, \ldots, NEHI$
$NEHJ_{nehi}$	Number of available systems of category *nehi*
$nehj$	Index to $NEHJ_{nehi}$; $nehj = 1, \ldots, NEHJ_{nehi},\ \forall\forall nehi = 1, \ldots, NEHI$
$e_{nehi,nehj}^{NEH}$	Generation efficiency of system *nehj* of category *nehi* (%)
$CST_{nehi,nehj}^{NEH}$	Installation cost of system *nehj* of category *nehi* (€)
$FU_{nehi,nehj,fuel}^{NEH}$	Parameter; equals 1, if system *nehj* of category *nehi* uses fuel *fuel*, else equals 0

Table A7. Cooling-only systems' parameters and data.

Parameters	Description
ECI	Number of available electrical cooling systems categories
eci	Index to ECI; $eci = 1, \ldots, ECI$
ECJ_{eci}	Number of available systems of category *eci*
ecj	Index to ECJ_{eci}; $ecj = 1, \ldots, ECJ_{eci},\ \forall eci = 1, \ldots, ECI$
$e_{eci,ecj}^{C}$	Generation efficiency of system *ecj* of category *eci* (%)
$CST_{eci,ecj}^{C}$	Installation cost of system *ecj* of category *eci* (€)

Table A8. Domestic hot water (DHW) supply-only systems' parameters and data.

Parameters	Description
EWI	Number of available electrical DHW systems' categories
ewi	Index to EWI; $ewi = 1, \ldots, EWI$
EWJ_{ewi}	Number of available systems of category *ewi*
ewj	Index to EWJ_{ewi}; $ewj = 1, \ldots, EWJ_{ewi},\ \forall ewi = 1, \ldots, EWI$
$e_{ewi,ewj}^{EW}$	Generation efficiency of system *ewj* of category *ewi* (%)
$CST_{ewi,ewj}^{EW}$	Installation cost of system *ewj* of category *ewi* (€)
$NEWI$	Number of available non-electrical DHW systems' categories
$newi$	Index to $NEWI$; $newi = 1, \ldots, NEWI$
$NEWJ_{newi}$	Number of available systems of category *newi*
$newj$	Index to $NEWJ_{newi}$; $newj = 1, \ldots, NEWJ_{newi},\ \forall newi = 1, \ldots, NEWI$

Table A8. *Cont.*

Parameters	Description
$e^{NEW}_{newi,newj}$	Generation efficiency of system *newj* of category *newi* (%)
$CST^{NEW}_{newi,newj}$	Installation cost of system *newj* of category *newi* (€)
$FU^{NEW}_{newi,newj,fuel}$	Parameter; equals 1, if system *newj* of category *newi* uses fuel *fuel*, else equals 0

Table A9. Combined heating–cooling systems' parameters and data.

Parameters	Description
$EHCI$	Number of available combined electrical heating-cooling systems' categories
$ehci$	Index to $EhCI$; $ehci = 1, \ldots, EHCI$
$EHCJ_{ehci}$	Number of available systems of category *ejci*
$ehcj$	Index to $EHCJ_{ehci}$; $ehcj = 1, \ldots, EHCJ_{ehci}$, $\forall ehci = 1, \ldots, EhCI$
$e^{HC}_{ehci,ehcj}$	Generation efficiency of system *ehcj* of category *ehci* (%)
$CST^{HC}_{ehci,ehcj}$	Installation cost of system *ecj* of category *eci* (€)

Table A10. Combined heating–DHW systems' parameters and data.

Parameters	Description
$EHWI$	Number of available combined electrical heating-DHW systems' categories
$ehwi$	Index to $EHWI$; $ehwi = 1, \ldots, EHWI$
$EHWJ_{ehwi}$	Number of available systems of category *ehwi*
$ehwj$	Index to $EHWJ_{ehwi}$; $ehwj = 1, \ldots, EHWJ_{ehwi}$, $\forall ehwi = 1, \ldots, EHWI$
$e^{EHW}_{ehwi,ehwj}$	Generation efficiency of system *ehwj* of category *ehwi* (%)
$CST^{EHW}_{ehwi,ehwj}$	Installation cost of system *ehwj* of category *ehwi* (€)
$NEHWI$	Number of available non-electrical combined heating-DHW systems' categories
$nehwi$	Index to $NEHWI$; $nehwi = 1, \ldots, NEHWI$
$NEHWJ_{nehwi}$	Number of available systems of category *nehwi*
$nehwj$	Index to $NEHWJ_{nehwi}$; $nehwj = 1, \ldots, NEHWJ_{nehwi}$, $\forall nehwi = 1, \ldots, NEHWI$
$e^{NEHW}_{nehwi,nehwj}$	Generation efficiency of system *nehwj* of category *nehwi* (%)
$CST^{NEHW}_{nehwi,nehwj}$	Installation cost of system *nehwj* of category *nehwi* (€)
$FU^{NEHW}_{nehwi,nehwj,fuel}$	Parameter; equals 1, if system *nehwj* of category *nehwi* uses fuel *fuel*, else equals 0

Table A11. Solar collectors' parameters and data.

Parameters	Description
A_{SLC}	Area of solar collector (m^2); here $A_{SLC} = 2$
$F_{S,SLC}$	Correction factor for shading of solar collector (%); here $F_{S,SLC} = 1$
U	Number of available solar collectors systems' categories
u	Index to U; $u = 1, \ldots, U$
B_u	Number of available solar collectors systems of category *u*
b	Index to B_u; $b = 1, \ldots, B_u$, $\forall u = 1, \ldots, U$
e^{SLC}_{ub}	Generation efficiency of system *b* of category *u* (%)
CST^{SLC}_{ub}	Installation cost of system *b* of category *u* (€/m^2)

Table A12. Fuel and emissions-related parameters and data [1].

Parameters	Description
$FUEL$	Number of fuels available for heating and DHW; here $FUEL = 2$
$fuel$	Index to $FUEL$; $fuel = 1, \dots, FUEL$; here $fuel = 1$ is oil and $fuel = 2$ is gas
F_{fuel}	Conversion factor of fuel $fuel$ to CO_2 emissions (kg of CO_2/kg of fuel); here $F_1 = 3.142$ and $F_{fuel} = 2.715$
LHP_{fuel}	Conversion factor of fuel $fuel$ to energy (MJ/kg of fuel); here $LHP_1 = 42.912$ and $LHP_2 = 49.788$
$F_{station}$	Emissions factor of electricity producing station (kg of CO_2/MJ); here $F_{station} = 0.295$
n_{el}	Return rate of electricity producing stations; here $n_{el} = 0.35$

[1] Parameter values have been adopted from [5].

Table A13. Parameters and data describing weather conditions at the building's location [1].

Parameters	Description
n	Month index; $n = 1, \dots, 12$; 1 corresponds to January, 2 to February, etc.
T_n	Duration of month n (s)
$\theta_{E,n}$	Average external temperature at building's location in month n (°C)
$I_{SL,wn,n}$	Solar radiation on window wn in month n (MJ/m^2/month)
$I_{SL,SLC,n}$	Solar radiation on solar collector in month n (MJ/m^2/month)
ρ_{air}	Air density at building's location (kg/m^3)
c_{air}	Air heat at building's location (J/kg°C)
V_{air}	Air volume (m^3)

[1] It is assumed that the building is located in the wider area of Athens, Greece [5].

Table A14. Parameters and data describing comfort-related user preferences and foreseen operational conditions of the building.

Parameters	Description
θ_{IH}	Internal design temperature during heating season (°C); here $\theta_{IH} = 21$
θ_{IC}	Internal design temperature during cooling season (°C); here $\theta_{IC} = 26$
HS_n	Parameter; equals 1, if heating is required for month n, else equals 0; here $HS_n = 1$ for $n = 1, 2, 3, 4, 10, 11, 12$ and $HS_n = 0$ for $n = 5, 6, 7, 8, 9$
CS_n	Parameter; equals 1, if cooling is required for month n, else equals 0; here $CS_n = 1$ for $n = 6, 7, 8, 9$ and $CS_n = 0$ for $n = 1, 2, 3, 4, 5, 10, 11, 12$
WS_n	Parameter; equals 1, if hot water supply is required for month n, else equals 0; here $WS_n = 1 \; \forall n = 1, 2, \dots, 12$
Q_{AINHG}	Average monthly heat gains (W); here $Q_{AINHG} = 8400$ [1]
Q_{dhwu}	Average energy requirements for hot water use (MJ/month); here $Q_{dhwu} = 425$ [1]

[1] Rough estimates assuming 4 inhabitants in the building [5].

Table A15. Decision variables.

Variable	Description
x_v^{DOR}	Doors choice; equals 1, if type v is selected, else equals 0
x_{st}^{WIN}	Windows choice; equals 1, if subtype t of type s is selected, else equals 0
x_w^{WAL}	Wall structure choice; equals 1, if structure w is selected, else equals 0
x_{wyc}^{mWAL}	Wall material choice; equals 1, if material c is selected for layer y of wall structure w, else equals 0
x_{wy}^{dWAL}	Thickness of material added in layer y of wall structure w (m)
x_d^{CEIL}	Ceiling structure choice; equals 1, if structure d is selected, else equals 0
x_{dfa}^{mCEIL}	Ceiling material choice; equals 1, if material a is selected for layer f of ceiling structure d, else equals 0

Table A15. *Cont.*

Variable	Description
x_{df}^{dCEIL}	Thickness of material added in layer f of ceiling structure d (m)
x_h^{FLO}	Floor structure choice; equals 1, if structure h is selected, else equals 0
x_{heg}^{mFLO}	Floor material choice; equals 1, if material g is selected for layer e of floor structure h, else equals 0
x_{he}^{dFLO}	Thickness of material added in layer e of floor structure h (m)
$x_{ehi,ehj}^{EH}$	Electrical heating system choice; equals 1, if system ehj of category ehi is selected, else equals 0
$x_{nehi,nehj}^{NEH}$	Non-electrical heating system choice; equals 1, if system $nehj$ of category $nehi$ is selected, else equals 0
$x_{eci,ecj}^{EC}$	Electrical cooling system choice; equals 1, if system ecj of category eci is selected, else equals 0
$x_{ewi,ewj}^{EW}$	Electrical DHW system choice; equals 1, if system ewj of category ewi is selected, else equals 0
$x_{newi,newj}^{NEW}$	Non-electrical DHW system choice; equals 1, if system $newj$ of category $newi$ is selected, else equals 0
$x_{ehci,ehcj}^{EHC}$	Electrical heating–cooling system choice; equals 1, if system $ehcj$ of category $ehci$ is selected, else equals 0
$x_{ehwi,ehwj}^{EHW}$	Electrical heating–DHW system choice; equals 1, if system $ehwj$ of category $ehwi$ is selected, else equals 0
$x_{nehwi,nehwj}^{NEHW}$	Non-electrical heating–DHW system choice; equals 1, if system $nehwj$ of category $nehwi$ is selected, else equals 0
x_{ub}^{SLC}	Solar collector choice; equals 1 if system b of category u is selected, else equals 0
$\mathbf{x}$	Vector of all decision variables x

Appendix B.2. Multi-Objective Decision Model

Minimize

$$g_1(\mathbf{x}) = \frac{Q^{HD}SEH_{el}}{n_{el}} + \sum_{fuel=1}^{FUEL}\left(Q^{HD}SEH_{nel,fuel}\right) + \frac{Q^{CD}SEC_{el}}{n_{el}} + \frac{Q^{WD}SEW_{el}}{n_{el}} + \sum_{fuel=1}^{FUEL}\left(Q^{WD}SEW_{nel,fuel}\right)$$

$$g_2(\mathbf{x}) = \left(Q^{HD}SEH_{el} + Q^{CD}SEC_{el} + Q^{WD}SEW_{el}\right)F_{station} + \sum_{fuel=1}^{FUEL}\left(Q^{HD}SEH_{nel,fuel} + Q^{WD}SEW_{nel,fuel}\right)\frac{F_{fuel}}{LHP_{fuel}}$$

$$g_3(\mathbf{x}) = COST_{DOR} + COST_{WIN} + COST_{WAL} + COST_{CEIL} + COST_{FLO} + COST_{HS} + COST_{CS} + COST_{WS} + COST_{HCS} + COST_{HWS} + COST_{SLC}$$

Subject to

$$Q^{HD} = \sum_{n=1}^{12} Q_n^{HD}$$

$$Q^{CD} = \sum_{n=1}^{12} Q_n^{CD}$$

$$Q^{WD} = \sum_{n=1}^{12} \left(WS_n DQ_{dDHW,n}\right)$$

$$SEH_{el} = \sum_{ehi=1}^{EHI}\sum_{ehj=1}^{EHJ_{ehi}}\left(\frac{x_{ehi,ehj}^{EH}}{e_{ehi,ehj}^{EH}}\right) + \sum_{ehci=1}^{EHCI}\sum_{ehcj=1}^{EHCJ_{ehci}}\left(\frac{x_{ehci,ehcj}^{EHC}}{e_{ehci,ehcj}^{EHC}}\right) + \sum_{ehwi=1}^{EHWI}\sum_{ehwj=1}^{EHWJ_{ehwi}}\left(\frac{x_{ehwi,ehwj}^{EHW}}{e_{ehwi,ehwj}^{EHW}}\right)$$

$$SEH_{nel,fuel} = \sum_{nehi=1}^{NEHI}\sum_{nehj=1}^{NEHJ_{nehi}}\left(\frac{x_{nehi,nehj}^{NEH}FU_{nehi,nehj,fuel}^{NEH}}{e_{nehi,nehj}^{NEH}}\right) + \sum_{nehwi=1}^{NEHWI}\sum_{nehwj=1}^{NEHWJ_{nehwi}}\left(\frac{x_{nehwi,nehwj}^{NEHW}FU_{nehwi,nehwj,fuel}^{NEHW}}{e_{nehwi,nehwj}^{NEHW}}\right) \forall fuel \in \{1,\ldots,FUEL\}$$

$$SEC_{el} = \sum_{eci=1}^{ECI} \sum_{ecj=1}^{ECJ_{eci}} \left(\frac{x_{eci,ecj}^{EC}}{e_{eci,ecj}^{EC}} \right) + \sum_{ehci=1}^{EHCI} \sum_{ehcj=1}^{EHCJ_{ehci}} \left(\frac{x_{ehci,ehcj}^{EHC}}{e_{ehci,ehcj}^{EHC}} \right)$$

$$SEW_{el} = \sum_{ewi=1}^{EWI} \sum_{ewj=1}^{EWJ_{ewi}} \left(\frac{x_{ewi,ewj}^{EW}}{e_{ewi,ewj}^{EW}} \right) + \sum_{ehwi=1}^{EHWI} \sum_{ehwj=1}^{EHWJ_{ehwi}} \left(\frac{x_{ehwi,ehwj}^{EHW}}{e_{ehwi,ehwj}^{EHW}} \right)$$

$$SEW_{nel,fuel} = \sum_{newi=1}^{NEWI} \sum_{newj=1}^{NEWJ_{newi}} \left(\frac{x_{newi,newj}^{NEW} FU_{newi,newj,fuel}^{NEW}}{e_{newi,newj}^{NEW}} \right) + \sum_{nehwi=1}^{NEHWI} \sum_{nehwj=1}^{NEHWJ_{nehwi}} \left(\frac{x_{nehwi,nehwj}^{NEHW} FU_{nehwi,nehwj,fuel}^{NEHW}}{e_{nehwi,nehwj}^{NEHW}} \right) \; \forall fuel \in \{1,\ldots,FUEL\}$$

$$Q_n^{HD} = \begin{cases} HS_n \left(\begin{array}{l} BLC(\theta_{IH} - \theta_{E,n})T_n + \rho_{air}c_{air}V_{air}(\theta_{IH} - \theta_{E,n}) - Q_{AINHG}T_n \\ - \sum_{wn=1}^{WN} \left(A_{wn}^{WIN} F_{F,wn} F_{S,wn} F_{CM,wn} I_{SL,wn,n} \sum_{s=1}^{S} \sum_{t=1}^{T_s} \left(x_{st}^{WIN} g_{st}^{WIN} \right) \right) \end{array} \right), & \text{if positive} \\ 0, & \text{else} \end{cases}$$

$$Q_n^{CD} = \begin{cases} CS_n \left(\begin{array}{l} \sum_{wn=1}^{WN} \left(A_{wn}^{WIN} F_{F,wn} F_{S,wn} F_{CM,wn} I_{SL,wn,n} \sum_{s=1}^{S} \sum_{t=1}^{T_s} \left(x_{st}^{WIN} g_{st}^{WIN} \right) \right) \\ + Q_{AINHG}T_n - BLC(\theta_{IC} - \theta_{E,n})T_n - \rho_{air}c_{air}V_{air}(\theta_{IC} - \theta_{E,n}) \end{array} \right), & \text{if positive} \\ 0, & \text{else} \end{cases}$$

$$BLC = \sum_{dr=1}^{DR} \left(A_{dr}^{DOR} b_{dr}^{DOR} \right) \sum_{v=1}^{V} \left(x_v^{DOR} U_v^{DOR} \right) + \sum_{wn=1}^{WN} \left(A_{wn}^{WIN} b_{wn}^{WIN} \right) \sum_{s=1}^{S} \sum_{t=1}^{T_s} \left(x_{st}^{WIN} U_{st}^{WIN} \right)$$
$$+ \frac{\sum_{wl=1}^{WL} \left(A_{wl}^{WAL} b_{wl}^{WAL} \right)}{\sum_{w=1}^{W} \left(x_w^{WAL} \left(\sum_{kwl=1}^{KWL_w} \left(\frac{d_{w,kwl}^{dWAL}}{kk_{w,kwl}^{mWAL}} \right) + \sum_{y=1}^{Y_w} \left(x_{wy}^{dWAL} \sum_{c=1}^{C_{wy}} \left(\frac{x_{wyc}^{mWAL}}{k_{wyc}^{mWAL}} \right) \right) \right) \right)}$$
$$+ \frac{\sum_{ce=1}^{CE} \left(A_{ce}^{CEIL} b_{ce}^{CEIL} \right)}{\sum_{d=1}^{D} \left(x_d^{CEIL} \left(\sum_{kcl=1}^{KCL_d} \left(\frac{d_{d,kcl}^{dCEIL}}{kk_{d,kcl}^{mCEIL}} \right) + \sum_{f=1}^{F_d} \left(x_{df}^{dCEIL} \sum_{a=1}^{A_{df}} \left(\frac{x_{dfa}^{mCEIL}}{k_{dfa}^{mCEIL}} \right) \right) \right) \right)}$$
$$+ \frac{\sum_{fl=1}^{FL} \left(A_{fl}^{FLO} b_{fl}^{FLO} \right)}{\sum_{h=1}^{H} \left(x_h^{FLO} \left(\sum_{kfl=1}^{KFL_h} \left(\frac{d_{h,kfl}^{dFLO}}{kk_{h,kfl}^{mFLO}} \right) + \sum_{e=1}^{E_h} \left(x_{he}^{dFLO} \sum_{g=1}^{G_{he}} \left(\frac{x_{heg}^{mFLO}}{k_{heg}^{mFLO}} \right) \right) \right) \right)}$$

$$DQ_{DHW,n} = \begin{cases} Q_{dhwu} - \dfrac{A_{SLC} I_{SL,SLC,n} F_{S,SLC} \sum_{u=1}^{U} \sum_{b=1}^{B_u} \left(x_{ub}^{SLC} e_{ub}^{SLC} \right)}{10^6}, & \text{if } Q_{dhwu} \geq \dfrac{A_{SLC} I_{SL,SLC,n} F_{S,SLC} \sum_{u=1}^{U} \sum_{b=1}^{B_u} \left(x_{ub}^{SLC} e_{ub}^{SLC} \right)}{10^6} \\ 0, & \text{else} \end{cases}$$

$$Q_{dSLC,n} = \frac{A_{SLC} I_{SL,SLC,n} F_{S,SLC} \sum_{u=1}^{U} \sum_{b=1}^{B_u} \left(x_{ub}^{SLC} e_{ub}^{SLC} \right)}{10^6}$$

$$COST_{DOR} = \sum_{dr=1}^{DR} \left(A_{dr}^{DOR} \right) \sum_{v=1}^{V} \left(x_v^{DOR} C_v^{DOR} \right)$$

$$COST_{WIN} = \sum_{wn=1}^{WN} \left(A_{st}^{WIN} \right) \sum_{s=1}^{S} \sum_{t=1}^{T_s} \left(x_{st}^{WIN} C_{st}^{WIN} \right)$$

$$COST_{WAL} = \sum_{wl=1}^{WL} \left(A_{wl}^{WAL} \right) \sum_{w=1}^{W} \left(x_w^{WAL} \left(\sum_{kwl=1}^{KWL_w} \left(d_{w,kwl}^{dWAL} CK_{w,kwl}^{mWAL} \right) + \sum_{y=1}^{Y_w} \left(x_{wy}^{dWAL} \sum_{c=1}^{C_{wy}} \left(x_{wyc}^{mWAL} C_{wyc}^{mWAL} \right) \right) \right) \right)$$

$$COST_{CEIL} = \sum_{ce=1}^{CE} \left(A_{ce}^{CEIL} \right) \sum_{d=1}^{D} \left(x_d^{CEIL} \left(\sum_{kcl=1}^{KCL_d} \left(d_{d,kcl}^{dCEIL} CK_{d,kcl}^{mCEIL} \right) + \sum_{f=1}^{F_d} \left(x_{df}^{dCEIL} \sum_{a=1}^{A_{df}} \left(x_{dfa}^{mCEIL} C_{dfa}^{mCEIL} \right) \right) \right) \right)$$

$$COST_{FLO} = \sum_{fl=1}^{FL} \left(A_{fl}^{FLO} \right) \sum_{h=1}^{H} \left(x_h^{FLO} \left(\sum_{kfl=1}^{KFL_h} \left(d_{h,kfl}^{dFLO} CK_{h,kfl}^{mFLO} \right) + \sum_{e=1}^{E_h} \left(x_{he}^{dFLO} \sum_{g=1}^{G_{he}} \left(x_{heg}^{mFLO} C_{heg}^{mFLO} \right) \right) \right) \right)$$

$$COST_{HS} = \sum_{ehi=1}^{EHI} \sum_{ehj=1}^{EHJ_{ehi}} \left(x_{ehi,ehj}^{EH} CST_{ehi,ehj}^{EH} \right) + \sum_{nehi=1}^{NEHI} \sum_{nehj=1}^{NEHJ_{nehi}} \left(x_{nehi,nehj}^{NEH} CST_{nehi,nehj}^{NEH} \right)$$

$$COST_{CS} = \sum_{eci=1}^{ECI} \sum_{ecj=1}^{ECJ_{eci}} \left(x_{eci,ecj}^{EC} CST_{eci,ecj}^{EC} \right)$$

$$COST_{WS} = \sum_{ewi=1}^{EWI} \sum_{ewj=1}^{EWJ_{ehi}} \left(x_{ewi,ewj}^{EW} CST_{ewi,ewj}^{EW} \right) + \sum_{newi=1}^{NEWI} \sum_{newj=1}^{NEWJ_{nehi}} \left(x_{newi,newj}^{NEW} CST_{newi,newj}^{NEW} \right)$$

$$COST_{HCS} = \sum_{ehci=1}^{EHCI} \sum_{ehcj=1}^{EHCJ_{ehci}} \left(x_{ehci,ehcj}^{EHC} CST_{ehci,ehcj}^{EHC} \right)$$

$$COST_{HWS} = \sum_{ehwi=1}^{EHWI} \sum_{ehwj=1}^{EHWJ_{ehwi}} \left(x_{ehwi,ehwj}^{EHW} CST_{ehwi,ehwj}^{EHW} \right) + \sum_{nehwi=1}^{NEHWI} \sum_{nehwj=1}^{NEHWJ_{nehwi}} \left(x_{nehwi,nehwj}^{NEHW} CST_{nehwi,nehwj}^{NEHW} \right)$$

$$COST_{SLC} = A_{SLC} \sum_{u=1}^{U} \sum_{b=1}^{B_u} \left(x_{ub}^{SLC} CST_{ub}^{SLC} \right)$$

$$\sum_{v=1}^{V} x_v^{DOR} = 1$$

$$\sum_{s=1}^{S} \sum_{t=1}^{T_s} x_{st}^{WIN} = 1$$

$$\sum_{w=1}^{W} x_w^{WAL} = 1$$

$$\sum_{c=1}^{C_{wy}} x_{wyc}^{mWAL} = x_w^{WAL}, \forall (y = 1, \ldots, Y_w, \forall w = 1, \ldots, W)$$

$$x_{wy}^{dWAL} \in \left[0, d_{\max,wy}^{WAL} \right], \forall (y = 1, \ldots, Y_w, \forall w = 1, \ldots, W)$$

$$\sum_{d=1}^{D} x_d^{CEIL} = 1$$

$$\sum_{a=1}^{A_{df}} x_{dfa}^{mCEIL} = x_d^{CEIL}, \forall (f = 1, \ldots, F_d, \forall d = 1, \ldots, D)$$

$$x_{df}^{dCEIL} \in \left[0, d_{\max,df}^{CEIL} \right], \forall (f = 1, \ldots, F_d, \forall d = 1, \ldots, D)$$

$$\sum_{h=1}^{H} x_h^{FLO} = 1$$

$$\sum_{g=1}^{G_{he}} x_{heg}^{mFLO} = x_h^{FLO}, \forall (e = 1, \ldots, E_h, \forall h = 1, \ldots, H)$$

$$x_{he}^{dFLO} \in \left[0, d_{\max,he}^{FLO} \right], \forall (e = 1, \ldots, E_h, \forall h = 1, \ldots, H)$$

$$\sum_{ehi=1}^{EHI} \sum_{ehj=1}^{EHJ_{ehi}} x_{ehi,ehj}^{EH} + \sum_{ehci=1}^{EHCI} \sum_{ehcj=1}^{EHCJ_{ehci}} x_{ehci,ehcj}^{EHC} + \sum_{ehwi=1}^{EHWI} \sum_{ehwj=1}^{EHWJ_{ehwi}} x_{ehwi,ehwj}^{EHW} + \sum_{nehi=1}^{NEHI} \sum_{nehj=1}^{NEHJ_{nehi}} x_{nehi,nehj}^{NEH} + \sum_{nehwi=1}^{NEHWI} \sum_{nehwj=1}^{NEHWJ_{nehwi}} x_{nehwi,nehwj}^{NEHW} = 1$$

$$\sum_{eci=1}^{ECI} \sum_{ecj=1}^{ECJ_{eci}} x_{eci,ecj}^{EC} + \sum_{ehci=1}^{EHCI} \sum_{ehcj=1}^{EHCJ_{ehci}} x_{ehci,ehcj}^{EHC} = 1$$

$$\sum_{ewi=1}^{EWI} \sum_{ewj=1}^{EWJ_{ehi}} x_{ewi,ewj}^{EW} + \sum_{ehwi=1}^{EHWI} \sum_{ehwj=1}^{EHWJ_{ehwi}} x_{ehwi,ehwj}^{EHW} + \sum_{newi=1}^{NEWI} \sum_{newj=1}^{NEWJ_{newi}} x_{newi,newj}^{NEW} + \sum_{nehwi=1}^{NEHWI} \sum_{nehwj=1}^{NEHWJ_{nehwi}} x_{nehwi,nehwj}^{NEHW} = 1$$

$$\sum_{u=1}^{U} \sum_{b=1}^{B_u} x_{ub}^{SLC} \leq 1$$

References

1. Costa-Carrapiço, I.; Raslan, R.; González, J.N. A systematic review of genetic algorithm-based multi-objective optimisation for building retrofitting strategies towards energy efficiency. *Energy Build.* **2020**, *210*, 109690. [CrossRef]
2. Hashempour, N.; Taherkhani, R.; Mahdikhani, M. Energy performance optimization of existing buildings: A literature review. *Sustain. Cities Soc.* **2020**, *54*, 101967. [CrossRef]
3. Wulfinghoff, D.R. *Energy Efficiency Manual*; Energy Institute Press: Wheaton, MD, USA, 1999.
4. Diakaki, C.; Grigoroudis, E.; Kolokotsa, D. Towards a multi-objective optimization approach for improving energy efficiency in buildings. *Energy Build.* **2008**, *40*, 1747–1754. [CrossRef]
5. Diakaki, C.; Grigoroudis, E.; Kabelis, N.; Kolokotsa, D.; Kalaitzakis, K.; Stavrakakis, G. A multi-objective decision model for the improvement of energy efficiency in buildings. *Energy* **2010**, *35*, 5483–5496. [CrossRef]
6. Diakaki, C.; Grigoroudis, E. Applying genetic algorithms to optimise energy efficiency in buildings. In *Multicriteria Decision Aid and Artificial Intelligence: Links, Theory and Applications*; Doumpos, M., Grigoroudis, E., Eds.; John Wiley & Sons Ltd: Hoboken, NJ, USA, 2013; pp. 309–333. [CrossRef]
7. Diakaki, C.; Grigoroudis, E.; Kolokotsa, D. Performance study of a multi-objective mathematical programming modelling approach for energy decision-making in buildings. *Energy* **2013**, *59*, 534–542. [CrossRef]
8. Ren, H.; Zhou, W.; Gao, W.; Wu, Q. A mixed-integer linear optimization model for local energy system planning based on simplex and branch-and-bound algorithms. In *Life System Modeling and Intelligent Computing*; Li, K., Fei, M., Jia, L., Irwin, G.W., Eds.; Springer: Berlin/Heidelberg, Germany, 2010; Volume 6328, pp. 361–371. [CrossRef]
9. Asadi, E.; Da Silva, M.G.; Antunes, C.H.; Dias, L. Multi-objective optimization for building retrofit strategies: A model and an application. *Energy Build.* **2012**, *44*, 81–87. [CrossRef]
10. Üçtug, F.G.; Yukseltan, E. A linear programming approach to household energy conservation: Efficient allocation of budget. *Energy Build.* **2013**, *49*, 200–208. [CrossRef]
11. Karmellos, M.; Kiprakis, A.; Mavrotas, G. A multi-objective approach for optimal prioritization of energy efficiency measures in buildings: Model, software and case studies. *Appl. Energy* **2015**, *139*, 131–150. [CrossRef]
12. Bayata, O.; Temiz, I. Developing a model and software for energy efficiency optimization in the building design process: A case study in Turkey. *Turk. J. Electr. Eng. Comput. Sci.* **2017**, *25*, 4172–4186. [CrossRef]
13. Shakouri, G.H.; Rahmani, M.; Hosseinzadeh, M.; Kazemi, A. Multi-objective optimization-simulation model to improve the buildings' design specification in different climate zones of Iran. *Sustain. Cities Soc.* **2018**, *40*, 394–415. [CrossRef]
14. Nielsen, A.N.; Jensen, R.L.; Larsen, T.S.; Nissen, S.B. Early stage decision support for sustainable building renovation—A review. *Build. Environ.* **2016**, *103*, 165–181. [CrossRef]
15. Siskos, J.; Despotis, D.K. A DSS oriented method for multiobjective linear programming problems. *Decis. Support. Syst.* **1989**, *5*, 47–55. [CrossRef]
16. Siskos, Y.; Yannacopoulos, D. UTASTAR: An ordinal regression method for building additive value functions. *Investig. Oper.* **1985**, *5*, 39–53.
17. Kreider, J.F.; Curtiss, P.; Rabl, A. *Heating and Cooling for Buildings, Design for Efficiency*; McGraw-Hill: New York, NY, USA, 2002.
18. Zopounidis, C.; Despotis, D.K.; Kamaratou, I. Portfolio selection using the ADELAIS multiobjective linear programming System. *Comput. Econ.* **1998**, *11*, 189–204. [CrossRef]
19. Lee, S.H.; Hong, T.; Piette, M.A.; Taylor-Lange, S.C. Energy retrofit analysis toolkits for commercial buildings: A review. *Energy* **2015**, *89*, 1087–1100. [CrossRef]

 *sustainability*

Article

Automatic Detection of Photovoltaic Farms Using Satellite Imagery and Convolutional Neural Networks

Konstantinos Ioannou [1,*] and **Dimitrios Myronidis** [2]

1 Hellenic Agricultural Organization "DEMETER", Forest Research Institute, Vasilika, 57006 Thessaloniki, Greece
2 Department of Forestry and Natural Environment, Aristotle University of Thessaloniki, 54124 Thessaloniki, Greece; myronid@for.auth.gr
* Correspondence: ioanko@fri.gr; Tel.: +30-231-046-1171 (ext. 225)

Abstract: The number of solar photovoltaic (PV) arrays in Greece has increased rapidly during the recent years. As a result, there is an increasing need for high quality updated information regarding the status of PV farms. This information includes the number of PV farms, power capacity and the energy generated. However, access to this data is obsolete, mainly due to the fact that there is a difficulty tracking PV investment status (from licensing to investment completion and energy production). This article presents a novel approach, which uses free access high resolution satellite imagery and a deep learning algorithm (a convolutional neural network—CNN) for the automatic detection of PV farms. Furthermore, in an effort to create an algorithm capable of generalizing better, all the current locations with installed PV farms (data provided from the Greek Energy Regulator Authority) in the Greek Territory (131,957 km^2) were used. According to our knowledge this is the first time such an algorithm is used in order to determine the existence of PV farms and the results showed satisfying accuracy.

Keywords: PV farms; deep learning; satellite imagery; CNN; automatic detection

Citation: Ioannou, K.; Myronidis, D. Automatic Detection of Photovoltaic Farms Using Satellite Imagery and Convolutional Neural Networks. *Sustainability* **2021**, *13*, 5323. https://doi.org/10.3390/su13095323

Academic Editor: Maria Malvoni

Received: 15 April 2021
Accepted: 7 May 2021
Published: 10 May 2021

Publisher's Note: MDPI stays neutral with regard to jurisdictional claims in published maps and institutional affiliations.

1. Introduction

During the last three decades mankind is witnessing an evolution in the energy sector as we notice a shift in energy production methods, from the usage of fossil fuels (petroleum, natural gas, coal, etc.) to more environmentally friendly methods. This is caused mainly due to the fact that a significant portion of the world's carbon dioxide production is a result of fossil fuels used for energy production [1–3].

However, as electricity consumption plays an important role for modern societies (and its usage cannot be reduced) other forms of energy production must be used in order to satisfy current and future energy demands [3–7].

Renewable energy methods can be considered as a viable solution for energy production and the reduction of CO_2 emissions. These methods include the usage of sustainable sources based on wind, water, biomass, solar and geothermal energy for energy production which are in general called renewable energy sources (RES) [8].

The exploitation of solar energy is considered as one of the most common types of RES. Solar panels are used for transforming energy from indecent sunlight, to electricity using solar cells based on the photovoltaic effect, thus they are also called photovoltaic (PV) panels [9]. Nowadays, massive arrays of PV panels (in the form of solar or PV farms) are used for energy production throughout the world. These farms energy production capability ranges from 1 to 2000 MW, in the case of mega projects covering thousands of hectares [10].

In Europe, PV farms account for 13% of the total RES production. Furthermore, solar power is the fastest-growing source: in 2008, it accounted for 1%. This means that the growth in electricity from solar power has been dramatic, rising from 7.4 TWh in 2008 to 125.7 TWh in 2019 [11].

In Greece, data provided by the Regulatory Authority for Energy (RAE) indicate that currently there are 9791 PV potential installations (farms) in a variety of stages (licensed investments, licensed installations, licensed production or under evaluation), currently producing 715.6 MW of electric energy.

The variety of the existing stages of PV farms is making difficult to track the infiltration of PV to the Greek market as in many cases the time period from the initial evaluation of the energy production license to production can be years. Financial difficulties, public reaction against the investment as well as technical difficulties can pause the entire installation process.

In this work we investigate a new method of collecting installed PV information which is potentially cheaper and faster than existing methods. The proposed approach uses an algorithm which can automatically detect the existing PV farms based on high resolution free to use satellite imagery, current RAE data for training and deep learning techniques. The entire methodology can be divided in two separate steps.

The first step involves the association of the data provided by RAE with satellite images. For the implementation of this step, we used an algorithm for automatically annotating the images and matching RAE data with satellite images in order to create two datasets. A high-resolution dataset and a low-resolution dataset.

The second step involve the usage of the output produced in the first step in order to train a deep learning (DL) algorithm to automatically detect the PV farm's locations. The algorithm apart from the determination of the locations can also help scientists to extract other information. As it is basically a data unaware algorithm, it can also provide information such as the effect of land use in the selection of PV farm locations, the effect of micrometeorology to the installation locations etc.

The proposed approach offers a series of benefits when compared with other data analysis methods. First it allows the scalability of the produced results as well as the automatic improvement of the data collection. Usage of higher resolution images will provide the user with better results. Thus, the user is free to use data which originate from a variety of sources even from Google Earth, with the best results however, to be expected with data from paid services such as LandSat [12,13].

Additionally, the implementation of the approach using a computer algorithm allows the automation of the process. The entire procedure is easy to use and can be executed multiple times in order to monitor the installation rate. The produced information can also help scientists to predict the level of energy produced as well as help the Government to initiate programs related with RES adoption and provide a valuable tool to enhance the decision-making process regarding the determination of potential installation sites [2,14]. Finally, the presented methodology can be easily adapted in order to monitor other types of RES and reproduced in other regions.

2. Literature Review

Computer applications, sensor networks as well as the Internet of Things are responsible for the creation of enormous amounts of data [15]. For this reason, new and innovative techniques must be applied in order to perform sufficient analysis of the accumulated data. Deep Learning is a part of machine learning (ML) methods based on the usage of artificial neural networks with representation learning (supervised, semi-supervised or unsupervised learning) [16].

Essentially DL is a methodology where many classifiers work together, and it is based on linear regression followed by activation functions. DL foundation relay on the same traditional statistical linear regression approach. The only difference is that there are many neural nodes in deep learning instead of only one node (in the case of linear regression). These nodes are known as neural network, and one classifier (a node) is known as perceptron. The network is organized in layers and each layer can have many hundreds or even thousands of nodes. Layers which are situated between the input and output layers constitute the hidden layer and accordingly the nodes which constitute this layer are known as hidden nodes. In contrary with traditional machine learning classifiers where the user must write complex hypothesis, in deep neural network applications the hypothesis is generated by the network itself, making it a powerful tool for learning nonlinear relationships effectively [16].

ML can be divided into two development phases, shallow learning (SL) and deep learning. The most widely spread SL methods include logistic regression, support vector machine (SVM) and Gaussian mixture models [17–26]. SL main disadvantage is that it cannot handle complex real-world problems such as voice and image recognition [16]. On the contrary DL specializes in solving problems such as image classification, voice recognition etc. For example, image classification of 1000 kinds of images provided a classification error rate of 3.5% which is higher than the accuracy of ordinary people [27].

Various DL algorithms were used for disease determination. Quiroz and Alferez [28] used DL image recognition of legacy blueberries in the rooting stage, planted in smart farms in Chile. For this reason, they used a convolutional neural network (CNN) to detect the presence of trays with living blueberry plants, the presence of trays without living plants and the absence of trays. The model produced results with 86% accuracy, 86% precision, 88% recall and 86% F_1 score.

Other researchers used DL for apple pathology image recognition and diagnosis [29]. For this reason, they trained a CNN that obtained a recall rate of 98.4% using error back propagation analysis of sampled elements. In the study of Liu et al. [30], DL was used for the identification of citrus cancer based on the AlexNet model, with an optimized network structure which could reduce the network parameters while maintaining the same level of accuracy. The results from the application showed that the recognition accuracy reached 98%. In the study of Amara et al. [31], DL was used for detecting two well-known banana diseases. For this reason, they used a deep CNN based on the LeNet architecture, with the results accuracy at 85.9%, precision accuracy 86.7%, recall 85.9% and F_1 score 86.3%.

DL was also used for other types of image recognition. Huang et al. [32] used DL for determining crack and leakage defects on metro shield tunnels which produced very good results with an identification error of 0.8%. Yang et al. [33] used a DL algorithm (in this case a modified AlexNet model) was used in order to determine wind turbine blade damage on images taken from an unmanned aerial vehicle. The model provided better results (97.1% average accuracy) when compared to the unmodified AlexNet model and support vector machine models. In [34], a DL approach was proposed for the classification of road surface conditions. For this, they used a CNN network and created a new activation function based on the rectified linear unit function. Their results showed a classification accuracy of 94.89% on the road state database. DL were also used to perform breast cancer classification. A new method called BDR-CNN-CGN was used to perform classification of breast cancer types, the results showed improved detection rates (accuracy 96.10%) compared to other neural network models [35]. A CNN was also used in order to perform COVID-19 diagnosis. The proposed CNN employed several new techniques such as rank-based average pooling and multiple-way data augmentation. Among the eight proposed models, the model named FGCNet performed better with performance percentage higher than 97% [36]. Finally, Malog et al. [37], used high resolution satellite imagery and deep forest algorithm in order to detect roof top installed photovoltaic arrays. Their data included imagery from an area of 135 Km^2 and the results showed 99.9% pixel-based detection accuracy and 90% object-based detection accuracy. Table 1 presents an overview of the aforementioned literature.

Table 1. State of the Art.

Reference	DL Usage
Ioannou et al., 2021	Application of Deep Learning in weather Data gathered from IoT devices
Dong et al., 2021	Deep learning methods
Freedman, D. 2009	A survey on deep learning and its applications
Mood, C. 2010; Kleinbaum and Klein, 2002; Hosmer et al., 2013	Applied Logistic Regression and applications
Soentpiet, R. 1999; Hearst et al., 1998; Steiwart and Christmann, 2008	Support Vector Learning, Support vector machines and applications
Schraudolph, N, 2002	A Gauss–Newton approximation of the Hessian from nonlinear least-squares problems to arbitrary loss functions
Li, Z., 2009	A deep investigation into new ways of applying computer technologies to biomedicine
Verbeek et al., 2003	Development of a greedy algorithm for application on unknown ratio mixtures of components
Gavali, P., Banu, J.S., 2019	DL for image classification using the CUDA Platform
Quiroz, Alferez, 2020; Tan et al., 2015; Liu et al., 2020; Amara et al., 2017	Application of DL for the recognition of fauna diseases
Huang et al., 2018; Yang et al., 2021; Cheng et al., 2019	Recognition of structural defects using DL
Zhang et al., 2021; Wang et al., 2021	Application of CC for medical purposes (cancer type determination, COVID-19 diagnoses, etc.)
Malog et al., 2016	Detection of residential PV arrays using DL and aerial imagery.

3. Materials and Methods

For the creation of the image data sets we used data provided by RAE as well as, data which are available from Apple Maps. Apple Maps is a free map service based on satellite data which are provided from DigitalGlobe. RAE data included a series of polygons (in Shape file form) which included all PV farm investments in Greece (Figure 1). The data were categorized depending on the status of the investment in:

- Investments with installation licenses;
- Investments with production licenses;
- Investments with operation licenses.

Each shape file was at first converted to GeoJSON format. GeoJSON is a geospatial data interchange format compatible with the GNU/General Public License (GPL) guidelines, based on JavaScript Object Notation (JSON). It defines several types of JSON objects and the manner in which they are combined to represent data about geographic features, their properties, and their spatial extents. GeoJSON uses a geographic coordinate reference system, World Geodetic System 1984, and units of decimal degrees [38].

A special PYTHON algorithm was written in order to match the polygons with base map data. The algorithm used a GNU/GPL library called jimutmap in order to read each polygon in GeoJSON form and create an image file. Thus, concluding the first step of the methodology. Jimutmap allows the user to select different zoom levels when annotating the data and create images of different resolutions.

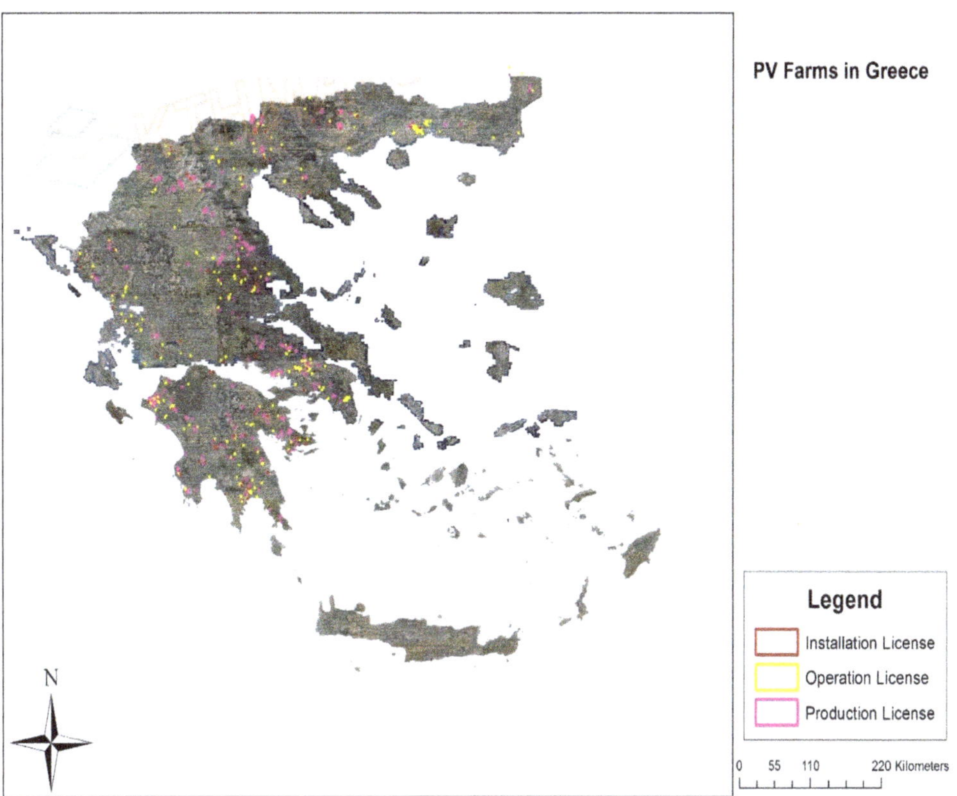

Figure 1. Map of Greece with PV farms in various phases (Basemap from Hellenic Cadastre).

In Figure 2, we can easily observe that the library user, can easily select the zoom level value, using the zoom variable, and thus determine the resolution of the images created (higher zoom level creates images with lower resolution). This is due to the fact that satellite imagery provided by free services has limited resolution. Additionally, the library allows the usage of multiple core threads in order to perform quicker the required annotations.

```
download_obj = api(min_lat_deg = 10,
max_lat_deg = 10.01,
min_lon_deg = 10.1,
max_lon_deg = 10.11,
zoom = 19,
verbose = False,
threads_ = 5,
container_dir = "myOutputFolder")
```

Figure 2. Code snippet from jimutmap library.

The second step included training a convolutional neural network to automatically detect the PV farm's locations. The CNN was developed using Google Collaboratory or Google Colab (GC) for short. GC is a product from Google Research allowing users to write and execute arbitrary PYTHON code using their browser, and is especially well suited to machine learning, data analysis. Additionally, it provides access to advanced cloud resources including the ability for the user to use graphics processor units (GPU's) and tensor processing units (TPU's). Unlike normal central processor units—CPU's (which are installed on all personal computers using the x64 architecture), GPU's are specialized electronic circuits designed to accelerate the creation and manipulation of images. Their highly parallel structure makes them more efficient than general-purpose (CPUs) for algorithms that process large blocks of data in parallel [39]. TPU's are artificial intelligence accelerator application-specific integrated circuits (ASICs) developed by Google specifically for neural network machine learning using TensorFlow a free and open-source software library for machine learning [40].

3.1. Convolutional Neural Networks

Convolutional neural networks (CNN) are inspired by the cat's cortex and were first proposed in the 1980s [41]. A CNN has similar structure with other multilayer neural networks, and it is comprised of layers. Each layer is composed of a number of two-dimensional planes and each plane has independent neurons. Sparse connections are used between layers, meaning that the neuron in each feature map only connects to the neurons in a small area in the upper map, in contrast with the traditional neural networks. The CNN structure depends mainly in the shared weight, the local experience field and the sub-collector to ensure the invariance of input data [42].

The following figure (Figure 3) presents the layout of a CNN. In this case the network is comprised from an input layer, four hidden layers and an output layer. This network was created for performing image processing. In more detail image recognition of characters written by hand. In this case the input layer is made up using 28×28 sensory nodes. This layer receives the images which have been approximately centered and normalized in terms of size. Afterwards the computational layouts alternate between convolution and subsampling as follows:

- The first hidden layer is responsible for the convolution. This layer consists of four feature maps, with each feature map consisting of 24×24 neurons. Each neuron is assigned a receptive field of 5×5 size.
- The second hidden layer is responsible for subsampling and local averaging. Like the previous layer, it also consists of four feature maps, but each feature map is now made up of 12×12 neurons. Each neuron has a receptive field of size 2×2, a trainable coefficient, a trainable bias, and a sigmoid activation function. The trainable coefficient and bias control the operating point of the neuron.
- The third hidden layer is responsible for the second convolution. It consists of 12 feature maps, with each feature map consisting of 8×8 neurons. Each neuron in this hidden layer may have synaptic connections from several feature maps in the previous hidden layer. Otherwise in operates in a manner similar to the first convolutional layer.
- The fourth hidden layer is responsible for performing a second subsampling and local averaging. It consists of 12 feature maps, but with each feature map in this case consisting of 4×4 neurons. Otherwise, it operates in a manner similar to the first sampling layer.
- Finally, the output layer is responsible for the final stage of convolution. This layer consists of 26 neurons, with each neuron assigned to one of 26 possible characters. As before each neuron is assigned a receptive field of size 4×4 [42].

The result of the previously described processes is the application of a bipyramidal effect. This means that with each convolutional or subsampling layer, the number of features maps is increased while the spatial resolution is reduced, compared to the corresponding previous layer.

CNN's first usage was for the identification of handwritten checks in banks, but they were incapable of recognizing large images. For this reason, ref. [43] developed LeNet-5 which was a classical model of convolutional neural network with low error rates (only 0.9% on the MNIST data-set).

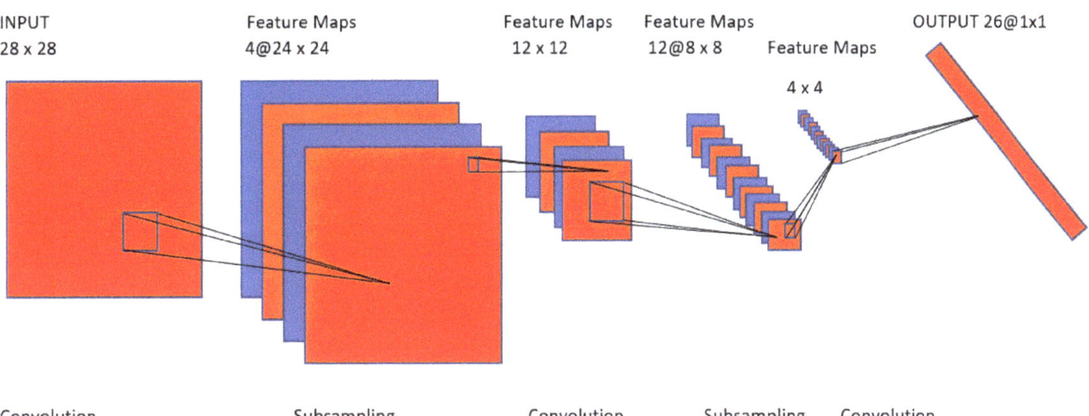

Figure 3. Convolutional Network for image processing.

The main bottleneck on the application of CNN is the long training time due to many hidden nodes on the networks. However, weight sharing which is a characteristic of the CNN allows parallel processing of weights if the proper infrastructure exists. Today as modern graphics processor units (GPU's) support parallel computing the application of CNN's is easier. In [44], a GPU algorithm was used in order to solve the ImageNet problem.

The CNN implemented for automatically detecting PV farms was based on Keras 2.3.0, a deep learning application programing interface written in PYTHON 3.7, running on top of the machine learning platform TensorFlow 2.4.1 supported by Google Colab. Keras was developed with a focus on enabling fast experimentation.

3.2. Building the Model

Keras supports various image classification models (Xception, ResNet, MobileNet, VGG, etc.). In this study we used the InceptionV3 model mainly because it performs significantly better than the other Keras Supported models [45]. The images that will be used were randomly divided in two categories, Training Images used for training and validating the model and evaluation images used for determining the network performance against new, unseen, images.

Before presenting the images to the network we perform a series of augmentations which will ensure that our model would never use twice the exact same picture thus, the model will try to overfit on the training data. For this reason, we used the image data preprocessing function of Keras. This function has a series of arguments for manipulating the training image datasets. The following arguments were used for the manipulation:

- Rotation range, rotates the images randomly;
- Height shift range, shifts the image along the X axis;
- Width shift range, shifts the image along the Y axis;
- Horizontal flip, flips the image across the X axis;
- Vertical flip, flips the image across the Y axis;
- Validation split, determines the fraction of images reserved from the training dataset for model validation;
- Zoom range, determines the zoom factor;
- Brightness range, modifies the image brightness level;
- Rescale, determines if the image is rescaled to specific dimensions;
- Shear range, determines the image distortion across an axis in order to create or rectify perception angle;
- Fill mode, determines the image location inside the canvas.

Continuing, we must determine the training epochs as well as the image batch size. Epochs refers to the number of times the network is trained through the entire dataset, whereas batch size determines the number of samples processed each time (before the model is updated).

In InceptionV3 we have the capability to use predefined training weights using the imagenet or initialize them randomly. Imagenet is an image database which is organized according to the WordNet hierarchy in which each node of the hierarchy is depicted by thousands of images [46]. The usage of this database is proven to significantly increase a CNN's performance [47]. Figure 4 displays the entire workflow of the model applied.

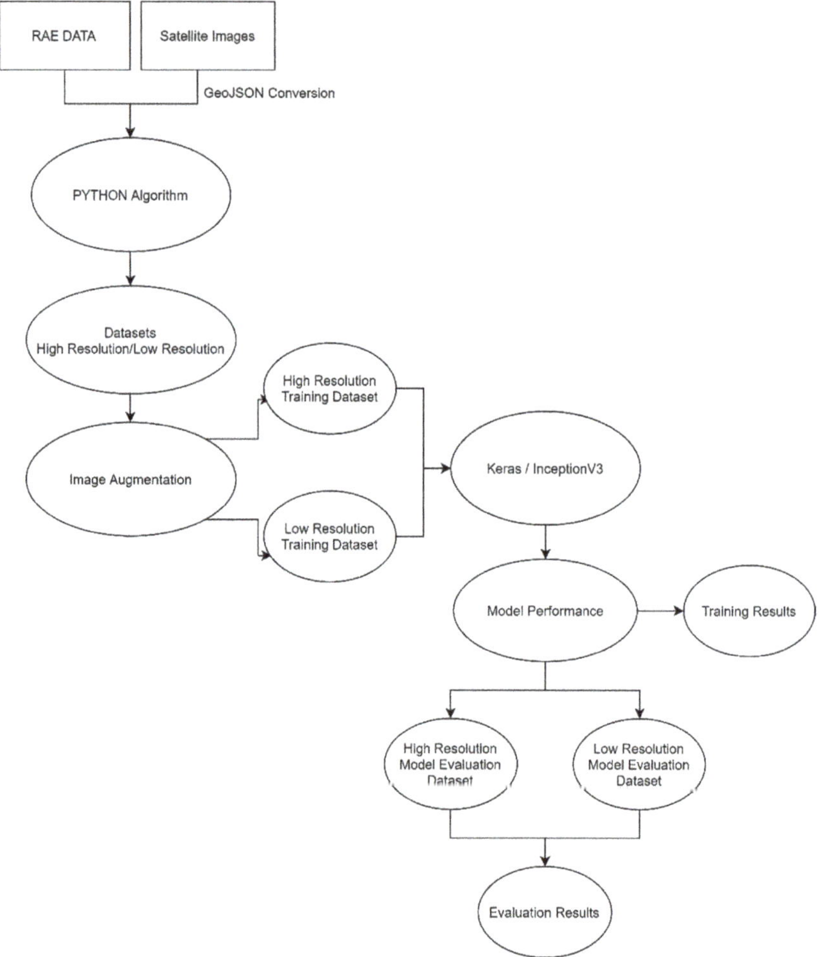

Figure 4. Methodology flowchart.

4. Results

The PYTHON algorithm used for extracting the images of PV farms created 570 images files. Of them, 220 where high-resolution images (approximately 1 MB each) and 350 where low-resolution (approximately 16 KB each). These images where divided randomly in Training and Evaluation datasets as show on Table 2.

Table 2. Image per dataset.

Resolution	Training Dataset	Evaluation Dataset
High Res	75	145
Low Res	250	100

Following that, the datasets where augmented using the image data processing function. The parameters used in this function are presented in Table 3.

Table 3. Image data processing parameters.

Parameter	Value
Rescale	1./255
Zoom_range	0.3
Rotation_scale	360
Width_shift_range	0.5
Shear_range	0.5
Horizontal_flip	True
Vertical_Flip	True
Brightness_range	0.6, 1.4
Fill_mode	Nearest
Vallidation_split	0.2
Shear Range	0.2
Fill mode	Nearest

Next, the images were imported to Keras and the InceptionV3 algorithm was applied, for 15, 20 and 25 epochs with a batch size of 15 using the ImageNet pre-trained weights. Batch size number was selected mainly because the number of the images used for training and validation is rather small. Generally, we use larger batch sizes when we have large datasets. The selection on the number of training epochs is based on the produced results (there is no guideline regarding the train period of a neural network). This means that if we notice overfitting in the results (meaning that the network cannot generalize properly), then we reduce training epochs.

Table 4 includes the results taken from the three training sessions applied. The results show the percentage of correct prediction using the training dataset and the validation dataset. From the table it is evident that the applied model does not provide better results when trained for more than 20 epochs, as it can also be seen in the graphical representation of the results in Figure 5. From Figure 5 it is also obvious, that the model performs erratically during the last validation session with large fluctuations during the validation of the model. This means that the model must have overfitted during training for 25 epochs.

Table 4. Training and validation results.

Accuracy	15 Epochs	20 Epochs	25 Epochs
Train Accuracy	94.23%	95.38%	87.31%
Validation Accuracy	90.77%	90.77%	86.15%

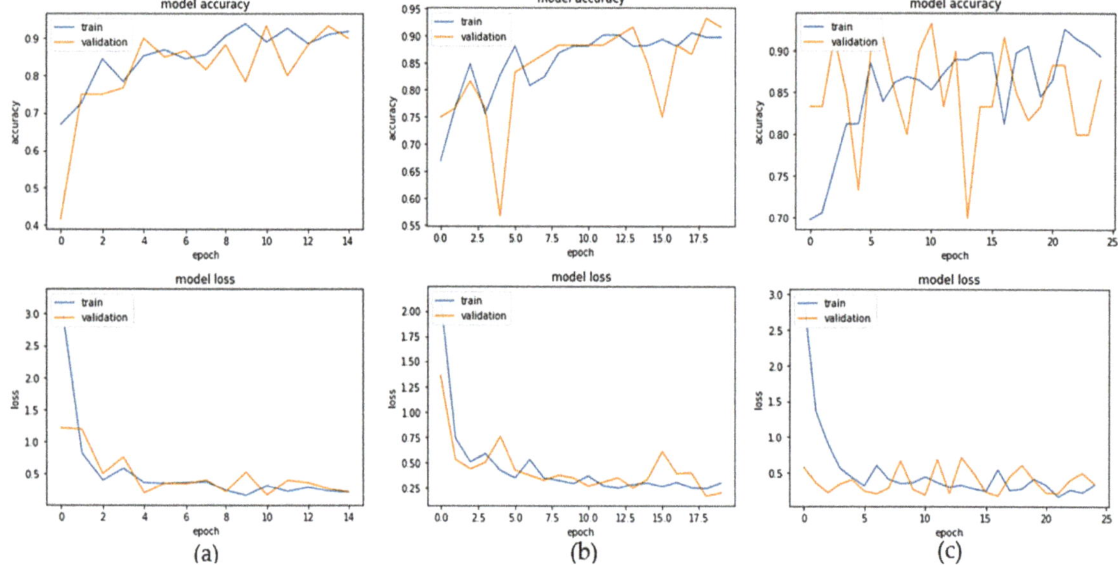

Figure 5. Model train and accuracy results for 15 epochs (**a**), 20 epochs (**b**) and 25 epochs (**c**).

Additionally, from the same figure it is also obvious that the model performs better when trained for 15 epochs (although the training performance in this session is slightly smaller compared to the performance during the next training session). As it can be seen in model accuracy section of the diagrams, the model validation line follows more closely the training line. Generally, models with a smaller curve fluctuation during accuracy elevation have better training convergence. Furthermore, model training is better when the two curves (train and validation) are closer. After training completion, the model is also tested against new data which were not used during train and validation sessions. The produced evaluation results are shown on Table 5 and Figure 6 which also prove that the model trained for 15 epochs provides the best overall predictions.

Table 5. Evaluation results.

Results	15 Epochs				20 Epochs				25 Epochs			
	Precision	Recall	F_1 Score	Support	Precision	Recall	F_1 Score	Support	Precision	Recall	F_1 Score	Support
Pv1	0.60	0.55	0.58	145	0.57	0.53	0.55	145	0.59	0.59	0.59	145
Pv2	0.42	0.47	0.44	100	0.38	0.42	0.40	100	0.40	0.39	0.39	100
Accuracy			0.52	245			0.49	245			0.51	245
Macro Avg	0.51	0.51	0.51	245	0.48	0.48	0.48	245	0.49	0.49	0.49	245
Weighted Avg	0.53	0.52	0.52	245	0.49	0.49	0.49	245	0.51	0.51	0.51	245

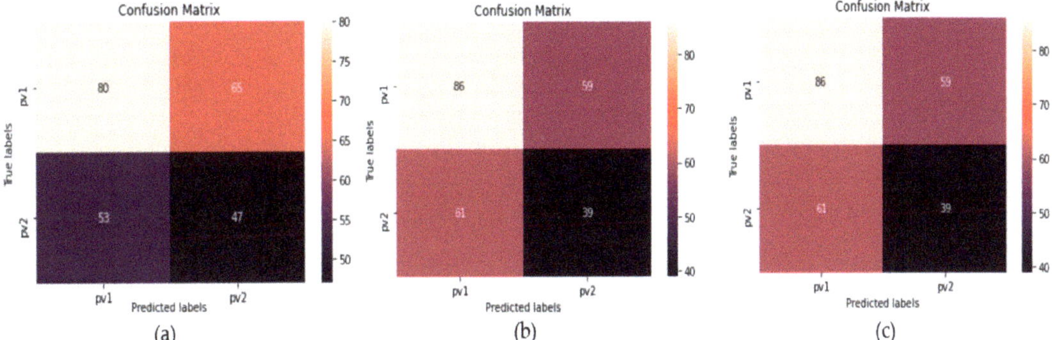

Figure 6. Confusion matrices for 15 epochs (**a**), 20 epochs (**b**) and 25 epochs (**c**).

On Table 5, Pv 1 refers to the high-resolution images' dataset, whereas Pv 2 refers to the low-resolution images' dataset. Precision is the ability of the classifier not to label as positive a sample that is negative. Or in other terms, precision is the number of correct results divided by the number of all returned results.

$$Precision = \frac{|\{relevant\ documents\} \cap \{retrieved\ documents\}|}{|\{retrieved\ documents\}|} \tag{1}$$

Recall is the ability of the classifier to find all the positive samples. Or in other terms, recall is the fraction of relevant documents that are successfully retrieved.

$$Recall = \frac{|\{relevant\ documents\} \cap \{retrieved\ documents\}|}{|\{relevant\ documents\}|} \tag{2}$$

F_1 score is a measure of the test's accuracy. It is the harmonic mean of the precision and recall:

$$F_1 = \frac{2}{recall^{-1} + precision^{-1}} \tag{3}$$

The worst value for this measure is 0 whereas the best is obtained when it equals to 1.

Accuracy is the weighted arithmetic mean of Precision and Inverse Precision (weighted by Bias) as well as the weighted arithmetic mean of Recall and Inverse Recall (weighted by Prevalence). Inverse Precision and Inverse Recall are simply the Precision and Recall of the inverse problem where positive and negative labels are exchanged. Higher accuracy values demonstrate better model performance.

Macro Average, computes the F_1 for each label and returns the average without considering the proportion for each label (in our case high- and low-resolution PV images) in the dataset. Weighted Average computes F_1 for each label (in our case high- and low-resolution PV images) and returns the average considering the proportion of each label to the dataset. Finally, support is the number of occurrences of the given class (or label) in the dataset.

The results on Table 4 indicate that the trained model produce's better results for 15 epochs in both datasets (high and low resolution).

5. Discussion

For most researchers, terms such as deep learning and machine learning seem interchangeable concerning the world of artificial intelligence. However, this approach is mistaken. Deep learning is a specialized subset of machine learning which, in turn, is a specialized subset of artificial intelligence. Deep learning describes algorithms that analyze data with a structure which is similar to how a human would draw to a conclusion. The only drawback in the application of DL is the requirement of incredibly vast amounts of data and the need for substantial computing power for its usage.

However, the application of deep learning algorithms nowadays is a necessity. The evolvement of Internet of Things has created multiple devices capable of collecting a variety of unstructured data, ranging from simple arithmetic values to images from satellites. Therefore, the need arises to evaluate this data and extract useful patterns. DL algorithms have no requirement for human intervention as the nested layers in the neural networks put data through hierarchies of different concepts, and eventually learn through their own errors. Therefore, the usage of DL algorithms can greatly help toward the process of collected data, mainly because these algorithms ignore the data types which are processing. Thus, they can (if trained properly) used for solving many problems, including image detection and classification.

This study presents a novel approach towards the problem of automatic recognition of PV farms. The recognition is based on the usage of satellite imagery and image classification techniques which until recently were used for other purposes (face recognition, flora and fauna species recognition, etc.). According to our research it is the first time that neural networks (in particular a CNN) was used for the automatic detection of PV farms. From the literature review we conducted, the only similar research used a CNN for the determination of small rooftop installed PV arrays, however we did not find any other similar research, which indicates that our approach is pioneering.

Furthermore, another novelty of our approach is that the used dataset's as well as the software (libraries, functions algorithms) used for the implementation of this research are freely available to the researchers, thus making our methodology easily replicable.

The results showed that (even though the original dataset was rather small) we can expect correct identification accuracy reaching 60% when using high resolution imagery and lower results in case we use lower resolution. From the confusion matrixes we can determine that for 15 epochs 127 correct identifications were performed, 125 correct identifications were performed for 20 epochs and 125 were also recognized correctly for 25 epochs.

However, the identification results can be further improved if we use larger datasets. Additionally, the results showed that, increase in the number of training epochs does not provide significant improvements. Table 5 presents the results showing that 15 training epochs can be considered adequate for the dataset used.

Finally, the application of the algorithm also proven that high resolution images perform significantly better even in smaller datasets compared to low resolution imagery. This result was not expected because we believed that increasing the number of low-res input data could compensate for the lower resolutions, mainly due to the fact that input data are characterized by a specific geometry.

The approach presented in this work can also be applied in the recognition of other types of RES, if trained properly. It can also be used in other cases where automatic image recognition is necessary. The results could be improved by using images provided from paid services (and therefore high resolution) and by using larger datasets. Further improvements can be achieved if the user performs some kind of image pre-processing on the dataset (edge detection, color corrections, etc.), or deeper networks (more hidden layers).

6. Conclusions

Image recognition can provide a valuable tool for monitoring the adaption rate of renewable energy sources. Modern deep learning methods are unaware of the processing data and therefore can be easily used in order to recognize the various forms of RES (wind turbines, PV panels, hydroelectric stations, etc.). However, there is a need for large datasets in order to train properly the algorithms. The existence of various satellite imagery services allows the user to collect these data in a variety of resolutions and create datasets which contain images of RES forms in a variety of installation environments, various angles, different weather and time. Therefore, it is possible to create a tool which will be capable of identifying them with increased accuracy.

This paper examined a first approach towards this goal. The dataset is based on the usage of PV farms in Greece and the results proved to be adequate given the size of the training dataset. As the years pass and more installations complete the algorithm can be trained again in order to increase its efficiency. Furthermore, advancements in computer technology and DL algorithms can also help towards this goal.

Finally, the combination of these algorithms with other types of software capable of calculating the annual solar energy output can help local and regional authorities to plan their energy policy. The methodology can also be used from the national authorities in an attempt to continuously monitor current RES status, determine the investment/adoption rate of RES in the various regions and regional units, and act as an overall tool for the application of national policy.

Author Contributions: Conceptualization, K.I.; methodology, K.I.; software, K.I.; validation, K.I. and D.M.; formal analysis, K.I.; resources, K.I. and D.M.; data curation, K.I.; writing—original draft preparation, K.I. and D.M. Both authors have read and agreed to the published version of the manuscript.

Funding: This research received no external funding.

Institutional Review Board Statement: Not applicable.

Informed Consent Statement: Not applicable.

Data Availability Statement: Data sharing not applicable.

Conflicts of Interest: The authors declare no conflict of interest.

References

1. Rafaj, P.; Barreto, L.; Kypreos, S. Combining policy instruments for sustainable energy systems: An assessment with the GMM model. *Environ. Model. Assess.* **2006**, *11*, 277–295. [CrossRef]
2. Konstantinos, I.; Georgios, T.; Garyfalos, A. A Decision Support System methodology for selecting wind farm installation locations using AHP and TOPSIS: Case study in Eastern Macedonia and Thrace region, Greece. *Energy Policy* **2019**, *132*, 232–246. [CrossRef]
3. Tsantopoulos, G.; Arabatzis, G.; Tampakis, S. Public attitudes towards photovoltaic developments: Case study from Greece. *Energy Policy* **2014**, *71*, 94–106. [CrossRef]
4. Rahman, S.M.; Miah, M.D. The impact of sources of energy production on globalization: Evidence from panel data analysis. *Renew. Sustain. Energy Rev.* **2017**, *74*, 110–115. [CrossRef]
5. Arto, I.; Capellán-Pérez, I.; Lago, R.; Bueno, G.; Bermejo, R. The energy requirements of a developed world. *Energy Sustain. Dev.* **2016**, *33*, 1–13. [CrossRef]
6. Andreoni, V. Energy Metabolism of 28 World Countries: A Multi-scale Integrated Analysis. *Ecol. Econ.* **2017**, *142*, 56–69. [CrossRef]
7. *International Energy Outlook 2020*; International Energy Agency: Paris, France, 2020.
8. Doukas, H.; Karakosta, C.; Psarras, J. RES technology transfer within the new climate regime: A "helicopter" view under the CDM. *Renew. Sustain. Energy Rev.* **2009**, *13*, 1138–1143. [CrossRef]
9. Ohl, R.S. Light-Sensitive Electric Device. US Patent Office: 2,402,662, 25 June 1946.
10. Wolfe, P. An overview of the world's largest solar power plants. *PV Magazine International.* 18 June 2019. Available online: https://www.pv-magazine.com/2019/06/18/an-overview-of-the-worlds-largest-solar-power-plants/ (accessed on 30 March 2021).
11. Eurostat Renewable Energy Statistics. Available online: https://ec.europa.eu/eurostat/statistics-explained/index.php/Renewable_energy_statistics (accessed on 30 March 2021).
12. Jayanthi, S.; Vennila, C. Performance improvement in satellite image classification using adaptive supervised multi-resolution approach. *Comput. Commun.* **2020**, *150*, 200–208. [CrossRef]
13. Liu, M.; Fu, B.; Xie, S.; He, H.; Lan, F.; Li, Y.; Lou, P.; Fan, D. Comparison of multi-source satellite images for classifying marsh vegetation using DeepLabV3 Plus deep learning algorithm. *Ecol. Indic.* **2021**, *125*, 107562. [CrossRef]
14. Ioannou, K.; Tsantopoulos, G.; Arabatzis, G.; Andreopoulou, Z.; Zafeiriou, E. A Spatial Decision Support System Framework for the Evaluation of Biomass Energy Production Locations: Case Study in the Regional Unit of Drama, Greece. *Sustainability* **2018**, *10*, 531. [CrossRef]
15. Ioannou, K.; Karampatzakis, D.; Amanatidis, P.; Aggelopoulos, V.; Karmiris, I. Low-Cost Automatic Weather Stations in the Internet of Things. *Information* **2021**, *12*, 146. [CrossRef]
16. Dong, S.; Wang, P.; Abbas, K. A survey on deep learning and its applications. *Comput. Sci. Rev.* **2021**, *40*, 100379. [CrossRef]

17. Freedman, D. *Statistical Models: Theory and Practice*; Cambridge University Press: Cambridge, UK, 2009.
18. Mood, C. Logistic Regression: Why We Cannot Do What We Think We Can Do, and What We Can Do About It. *Eur. Sociol. Rev.* **2010**, *26*, 67–82. [CrossRef]
19. Kleinbaum, D.; Klein, M. *Logistic Regression: Why We Cannot Do What We Think We Can Do, and What We Can Do about It*; Springer: Berlin/Heidelberg, Germany, 2002.
20. Hosmer, W.; Lemeshow, S.; Sturdivart, X. *Applied Logistic Regression*; John Wiley & Sons: Hoboken, NJ, USA, 2013.
21. Soentpiet, R. *Advances in Kernel Methods: Support Vector Learning*; MIT Press: Cambridge, MA, USA, 1999.
22. Hearst, M.A.; Dumais, S.T.; Osuna, E.; Platt, J.; Scholkopf, B. Support vector machines. *IEEE Intell. Syst. Appl.* **1998**, *13*, 18–28. [CrossRef]
23. Steinwart, I.; Christmann, A. *Support Vector Machines*; Springer Science & Business Media: Berlin/Heidelberg, Germany, 2008.
24. Schraudolph, N.N. Fast Curvature Matrix-Vector Products for Second-Order Gradient Descent. *Neural Comput.* **2002**, *14*, 1723–1738. [CrossRef]
25. Li, Z. *Encyclopedia of Biometrics I-Z*; Springer Science & Business Media: Berlin/Heidelberg, Germany, 2009.
26. Verbeek, J.J.; Vlassis, N.; Kröse, B. Efficient Greedy Learning of Gaussian Mixture Models. *Neural Comput.* **2003**, *15*, 469–485. [CrossRef] [PubMed]
27. Gavali, P.; Banu, J.S. Chapter 6—Deep Convolutional Neural Network for Image Classification on CUDA Platform. In *Deep Learning and Parallel Computing Environment for Bioengineering Systems*; Arun, K.S., Ed.; Academic Press: Cambridge, MA, USA, 2019; pp. 99–122. ISBN 978-0-12-816718-2.
28. Quiroz, I.A.; Alférez, G.H. Image recognition of Legacy blueberries in a Chilean smart farm through deep learning. *Comput. Electron. Agric.* **2020**, *168*, 105044. [CrossRef]
29. Tan, W.; Zhao, C.; Wu, H.; Gao, R. A deep learning network for recognizing fruit pathologic images based on flexible momentum. *Nongye Jixie Xuebao/Trans. Chin. Soc. Agric. Mach.* **2015**, *46*, 20–25. [CrossRef]
30. Liu, Z.; Xiang, X.; Qin, J.; Ma, Y.; Zhang, Q.; Xiong, N.N. Image Recognition of Citrus Diseases Based on Deep Learning. *Comput. Mater. Contin.* **2020**, *66*, 457–466. [CrossRef]
31. Amara, J.; Bouaziz, B.; Algergawy, A. A Deep Learning-based Approach for Banana Leaf Diseases Classification. In *Datenbanksysteme für Business, Technologie und Web (BTW 2017)—Workshopband*; Mitschang, B., Nicklas, D., Leymann, F., Schöning, H., Herschel, M., Teubner, J., Härder, T., Kopp, O., Wieland, M., Eds.; Gesellschaft für Informatik: Bonn, Germany, 2017; pp. 79–88.
32. Huang, H.W.; Li, Q.T.; Zhang, D.M. Deep learning based image recognition for crack and leakage defects of metro shield tunnel. *Tunn. Undergr. Space Technol.* **2018**, *77*, 166–176. [CrossRef]
33. Yang, X.; Zhang, Y.; Lv, W.; Wang, D. Image recognition of wind turbine blade damage based on a deep learning model with transfer learning and an ensemble learning classifier. *Renew. Energy* **2021**, *163*, 386–397. [CrossRef]
34. Cheng, L.; Zhang, X.; Shen, J. Road surface condition classification using deep learning. *J. Vis. Commun. Image Represent.* **2019**, *64*, 102638. [CrossRef]
35. Zhang, Y.-D.; Satapathy, S.C.; Guttery, D.S.; Górriz, J.M.; Wang, S.-H. Improved Breast Cancer Classification Through Combining Graph Convolutional Network and Convolutional Neural Network. *Inf. Process. Manag.* **2021**, *58*. [CrossRef]
36. Wang, S.-H.; Govindaraj, V.V.; Górriz, J.M.; Zhang, X.; Zhang, Y.-D. Covid-19 classification by FGCNet with deep feature fusion from graph convolutional network and convolutional neural network. *Inf. Fusion* **2021**, *67*, 208–229. [CrossRef]
37. Malof, J.M.; Bradbury, K.; Collins, L.M.; Newell, R.G. Automatic detection of solar photovoltaic arrays in high resolution aerial imagery. *Appl. Energy* **2016**, *183*, 229–240. [CrossRef]
38. Butler, H.; Daly, M.; Doyle, A.; Gillies, S.; CiteHagen, S.; Schaub, T. Internet Engineering Task Force RFC: 7946. 2016. Available online: https://datatracker.ietf.org/doc/html/rfc7946 (accessed on 30 March 2021).
39. Harding, S.; Banzhaf, W. Fast genetic programming on GPUs. In Proceedings of the 10th European Conference on Genetic Programming; Ebner, M., O'Neill, M., Ekárt, A., Vanneschi, L., Esparcia-Alcázar, I., Eds.; Springer: Berlin/Heidelberg, Germany, 2007; pp. 90–101.
40. Abadi, M.; Barham, P.; Chen, J.; Chen, Z.; Davis, A.; Dean, J.; Devin, M.; Ghemawat, S.; Irving, G.; Isard, M.; et al. TensorFlow: A system for large-scale machine learning. In Proceedings of the 12th {USENIX} Symposium on Operating Systems Design and Implementation ({OSDI} 16), Savannah, GA, USA, 2–4 November 2016; USENIX Association: Berkeley, CA, USA, 2016; pp. 265–283.
41. Hubel, D.H.; Wiesel, T.N. Receptive fields, binocular interaction and functional architecture in the cat's visual cortex. *J. Physiol.* **1962**, *160*, 106–154. [CrossRef]
42. Haykin, S. *Neural Networks A Comprehensive Foundation*, 2nd ed.; Subsequent Edition (1 July 1998); Prentice Hall: Hoboken, NJ, USA, 1999; ISBN 978-0132733502.
43. LeCun, Y.; Boser, B.; Denker, J.; Henderson, D.; Howard, R.; Hubbard, W.; Jackel, L. Handwritten Digit Recognition with a Back-Propagation Network. In Proceedings of the Advances in Neural Information Processing Systems; Touretzky, D., Ed.; Morgan-Kaufmann: Burlington, MA, USA, 1990; Volume 2.
44. Krizhevsky, A.; Sutskever, I.; Hinton, G.E. ImageNet classification with deep convolutional neural networks. *Commun. ACM* **2017**, *60*, 84–90. [CrossRef]
45. Szegedy, C.; Vanhoucke, V.; Ioffe, S.; Shlens, J.; Wojna, Z. Rethinking the Inception Architecture for Computer Vision. *CoRR* **2015**. [CrossRef]

46. Yang, K.; Qinami, K.; Fei-Fei, L.; Deng, J.; Russakovsky, O. Towards fairer datasets. In *Proceedings of the 2020 Conference on Fairness, Accountability, and Transparency*; ACM: New York, NY, USA, 2020; pp. 547–558.
47. Russakovsky, O.; Deng, J.; Su, H.; Krause, J.; Satheesh, S.; Ma, S.; Huang, Z.; Karpathy, A.; Khosla, A.; Bernstein, M.; et al. ImageNet Large Scale Visual Recognition Challenge. *Int. J. Comput. Vis.* **2015**, *115*, 211–252. [CrossRef]

Review

Social Acceptance of Carbon Capture and Storage (CCS) from Industrial Applications

Katja Witte

Wuppertal Institute for Climate, Environment and Energy, Division Future Energy and Industry Systems, Doeppersberg 19, 42103 Wuppertal, Germany; katja.witte@wupperinst.org; Tel.: +49-202-2492-218

Abstract: To limit global warming, the use of carbon capture and storage technologies (CCS) is considered to be of major importance. In addition to the technical–economic, ecological and political aspects, the question of social acceptance is a decisive factor for the implementation of such low-carbon technologies. This study is the first literature review addressing the acceptance of industrial CCS (iCCS). In contrast to electricity generation, the technical options for large-scale reduction of CO_2 emissions in the energy-intensive industry sector are not sufficient to achieve the targeted GHG neutrality in the industrial sector without the use of CCS. Therefore, it will be crucial to determine which factors influence the acceptance of iCCS and how these findings can be used for policy and industry decision-making processes. The results show that there has been limited research on the acceptance of iCCS. In addition, the study highlights some important differences between the acceptance of iCCS and CCS. Due to the technical diversity of future iCCS applications, future acceptance research must be able to better address the complexity of the research subject.

Keywords: carbon capture; acceptance; public perception; industrial applications; literature review; knowledge; awareness; communication

check for **updates**

Citation: Witte, K. Social Acceptance of Carbon Capture and Storage (CCS) from Industrial Applications. *Sustainability* **2021**, *13*, 12278. https://doi.org/10.3390/su132112278

Academic Editors: Georgios Tsantopoulos and Evangelia Karasmanaki

Received: 25 September 2021
Accepted: 3 November 2021
Published: 7 November 2021

Publisher's Note: MDPI stays neutral with regard to jurisdictional claims in published maps and institutional affiliations.

1. Introduction

To limit global warming to 1.5 °C, the use of carbon capture and storage technologies (CCS) is considered to be of major importance [1–5]. In international parlance, CCS stands for a mix of technological processes for CO_2 capture and storage. These are large-scale processes in which carbon dioxide (CO_2) is captured from huge CO_2 point sources. The captured CO_2 is transported via pipeline, ship, or heavy transport and then either reused or injected underground into a suitable geological formation (onshore or offshore) [6].

The use of CO_2 capture processes is feasible both in fossil-fired power plants for electricity generation and in energy-intensive industrial processes (for example, steel or cement plants) and could enable a significant reduction in CO_2 emissions in these applications. According to the International Energy Agency [7], fossil-fired power plants accounted for about 42.5% of total global CO_2 emissions in 2013. In comparison, the share of CO_2 emissions caused by industrial activities was around 25%.

In recent years, the discussion around CCS has increasingly focused on its use in the context of industrial facilities (in the following, the term "industrial CCS" is referred to as iCCS). This is mainly because the technical options for the extensive reduction of CO_2 emissions in the area of energy-intensive industries without the use of iCCS are not sufficient to achieve the targeted GHG neutrality in the industrial sector. Ref. [4] However, what exactly distinguishes the term iCCS from the classic CCS application? Fossil fuels are an essential input to the production process of the steel, cement, lime and chemical industries, the so-called energy-intensive industries. These fuels are used in the industries for their chemical and physical properties rather than as a primary energy source for power generation, as is the case with CCS [8]. However, unlike electricity generation, it is not possible to replace fossil fuels with renewable energy sources to reduce emissions. This literature review focuses explicitly on the application of CCS to these industrial processes.

The debate to date on the commercial introduction of CCS in fossil-fired power plants (abbreviated below as CCS) has made it clear that numerous other factors are relevant in addition to purely technical and economic indicators. On the part of policymakers, there is a need for a reliable agreement and strategy on the future role of CCS, taking into account international developments around CCS as well as other technological climate protection paths. This will create planning and legal certainty for industry and society and enable the early development of CO_2 infrastructure.

Another essential factor, which is the focus of this publication, is the social perception of iCCS technologies and the possible assessment of their future acceptance. Previous research on CCS acceptance has made it clear that CCS technologies may meet with strong opposition, especially in regions where the applications have been tested or were intended to be deployed on a long-term, permanent basis [9,10]. For example, in Germany and the Netherlands, some projects to explore potential CO_2 storage formations were abandoned early, primarily due to massive opposition from local communities [11,12]. Since the early 2000s, the number of scientific publications on the acceptance of CCS has continuously increased (see also Section 3). The perception and acceptance of CCS is strongly dependent on the respective country [13] and due to the low level of knowledge about CCS [14,15], it remains difficult to make valid predictions about how specific local attitudes towards CCS might develop.

This study is the first literature review to address the acceptance of industrial CCS (iCCS). The objectives of this study are fourfold. First, it examines the extent to which iCCS acceptance has already been empirically studied. Second, an analytical framework is proposed to systematically review the existing literature. Third, factors that influence iCCS acceptance are identified and discussed based on the review. Fourth, the results on the acceptance of iCCS are compared with the acceptance of CCS in the context of fossil-fired power plants. The assumption is that the attitude of society towards iCCS differs from the attitude towards CCS along individual process steps and value chains. In this regard, first scientific findings are emerging [16,17]. It is unclear in which direction these attitude differences tend.

This study's results should not only contribute to the scientific discussion and further development of the research field, but also hopefully feed into the ongoing practical iCCS discourse in industry and politics. At the international level, there are already associations of industry players testing different technical use cases for iCCS in the form of pilot projects, for example the European Cement Research Academy (ECRA). In some industrial processes, the capture of CO_2 emissions is already practiced today, and currently the first projects are underway worldwide in different sectors, such as chemicals (Illinois Industrial), iron and steel (Abu Dhabi Phase 1), and hydrogen (QUEST) [18]. The results of this literature review should also provide indications of possible communication and empowerment needs on the part of the general public and at the same time enable the more technology-based scientific disciplines to place their developments on iCCS in a broader societal context.

In order to be able to better classify the present analysis, the technological component of the research object should first be explained in more detail. For a better understanding of this, Renn's classification [19] of the three areas of technology and their acceptance parameters is helpful. He distinguishes between (1) products—everyday and leisure technology; (2) technology in working life; and (3) external, large-scale and risky technology. The three technology areas differ in terms of their acceptance testing criteria. In the case of current acceptance research on carbon capture and usage (CCU), for example, the focus is often on the concrete evaluation of an end product, which can often be explained in terms of buying or not buying, manageability, long-term durability or direct physical risks (although the research approach here is also broader, for example [20–22]). In the context of the present analysis, all scientific publications dealing with acceptance research on concrete end products (e.g., mattresses, fuels) of CCU technologies were explicitly excluded. This also appears consistent with [23], who clarify that CO_2 utilization is often compared and contrasted with CCS; however, they are two different technology pathways so it is necessary to address and evaluate these technologies separately. Since the subject of the present analysis is the broader society, technology area 2, which deals with technology in the workplace and thus targets "employees", can also be excluded. Following the exclusion principle, only studies dealing with iCCS as an external, large-scale and risky technology (area 3) were analyzed here. For this technology area, the test criteria of acceptability are, for example, societal interests, rights, responsibilities, and legitimacy issues. The focus of this review is therefore on technology pathways that capture CO_2 on a large scale and transport it for further purposes without further differentiating whether and how the CO_2 is further used.

This paper is structured as follows. First, Section 2 presents the selection of articles analyzed, the methodological approach and the acceptance factors for CCS already identified in the scientific literature, which are also used here as analysis dimensions. The results of the content analysis are explained in detail in Section 3. In the Discussion (Section 4), we present which of the identified acceptance factors for iCCS can be considered crucial for the further development of iCCS and which scientific implications the results induce. The conclusions in Section 5 illustrate some rough propositions for relevant groups of actors dealing with issues of societal acceptance on iCCS in the future.

2. Materials, Methods and Acceptance Factors

In order to assess the state of scientific research in the field of acceptance of industrial CCS, a content analysis of scientific articles was conducted. Only articles published in English between 2012 up to and including the end of 2020 were included. This time period was chosen because, to the best of the author's knowledge, no articles were published before 2012 that approached this topic. Thus, the chosen period of analysis seemed sufficient to generate as complete an overview as possible of the state of the scientific literature on this topic.

2.1. Selection of Articles

Articles were identified using two online databases. First, the online database of the publisher Elsevier (sciencedirect.com), a full-text database with an inventory of more than 16 million articles and book chapters [24]. Although documents from other scientific publishers are not included, Elsevier is one of the top 5 publishers in the world with over 2000 journals published [25]. Second, the online database was used through scholar.google.com. Google's search engine presents only scientific literature; that is, books or papers from professional journals [26]. Using these two most popular online databases, it was possible to generate the largest possible proportion of scientific literature on the topic of iCCS acceptance.

Only scientific papers, book and conference contributions that could be generated by keyword searches via the two online databases were included in the analysis. In addition, one master's thesis was evaluated that was identified via the online database scholar.google.com and appeared to be relevant. No other dissertations or master's or bachelor's theses were systematically searched for.

Items were identified from November 2020 to 16 January 2021. The following search terms were used to select the technology:

- carbon capture and storage;
- carbon capture;
- CCS;
- carbon capture and storage industry;
- carbon capture industry;
- CCS industry.

The technical search terms were each combined with the following acceptance-related terms:

- acceptance;
- acceptability;
- perceptions;

- attitudes;
- public opinion.

Using a combination of search terms, between 4099 (maximum at sciencedirect.com) and 16,900 (maximum at scholar.google.com) articles were identified in the two online databases. Only articles that explicitly address the topic of industrial CCS were to be included (see Section 1 for narrowing criteria). For further identification of these articles from the existing material, the so-called PRISMA criteria were followed [27]. Based on this procedure, a complete search strategy for one of the databases used is presented below. The presentation is intended to create the prerequisite for the best possible reproducibility of the search.

The search strategy described here as an example refers to the online database scholar.google.com. As previously described, the initial selection was made according to the search terms presented above. With the search term "carbon capture industry acceptance", approximately 16,900 articles were identified on 16 January 2021 (initial access on 8 November 2020). In advance, the search of the articles was restricted to the years from 2012 to 2020 inclusive in the menu under "select period". Subsequently, the search result was sorted by relevance (an option offered by the online database in the menu). The individual short descriptions of the list of results on the homepage were read (not clicked on) and checked to see if all individual search terms were included in the respective text descriptions. This was an indication that all search terms were actually included in the respective target article. In addition, it was checked whether the keywords appeared in the desired context. If, for example, the term "industry" was linked to "coal industry" and the title also indicated that the article was exclusively about CCS as a low-carbon technology for energy generation, the article was excluded from further analysis. The matches identified in this way were further checked for accuracy of fit by reading the respective abstract or, if this did not appear to be sufficient for assessing accuracy of fit, the conclusions.

All hits identified in this way were then included in the pool for further analysis. During the course of the search, it became apparent that after approximately the fourth to fifth page of results on the homepage, the articles listed no longer appeared relevant for the analysis due to missing keywords in the short text. Additional tools from scholar.google.com were used to further identify relevant articles. The option "cited by" lists all articles in which the original hit was cited. A check of these articles was performed according to the criteria already mentioned. The option "related articles" was also used. Using these options, few additional articles could be identified. In addition, an "alert" was created, which was used to automatically notify the author via email when new articles with the given keywords appeared. This option appeared valuable in generating articles that did not appear until the end of the analysis period. To ensure that all articles published by the end of 2020 were identified, a final search query took place in mid-January 2021. The search query at sciencedirect.com followed the same procedure and selection criteria. Beyond the use of the two online databases, a few articles were identified via the references or sources of the articles already identified and read in the course of the evaluation and included in the analysis pool. Using these procedures, a total of 67 articles were identified and included in the closer analysis.

All 67 articles were then read completely. Of these, 42 articles were excluded. There were two main reasons for articles to be excluded:

- Some articles only hinted at possible acceptance conditions for iCCS in their conclusions. A presentation of these references to acceptance seemed mostly comprehensible, but since they could not be sufficiently derived empirically from the study results, the articles were not considered for further analysis.
- Other articles, as part of their methodological approach, focused only on the use of CO_2 (CCU) and did not differentiate by source (industrial capture or capture in the context of electricity generation).

Ultimately, 25 articles met the criteria to be included. It can be assumed that a large part of the relevant literature was identified.

2.2. Methodical Approach

A qualitative content analysis of 25 articles was carried out using the MAXQDA software. The software allows qualitative data and text analyses and is internationally established in the field of science. For content analysis, a deductive category system was developed (referred to as "analysis dimensions" in the following). It was derived from the previous state of attitude and acceptance research on CCS. During the coding process, some of the analysis dimensions were adapted and the possibility was left open to inductively generate new dimensions, in accordance with the approach of [28]. The individual dimensions or acceptance factors are discussed in more detail in the following subsection.

2.3. Acceptance Factors from the Field of CCS

A wealth of individual studies, results, and initial overview studies are available on the perception, attitude, and acceptance of CCS [29,30]. The first studies on the subject appeared from 2002 [31–33]. In the literature up to 2015, publications on the acceptance of CCS focus mainly on the use of the technology in the context of fossil power generation. Therefore, a considerable number of factors determining the acceptance of CCS have been proposed, many of which are commonly used to explain the acceptance of new technologies. There is not a consensus on the one model best suited to predict CCS or technology acceptance [29], although there are publications that present a technology acceptance framework [34] or provide a model approach for selected factors [20,35,36]. Most studies, as mentioned, examine the determining factors along specific research questions that can be categorized into some thematic groups. These groups of topics mainly include (a) general acceptance analyses "of the general public" in one country or in several countries; (b) analyses of real-life-projects across different groups of actors, including the local society; (c) analyses on communication and participation of CCS; and (d) analyses on specific process steps of CCS, especially storage. In recent years, since 2015, more studies have been added on the topic of CCU [20–23,37–41], which can be assigned to the abovementioned group of topics and perhaps also represent a research unit in their own right (cf. chapter 1). However, these factors have predominantly become established and are repeatedly used as a starting point for new research studies and questions. Additionally, for the analysis of the articles identified here for the topic area of industrial CCS, analysis dimensions were generated on the basis of the acceptance factors just mentioned or the state of science (cf. Table 1, here especially the factors from 1 to 8) (a similar set of influencing factors can also be found in the acceptance research on the energy transition [42]). After the initial review of the articles (relevance check), additional dimensions that seemed useful for analyzing the acceptance of industrial CCS were added (compare factors 9 to 11).

Table 1. Analysis dimensions of iCCS acceptance within the framework of the review.

No	Potential Acceptance Factors	Explanation	Source [1]
1	Perceived benefits	What personal/societal benefits are associated with iCCS? (social benefits include environmental benefits)	[13,16,43–45]
2	Perceived risks	What personal/societal benefits are associated with iCCS (including possible costs)?	[13,16,31,44]
3	Values/attitudes	Can certain patterns of attitudes be identified that have an influence on the acceptance of iCCS?	[34,44,46]
4	Regional factors	What contribution do regional factors make to the evaluation of iCCS technology? For example, are citizens' previous experiences with potential iCCS companies or local storage options decisive?	[11,12,47,48]

Table 1. *Cont.*

No	Potential Acceptance Factors	Explanation	Source [1]
5	Trust	How important is trust in iCCS actors for acceptance? What are the reasons for a lack of trust?	[10,41,49–51]
6	Knowledge/awareness	How does the level of knowledge about iCCS influence the evaluation of the technology? Are initial perceptions of iCCS also important for acceptance?	[52,53]
7	Communication/participation	What is the need for participatory instruments/communication concepts for the implementation of iCCS? Which communication strategy do companies pursue for marketing/which actors do they involve?	[54–58]
8	Socio-demographic factors	Can different socio-demographic factors induce distinguished iCCS perceptions?	[44,47,59,60]
9	Perceived differences to iCCS in the power plant sector	Are there significant differences between the acceptance of CCS in the power plant sector and for industrial applications?	[16,17,41,61–63]
10	Evaluation according to process step	How is the use of iCCS evaluated along the value chain stages (from investment to capture/transport to CO_2 storage and possible reuse)? How is iCCS assessed in the context of other carbon abatement technologies and pathways?	[14,17,41,64]
11	Regulatory/political aspects	How can a lack of regulatory frameworks, political support and unresolved/complex approval procedures influence iCCS acceptance?	[14,65–67]

[1] It should be noted that the sources cited in the table are only a small excerpt of possible sources that have dealt with the topic. A comprehensive presentation of studies that have produced results on the respective dimensions of analysis is not intended here. Moreover, the assignment of sources is not exclusive because the respective studies often explored several categories of analysis. In this respect, relevant sources were also assigned to more than one analysis category.

In the following, the results of the evaluated articles are presented along the acceptance factors described in Table 1. In addition to a presentation of the characteristic features, such as methodology used, year of publication and technology path, the analysis clarifies which influencing factors were assumed and investigated to explain the acceptance of industrial CCS. In Section 4 (Discussion), these results are then reflected on and classified in the context of the entire acceptance research on CCS so that first insights can be gained on whether the acceptance factors on iCCS differ from the previous ones, in which areas they differ, if any, and whether new factors have been added.

3. Results

3.1. Characteristics of the Analyzed Articles

To place the iCCS publications in the overall context of all publications on the topic of CCS acceptance, it should be mentioned in advance that until circa 2014 the number of scientific publications on the acceptance of CCS increased steadily [29]. Between 2015 and around 2018, the number of publications on the topic of CCS acceptance then remained at a lower level than in the years between 2010 and 2014 [30]. Up to this point, publications on the acceptability of CCS focused on the use of the technology in the context of fossil fuel power generation. Triggered by the Paris Agreement 2015 [2], which highlighted the urgency of limiting global warming to as close to 1.5 °C as possible, as well as a number of other publications [1,3–5], as described in Section 1, the discussion about CCS has continuously broadened and has more often focused on technology pathways that are not directly related to fossil energy production. Since then, there has also been an increasing number of scientific publications dealing with the acceptance of different technology paths of CCS.

The articles analyzed here were published between 2012 and 2020. Table 2 illustrates the year of publication of the articles in combination with the selected technology path.

Table 2. Theme clusters of iCCS acceptance in combination with year of publication [13–17,30,41,61–78].

Technology Path	2012–2015	2016	2017	2018	2019	2020	2021 [1]
iCCS without further specification		Haug et al. [64], Broecks et al. [63]	Pihkola et al. [69]	Xenias et al. [68], Kashintseva et al. [67], Ilinova et al. [70], Thomas et al. [71], van Os [72]	Tcvetkov et al. [30], Whitmarsh et al. [13], Serdoner [73]	Swennenhuis et al. [65], Boomsma et al. [74]	
Evaluation of different technology pathways (variation of source, transport, storage)	De Best-Waldhober et al. [17], Wallquist et al. [16], Dütschke et al. [61]					Offermann-van Heek et al. [41]	
iCCS with focus on CO_2-storage				Gough et al. [14]			
iCCS as low carbon technology for energy-intensive industry (cement, steel)					Aursland et al. [66]		Williams et al. [62]
Bioenergy with CCS (BECCS)			Kojo et al. [75]		Haikola et al. [76]	Rodriguez et al. [77]	
iCCS with reference to hydrogen applications					Alcalde et al. [78]		Glanz et al. [15]
Total	2/1	2	2	6	6	4	2

[1] These two articles have already been published in mid-January 2021. Due to their relevance, the author decided to include them before completing this article at the end of January. No other articles from 2021 were included in the analysis.

As shown in Table 2, by the end of 2020, most articles on iCCS were published in 2018 and 2019 (n = 6 in each year). A slight majority of the 25 articles (n = 13) use the terminology "industrial CCS" (compare row 1 Table 2), but do not further explain which technological concept of iCCS technologies is involved in the definition or within the operationalizations. This is not surprising, as the technological applications of iCCS are highly complex along the process steps and the different value chains that may be involved.

To address this complexity, four of the studies provided their participants with a selection of different realistic CCS technology pathways to evaluate (compare row 2 Table 2), which at least allowed for a more differentiated view according to different CO_2 sources, such as the evaluation of CO_2 capture in a chemical plant [41]. Since 2019, there has been an increase in acceptance studies investigating the impact of specific industrial CCS applications, such as from cement or steel plants or for the BECCS sector. These studies are often linked to specific project proposals, for example the ALIGN project (It is expected that further scientific publications on the acceptance of iCCUS will be published in 2021 from research projects that have been and will be funded within the framework of Horizon 2020 of the European Commission, such as the ALIGN-CCUS and STRATEGY CCUS projects) [74], and concentrate on regions with industrial clusters that are significant geologically and in terms of their industrial structure with regard to the development of iCCS and are already being scientifically researched in part (compare lines 4 to 6, Table 2).

The analyzed articles on iCCS acceptance come from a total of 15 different countries, of which European countries represented 13—an overwhelming majority. The following European countries were involved in the preparation of the articles: United Kingdom = 7; The Netherlands and Germany = 4 each; Norway = 3; Finland and Sweden = 2 each; and Austria, Belgium, Lithuania, Portugal, Romania, Spain and Switzerland with one article each. Five of the European articles involved more than one country. As mentioned at the beginning, previous studies on the acceptance of CCS have made clear that protests and risk perceptions on CCS have formed along exploration plans and projects, especially in Europe—particularly in the Netherlands [12] and Germany [11].

In this respect, if an iCCS strategy is to be pursued on the political level in the long term, these countries seem to have a particular interest in predicting future developments regarding the acceptance of iCCS. For Great Britain, the situation is similar; here, according to [79], 17.2% would "probably not use" or "definitely not use" CCS technologies according to a representative survey. A further three articles come from Russia and another one from the United States of America. According to [30], Russia has a special interest in the use of enhanced oil recovery (EOR) technology, which requires a lot of CO_2, and therefore is considering CCS as a future option to develop this technology.

The relevant articles on the acceptance of iCCS were published in a wide range of journals. In total, the 25 articles come from 15 different journals. *The International Journal of Green-house Gas Control* accounts for 8 articles—by far the most. This is followed by the journals *Energy Procedia* and *Journal of Cleaner Production*, with 2 publications each on the topic. One of the analyzed articles is a Master's thesis, which was written at the University of Graz and cannot be assigned to any journal [73].

Different theoretical concepts and approaches were used in the articles included. Twelve of the analyzed articles on iCCS acceptance do not mention any theoretical concepts. The concept of Wüstenhagen [80] to classify three different dimensions of social acceptance is mentioned and applied in two articles. Studies that focus their analysis more on the regional or project level often include actor and communication-related approaches, such as the theory of public engagement in [68], the social licence to operate (SLO) in [14,74], the end-to-end stakeholders involvement approach in [67], the concept of procedural fairness in [62], the concept of media agenda-setting in [75], the stakeholder theory for management in [70] and the cognitive theory of shifting coalitions in [73].

In addition, the articles mention social-psychological concepts that illuminate social behavior even more against the background of cultural aspects and certain values, such as the theory of planned behavior in [30] and, in the context of the Master's thesis, the concept of the Ethical landscape of CCS, the theory of worldviews and the cultural theory to specify belief systems in [73]. Two of the analyzed articles reflect their findings on iCCS acceptance to the whole debate on energy system transformation using the just transition approach [65,78] or the multidimensional research concept as in [15].

A complete table of the analyzed articles with the categories "first author", "year of publication", "method(s) used", "country", "iCCS-related technology", and "important statement in relation to iCCS" is provided in the Appendix A (Table A1: Overview of the analyzed articles).

3.2. Key Findings along the Dimensions of Analysis as well as Additional Insights

In the following, the main results of the analyzed articles are presented along the analysis dimensions shown in Table 1.

3.2.1. Perceived Benefits

The results of the studies analyzed have identified some benefits that appear to be associated with the use of iCCS and thus may have a positive impact on social acceptance. These benefits include the possibility of creating local and national value through iCCS projects [64].

For example, the municipality of Porsgrunn in Norway considers iCCS important in legitimizing industry in the region and thus sustaining related jobs in the long term [64]. Additionally, ref. [71] sum up that the potential of iCCS can protect and rejuvenate historical employment patterns and this opportunity makes iCCS an attractive option for an area. This is also important to counteract the out-migration of the local population that threatens to occur if established industries go away [64]. Beyond protecting existing jobs, ref. [71] make the argument that providing infrastructure for iCCS can also create additional employment opportunities in the region. Consistent with this, communities hosting CCS projects would benefit economically from the jobs and revenue that the industry would provide [13].

In addition, regional clusters containing multiple capture projects can benefit from shared CO_2 transport and storage infrastructure to maximize value, share investment decisions and operating costs, and thus reduce development costs [78]. Thus, ref. [64] postulate benefits from mergers of larger regional clusters for iCCS (across national borders). For example, in their study, they identified the notional "Skagerrak Cluster" for the countries of Norway, Sweden, and Denmark, which identifies some key geographic features that have good conditions for establishing iCCS technology (similar to the northeast region of Scotland). The advantages come from the possibility of storing the CO_2 offshore, with emission sources relatively close to the sea. According to [64], the relevance of looking more closely at the Skagerrak cluster provides valuable input for evaluating acceptance and communication challenges for other iCCS clusters in the Nordic region. These benefits of iCCS overall can be linked to increasing the economic viability of both the technology itself and the region in question, these are benefits that [30,70] also highlight in their study.

However, not only is the preservation or renewal of existing economic structures identified as a benefit of iCCS, but the technologies should also serve to promote and profile municipalities and regions as environmental and technological leaders, ultimately to develop new industrial activities [64]. In this context, there is also talk of a potential image boost for iCCS industries and regions [62]. For example, refs. [75,77] argue the relevance of developing and deploying BECCS, a technology pathway discussed as an advantage for forest-rich countries such as Finland [75] and which holds the potential to establish itself as a "first mover" [77]. Without BECCS it would be a challenge to meet emission targets, but with BECCS Finland could gain advantages by saving and trading emission rights [75] (see also Section 3.2.11).

Regarding the impact of environmental effects (reduction of CO_2 emissions, slowing of climate change) and their classification as a benefit for the acceptance of iCCS, there are different results in the analyzed studies. Some study results suggest that attributing the benefits of iCCS to improving the regional and global environmental situation can create an advantage for the perception of acceptance [15,30,70,75]. Similarly, the results of a representative study in Canada, the USA, the UK, the NL, and Norway illustrate that iCCS can help mitigate climate change and support the economy according to the respondents in [13], which could be interpreted as a benefit for the technology. However, the same study also highlighted that framing CCS as dealing with 'waste' (in conjunction with CO_2 reuse) seems to be more persuasive in encouraging support than framing it in terms of climate or economic benefits. The authors of [74] critically note that the siting of new or expanded iCCS facilities is more likely to be associated with national and international benefits, for example achieving energy and climate goals and economic revenues (on this also see [70]), and that the apparent benefit to local communities may turn out to be a potential burden, for example through subjectively perceived risks. Such a perceived imbalance between (negative) local impacts and national or global benefits would pose a challenge when it comes to public response to iCCS technologies [74]. Hence, currently there is no consistent evidence from the scientific community as to whether iCCS is perceived as a mitigation option for CO_2, and thus as a climate technology, and whether this has a positive or negative effect on the perception of the benefits of the technology. Moreover, such a perception is certainly also dependent on many regional factors.

For completeness, here are the five main benefits of CCS industrial projects according to [70]: (1) reduction of negative impacts on the environment, (2) contribution to socio-economic development of regions and territories, (3) attractive direction for socially responsible investments, (4) support for sustainable development of companies involved in CCS projects, (5) use of CO_2 for purposes such as improving oil recovery by oil and gas companies, increasing energy efficiency of industrial companies.

The analysis of perceived benefits gives the impression, as also indicated by [30] and previous studies on the benefits and risks of CCS, that benefit perception may exert a stronger influence on iCCS acceptance than risk perception.

3.2.2. Perceived Risks

According to the studies analyzed, the use of iCCS technologies is associated with various societal risks that can have a negative impact on acceptance. These include perceived risks at the local level, for subsequent generations and for ecological and economic systems, but also risks for making political decisions that do not contribute to improving climate protection in the long term. The most frequently mentioned risk perceptions in the studies relate to negative health impacts, especially for people living near CO_2 storage and transport infrastructure [62].

The local impacts of iCCS are particularly addressed here [68], and with it the accompanying sense of unfair treatment of those who suffer disadvantages [30,74]. It is believed that iCCS could become locally entrenched as a "risky technology" in the perception of local and regional populations [15], especially if CO_2 storage occurs on land [77]. Hazards are expected from possible CO_2 leakage and seismic risks [15,75,77]. The perception would not improve even if already existing infrastructure were used [15]. The same applies to the CO_2 transport route; here, too, leakages and unforeseen risks are feared by the population [15]. In addition, several stakeholders in Germany expected so-called spillover effects, which occur when already existing rejections of CO pipelines are transferred to CO_2 pipelines on the grounds that these transport options are not sufficiently differentiated in society [15].

In this context, the fear of a lack of acceptance of responsibility on the part of politics and industry [71,77] and the societal desire to avoid uncertainties are mentioned [30], especially when it comes to long-term monitoring of CO_2 infrastructure, which is primarily intended to ensure the protection of future generations [71,73]. In addition to health risks from the use of iCCS, ecological risks were also mentioned in the analyzed articles [15,75,76], which can have an unfavorable impact on acceptance. For example, interventions in the ecological system through the construction of new CO_2 infrastructure can permanently endanger the environment [15]. In addition, one study expressed fears about the possible effects of stored CO_2 in the seabed [73], which could, for example, affect the fauna and flora of nearby coastal regions and lead to catastrophic consequences there [71]. At the same time, the use of iCCS technologies was interpreted as a standstill for other climate protection measures in industry that would lead to lock-in effects of unsustainable corporate practices [73]. However, the results on the perception of iCCS technologies are partly contradictory; on the other hand, there is apparently the concern that without their use, no adequate emission reductions for the climate can be achieved by energy-intensive industries [62] (which can ultimately be seen as an advantage for iCCS).

In addition to these societal risks, the studies also mentioned some personal risks that may be decisive with regard to the perception of iCCS. These include, in particular, the previously mentioned perceived health risks, which could lead to a strong rejection of iCCS technologies, especially on the part of the local population [13,30,71]. Personal risks may also be perceived in conjunction with the economic factors of iCCS. For example, the results of the analyzed studies illustrate that the factor of employment can be perceived as both a personal risk and a benefit [14,65] for people in a region in the context of iCCS. For example, one study expressed concerns that iCCS may impose costs that are then offset by, for example, lower employment levels in iCCS operations. On the other hand, the introduction of the technologies could create new areas of work and if steps were taken to retrain and employ industrial workers within the iCCS sector, this would be a benefit [71]. However, there has been an equal concern that there may be inflation of products through use with iCCS and in the long run this effect will contribute to industrial companies becoming uncompetitive in the global market and may lead to local plant closures [65].

3.2.3. Preferences/Values

In the context of the studies analyzed, a variety of values and attitudes were explained that can have an influence on the acceptance of iCCS. These broadly include cultural identity, the closely related moral concepts of a society, environmental awareness, the perceived influence of iCCS on people's living conditions and attitudes toward technological developments and industry.

According to the study by [13], nationality is the strongest predictor of support for iCCS. Closely related to nationality is the cultural identity of a country. Thus, a study explained that compensation services to communities [74] must take into account the cultural as well as the social context [14,30,62]. Here, it is especially important that sacred values such as human safety are not mixed against a secular value, for example, by accommodating a hazardous facility in exchange for monetary compensation [74]. Certain normative ideas and moral values are also obviously advantageous for the development of a positive attitude towards iCCS [63,76]. Insofar as the use of iCCS can compensate for possible inequalities in society [65], for example, by allowing regions with a high proportion of energy-intensive industries to hold on to their economies to some extent or to operate them in a climate-friendly manner through iCCS, this represents an advantage for the perception of iCCS [64]. However, such perspectives do not go hand in hand with the moral notion that iCCS is interpreted as an intrusion into the subsurface "wilderness" or that BECCS is morally indefensible due to the still unclear availability of biomass, as stated in [71]. A view that, according to [71], occurs among those with strongly ecological values. According to [71], iCCS can only contribute to justice in society where a common understanding of cultural, natural and socio-economic systems prevails.

The influence of environmental awareness on the acceptance of iCCS is still evaluated very differently. Thus, ref. [13] clarify that a high environmental awareness can lead to a low acceptance of iCCS as the technology is seen as less important for coping with climate change than other technological options [63]. Whereas BECCS technologies seem to get a better rating in [71] compared to CO_2 capture from further industrial processes (here certain views of environmental awareness do not seem to be in conflict with the moral risks of BECCS mentioned above). Either way, BECCS is obviously viewed positively here because it is more likely to be associated with natural processes through the use of biomass [16]. However, if iCCS technologies are placed in the larger context of addressing climate change, where the technologies are embedded as part of an overall strategy to reduce CO_2, their perception as an environmentally conscious technology may change if necessary [13,65]. Here, the urgency to address climate change postulated in recent years seems to have become a helpful vehicle for improving society's perception of iCCS technologies [63]. Another step towards valuing iCCS as an environmental technology focuses on the perception of CO_2 as a significant resource [64] rather than a waste product (see Section 3.2.1) or iCCS as a socially desired argument to support energy intensive industries in the context of political decarbonization intentions [53].

It remains open whether, far from being environmentally conscious, people can develop a positive perception of iCCS out of a certain technological affinity. The authors of [30] present a study in which people with a positive attitude toward gas infrastructure development are more supportive of iCCS than people without this attitude. In addition to environmental awareness and technological affinity, the perceived impact of iCCS on people's concrete living conditions is also likely to be significant in assessing acceptance [68]. For example, results from a focus group [71] illustrate people's fears that a life based on the renewable energy technology system may be very regimented and "robotic" and that this development may negatively affect previously valued lifestyles. In light of these considerations, the use of iCCS technologies is evaluated in a different context; in which through them traditional ways of life can be maintained for longer, which is evaluated as quite positive [70]. The authors of [13] also found in their study that people with energy-intensive lifestyles were more likely to prefer iCCS than others because they too could maintain their lifestyles while not being accused of promoting climate change.

The general attitude of the population toward the industry could also be an indicator for the future acceptance of iCCS. This is an aspect that will be discussed in more detail in the following section, as it is very closely linked to questions of the regional affiliation of the public.

3.2.4. Regional Factors

In this section, we will focus on the factors that can exclusively determine the regional characteristics and conditions for the development of iCCS acceptance (independent of other factors such as trust, knowledge, and communication, which can also influence the regional perception of iCCS). These factors on regional specificity include the specific history of an area and the regional perception of iCCS technologies in the context of other developments, such as the economic activities and geological conditions of the region.

The results of the studies analyzed suggest that despite the processes of deindustrialization in advanced capitalist economies, deeply rooted cultural narratives of industrial modernity and manufacturing employment remain powerful markers of identity and social progress [64,71]. In regions with an industrial heritage, where the local public feels connected to industry, this identity is particularly high [74]. Regional populations appreciate it when industrial actors inform them and involve them in their activities and plans to give them a sense of belonging and identity [66,74]. It is becoming apparent that people in such regions are concerned that these industries remain fully intact and are becoming sustainable [13,62].

Ref. [14] contribute to this thesis, for example, with the study of Teesside (UK). Teesside is a conurbation with a strong industrial base that residents rely on. Ref. [74] also assume that people in such regions are more positive about iCCS development than people who are less rooted in their industrial heritage. For example [66], describes that the Norcem industry began producing cement as early as 1919 and quickly became a major player in the economic life of the region. Ref. [64] emphasize the aspect of habituation. If people are used to industrial activities, especially when industry has operated in the area for decades, this has a positive effect on trust towards local industry and politics. For example, residents in northern regions are also accustomed to transporting products that are considered more dangerous than CO_2, such as ammonia.

Ref. [13] assume that areas where iCCS plants are likely to be built are typically those locations where (analogous) industry already exists. Subjective familiarity with such an industry could also serve to reduce the perceived risks associated with new infrastructure, leading to greater acceptance (or tolerance) of iCCS within regions. Fundamentally, according to [74], there is a need to understand local social realities, such as understanding what a particular place means to the local public, as well as how iCCS technology can impact this meaning at an early stage of the projects.

However, refs. [15,30,67] also emphasize that past economic activities, for example, when coal mines are present in the region or there have been incidents with health impacts for local residents, can have a lasting negative effect on the implementation of new projects. For example, the explosion of a gas pipeline in Belgium in 2004 increased public concern about the perceived reliability of CO_2 transport [30] (see also [15] regarding the CO pipeline in Section 3.2.2).

Another crucial factor for the regional acceptance of iCCS seems to be the specific perception of actors and issues related to a (possible) project. Ref. [74] suggests that this debate is also in the literature on the so-called social license to operate (SLO): "SLO refers to the informal permission granted to industry by the local community and wider society to develop a technology; in the context of CCS, SLO has been recognized as very preliminary and fragile". The following factors are summarized for achieving an SLO by [74] and are supplemented here by the results of other studies:

- Weighing the costs and benefits to the community, based on the particular characteristics of the project (see also [13]). Here, the ability of iCCS to protect jobs was identified as one of the key benefits. These benefits can be felt even more strongly for iCCS as it both protects employment in existing industries and provides infrastructure that can attract new investment and employment opportunities [13,66,71];

- Creation of socio-political legitimacy; that is, whether an industry and all other (interest) groups act fairly, respect local lifestyles, and, in sum, the community plays a role and is involved (see also [13]). This can also include industry engagement with the local public, which is seen as the "key vehicle for achieving social license" by [81]. Part of this engagement can be compensation measures offered to the community [74];
- Creation of interactional trust; in which all participants engage in a mutual dialogue (in relation to communication, compare also Section 3.2.7);
- Establishing an institutionalized trust in which a lasting relationship with community representatives is established, taking into account mutual interests. This dialogue also includes the industry's ongoing efforts to address environmental challenges, including iCCS—see also [64].

In addition to the factors already mentioned, the studies identified further aspects that may have an influence on the regional acceptance of iCCS; these include the specific economic situation and the geological conditions of a region. These have already been discussed in more detail in Section 3.2.1 on the perceived benefits of iCCS and will not be repeated here.

3.2.5. Trust

In almost all analyzed studies (n = 23), the topic "trust" was treated as a crucial acceptance factor for iCCS. Ref. [74] conclude that research indicates that trust in developers and other stakeholders is a critical factor influencing public response to a development such as iCCS as a whole, as well as at the community level. Within the studies analyzed, the trust factor is predominantly discussed in the context of regional processes and stakeholders on iCCS. Some stakeholder groups enjoy more trust among the population than others. These groups include in particular (environmental) non-governmental organizations (ENGOs) and local stakeholders, for example politicians and investors, who are considered to represent local and civic interests [15]. These groups of people are thus seen as having a certain degree of integrality. Whereas [62] notes that in the context of a focus groups in Wales (United Kingdom), a distrust of both a major steel producer and the government at all levels was mentioned based on a lack of integrity and competence. According to [14], perceptions of trust in key institutions depend on the track record of those institutions in managing past industrial processes.

Local authorities seem to have a special role to play here in developing a deeper commitment, as they can act as facilitators for the deployment of iCCS [65]. The importance of the position of the municipality towards CCS projects has been shown in previous studies. In Barendrecht in the Netherlands, the local government rejected a proposed CCS project because they feared negative impacts on public health and a decline in property values [64]. Accordingly, it is important that the community, including the people who live there, feel that the continued efforts of industry to build technology like iCCS is also directed toward solutions to environmental challenges [64]. This is where community familiarity with industry relevant to CCS implementation may also be important [64]. Moreover, ref. [13] argues that subjective familiarity with such an industry may serve to reduce the perceived risks associated with new infrastructure, leading to greater acceptance of iCCS within the intended communities.

At the same time, gaining public trust is an extremely lengthy and labor-intensive process that is highly dependent on experience in the interaction between laypersons and project stakeholders [30]. It is also important to avoid violating trust as much as possible, as it can be difficult to rebuild and can also cause negative spillover effects on perceptions of other technologies and projects [14]. Distrust can have an effect in different areas, on the one hand with regard to the competence of the responsible persons (competence-based distrust), especially when it comes to the implementation of a complex infrastructure project such as iCCS technology [62]. On the other hand, distrust can also relate to procedural fairness in the participation process (integrity-based distrust [62]; compare also the comments on socio-political legitimacy in Section 3.2.4). According to [74], without a more comprehensive public involvement strategy, the question remains whether this is sufficient to build a sense of trust towards the developer.

3.2.6. Knowledge/Awareness

As expected, none of the studies analyzed provide any information on what the state of public perception and knowledge of iCCS technologies is. However, the results of [13] show that public awareness of CCS (without concreteness to iCCS) remains low (here for Canada, the Netherlands, Norway, the UK and the US) and this result is also in line with previous research. However, in deciding whether to accept or reject CCS, the general level of knowledge and awareness plays an important role, as illustrated by the presentations from Tcvetkov's literature review on CCS [30]. Stakeholders interviewed by [15] in the ELEGANCY project rate public knowledge about CCS as rather low and perceive that iCCS technologies are not yet present in the current public discussion due to low market penetration. The results of [61] in the context of an experiment suggest that iCCS is viewed more positively by those who claim to have more knowledge about iCCS and that they are also likely to show a higher interest in the technology. Additionally, ref. [41] found that higher information levels can fundamentally change the evaluation of CO_2 capture options (for example air capture or from chemical plants).

The study [64] emphasizes that the local population in Porsgrunn (Norway) is not only used to industrial activities, but is also likely to have concrete experience with iCCS activities. There is a sense that the local population is positive about the proactive approach to managing CO_2 emissions, and this assumes that there is some level of knowledge about iCCS locally. Beyond this level of knowledge about iCCS, ref. [77] clarified that industries also have an interest in iCCS technologies becoming more widely known. For example, to market BECCS, public knowledge of low-carbon technologies is a possible positive aspect. The reasoning is that customer demands for negative emissions make investment decisions easier for industries because they can integrate iCCS technologies as part of their sustainability strategy. According to [65], however, even key stakeholders such as trade unions and environmental organizations lack evidence-based information on the iCCS capabilities of carbon-intensive industries. Ref. [73] also assumes that environmental organizations (related to Europe) lack the necessary resources to acquire knowledge about different iCCS technology options in detail. This lack of capacity also contributes to the apparent lack of official positions on issues such as iCCS until 2018 [73].

Beyond just awareness and knowledge of iCCS, the studies address the need for contextual knowledge. For example, ref. [72] suspects that there will be a more positive perception of iCCS as people become more aware of their individual climate impacts. Thus, some of the stakeholders interviewed in the study of [15] also see a general lack of societal acceptance regarding energy technologies and large-scale infrastructure, attributed in part to a lack of knowledge. Perception of global warming issues, understanding of the role of humans in this process, and developing an objective view of the prospects of low-carbon technologies, including CCS, depend on the education of respondents [26,30]. Therefore, implementation of an educational strategy for sustainable development should be considered, which starts at school and could be part of a national "green" policy. Ref. [71] clarified in their study that with the level of knowledge about iCCS and the integration of the technologies into a higher-level thematic context, the initially perceived assessment of iCCS can change once again. If iCCS is initially interpreted as a potential threat to natural systems, subsequent presentations and scenario discussions led to a gradual shift in how participants interpreted iCCS. Similarly, ref. [62] clarifies that participants in two focus groups on the Port Talbot steel mill development acquired contextual knowledge to evaluate iCCS. For example, they express concerns that if iCCS makes steel more expensive, the Welsh steel industry could lose out to foreign competitors who continue to produce emissions-intensive steel at the lowest price. If nothing else, these findings illustrate that awareness of iCCS does not immediately predict public acceptance of a project [30]. Ref. [66] also note that regardless of the depth of their insight and knowledge, people will acquire subjective perceptions about iCCS. Ref. [30] sees consolidating government, industry and NGO efforts as one of the key challenges to improving public perceptions of CCS.

3.2.7. Communication/Participation

The discussion of CCS communication and participation in the articles analyzed is extensive and is therefore presented in the form of a table (Table 3). Ref. [68] suggests that the CCS community is generally aware of the range of factors that influence public engagement. Whether this range changes significantly for communication about iCCS cannot be adequately answered using the available results. Ref. [74] illustrates that effective public engagement will be key to successful iCCS implementation. With this comes the need to further explore how to most effectively engage with the local public.

Table 3. Overview of the acceptance factor "communication/participation" of iCCS (who/what/how).

Who should communicate?
Persons of trustPersons within the scope of their respective expertise
Qualified project team
Entire community of interest (to be defined on a case-by-case basis)
Inclusion of new players, e.g., business and trade associations, companies along the entire value chain
What should be communicated?
iCCS narrative embedded in the overall context of sustainability
Urgency to combat climate change
Framing of iCCS as environmental technology (where there is no alternative)
Discussion of alternative technologies
Integration into norms and values of society
Costs in the context of the overall energy transition
Economic advantages and disadvantages
Set economic consequences in relation to ecological ones
Infrastructure challenges/use of existing infrastructure
Presentation of project experiences incl. risk analyses
Integration into current political context
Liabilities/standards/regulatory framework/securityRole of iCCS for global economy/international cooperation
How to communicate?
Develop an empowerment and communication strategy and plan
Take into account the main principles of public participation
Meaningful voice during decision-making processes
Establish continuity in communication
Fairness/greatest possible transparency/inclusion of all/neutral/clear/high quality
Creation of problem-oriented knowledge, e.g., FCDP
Include local needs and contexts/site characterization.
Consider community compensation
Use of classic media, such as brochures, local media
Facilitate face to face exchange, e.g., local activities and events
Use of digital media

The chosen order of the factors does not represent a weighting.

In this context, it seems important to mention again the aspect of [74], which emphasizes a certain flexibility in dealing with iCCS projects, as specific concerns and needs may change over time in different regions. Here, regular adjustments of the implementation strategy of iCCS projects have to be taken into account.

3.2.8. Socio-Demographic Factors

The analysis of the influence of socio-demographic factors on the acceptance of iCCS from the available studies does not reveal any meaningful trend. According to [67], for example, the acceptance of iCCS among women is about three times higher than among men (in selected European countries). Additionally, according to [13], men (as well as older people and people with high incomes) showed lower support for iCCS (but only after reading the message on CCS and possible lifestyle change). In contrast, ref. [30] presents findings in which men show more tolerant perceptions of CCS risks when the economic potential is present, while women are more concerned about safety. Additionally, as mentioned earlier in the context of a country's cultural identity (see Section 3.2.3), nationality represents the strongest predictor of support for iCCS [13].

All other results on the influence of the socio-demographic factor do not explicitly refer to iCCS technologies and therefore do not find any further explanation here.

3.2.9. Perceived Differences between CCS and iCCS

In the following, the question is addressed whether significant differences between the acceptance of CCS from fossil-fired power generation plants and the acceptance of iCCS from industrial processes can be derived from the results of the analyzed studies. There are a number of initial results on this, but they target different technology pathways and are therefore hardly comparable. First, ref. [30] suggests that CCS technologies received general support from respondents in a survey, but when it comes to specific options for implementation, for example as part of gas and coal-fired power plants, initial public preferences may be negated. Additionally, according to [71], focus group participants articulate more positive visions for iCCS and BECCS than for coal CCS. They affirm support for growth through iCCS in manufacturing industries, as this is highly desired by society. Additionally, ref. [15] assume that iCCS will have higher social acceptance than CCS. Beyond this more economic aspect, ref. [68] represents the need to significantly broaden the iCCS discussion to include heavy industry and processes outside of power generation. This was seen as necessary to counter the traditional arguments of environmental groups that reject CCS because of its ability to re-generate electricity. In addition, initial studies compare the acceptance of iCCS with the acceptance of gas-fired power plants. For example, ref. [16] show in their experiment that BECCS plants receive higher approval than those using conventional gas. Interestingly, as perceptions of BECCS improve, so does the willingness of one's community to accept CO_2 storage. Ref. [17] also found that large-scale plants converting gas to hydrogen (H2) with CCS tend to be viewed negatively by most respondents. Basically, ref. [71] assumes that fossil CCS is considered unacceptable by the local population, while other CCS options, like iCCS, remain feasible.

3.2.10. Evaluation of iCCS for Different Process Steps

iCCS technologies encompass many different technological concepts and potential target applications. The results presented below are intended to illustrate the acceptance of iCCS along the stages of different value chains and the underlying factors. It should be mentioned at the outset that the studies analyzed did not examine in detail the possible effect of the technical feasibility of different iCCS technologies on iCCS acceptance.

The following findings are available on the CO_2 source and the capture process step:

- BECCS: as briefly indicated before, BECCS is preferred to fossil-based CCS. According to [76], the technological approach has reached a stage of normalization in the debate, at least in the scientific discourse, after several years of intense criticism, and has become a self-evident aspect of climate change discourse. Especially for countries with a strongly biomass-based economy, such as Finland, BECCS seems to generate benefits [75]. With reference to [71], CCS was seen as a more intuitive and natural process when linked to managed forestry and the carbon cycle. Similarly, ref. [41] presents the use of biogas plants as a source of CO_2 as a promising option for industry and policy makers to achieve a socially acceptable form of carbon capture. Environmental organizations such as Greenpeace and Biofuelwatch disagree here, according to [76], emphasizing problems with agricultural production and water scarcity in the context of BECCS. This aspect is also critically addressed in the Convention on Biological Diversity from 2019 [82]. This is because significant negative impacts on biodiversity and food security are expected as a result of the extensive land use changes caused by the consistent use of bioenergy, including BECCS. It remains to be seen what effect this position can have in terms of shaping public opinion. However, ref. [13] assume that BECCS is more supported than shale gas, underground coal gasification, and the application of CCS in heavy industry.

- Post-combustion capture: while the process can be retrofitted into existing energy infrastructure, it does not promise economic feasibility due to low efficiency and increases the need for fossil fuels, thus having a comparatively high environmental impact. For these reasons, the process is generally not considered beneficial from the perspective of interviewed stakeholders [69]. In contrast to oxy-fuel technology, post-combustion requires larger constructional measures and entails a visible and significant change to the existing plant. Therefore, acceptance-relevant aspects may occur due to construction sites and changes in the landscape [15].
- Direct air capture (DAC): according to [41], capturing CO_2 from ambient air is not an accepted option among the public, especially when detailed information on efficiency and energy requirements is available.
- CO_2 capture from chemical plants: the results of a study by [41] show that providing technically correct and comprehensible information has the potential to completely revise previous negative opinions of study participants. The prerequisite is that it is explained transparently that the capture of CO_2 from a chemical plant is highly efficient and has a lower environmental impact compared to other alternatives. Initially negative reactions can thus be transformed into positive acceptance ratings.

The following findings are available on the acceptance of the CO_2 transport process step:

- Rejection of CO_2 pipelines: Respondents' judgments in an experiment by [16] were most influenced by the pipeline factor, to a lesser extent by the plant factor, and least by the storage location factor (there are a variety of contrary results on this). However, people seem unwilling to live near a pipeline (respondents from Switzerland), although they would prefer a CO_2 pipeline to a gas pipeline. Field testing of geological storage in densely populated areas may therefore consider avoiding pipeline transport to increase the likelihood of public acceptance [13].
- Use of existing infrastructure: ref. [41] make clear in their study that CO_2 transport by truck and a mix of trucks and pipelines are not preferred by the participants. In particular, the negative ecological effects expected for the construction of new infrastructure packages are mentioned here. Instead, it is recommended to examine the potential of using the existing infrastructure for alternative fuel production. A further step would even be the avoidance of CO_2 transports by spatially linking CO_2 capture and fuel production—an option that should be examined in terms of acceptance.

The following findings are available on the acceptance of CO_2 use:

- Methanol production: according to [30], the most preferred way to use CO_2 is methanol production, while the CCS-EOR process chain is perceived as one of the worst alternatives, second only to CCS without the link to the beneficial use of CO_2.
- Chemical looping and CO_2 removal from calcination processes: these have shown potential according to [69] in the study area of Finland, especially in small CCU applications and in some cases also in CHP production. Opportunities to recycle the captured carbon could help solve the economic feasibility problem due to lower transportation and storage costs and potential revenue from recycling. Whether optimizing economic feasibility may also have an effect on public perception is not addressed.

- CO_2-based fuel production: ref. [41] make clear that the public is less interested in the process step of CO_2-based fuel production and efficiency improvements in chemical production, but rather in the processes of CO_2 capture and transport.
- H_2/CCS value chain: ref. [14] represent that the H2 part of this joint value chain is more socially accepted than the CCS part. Nevertheless, the type of H_2 (green, blue, conventional) is also estimated to be relevant for acceptance. They also hypothesize that only established larger industries can address these infrastructure issues, but that the trust on the ground, where the (re)construction of the infrastructure takes place, is more likely to be given to local stakeholders.

The following findings are available on the acceptability of CO_2 storage in conjunction with iCCS:

- Onshore storage: ref. [16] suggest avoiding the NIMBY (not in my backyard) effect in field trials of CO_2 storage using BECCS as the CO_2 source. It is likely that the source of the CO_2 is critical to the acceptance of the storage site.
- Offshore storage: Haug's results show that the possibility of the offshore storage of CO_2 could be a clear advantage for the Nordic regions for the establishment of an iCCS economy [64]. As an example, the municipality of Porsgrunn in Norway, whose positive attitude towards existing and potential iCCS activities may result from the option of offshore storage, should be mentioned once again. The Sleipner project in the North Sea was also realized without much public controversy, and ref. [64] suggest that this could also be a result of the offshore location. In sum, the off-shore option could be a great advantage for the Nordic region, but it is important to note that it must also gain the consent of the stakeholders in the use of the sea and that there is no guarantee of acceptance if these stakeholders are neglected [64].
- Geological and infrastructural prerequisites: Countries with an interest in establishing an iCCS economy should carefully examine their geological prerequisites. According to [75], CO_2 storage is an open question in Finland, as the country lacks potential geological formations for it, which also underscores the importance and cost of CO_2 transport [75]. Russia, on the other hand, has extensive area and therefore allows CO_2 storage at a considerable distance from industrial centers and residential areas, which could potentially weaken stakeholder opposition to the projects [70]. Another option, he said, is to look at reusing existing infrastructure for CO_2 storage, as proposed in the Acorn project. Significant cost savings can be achieved through this approach, and this also represents a societal approach to enable broader CCS deployment [78]. For example, existing CO_2 transport and storage infrastructure could be shared by multiple capture projects to maximize value, simplify investment decisions, share operating costs, and thus reduce development costs.

Finally, it should be summed up here that several studies consider the acceptance of iCCS along the different process steps and value creation stages to be possible. An important approach to developing iCCS acceptance, initially primarily from an economic perspective, is the pursuit of a cluster and network approach [14,41,62,64,67,78], which is already emerging as a trend in practice (see Section 3.2.1 for a more detailed discussion).

3.2.11. Regulatory/Political Aspects

This literature review also noted circumstantial evidence suggesting that a lack of regulatory frameworks, political support, and missing or complex approval processes may influence iCCS adoption.

The findings highlight a fundamental need for strong regulation and policy on iCCS, both to leverage the skills and experience of the private sector and to maintain the common good and public interest [65]. For example, a UK opinion poll cited by [62] found that a majority (74%) of adults support policies to regulate heavy industry to ensure emissions reductions in the sector. Focus group participants from a region of Scotland that has historically been closely associated with energy-intensive industry (Port Talbot steelworks) assume that there will be stricter emissions legislation for these industries in the long term, and therefore refer to iCCS as an "inevitable" option [62]. This would imply that expectations of stricter emissions legislation in the future from national and EU levels alone can convince people that iCCS is inevitable in the future. On the other hand, the participants of this study also valued the European Union as an important partner for the implementation of iCCS technologies [60], especially by providing the necessary funding. In this context, ref. [69] also mention the funding for the development of the necessary CO_2 transport infrastructure.

Ref. [30] go one step further and assume that an important factor for further iCCS development is international cooperation. On the one hand, so that individual countries can embed and position their iCCS policies internationally [14], and on the other hand, international cooperation would make it possible to combine national efforts, create favorable conditions for project proposals and adopt successful experiences of other countries. Thus, it would be necessary to create a political context that can strengthen public trust due to the importance of collaborative decision-making [30]. Local and regional networks alone would be insufficient to influence national policy [14,63]. In addition, ref. [65] describe that there would be limited public communication of an iCCS project proposal if political uncertainties prevail. For this, it is also important to have political long-term strategies that create reliability, for example, regarding BECCS technology and its integration into the European Union Emissions Trading Scheme (ETS-EU) [77]. This integration would be important for Finland, for example. Without BECCS, it would be challenging to meet emissions targets, but with BECCS, Finland could gain benefits by saving and trading emissions allowances [75]. The need for ETS-EU was frequently mentioned in the analyzed studies, but mostly by industrial actors and other experts [69,75,77].

4. Discussion

The present study is the first literature review to address the acceptance of iCCS. The objective of this study was fourfold. Firstly, it is examined to what extent the acceptance of iCCS is already being empirically investigated. Secondly, an analytical framework is proposed in order to systematically review the existing literature. Thirdly, based on the review, factors influencing the acceptance of iCCS are identified and discussed. Fourthly, results for the acceptance of iCCS are compared to CCS, highlighting some important differences between the two areas of application.

First, the results show that there is still only limited research on the acceptance of iCCS. Between 2012 and 2020, 25 scientific articles were published on the subject, with very different and incomparable methodological tools and research questions.

Secondly, during the evaluation process, it became apparent that the analytical framework transferred from CCS acceptance research, with its well-established dimensions (cf. Table 1), was sufficient to systematically gather the results from the articles. The research findings of the analyzed articles could be assigned to one or more dimensions, such as findings on local aspects (as suggested by Table A1 in the Appendix A, see column "Important statement related to iCCS"). Influencing variables that emerged in the analyzed articles and initially deviated from the established factors for CCS acceptance research (for example, the employment factor) could be assigned to the existing dimensions by the author during the evaluation. Accordingly, no further factors were inductively added to the analytical framework established in Section 2. As a result, many factors explaining the acceptance of CCS seem to be decisive for the acceptance of iCCS as well. However, it became apparent that the weighting and the expressions of acceptance factors to iCCS appears to vary compared to CCS, as shown in the following. Moreover, only tentative trends for the acceptance of iCCS can be derived from the studies analyzed. It remains unclear whether iCCS applications are more likely to be accepted or rejected by society in the future. Moreover, from a scientific point of view, a methodological concept for analyzing iCCS acceptance is still lacking, even though the factors considered here already provide a good starting point for operationalizing the research subject. Given the wide range of technological options and the resulting societal implications, this task also appears to be non-trivial.

The discussion of objectives 3 (factors influencing iCCS) and 4 (differences of CCS and iCCS) of this content analysis are now discussed in conjunction.

More specifically, acceptance at the regional level, for example, appears to depend even more significantly on the perceived societal benefits that people associate with iCCS. The potential to maintain and increase local employment through the use of iCCS applications was frequently mentioned [13,64,66,71]. This represents a difference from the debate in perceptions of the societal benefits of CCS. In sum, it appears as if the population expects the safeguarding or even increasing of economic performance in their local environment with the use of iCCS. Previous research findings illustrate that societal benefits have either the same or slightly higher explanatory power for CCS acceptance than societal risks [31,35,47,83] (see Table 1). Whether this is also valid for the acceptance of iCCS remains to be investigated.

What is clear is that both factors will also be very significant in the context of iCCS. Subjectively perceived risk associated with CO_2 storage has been a crucial factor in explaining local and regional resistance in the context of CCS technologies [11,16,44]. It is different from factual risk in this regard as [53] illustrated with their approach to misconceptions. The fact is that CO_2 pipelines are state of the art and have been operating in the United States for example since the 1970s. Additionally, no significant research and development budgets are being spent on CO_2 transport and the associated potential risks worldwide. In contrast, geological storage of CO_2 has been the subject of intensive research and development work internationally for many years, even though CO_2 storage is already being successfully operated in many countries [84]. Here, the exploration methods for CO_2 storage, the procedures for storage monitoring, the competition with other storage utilization options, the impact on geothermal energy utilization, and the theoretically possible effects on drinking water supplies are often the subject of interest [85]. In sum, the question is not so much whether CO_2 storage is fundamentally possible, but under what conditions it is as safe as possible. Besides these science-based facts of technical and environmental aspects, the subjectively perceived risk factor will be important in the context of iCCS acceptance, as many of the studies analyzed have made clear [13–15,30,62,65,68,71,73–77]. It seems that in this context the aspect of fair distribution of risks and benefits has to be more in focus than in the context of the CCS debate. If, in the future, the benefits associated with the use of iCCS are perceived by the population primarily at the global level in the context of climate protection and the local population gains the impression that, in contrast, they are more likely to be confronted with the disadvantages of iCCS applications, this would probably be a barrier to the development of acceptance. In relation to the perception of an equitable distribution of risks and benefits, the explicit understanding of the benefits associated with iCCS for a region therefore seems to be of importance. This starting point of an unequal distribution of risks and benefits in the context of the future deployment of iCCS offers a possible field of action, both for research and for the implementation of practical iCCS projects. The previous research approaches of possible compensation benefits in the context of CCS will be examined here for their transferability and applicability.

Furthermore, the factor trust, which was evaluated in most studies as an important tipping point for or against the acceptance of iCCS (see [13–15,30,62,64,65,74]), should be further investigated. It became clear that in the development of local iCCS projects, trust in the stakeholders involved becomes especially important when it comes to large infrastructure measures related to CO_2 transport [13,16,41,62]. It seems that the process step of transport has become critical to the CCS debate, even though the negative sign in the assessment of CO_2 pipelines does not seem to have changed. With respect to transport infrastructure, more knowledge is still needed on the acceptance of iCCS. It is unclear whether, for example, the "joint" use of infrastructure or the use of existing infrastructure by industry clusters or hubs leads to an improvement in the acceptance of iCCS. Another research question could be whether the CO_2 source has an influence on the acceptance of CO_2 transport, for example if the source is associated with an industry that is deeply rooted in the local society and contributes to its identity.

In this context, the role of framing or a possible narrative for iCCS (and also the greenhouse gas CO_2) implementation should also be further explored. The studies have illustrated that framing iCCS as a climate change mitigation technology can lead to both positive and negative acceptance tendencies [13,16,53,63–65,71]. There does not seem to be a determination yet as to whether or not iCCS technologies are perceived by society as a climate change technology. This framing was hardly conceivable in the context of the CCS debate since there were sufficient technological alternatives for sustainable generation of electricity through the use of renewable energy technologies. Anyway, it is clear that iCCS operators would benefit from such "green" framing of iCCS applications, especially in marketing potential products along the value chain. This framing approach, based on a rather economically oriented marketing strategy, would certainly fall short. Ultimately, there is an obvious need for a more overarching narrative that takes into account both the aspect of sustainability and the reduction of CO_2, as well as economic issues that not only affect individual technology paths, but in sum relate to the economic viability of an entire region. As a consequence, this would mean embedding iCCS in a discourse around sustainable structural change. After all, regions with energy-intensive economic sectors are particularly affected by the challenge of structural change.

The articles analyzed have also made it clear that the factors of social "values and attitudes" can be significant for the acceptance of iCCS [13,14,30,62,64,65,71,74,76]. In this context, further research is particularly needed on the question of whether a certain environmental awareness has a positive or negative influence on the perception of iCCS. Compared to the CCS context, the clarification of this research question seems to be much more complex due to the many different possible applications of iCCS. In the context of CCS acceptance, existing studies indicate that people with high environmental awareness tend to evaluate the technology negatively [49,86]. Some authors mentioned that, triggered by the Paris Agreement, the absolute urgency of the transformation to a sustainable economy and way of life has now arrived in the perception of society. In light of this urgency, the evaluation of iCCS could also be developed in a more positive direction [63,65]. Again, there are only assumptions and no evidence-based findings yet. Interestingly, in one study, this urgency emerged as a driver of iCCS acceptance. This happens when this urgency is interpreted by society as a threat to their current lifestyles and cherished habits for everyday life, and iCCS is perceived as an option to hold on to these habits without regret [71]. This approach could also be a starting point for new research questions on the acceptance of iCCS. In addition to this urgency, another aspect could influence the perception of iCCS in the future. For example, the Global Assessment Report on Biodiversity and Ecosystem Services (IPBES) [87] does not exclude CCS (and thus iCCS) as a measure to mitigate negative impacts on biodiversity (see Glossary). Consequently, iCCS could be considered not only an option to reduce global CO_2 emissions, but also a generally accepted measure to avoid or limit potential negative impacts on biodiversity. If such a perception is perpetuated among individuals with a high level of environmental awareness, this aspect could be interpreted as an advantage for the use of iCCS and possibly have a positive impact on social acceptance. Whether this assumption is well-founded needs to be explored in future studies on the acceptance of iCCS.

5. Conclusions

The IEA [88] estimates that iCCS in the cement, iron and steel, and chemicals sectors will need to deliver around $28GtCO_2$ of emission reductions between now and 2060 to meet the climate target of the Paris Agreement. To achieve these reduction goals globally, strategies for robust and timely market introduction of iCCS technologies need to be developed. For such a market introduction of iCCS, social acceptance is of particular importance in addition to technical-economic and environmental indicators, as the example of CCS has illustrated.

In the studies analyzed, a large number of indications for the design of a communication strategy were derived, largely on the basis of the findings from CCS acceptance research as well as on the basis of all the research on energy transformation (see Table 3). In view of the abovementioned abundance of requirements for such an iCCS communication, the question arises as to which institution is capable of organizing such a permanent and trust-based process and in which larger thematic context this communication can be embedded? This appears to be a difficult question to answer, especially against the background of often missing political strategies and the related regulatory frameworks on the national level. The present literature analysis shows on the one hand which starting points for the market introduction of iCCS exist so far from social science research for political and economic actors and on the other hand which research efforts are still required.

Funding: This research received no external funding.

Data Availability Statement: The articles used in the literature analysis can be accessed via the respective publishers (for more details, see the bibliography).

Conflicts of Interest: The author declares no conflict of interest.

Appendix A

Table A1. Overview of the analyzed articles.

First Author (Year of Publication) [Reference]	Method	Country	iCCS-Related Technology	Important Statement in Relation to iCCS
Alcalde et al. (2019) [78]	Evaluation of ACT Acorn findings and review of scholarly/industrial literature	UK	Complete iCCS value chain (Acorn Project)	Seven key elements for iCCS projects: Infrastructure reuse, storage development plan, low-carbon build-out options, full-chain development plan, policy support, just transition, public engagement, and knowledge exchange.
Aursland et al. (2019) [66]	Case study with local residents and Norcem employees (n = 15, face-to-face)	NO	CO_2 capture from the cement industry	Positive image of cement company conducive to acceptance, effects on local employment and environment perceived as benefits. However, also concern whether project affects local living conditions.
Boomsma et al. (2020) [74]	Literature review from academic literature (non-systematic, n = N/A) and publicly available documents (n = 25)	Academic literature: international; public documents (DE = 7, NL = 4, RO = 5, UK = 9)	No specific iCCS technique defined (focus on community compensation)	When implementing iCCS projects, it is important to understand local social conditions and examine what impact they have. Sites where the local public feels connected to the industry may be more positive about iCCS development. Compensation for communities needs to be integrated into broader public involvement strategies.
Broecks et al. (2016) [63]	Quantitative online survey representative for NL (n = 920) and discrete choice experiment	NL	No specific iCCS technique defined (=industrial applications)	"Industrial applications" is the most convincing pro-argument for CCS, followed by "dispose of CO_2 garbage", "safety of natural gas fields". Arguments on climate change are less convincing.

Table A1. *Cont.*

First Author (Year of Publication) [Reference]	Method	Country	iCCS-Related Technology	Important Statement in Relation to iCCS
de Best-Waldhober et al. (2012) [17]	Quantitative study (ICQ [1]) representative for NL (n = 971)	NL	Large plants where gas is converted into hydrogen with CCS	iCCS option rated lower compared to other energy production/mitigation options (except nuclear).
Dütschke et al. (2015) [61]	Quantitative online experimental survey design representative for DE (n = 1.672), assessment of 18 scenarios	DE	Industry and biomass power plant as CO_2 source	CCS scenarios that include either an energy-intensive industry or a biomass power plant as a source of CO_2 are perceived more positively than scenarios in which the CO_2 is captured from a coal-fired power plant. Rating of the respective CO_2 source as the strongest predictor.
Glanz et al. (2021) [15]	Qualitative explorative stakeholder interviews (n = 10)	DE	Hydrogen and carbon capture and storage infrastructure/ chain	Restricting the use of CCS for certain applications (industry, bioenergy) represent trade-offs that are supported by various stakeholder groups and offer a balance of environmental and economic arguments. Assumption: only large industries can address iCCS/H2 and its infrastructure challenges, but local trust is given to other stakeholders.
Gough et al. (2018) [14]	Mixed-methods approach: stakeholder interviews (n = 12) and two focus groups (n = 8 each group) with lay public	UK	iCCS with focus on CO_2 storage	Success of iCCS activities in a community dependent on social context, trust in key actors, track record of previous industrial processes. Hurdles related to procedural justice.
Haikola (2019) [76]	Qualitative analysis of (popular) science and news media from 2008—2018 (n= ca. 800)	International	BECCS	Scientific discussion about BECCS is becoming more neutral due to the time pressure to take action on climate protection. Debate moves away from the question of moral hazard and focuses instead on the need to act.
Haug et al. (2016) [64]	Interviews with municipalities (n = N/A [2]) and literature review	DK, NO, SE	No specific iCCS technique defined	Communities can consider iCCS as an advantage for regional value creation. Positive evaluation if local population is used to industrial activities and has concrete iCCS experience. Potential for offshore storage in a region is evaluated as an advantage.
Ilinova et al. (2018) [70]	Case studies (n = N/A), stakeholder management tools, and a checklist method	International	No specific iCCS technique defined	Most attention in CCS project planning/implementation should be focused on industrial companies/investors, government and society. CCS projects are mostly local projects; however, they are implemented in the context of national and even international interests. Therefore, the circle of stakeholders is large and establishing a constructive dialogue with all proves to be a difficult task.

Table A1. *Cont.*

First Author (Year of Publication) [Reference]	Method	Country	iCCS-Related Technology	Important Statement in Relation to iCCS
Kashintseva et al. (2018) [65]	Empirical model based on representative online survey (n = 564)	CZ, DE, IT, NL, PL, SK, UK	No specific iCCS technique defined (iCCS products and technologies)	Increase of iCCS sites, including those in the neighboring regions and countries, leads to the increase of negative consumer attitudes to iCCS and renewable energy policies. NIMBY effect is considered relevant.
Kojo et al. (2017) [75]	Quantitative longitudinal analysis of newspaper articles from 1996–2015 (n = 282)	FI	No specific iCCS technique defined (pertains to BECCS)	Agenda setting of the media regarding CCS is strongly dependent on real plant projects and communication measures of industrial actors. iCCS actors are not yet involved in communication in Finland. Business models are missing, costs are overestimated, a debate specifically about possible international developments is missing.
Offermann-van Heek et al. (2020) [41]	Quantitative online survey representative for DE (n = 300) and best-/worst-case scenarios	DE	DAC, biogas and chemical plant	Capture and transport process step more relevant to public than further use of CO_2, use of existing infrastructure conducive to acceptance, CO_2 use from BECCS and chemical plants viewed positively, DAC not an accepted option.
Pihkola et al. (2017) [69]	PESTEL [3] framework (analysis macro-environment of industries), stakeholder interviews (n = 12) from 2011–2012, media analyses (n = N/A), literature reviews	FI	No specific iCCS technique defined (pertains to BECCS)	iCCS needs a regulatory framework and political support, especially for the development of infrastructure. More systematic and differentiated consideration of iCCS applications is required for Finland. BECCS/CCU is seen as an opportunity for iCCS due to the central role of the Finnish energy-intensive industry.
Rodriguez et al. (2020) [77]	Qualitative inductive interviews with company representatives (n = 20)	FI, SE	BECCS	BECCS is technically feasible; what remains unclear is who will create a financially viable business case and establish supporting policies, as well as who will build the necessary transportation and storage infrastructure. In addition, customer requirements for negative emissions are still lacking.
Serdoner (2019) [73]	Qualitative interviews (n = 3) with representatives of EU environmental organizations, analysis of their public relations activities and literature review	EU	No specific iCCS technique defined	Positions of ENGOs operating in Europe on iCCS are closely related to previous debates on the application of the same technology in the power sector. Previous experience has led ENGO to approach the technology with skepticism and caution. They are either neutral toward iCCS or opposed to it.

Table A1. *Cont.*

First Author (Year of Publication) [Reference]	Method	Country	iCCS-Related Technology	Important Statement in Relation to iCCS
Swennenhuis et al. (2020) [65]	In-depth semi-structured interviews (n = 25) with regional stakeholders and workshops (UK)	NO, NL, UK	No specific iCCS technique defined	Narrative that iCCS is deployed for benefit of citizens/communities/workers and not in support of private sector, policy that leverages private sector capabilities without setting aside the public interest, need for deeper engagement with local governments that act as facilitators for iCCS deployment.
Tcvetkov et al. (2019) [30]	Literature review from 2002–2018 (n = 135)	international	No specific iCCS technique defined	Development of a regulatory framework to control the industry, important for public trust. Public preferences regarding capture plants are explained by problems with existing energy infrastructure. Public trust in environmental arguments of industry lower compared to NGOs, arguments of industry about economic aspects of project implementation are better perceived than by NGOs.
Thomas et al. (2018) [71]	Two qualitative deliberative workshops with local population (n= 12 each)	UK	Industrial CCS and BECCS	Depending on the context, iCCS may be perceived as a threat or a support to local social and economic interdependence. As a threat, for example, through costs that could harm employment in local industries, as a benefit through protecting and at the same time rejuvenating historical employment patterns through iCCS.
van Os (2018) [72]	Interview with Peter van Os	NL	Complete iCCS value chain (ALIGN CCUS Project)	Assumption that there will be a more positive perception of CCUS as the public becomes more aware of their individual impacts on climate. Uncertainties related to the cost of implementing CCUS, costs will decrease as implementation of CCUS technology progresses
Wallquist et al. (2012) [16]	Online Experiment (n = 139)	CH	BECCS	CO_2 source decisive for acceptance of storage site, avoidance of CO_2 pipeline transport in densely populated areas, avoidance of the NIMBY effect through the use of BECCS.
Whitmarsh et al. (2019) [13]	International experimental online study (n = 5.406), national and local samples	CA, NL, NO, UK, US	No specific iCCS technique defined	Bioenergy with CCS is more supported, while shale gas, underground coal gasification, and heavy industry with CCS are less supported. Areas where CCS facilities are likely to be built are typically locations where (analogous) industry already exists. Subjective familiarity with this industry could serve to reduce perceived risks associated with new infrastructure.

Table A1. *Cont.*

First Author (Year of Publication) [Reference]	Method	Country	iCCS-Related Technology	Important Statement in Relation to iCCS
Williams et al. (2021) [62]	Two qualitatively designed focus groups with citizens (n = 11 and n = 10)	UK	iCCS in the steel industry	Community could endorse use of iCCS if developer/government collaborate from local to national level, provide transparent dialogue process that supports community trust in intent, integrity, and competence of implementing organizations.
Xenias et al. (2018) [68]	Mixed-methods approach: interviews (n = 13) and online survey (n = 99) with experts	Interviews: NO, NL, UK; Online survey: DE, NL, NO, UK, others	No specific iCCS technique defined	Need to expand CCS discussion to heavy industry, iCCS benefits at global level and greater risks at local level, learning from public engagement research literature

[1] ICQ = Information-Choice Questionnaire; [2] N/A = not available; [3] PESTEL = Political, economic, social, technological, environmental, legal.

References

1. IPCC. Fifth Assessment: Report of the Intergovernmental Panel on Climate Change. In *Contribution of Working Group III*; Cambridge University Press: Cambridge, UK, 2014.
2. UNFCCC. Addendum Contents Part two: Action taken by the Conference of the Parties at its twenty-first session. In *Proceedings of the Conference of the Parties on Its Twenty-First Session*; Paris, France, 30 November–13 December 2015, UNFCC: Bonn, Germany, 2015.
3. IEA. *20 Years of Carbon Capture and Storage. Accelerating Future Deployment*; International Energy Agency: Paris, France, 2016.
4. IEA. *World Energy Outlook 2018*; International Energy Agency: Paris, France, 2018; ISBN 978-92-64-30677-6.
5. IPCC. *Global Warming of 1.5 °C. An IPCC Special Report on the Impacts of Global Warming of 1.5 °C above Pre-Industrial Levels and Related Global Greenhouse Gas Emission Pathways, in the Context of Strengthening the Global Response to the Threat of Climate Change, Sustainable Development, and Efforts to Eradicate Poverty*; Intergovernmental Panel on Climate Change: Geneva, Switzerland, 2018; Available online: https://www.ipcc.ch/site/assets/uploads/sites/2/2019/06/SR15_Full_Report_High_Res.pdf (accessed on 19 December 2020).
6. Bui, M.; Adjiman, C.S.; Bardow, A.; Anthony, E.J.; Boston, A.; Brown, S.; Fennell, P.S.; Fuss, S.; Galindo, A.; Hackett, L.A. Carbon capture and storage (CCS): The way forward. *Energy Environ. Sci.* **2018**, *11*, 1062–1176. [CrossRef]
7. IEA—International Energy Agency. *World Energy Outlook 2015*; International Energy Agency: Paris, France, 2015; Available online: https://www.globalccsinstitute.com/wp-content/uploads/2019/08/Introduction-to-Industrial-CCS.pdf (accessed on 19 December 2020).
8. Global CCS Institute. *The Global Status of CCS. Special Report: Introduction Industrial Carbon Capture and Storage*; Global CCS Institute: Melbourne, Australia, 2016.
9. Dütschke, E.; Schumann, D.; Pietzner, K. Chances for and Limitations of Acceptance for CCS in Germany. In *Geological Storage of CO₂—Long Term Security Aspects, Advanced Technologies in Earth Sciences*; Springer: Cham, Switzerland, 2015; pp. 229–245.
10. Terwel, B.W. Public participation under conditions of distrust: Invited commentary on 'Effective risk communication and CCS: The road to success in Europe'. *J. Risk Res.* **2015**, *18*, 692–694. [CrossRef]
11. Schumann, D.; Duetschke, E.; Pietzner, K. Public perception of CO_2 offshore storage in Germany: Regional differences and determinants. *Energy Procedia* **2014**, *63*, 7096–7112. [CrossRef]
12. Terwel, B.W.; ter Mors, E.; Daamen, D.D.L. It's not only about safety: Beliefs and attitudes of 811 local residents regarding a CCS project in Barendrecht. *Int. J. Greenh. Gas Control* **2012**, *9*, 41–51. [CrossRef]
13. Whitmarsh, L.; Xenias, D.; Jones, C.R. Framing effects on public support for carbon capture and storage. *Palgrave Commun.* **2019**, *5*, 17. [CrossRef]
14. Gough, C.; Cunningham, R.; Mander, S. Understanding key elements in establishing a social license for CCS: An empirical approach. *Int. J. Greenh. Gas Control* **2018**, *68*, 16–25. [CrossRef]
15. Glanz, S.; Schönauer, A.-L. Towards a Low-Carbon Society via Hydrogen and Carbon Capture and Storage: Social Acceptance from a Stakeholder Perspective. *J. Sustain. Dev. Energy Water Environ. Syst.* **2021**, *9*, 9. [CrossRef]
16. Wallquist, L.; Seigo, S.L.; Visschers, V.H.M.; Siegrist, M. Public acceptance of CCS system elements: A conjoint measurement. *Int. J. Greenh. Gas Control* **2012**, *6*, 77–83. [CrossRef]

17. De Best-Waldhober, M.; Daamen, D.; Ramirez, A.R.; Faaij, A.; Hendriks, C.; de Visser, E. Informed public opinion in the Netherlands: Evaluation of CO_2 capture and storage technologies in comparison with other CO_2 mitigation options. *Int. J. Greenh. Gas Control* **2012**, *10*, 169–180. [CrossRef]

18. 2000. Available online: https://CO2re.co/FacilityData (accessed on 20 December 2020).

19. Renn, O. Technikakzeptanz: Lehren und Rückschlüsse der Akzeptanzforschung für die Bewältigung des technischen Wandels. *Tech. Theor. Prax.* **2005**, *14*, 29–38. [CrossRef]

20. Arning, K.; Heek, J.O.-V.; Sternberg, A.; Bardow, A.; Ziefle, M. Risk-benefit perceptions and public acceptance of Carbon Capture and Utilization. *Environ. Innov. Soc. Transit.* **2020**, *35*, 292–308. [CrossRef]

21. Arning, K.; Zaunbrecher, B.S.; Sternberg, A.; Bardow, A.; Ziefle, M. Blending Acceptance as Additional Evaluation Parameter into Carbon Capture and Utilization Life-Cycle Analyses. In *SMARTGREENS—7th International Conference on Smart Cities and Green ICT Systems*; Science and Technology Publications SciTePress: Setubal, Portugal, 2018; pp. 34–43.

22. Heek, J.O.-V.; Arning, K.; Linzenich, A.; Ziefle, M. Trust and distrust in Carbon Capture and Utilization industry as relevant factors for the acceptance of carbon-based products. *Front. Energy Res.* **2018**, *6*, 73. [CrossRef]

23. Jones, C.R.; Olfe-Kräutlein, B.; Kaklamanou, D. Lay perceptions of Carbon Dioxide Utilisation technologies in the United Kingdom and Germany: An exploratory qualitative interview study. *Energy Res. Soc. Sci.* **2017**, *34*, 283–293. [CrossRef]

24. Available online: https://www.elsevier.com/solutions/sciencedirect (accessed on 2 December 2020).

25. Ware, M.; Mabe, M. *The STM Report: An Overview of Scientific and Scholarly Journal Publishing*; International Association of Scientific, Technical and Medical Publishers: The Hague, The Netherlands, 2015.

26. Available online: https://link.springer.com/article/10.1007/s11192-018-2958-5 (accessed on 2 December 2020).

27. Liberati, A.; Altman, D.G.; Tetzlaff, J.; Mulrow, C.; Gøtzsche, P.C.; Ioannidis, J.P.A.; Clarke, M.; Devereaux, P.J.; Kleijnen, J.; Moher, D. The PRISMA Statement for Reporting Systematic Reviews and Meta-Analyses of Studies That Evaluate Health Care Interventions: Explanation and Elaboration. *PLoS Med.* **2009**, *62*, e1–e34.

28. Gläser, J.; Laudel, G. *Experteninterviews Und Qualitative Inhaltsanalyse Als Instrumente Rekonstruierender Untersuchungen*, 4th ed.; Lehrbuch; VS Verlag Für Sozialwissenschaften: Wiesbaden, Germany, 2010; ISBN 978-3-531-17238-5.

29. L'Orange Seigo, S.; Dohle, S.; Siegrist, M. Public perception of carbon capture and storage (CCS): A review. *Renew. Sustain. Energy Rev.* **2014**, *38*, 848–863. [CrossRef]

30. Tcvetkov, P.; Cherepovitsyn, A.; Fedoseev, S. Public perception of carbon capture and storage: A state-of-the-art overview. *Heliyon* **2019**, *5*, e02845. [CrossRef]

31. Gough, C.; Taylor, I.; Shackley, S. Burying Carbon under the Sea: An Initial Exploration of Public Opinions. *Energy Environ.* **2002**, *13*, 883–900. [CrossRef]

32. Haug, P. Public acceptance: A wake-up call from the people of Europe. In *World Nuclear Association Annual Symposium*; World Nuclear Association: London, UK, 2002; pp. 4–6.

33. Huijts, N. Public Perception of Carbon Dioxide Storage. The Role of Trust and Affect in Attitude Formation. Master's Thesis, Eindhoven University of Technology, Eindhoven, The Netherlands, 2003.

34. Huijts, N.M.A.; Molin, E.J.E.; Steg, L. Psychological factors influencing sustainable energy technology acceptance: A review-based comprehensive framework. *Renew. Sustain. Energy Rev.* **2012**, *16*, 525–531. [CrossRef]

35. Kraeusel, J.; Möst, D. Carbon Capture and Storage on its way to large-scale deployment: Social acceptance and willingness to pay in Germany. *Energy Policy* **2012**, *49*, 642–651. [CrossRef]

36. Warren, D.C.; Carley, S.; Krause, R.M.; Rupp, J.A.; Graham, D. Predictors of attitudes toward carbon capture and storage using data on world views and CCS-specific attitudes. *Sci. Public Policy* **2014**, *41*, 821–834. [CrossRef]

37. Arning, K.; Heek, J.O.-V.; Linzenich, A.; Kaetelhoen, A.; Sternberg, A.; Bardow, A.; Ziefle, M. Same or different? Insights on public perception and acceptance of carbon capture and storage or utilization in Germany. *Energy Policy* **2019**, *125*, 235–249. [CrossRef]

38. Arning, K.; van Heek, J.; Ziefle, M. Risk perception and acceptance of CDU consumer products in Germany. *Energy Procedia* **2017**, *114*, 7186–7196. [CrossRef]

39. Jones, C.R.; Radford, R.L.; Armstrong, K.; Styring, P. What a waste! Assessing public perceptions of Carbon Dioxide Utilisation technology. *J. CO2 Util.* **2014**, *7*, 51–54. [CrossRef]

40. Jones, C.R.; Olfe-Kräutlein, B.; Naims, H.; Armstrong, K. The social acceptance of carbon dioxide utilisation: A review and research agenda. *Front. Energy Res.* **2017**, *5*, 11. [CrossRef]

41. Heek, J.O.-V.; Arning, K.; Sternberg, A.; Bardow, A.; Ziefle, M. Assessing public acceptance of the life cycle of CO_2-based fuels: Does information make the difference? *Energy Policy* **2020**, *143*, 111586. [CrossRef]

42. Gölz, S.; Wedderhoff, O. Explaining regional acceptance of the German energy transition by including trust in stakeholders and perception of fairness as socio-institutional factors. *Energy Res. Soc. Sci.* **2018**, *43*, 96–108. [CrossRef]

43. Wennersten, R.; Sun, Q.; Li, H. The future potential for Carbon Capture and Storage in climate change mitigation—an overview from perspectives of technology, economy and risk. *J. Clean. Prod.* **2015**, *103*, 724–736. [CrossRef]

44. Seigo, S.L.; Arvai, J.; Dohle, S.; Siegrist, M. Predictors of risk and benefit perception of carbon capture and storage (CCS) in regions with different stages of deployment. *Int. J. Greenh. Gas Control* **2014**, *25*, 23–32. [CrossRef]

45. Siegrist, M.; Cvetkovich, G.; Roth, C. Salient value similarity, social trust, and risk/benefit perception. *Risk Anal.* **2000**, *20*, 353–362. [CrossRef]

46. Zhang, L.; Wang, J.; You, J. Consumer environmental awareness and channel coordination with two substitutable products. *Eur. J. Oper. Res.* **2015**, *241*, 63–73. [CrossRef]
47. Krause, R.M.; Carley, S.; Warren, D.C.; Rupp, J.A.; Graham, D. "Not in (or under) my backyard": Geographic proximity and public acceptance of carbon capture and storage facilities. *Risk Anal.* **2014**, *34*, 529–540. [CrossRef]
48. Ter Mors, E.; Terwel, B.W.; Zaal, M.P. Can monetary compensation ease the siting of CCS projects? *Energy Procedia* **2014**, *63*, 7113–7115. [CrossRef]
49. Yang, L.; Zhang, X.; McAlinden, K.J. The effect of trust on people's acceptance of CCS (carbon capture and storage) technologies: Evidence from a survey in the People's Republic of China. *Energy* **2016**, *96*, 69–79. [CrossRef]
50. Lock, S.J.; Smallman, M.; Lee, M.; Rydin, Y. "Nuclear energy sounded wonderful 40 years ago": UK citizen views on CCS. *Energy Policy* **2014**, *66*, 428–435. [CrossRef]
51. Midden, C.J.H.; Huijts, N.M.A. The Role of Trust in the Affective Evaluation of Novel Risks: The Case of CO_2 Storage. *Risk Anal.* **2009**, *29*, 743–751. [CrossRef]
52. Curry, T.E.; Ansolabehere, S.; Herzog, H. *A Survey of Public Attitudes towards Climate Change and Climate Change Mitigation Technologies in the United States: Analyses of 2006 Results*; Massachusetts Institute of Technology: Cambridge, MA, USA, 2007.
53. Wallquist, L.; Visschers, V.H.M.; Siegrist, M. Impact of Knowledge and Misconceptions on Benefit and Risk Perception of CCS. *Environ. Sci. Technol.* **2010**, *44*, 6557–6562. [CrossRef] [PubMed]
54. Sacuta, N.; Anderson, K. Creating core CCS messages: Focus Group Testing and Peer Review of Questions and Answers from the IEAGHG Weyburn-midale CO_2 Monitoring and Storage Project. *Energy Procedia* **2014**, *63*, 7061–7069. [CrossRef]
55. Ashworth, P.; Wade, S.; Reiner, D.; Liang, X. Developments in public communications on CCS. *Int. J. Greenh. Gas Control* **2015**, *40*, 449–458. [CrossRef]
56. Brunsting, S.; Upham, P.; Dütschke, E.; De Best Waldhober, M.; Oltra, C.; Desbarats, J.; Riesch, H.; Reiner, D. Communicating CCS: Applying communications theory to public perceptions of carbon capture and storage. *Int. J. Greenh. Gas Control* **2011**, *5*, 1651–1662. [CrossRef]
57. Daamen, D.D.; Terwel, B.W.; Ter Mors, E.; Reiner, D.M.; Schumann, D.; Anghel, S.; Boulouta, I.; Cismaru, D.M.; Constantin, C.; de Jager, C.C. Scrutinizing the impact of CCS communication on opinion quality: Focus group discussions versus Information-Choice Questionnaires: Results from experimental research in six countries. *Energy Procedia* **2011**, *4*, 6182–6187. [CrossRef]
58. Ter Mors, E.; Terwel, B.W.; Daamen, D.D.; Reiner, D.M.; Schumann, D.; Anghel, S.; Boulouta, I.; Cismaru, D.M.; Constantin, C.; de Jager, C.C. A comparison of techniques used to collect informed public opinions about CCS: Opinion quality after focus group discussions versus information-choice questionnaires. *Int. J. Greenh. Gas Control* **2013**, *18*, 256–263. [CrossRef]
59. Hope, A.L.B.; Jones, C.R. The impact of religious faith on attitudes to environmental issues and Carbon Capture and Storage (CCS) technologies: A mixed methods study. *Technol. Soc.* **2014**, *38*, 48–59. [CrossRef]
60. Carley, S.R.; Krause, R.M.; Warren, D.C.; Rupp, J.A.; Graham, J.D. Early Public Impressions of Terrestrial Carbon Capture and Storage in a Coal-Intensive State. *Environ. Sci. Technol.* **2012**, *46*, 7086–7093. [CrossRef] [PubMed]
61. Dütschke, E.; Wohlfarth, K.; Höller, S.; Viebahn, P.; Schumann, D.; Pietzner, K. Differences in the public perception of CCS in Germany depending on CO_2 source, transport option and storage location. *Int. J. Greenh. Gas Control* **2016**, *53*, 149–159. [CrossRef]
62. Williams, R.; Jack, C.; Gamboa, D.; Shackley, S. Decarbonising steel production using CO_2 Capture and Storage (CCS): Results of focus group discussions in a Welsh steel-making community. *Int. J. Greenh. Gas Control* **2021**, *104*, 103218. [CrossRef]
63. Broecks, K.P.; van Egmond, S.; van Rijnsoever, F.J.; Verlinde-van den Berg, M.; Hekkert, M.P. Persuasiveness, importance and novelty of arguments about Carbon Capture and Storage. *Environ. Sci. Policy* **2016**, *59*, 58–66. [CrossRef]
64. Haug, J.K.; Stigson, P. Local acceptance and communication as crucial elements for realizing CCS in the Nordic region. *Energy Procedia* **2016**, *86*, 315–323. [CrossRef]
65. Swennenhuis, F.; Mabon, L.; Flach, T.A.; de Coninck, H. What role for CCS in delivering just transitions? An evaluation in the North Sea region. *Int. J. Greenh. Gas Control* **2020**, *94*, 102903. [CrossRef]
66. Aursland, K.; Jordal, A.B.K. Case study of communication and social perceptions towards CCS in the cement industry. In Proceedings of the 14th Greenhouse Gas Control Technologies Conference, Melbourne, Austria, 21–26 October 2018.
67. Kashintseva, V.; Strielkowski, W.; Streimikis, J.; Veynbender, T. Consumer attitudes towards industrial CO_2 capture and storage products and technologies. *Energies* **2018**, *11*, 2787. [CrossRef]
68. Xenias, D.; Whitmarsh, L. Carbon capture and storage (CCS) experts' attitudes to and experience with public engagement. *Int. J. Greenh. Gas Control* **2018**, *78*, 103–116. [CrossRef]
69. Pihkola, H.; Tsupari, E.; Kojo, M.; Kujanpää, L.; Nissilä, M.; Sokka, L.; Behm, K. Integrated sustainability assessment of CCS–identifying non-technical barriers and drivers for CCS implementation in Finland. *Energy Procedia* **2017**, *114*, 7625–7637. [CrossRef]
70. Ilinova, A.; Cherepovitsyn, A.; Evseeva, O. Stakeholder Management: An Approach in CCS Projects. *Resources* **2018**, *7*, 83. [CrossRef]
71. Thomas, G.; Pidgeon, N.; Roberts, E. Ambivalence, naturalness and normality in public perceptions of carbon capture and storage in biomass, fossil energy, and industrial applications in the United Kingdom. *Energy Res. Soc. Sci.* **2018**, *46*, 1–9. [CrossRef]
72. Van Os, P. Accelerating low carbon industrial growth through carbon capture, utilization and storage (CCUS). *Greenh. Gases Sci. Technol.* **2018**, *8*, 994–997. [CrossRef]

73. Serdoner, A. European Environmental Advocacy Coalitions: Environmental Non-Governmental Perceptions of Carbon Capture and Storage in Energy Intensive Industries. Masters's Thesis, University of Graz, Graz, Austria, 2019.
74. Boomsma, C.; ter Mors, E.; Jack, C.; Broecks, K.; Buzoianu, C.; Cismaru, D.M.; Peuchen, R.; Piek, P.; Schumann, D.; Shackley, S. Community compensation in the context of Carbon Capture and Storage: Current debates and practices. *Int. J. Greenh. Gas Control* **2020**, *101*, 103128. [CrossRef]
75. Kojo, M.; Innola, E. Carbon Capture and Storage in the Finnish Print Media. *Risk Hazards Crisis Public Policy* **2017**, *8*, 113–146. [CrossRef]
76. Haikola, S.; Hansson, A.; Anshelm, J. From polarization to reluctant acceptance–bioenergy with carbon capture and storage (BECCS) and the post-normalization of the climate debate. *J. Integr. Environ. Sci.* **2019**, *16*, 45–69. [CrossRef]
77. Rodriguez, E.; Lefvert, A.; Fridahl, M.; Grönkvist, S.; Haikola, S.; Hansson, A. Tensions in the energy transition: Swedish and Finnish company perspectives on bioenergy with carbon capture and storage. *J. Clean. Prod.* **2020**, *280*, 124527. [CrossRef]
78. Alcalde, J.; Heinemann, N.; Mabon, L.; Worden, R.H.; de Coninck, H.; Robertson, H.; Maver, M.; Ghanbari, S.; Swennenhuis, F.; Mann, I. Acorn: Developing full-chain industrial carbon capture and storage in a resource- and infrastructure-rich hydrocarbon province. *J. Clean. Prod.* **2019**, *233*, 963–971. [CrossRef]
79. Yu, H.; Reiner, D.; Chen, H.; Mi, Z. *A Comparison of Public Preferences for Different Low-Carbon Energy Technologies: Support for CCS, Nuclear and Wind Energy in the United Kingdom*; Energy Policy Research Group (EPRG) Working Paper 1810, Cambridge Working Paper in Economics 1826; JSTOR: Cambridge, UK, 2018; Available online: https://www.jstor.org/stable/pdf/resrep30335.pdf (accessed on 6 November 2020).
80. Wüstenhagen, R.; Wolsink, M.; Bürer, M.J. Social acceptance of renewable energy innovation: An introduction to the concept. *Energy Policy* **2007**, *35*, 2683–2691. [CrossRef]
81. Dare, M.; Schirmer, J.; Vanclay, F. Community engagement and social licence to operate. *Impact Assess. Proj. Apprais.* **2014**, *32*, 188–197. [CrossRef]
82. Convention on Biological Diversity. *Biodiversity and Climate Change*; No. CBD/SBSTTA/23/3; UNEP: Montreal, QC, Canada, 2019; Available online: https://www.cbd.int/doc/c/326e/cf86/773f944a5e06b75dfc5866bf/sbstta-23-03-en.pdf (accessed on 1 November 2021).
83. Tokushige, K.; Akimoto, K.; Tomoda, T. Public acceptance and risk-benefit perception of CO_2 geological storage for global warming mitigation in Japan. *Mitig. Adapt. Strateg. Glob. Change* **2007**, *12*, 1237–1251. [CrossRef]
84. Liebscher, A.; Münch, U. *Geological Storage of CO_2—Long Term Security Aspects*; Springer International Publishing: Geotechnologien, Switzerland, 2015.
85. Markewitz, P.; Zhao, L.; Robinius, M. Technologiebericht 2.3 CO_2—Abscheidung und Speicherung (CCS), Wuppertal, Karlsruhe, Saarbrücken: Wuppertal Institut Für Klima, Umwelt Und Energie. 2017. Available online: https://epub.wupperinst.org/files/7051/7051_CCS.pdf (accessed on 1 November 2021).
86. Pietzner, K.; Schumann, D.; Tvedt, S.D.; Torvatn, H.Y.; Næss, R.; Reiner, D.M.; Anghel, S.; Cismaru, D.; Constantin, C.; Daamen, D.D. Public awareness and perceptions of carbon dioxide capture and storage (CCS): Insights from surveys administered to representative samples in six European countries. *Energy Procedia* **2011**, *4*, 6300–6306. [CrossRef]
87. Brondízio, E.S.; Settele, J.; Díaz, S.; Ngo, H.T. *IPBES: Global Assessment Report of the Intergovernmental Science-Policy Platform on Biodiversity and Ecosystems Services*; IPBES Secretariat: Bonn, Germany, 2019.
88. Available online: https://iea.blob.core.windows.net/assets/0d0b4984-f391-44f9-854f-fda1ebf8d8df/Transforming_Industry_through_CCUS.pdf (accessed on 1 November 2021).

*sustainability*

Article

Selection of Renewable Energy in Rural Area Via Life Cycle Assessment-Analytical Hierarchy Process (LCA-AHP): A Case Study of Tatau, Sarawak

Cyril Anak John [1], Lian See Tan [1,*], Jully Tan [2], Peck Loo Kiew [1], Azmi Mohd Shariff [3] and Hairul Nazirah Abdul Halim [4]

[1] Malaysia-Japan International Institute of Technology, Universiti Teknologi Malaysia, Jalan Sultan Yahya Petra, Kuala Lumpur 54100, Malaysia; cyriljohn19@gmail.com (C.A.J.); plkiew@utm.my (P.L.K.)

[2] School of Engineering, Monash University Malaysia, Jalan Lagoon Selatan, Bandar Sunwa 47500, Malaysia; tan.jully@monash.edu

[3] Chemical Engineering Department, Universiti Teknologi PETRONAS, Seri Iskandar 32610, Malaysia; azmish@utp.edu.my

[4] Faculty of Chemical Engineering Technology, Universiti Malaysia Perlis, Kompleks Pusat Pengajian Jejawi 3, Arau 02600, Malaysia; hairulnazirah@unimap.edu.my

* Correspondence: tan.liansee@utm.my; Tel.: +603-22031311

Citation: John, C.A.; Tan, L.S.; Tan, J.; Kiew, P.L.; Shariff, A.M.; Abdul Halim, H.N. Selection of Renewable Energy in Rural Area Via Life Cycle Assessment-Analytical Hierarchy Process (LCA-AHP): A Case Study of Tatau, Sarawak. *Sustainability* **2021**, *13*, 11880. https://doi.org/10.3390/su132111880

Academic Editors: Georgios Tsantopoulos and Evangelia Karasmanaki

Received: 28 September 2021
Accepted: 25 October 2021
Published: 27 October 2021

Publisher's Note: MDPI stays neutral with regard to jurisdictional claims in published maps and institutional affiliations.

Abstract: With a growing global population and energy demand, there is increasing concern about the world's reliance on fossil fuels, which have a negative impact on the climate, necessitating the immediate transition to a cleaner energy resource. This effort can be initiated in the rural areas of developing countries for a sustainable, efficient and affordable energy source. This study evaluated four types of renewable energy (solar, wind, biomass, and mini-hydro energy) using the integrated Life Cycle Assessment (LCA) and Analytical Hierarchy Process (AHP) approaches to select the best renewable energy source in Tatau, Sarawak. The criteria under consideration in this study included the environment, engineering and economics. The LCA was used to assess the environmental impact of renewable energies from gate-to-grave boundaries based on 50 MJ/day of electricity generation. The AHP results showed that solar energy received the highest score of 0.299 in terms of the evaluated criteria, followed by mini-hydro, biomass and wind energy, which received scores of 0.271, 0.230 and 0.200, respectively. These findings can be used to develop a systematic procedure for determining the best form of renewable energy for rural areas. This approach could be vital for the authorities that are responsible for breaking down multi-perspective criteria for future decision making in the transition into renewable energy.

Keywords: renewable energy; life cycle assessment; analytical hierarchy process; multi-perspective criteria

1. Introduction

Petroleum crude oil and natural gas are the main energy sources in Malaysia [1]. The overall conventional fuel business, on the other hand, has deteriorated due to price instability, supply insufficiency, and the environmental damage it causes, thereby ushering us into the inevitable era of renewable energy [2]. In 2017, renewable energy only accounted for approximately 5.8% of Malaysia's total energy consumption [1]. The Malaysian government has set a target of 20% renewable energy, in terms of total electricity generation, by 2025 [3]. According to Abdullah et al. [4], Malaysia has a wide range of opportunities and potential for focusing on renewable energy, particularly solar, wind, hydro, biogas and biomass. However, realizing this potential would necessitate an immense effort from the government in terms of providing incentives as well as developing and implementing systemically effective policies.

It was estimated that 4% of the region in Sabah and 15% of that in Sarawak still have no access to electricity. In response to this, the Malaysian government has set a goal of providing modern energy facilities to as many people as possible, particularly in the remote parts of Sarawak and Sabah [5]. Due to the geographic profile of such places, increasing the grid's electricity supply is a challenge. Power distribution is uneconomic due to uneven terrain and dense forest. High transmission loss is also an issue, implying that a grid power supply in remote areas is not possible [6]. On the contrary, off-grid electricity, which can be generated using renewable energy sources such as solar, wind, or hydro technologies, can be used to power remote areas. The available resources can boost rural electrification capacity and benefit the villagers as an economic strategy and a sustainable source of energy.

In rural areas, the local electrical authorities commonly opted for diesel-powered generators or, most recently, the hybrid-solar system as a quick and short-term fix to supply electricity to essential facilities such as remote schools, clinics, administrative offices, and small villages for a limited period of time per day [7]. Nonetheless, the Sarawak state government deserves credit for increasing the state-level electricity coverage from 79.2% to 90% between 2009 and 2015 [8]. To avoid this rural electrification initiative failing on a long-term basis, extensive planning in terms of implementation, technical and social difficulties must be done [9]. Therefore, energy planning analysis should be done in a more holistic manner, such as integrating the methods of the Life Cycle Analysis (LCA) and Multi-Criteria Decision Making (MCDM) [10]. This would allow for a more comprehensive analysis, as well as the development of a new analytical tool to replace the conventional cost-benefit or techno-economic analysis.

The practical application of the LCA to product or process design and development in order to reduce aggregate environmental impacts is gaining traction, either through the modification of some input variables or a scenario analysis. Based on the LCA analysis of supercritical carbon dioxide extraction of caffeine from coffee beans, De Marco et al. [11] claimed that when a portion of the electricity at the grid was replaced with electricity produced by photovoltaic (PV) panels, the environmental impact could be reduced by 176% in terms of human health, 10.3% in terms of ecosystem diversity, and 16.1% in terms of resource availability. On the other hand, based on the LCA analysis conducted by Gallucci et al. [12], the authors reported that using PV energy as a renewable energy source in the production of hollow glass containers for food packaging was able to significantly reduce the global warming potential.

Likewise, MCDM has been implemented in recent years to evaluate many solutions to real-world problems relating to policy and strategy [13]. Hassan et al. [14] used a multi-criteria decision-making tool in the form of the Analytical Hierarchy Process (AHP) in order to analyze renewable generation sources in Saudi Arabia. A similar approach was also taken by Algarín et al. [15] in evaluating the renewable energy sources of rural areas in the Caribbean region of Colombia. A study was done by Das [16] that integrated the AHP and Quality Function Deployment (QFD) methods in order to determine the most viable renewable energy source for the state of Maharashtra. Hilorme et al. [17] developed a decision-making model for introducing energy-saving technologies based on the AHP. Zhang et al. [18] studied the economic development of the biomass energy industry in the Heilongjiang province based on the AHP. Nevertheless, the application of the combined LCA-MDCM methodologies for the analysis of renewable energy in an Asian context is limited, with Ren et al. [19] evaluating the sustainability of renewable fuel production, i.e., bioethanol. Hence, in this study, a robust systematic evaluation of renewable energy systems, using the integrated LCA-AHP methodologies, was expanded for the analysis of the Asian regions, particularly Malaysia, in order to promote the sustainable development of zero-carbon technology.

In recent years, several studies have been conducted to assess the current state of renewable energy in Malaysia. According to Hannan et al. [5], the rural electrification effort in Malaysia requires more attention in order to contribute to the country's future energy security and sustainability. Based on a study by Basri et al. [20], the abundance of renewable energy resources in Malaysia could potentially produce a stable supply of renewable energy. Despite this, there is no clear approach for choosing the most suitable type of renewable energy to meet the complicated economic, social, and environmental requirements.

Unsystematic decision making in the determination of the best renewable energy that can satisfy the needs of each individual rural location could hamper the successful implementation of the rural electrification initiative. Various elements, including environmental, engineering and economic factors, must be considered in order to overcome this. The application of the Life Cycle Assessment-Analytical Hierarchy Process (LCA-AHP) as a decision-making tool would allow for a comprehensive evaluation that could take such considerations into account. Tatau in Bintulu, Sarawak, was chosen as the case study for this study.

This research aims to characterize pollutant emissions and to determine the cost of generating electricity using different renewable energy systems. The environmental impact of different renewable energy systems was determined using the LCA method while the best renewable energy for Tatau, Bintulu from a combined engineering, environmental and economic perspective was evaluated using the AHP method. The findings of this study would provide a systematic process for determining the most appropriate renewable energy system for the rural area of Tatau, Sarawak. This approach will be critical for the authorities that are responsible for breaking down the multi-perspective criteria that may be used in the future system design.

2. Materials and Methods

2.1. Life Cycle Assessment (LCA)

The LCA was used to determine the environmental impact of the renewable energy sources. In this study, the LCA data for each renewable energy system was extracted from sources in the literature in order to reflect the global warming potential (GWP) and acidification potential (AP) as their respective environmental factor score. GWP is correlated to greenhouse gas emissions, and it is an indication of the system's potential contribution to climate change. Meanwhile, the AP could indicate the environmental impact of the system as it relates to the acidification of water bodies and soil [21]. This LCA approach, which was based on ISO 14040 and ISO 14044 [22], was comprised of four steps, as shown in Figure 1. The first step was to define the goal and scope of the project. This measure determined the objective, system boundaries, functional unit and assumptions. Then there was the life cycle inventory (LCI), which involved data collection from all stages within the life cycle boundary, including the input, intermediate processes, and output. The third step was the life cycle impact assessment (LCIA), whereby the potential impact on the environment by the system was evaluated. Lastly, the data was interpreted based on the goal and scope definitions, as well as the LCI and LCIA data. The vital points were assessed and suggestions for future improvements were made.

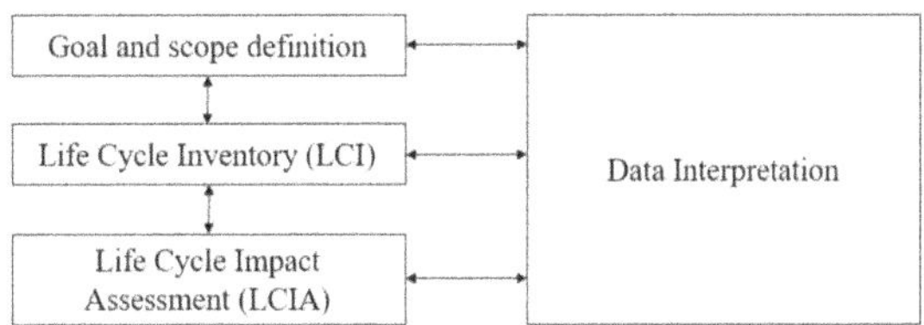

Figure 1. Framework of the life cycle assessment (LCA).

2.1.1. Goal and Scope Definition

This study focuses on the potential electrification of the rural area of Tatau, Bintulu using renewable energy. The goal of this assessment was to determine the overall impact of selecting the different renewable energy systems that were evaluated in this study, which included solar, wind, biomass and mini-hydro energy, for every 50 MJ/day of energy output. This is equivalent to a total of 13.89 kWh/day of electricity, which was sufficient to power an estimated 25 houses in Tatau, Sarawak, with an average of 2 MJ/day per household of the rural community [23]. Tatau is a district in Bintulu, Sarawak, with a total land size of 4945.80 km^2 and a population of approximately 25,000 people. The economic background of Tatau mostly involves the timber and agricultural industries. The rural areas of Tatau are only travelable via timber routes and palm estate paths, one of them being Kakus road [24].

Jong et al. [25] found that Tatau has an abundance of potential for renewable energy due to its strategic location relative to the renewable resources, reasonable distance to road access, considerable population and mild land slope. However, they only focused on evaluating the potential of renewable energy for Tatau, and several other locations in Sarawak, based on geographical data. Therefore, the evaluation that was done in this study aimed to further evaluate the potential renewable energy sources that could be sustainable for rural areas in terms of engineering feasibility, environmental impact, and economic feasibility.

For the four renewable energy systems under evaluation, the environmental impact was assessed from the point of manufacture (gate) to the point of end-of-life (grave). This included the stages of component manufacturing, construction, operation, maintenance, and finally, disposal. Figure 2 illustrates the system boundaries for the renewable energy systems.

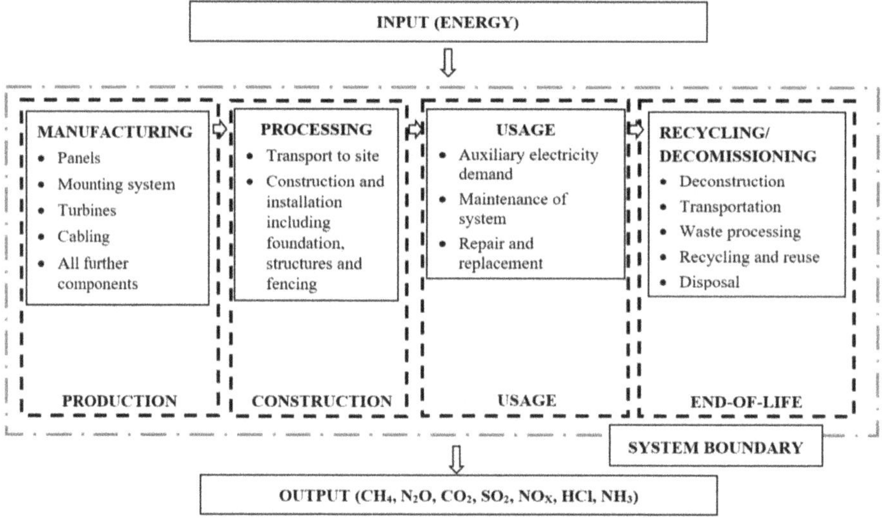

Figure 2. System boundaries for all alternatives in this study.

The energy consumptions and the materials consumed throughout the system boundaries for each renewable energy system were used as inputs in this study. As shown in Figure 2, the system boundary is made up of four stages of processes: production, construction, usage and end-of-life. The production stage of renewable energy started with the extraction of raw materials for the manufacturing of components and their assembly. This includes the assembly of panels, mounting systems, cables, and other components that are required to generate sufficient amounts of electricity. At this stage, only the main materials and quantities required for the production were defined.

Meanwhile, in the construction stage, local land routes and maritime deliveries were considered as the means of system transportations to the site. This study did not take into account transportation from outside of the country, as the energy system could be procured from any country, but the entry route to Tatau is consistent, which is through its port. According to Google Maps, the distance for local land transportation from Bintulu's nearest port to the rural area of Tatau was 63 km. The energy consumption during the construction and installation of the foundation, structures and fencing were also included in this stage.

The usage stage of the system involved the demand for auxiliary electricity when necessary. Scheduled maintenance of the system was included as it was vital to inspect its performance and maintain the system's efficiency. This stage also took into consideration any necessary repairs or replacements during the course of its use.

The final phase of the system was the end-of-life stage, which included deconstruction, transportation, recycling and reuse where applicable, and waste processing. The aim was to evaluate the impact of waste recycling and disposal on the environment. It was estimated that the system has a lifespan of 25 years.

- The data for this LCA analysis were extracted from reviews in the literature and publicly available databases. The data were scaled to 1 kWh of electricity produced for all stages before being normalized to 13.89 kWh, which is the functional unit of this study. The following assumptions were made for the inventory data collection: Only electricity was included as the input for this study. Other materials were not considered as alternatives would have required the use of exclusive materials for the manufacturing of components [26].
- Electricity used during the production stage was assumed to be generated using natural gas, which had the specific natural energy of 53.6 MJ/kg [27] and an efficiency of 55.13% [28]. During the usage stage, self-generated electricity was used for auxiliary electricity demand.
- The transportation stage only included land transportation and did not include sea or air transportation.
- Only output and pollutants, i.e., methane (CH_4), nitrous oxide (N_2O), carbon dioxide (CO_2), sulfur dioxide (SO_2), nitric oxide (NO_x), hydrogen chloride (HCl) and ammonia (NH_3), which were related to GWP and AP, were taken into consideration in the LCI.

2.1.2. Life Cycle Impact Assessment (LCIA) Scope Definition

The aim of this step was to extract the relevant environmental indicator from the results of the inventory analysis. The impact classification included global warming potential (GWP) and acidification potential (AP). The characterization factors were taken from the literature [26,29–31].

2.2. Simulation of HOMER Pro

This study used HOMER Pro version 3.11.2 simulation software to design an optimize a renewable-energy-electrification system for the case study area, i.e., Tatau, Sarawak. The information and details of the actual location served as the input data for the simulation. The details included the renewable resources of the alternatives evaluated, which were solar irradiation, wind speed, biomass resource, and stream flow with regards to the load profile of the case study area. The information was obtained from the coordinates of Tatau, Sarawak. Next, the system was designed based on the specifications and costs of the components that were obtained from sources in the literature and previous project data. The results from the HOMER Pro simulation provided the relevant costs required, which included the capital cost as well as the operation and management costs [32]. The flow chart of the use of HOMER for the optimization process of the proposed renewable energy system is shown in Figure 3.

In terms of load profiles, the average electric consumption for the case study area of Tatau, Sarawak was assumed to be 50 MJ/day or 13.89 kWh/day, based on the average household energy consumption. It is also worth noting that the assumed load profile was based on a working day during the dry season of the year. The energy consumption profile may differ from the input load profile during other seasons.

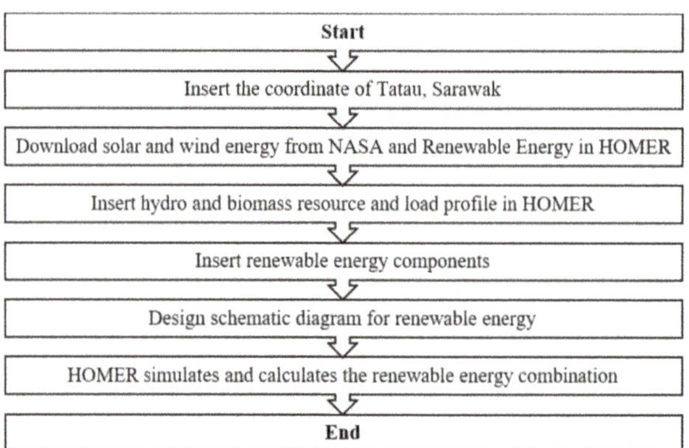

Figure 3. Flowchart of HOMER Pro simulation method.

2.3. Analytical Hierarchy Process (AHP)

The initial phase of the AHP involved the division of the multiplex problem into several levels of a hierarchy [33]. The topmost level of this hierarchy represented the main goal of a decision maker. The second level represented the evaluated criteria, and the bottom level corresponded to the alternatives that were under consideration. In some cases, sub-criteria can be considered under the main decision criteria in order to incorporate additional problem-specific decision levels. Following that, pairwise comparison was performed by weighing and ranking the priorities and alternatives. Saaty [34] advocated for the use of measurements on a scale of 1 to 9 and the eigenvector approach for this comparison. This pairwise comparison could be executed on both the quantitative and qualitative characteristics of the alternative energy sources.

The outcome of these steps helped to forecast the impact of each alternative on the overall goal of the hierarchy decision. It also helped to distinguish competing criteria and eventually rank them according to their priorities. Following that, the data were examined in order to identify inconsistencies in the judgements made. As the result obtained may have been subjective, this consistency check was an important step in the AHP method [14]. The final stage of this approach was to evaluate the scores of each criterion, sub-criterion, and lastly, alternative.

AHP Model

A hierarchical structure was developed in this study which incorporated four levels: goal, main criteria, sub-criteria and alternatives.

(a)　Level 0: Goal To determine the best renewable energy system for Tatau, Sarawak.
(b)　Level 1: Main Criteria Main criteria in this study were the environment, engineering and economy.
(c)　Level 2: Sub-criteria The sub-criteria in this study were the land requirements, environmental impact (global warming potential (GWP) and acidification potential (AP)), resource availability, efficiency of the system, technology maturity, capital cost and operating and management costs.
(d)　Level 3: Alternatives The alternatives being assessed in this study were solar, wind, biomass and mini-hydro energy system. The layout of this hierarchy is represented in Figure 4.

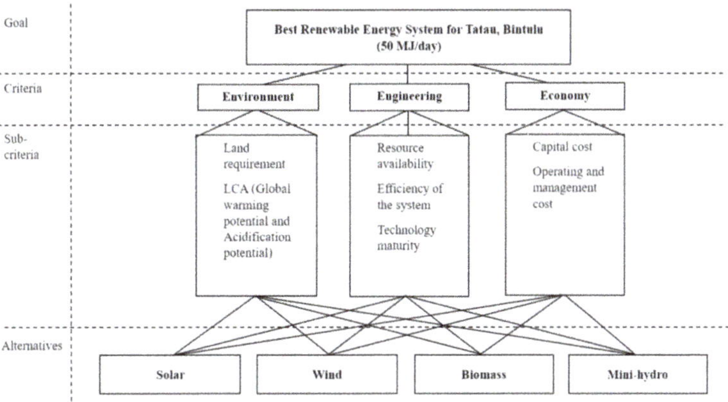

Figure 4. AHP model for this study.

3. Results

3.1. Environmental Impacts of Renewable Energy Alternatives

Table 1 shows the pollutants emitted by the renewable energy sources that were evaluated in this study, namely solar, wind, biomass, and mini-hydro energy. The pollutants that contributed to the impact assessed in the LCA boundary of this study, which were GWP and AP, were included in the results. The pollutant emissions were classified into four stages within the gate-to-grave boundary of manufacturing, construction, usage and end-of-life.

Table 1. Pollutant emissions of solar energy system [35,36]; wind energy system [35,37,38]; biomass energy system [39,40]; and mini-hydro energy system [41,42].

Energy System	Type of Pollutants (kg/kWh)	Manufacturing ($\times 10^{-2}$)	Construction ($\times 10^{-2}$)	Usage ($\times 10^{-2}$)	End-of-Life ($\times 10^{-2}$)
Solar	CO_2	156.14	26.80798	0.00	11.03
	CH_4	346.98	59.55	0.00	24.52
	N_2O	80.50	13.82	0.00	5.69
	SO_2	80.50	13.82	0.00	5.69
	NOx	27,271.27	4680.52	0.00	1927.15
	HCl	232.61	39.92	0.00	16.44
	NH_3	56.95	9.77	0.00	4.02
Wind	CO_2	18.82	0.24	0.17	0.18
	CH_4	44.46	0.56	0.39	0.43
	N_2O	0.37	0.01	0.00	0.00
	SO_2	3799.77	48.20	33.23	36.51
	NOx	2991.72	37.95	26.16	28.77
	HCl	19.24	0.24	0.17	0.19
	NH_3	2.89	0.04	0.03	0.03
Biomass	CO_2	1.21	0.03	1.64	0.03
	CH_4	1.81	0.04	2.46	0.04
	N_2O	3.54	0.08	4.79	0.08
	SO_2	1091.65	25.04	1479.81	25.04
	NOx	10,832.53	248.46	14,684.22	248.464
	HCl	209.93	4.82	284.58	4.82
	NH_3	587.81	13.48	796.82	13.48
Mini-hydro	CO_2	10.42	5.56	1.39	0.02
	CH_4	21.89	11.68	2.92	0.04
	N_2O	0.42	0.22	0.06	0.00
	SO_2	1010.25	538.80	134.70	1.68
	NOx	2139.34	1140.98	285.25	3.57
	HCl	5.94	3.17	0.79	0.01
	NH_3	2.38	1.27	0.32	0.00

Based on the aggregated pollutants, the GWP and AP of the renewable energy alternatives are shown in Figure 5. The results showed that solar energy had the greatest impact in terms of GWP and AP, followed by biomass energy and wind energy. Mini-hydro energy exhibited the lowest environmental impact of the four renewable energy sources that were evaluated. Figure 6 shows the percentage of environmental impact contribution in order to further analyze which stage within the gate-to-grave scope was responsible for the GWP and AP emission levels. A significant portion of the aggregated pollutants from solar and wind energy came from the manufacturing stage of the system. According to Mulvaney [43], this was due to the high energy consumption of the solar panel manufacturing process in particular. The processing of raw silicon requires a huge amount of energy as the process involves high temperatures that contribute significantly to carbon emissions. Similarly, the manufacturing phase of the wind energy system requires heating and cooling processes for the fabrication of turbines [44,45]. While the manufacturing stage contributed less to the environmental impact of mini-hydro energy, the construction stage accounted for a significant portion of the pollutants in this system. Concrete production and the transportation of rocks for the construction of dams and tunnels were among the major contributors to the pollutant emissions of a mini-hydro energy system [46]. Biomass energy, on the other hand, was found to emit a higher percentage of pollutants during the usage stage when compared to the other evaluated stages in the system boundary. This could be due to the release of pollutants during the operation of the system as a result of biomass combustion [40].

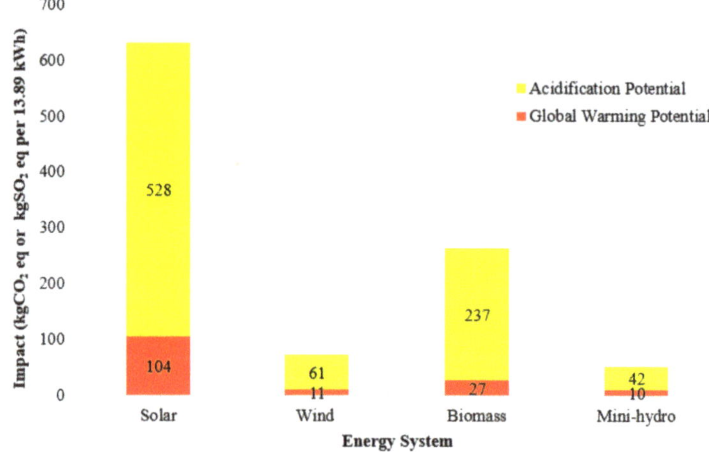

Figure 5. Impact assessment results of the renewable energy alternatives.

3.2. Cost for Electricity Generation

Based on the HOMER Pro Simulation, it was found that solar and wind energy demanded the highest costs in terms of building and operating the energy system, with an estimated total of US $14,821.01 and US $14,626.00, respectively. The capital costs for both energy systems were significantly higher due to the expensive materials needed to manufacture the energy systems [47,48]. It was also noted that, according to the simulation, the operational and maintenance costs of a biomass energy system was the highest, at approximately US $5,447.09. This was due to the cost of replacing the electrical generator over the course of a year. The replacement was necessary to maintain the efficiency of the system in meeting the electrical load demand [49]. Table 2 shows the summary of the costs needed for renewable energy alternatives in this study, which were simulated using the HOMER Pro simulation software version 3.11.2.

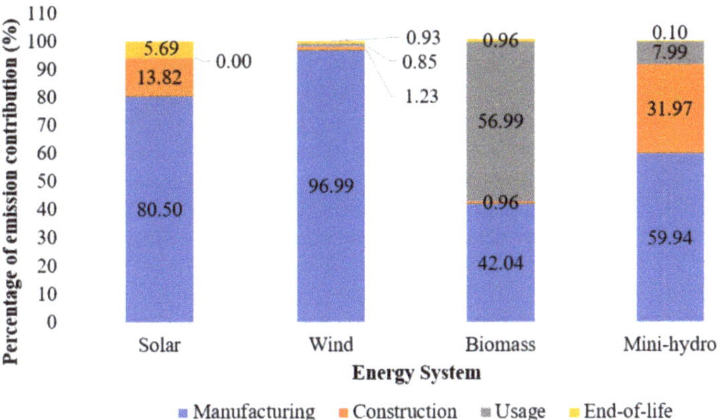

Figure 6. Percentage of emission contribution.

Table 2. Summary of cost * needed for renewable energy alternatives.

Expenditure	Solar Energy	Wind Energy	Biomass Energy	Mini-Hydro Energy
Capital cost (US$)	11,618.67	12,337.18	841.75	5,782.83
Operational and maintenance cost (US$)	3,202.34	2,288.82	5,447.09	773.38
Total (US$)	14,821.01	14,626.00	6288.84	6556.21

* The currency exchange rate used was US$1 = RM4.16.

3.3. Analytical Hierarchy Process (AHP)

The results of the environmental impact study and the costs from LCA and HOMER Pro simulation from the previous section yielded the score for the GWP, AP, capital cost, and operational and maintenance cost sub-criteria of the AHP model in Figure 4. In this section, the remaining sub-criteria data were analyzed based on sources in the literature from various studies. The results of various studies were compared to determine the data deviation and average value. Figure 7 shows a summary of the land requirements for each renewable energy alternative that was investigated in this study.

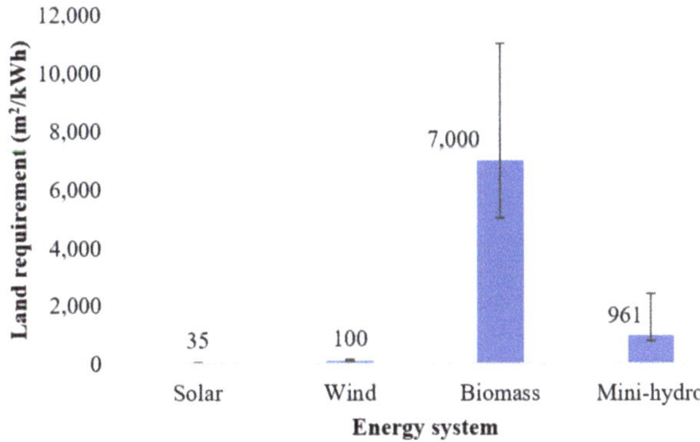

Figure 7. Land requirement of the renewable energy alternatives [50–57].

Biomass energy required the most land in terms of site area, with an average of 7000 m^2/kWh. This was because the biomass resource required a large receiving and processing area [58]. Other studies revealed highly variable land sizes due to the use of different technologies with varying equipment sizes to process the biomass resource.

As a result of this finding, it was observed that solar energy was the most sustainable renewable energy in terms of land requirement, as it required the least amount of land to build the energy system. This was largely due to the fact that the components of the energy system were small and did not take up much space [59]. Figure 8 shows the availability of resources for all renewable energy alternatives. The results showed that solar energy was leading in terms of its resource availability in Malaysia, with a generation potential as high as 6500 MW due to the high annual solar irradiance of the country [60]. This was followed by mini-hydro energy, with the capacity of 28.9 MW. The high annual rainfall and river flow in the proximity of the case study area were deemed as advantageous for the mini-hydro energy system [61].

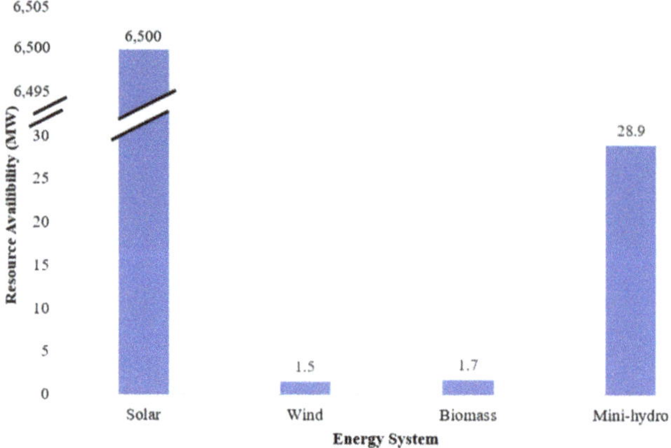

Figure 8. Resource availability of the renewable energy alternatives [32,58,62,63].

In terms of resource availability, solar energy was the most suitable type of renewable energy with the highest generation potential. Wind energy, on the other hand, was the least suitable as the wind speed in the case study area was not sufficient to meet the electrical load demand. This was compounded by the high variability of wind speeds throughout the year in Malaysia, which was not ideal for the output efficiency [48].

Technological maturity was the next sub-criterion under the engineering perspective. Figure 9 shows the maturity of renewable energy alternatives in Malaysia based on the total number of projects completed in the past [59,64–66]. The number of projects considered in this study referred to projects that were initiated under the Malaysian Government's Small Renewable Energy Programme (SREP) in order to promote small-scale renewable electricity in the country. The result showed that solar energy had the most projects in Malaysia in the past, with a total of 38 successful projects. This indicated that solar energy was the most established and reliable technology, which was also supported by Tang et al. [67], who found that this energy system had a high installation capacity compared to the other alternatives. Meanwhile, wind energy was considered to be the least matured technology with a low number of projects in Malaysia. This was due to numerous projects breaking down during operation, which raised concerns about their reliability and long-term prospects [68].

Another sub-criterion under the engineering criteria was the efficiency of the system. Figure 10 shows the efficiency of the renewable energy alternatives. When compared to other resources, mini-hydro energy had the highest output efficiency with an average 67% efficiency. This was largely due to the availability of water flow throughout the day. The high annual rainfall and river flow also contributed to this, as the abundance of resources benefitted the output efficiency of the energy system [61]. On the other hand, solar energy displayed the lowest efficiency at 11% compared to other alternatives. This was because of the low purity of the materials used in photovoltaic cells, which resulted in the low overall efficiency of the solar energy system [69].

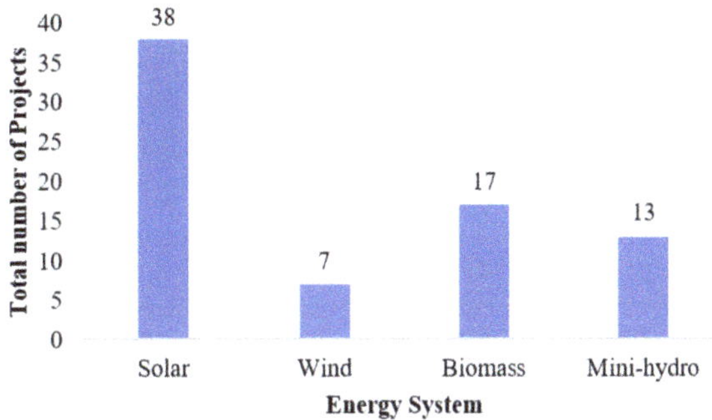

Figure 9. Technology maturity of the renewable energy alternatives [59,64–66].

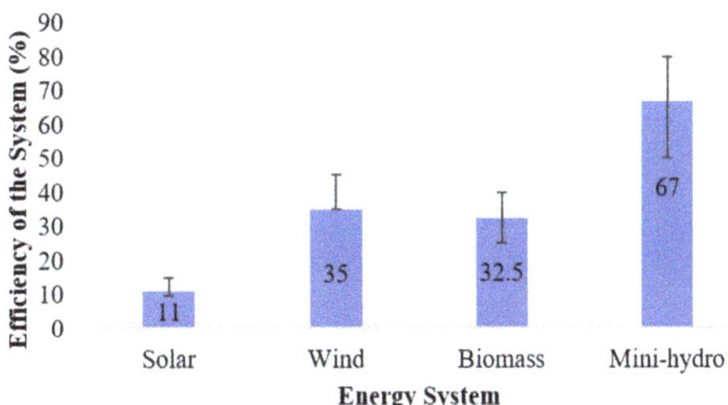

Figure 10. Efficiency of the renewable energy alternatives [70–78].

Following the data collection for the sub-criteria of the AHP model, the importance scores for the criteria and sub-criteria were compiled and normalized. This step was conducted based on the AHP model depicted in Figure 4 from the top level, which was the goal, to the bottom layer of alternatives.

(a)   Level 0: Goals To determine the best renewable energy for Tatau, Sarawak.
(b)   Level 1: Main Criteria Since the relative weight and score for the criteria were extracted from literature sources with equivalent goals, the pairwise comparison was disregarded at this level. The importance score was derived from a review of the literature. Figure 12 presents the normalized scores that were used to fit the values into the project model.
(c)   Level 2: Sub-criteria Figure 11 shows the importance scores for each sub-criterion and the overall importance score for sub-criteria that corresponded to the main criterion, as extracted from the literature sources. All scores were normalized to fit into the AHP model.
(d)   Level 3: Alternatives Table 3 tabulates the definition of the importance score and the literature source for each of the environmental, engineering and economic criteria, respectively.

In terms of land requirement, the larger the land area required to build the energy system, the lower the importance score. In this case, solar energy was given the highest priority over the others. The calculation of the score was based on Saaty's importance score of 1–9. The pairwise comparison data were then tabulated and normalized in order to produce the priority vectors shown in Table 4.

Table 3. Definition of importance score and literature sources for environmental, engineering and economic sub-criteria.

Criteria	Sub-Criteria	Definition of Importance Score	Data Source
Environmental	Land requirement	Larger land required indicates lower importance score (lower environmental sustainability)	Solar energy: 35 m^2/kWh [51]
			Wind energy: 100 m^2/kWh [53]
			Biomass energy: 7000 m^2/kWh [52]
			Mini-hydro energy: 961 m^2/kWh [51]
	GWP and AP	Higher impact value indicates lower importance score (lower environmental sustainability)	LCA
Engineering	Resource availability	Higher generation potential indicates higher importance score (higher engineering sustainability)	Solar energy: 6500 MW [62] Wind energy: 1.5 MW [4]
			Biomass energy: 1.7 MW [63]
			Mini-hydro energy: 28.9 MW [32]
	Efficiency of the system	Higher efficiency indicates higher importance score (higher engineering sustainability)	Solar energy: 11% [62]
			Wind energy: 35% [71]
			Biomass energy: 32.5% [72]
			Mini-hydro energy: 67% [70]
	Technology maturity	Higher number of past projects indicates higher importance score (higher engineering sustainability)	Solar energy: 38 projects [59]
			Wind energy: 7 projects [65]
			Biomass energy: 17 projects [66]
			Mini-hydro energy: 13 projects [64]
Economic	Capital cost	Higher cost indicates lower importance score (lower economical sustainability).	HOMER Pro results
	Operation and maintenance cost		

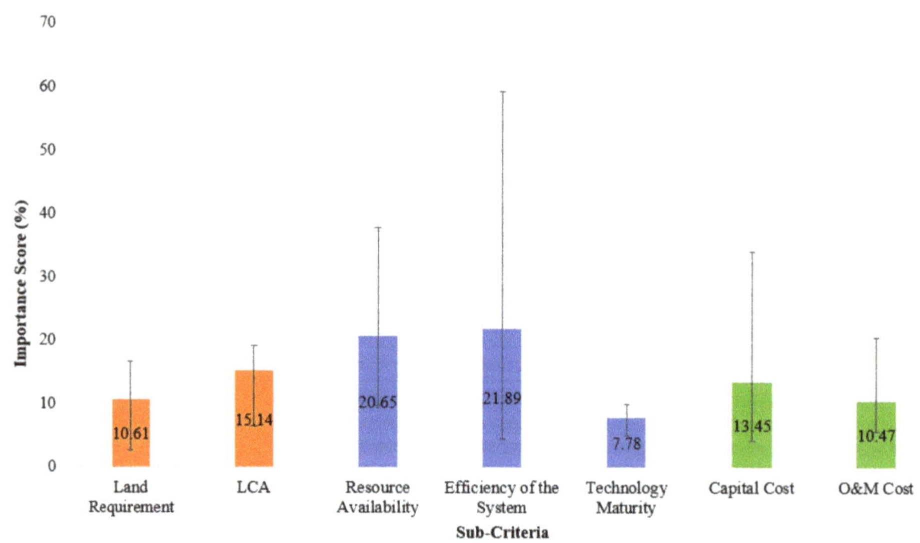

Figure 11. Overall importance score for sub-criteria [14,15,79–83].

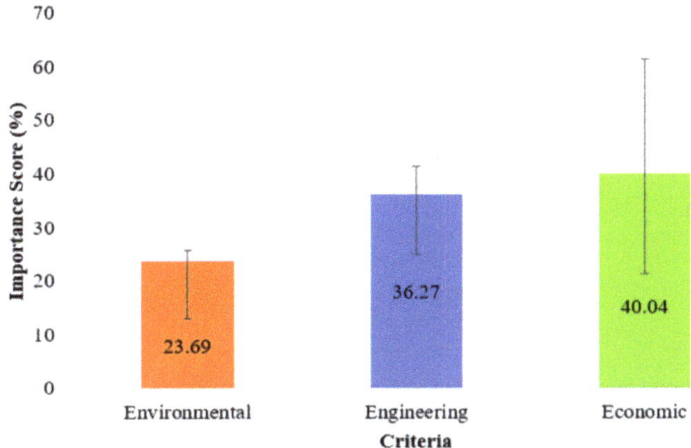

Figure 12. Importance score for main criteria [14,15,79–81].

Table 4. Priority vector with regards to land requirement sub-criterion.

Energy System	Priority Vector
Solar Energy	0.312
Wind Energy	0.284
Biomass Energy	0.180
Mini-hydro Energy	0.225

The same techniques were used for other sub-criteria in the AHP model. Following that, the linear multiplication of the priority weightings for each segment within all levels for each alternative was conducted in order to determine the final importance score with regards to the AHP model's goal of determining the best renewable energy source in Tatau, Sarawak. Table 5 shows the overall importance scores for all renewable energy alternatives and Table 6 tabulates the final importance scores of all energy sources with respect to the goal of the AHP model.

Table 5. The overall important scores for all renewable energy alternatives.

Level 1		Level 2		Level 3			
Criteria	Importance Score	Sub-Criteria	Importance Score	Solar Energy	Wind Energy	Biomass Energy	Mini-hydro Energy
Environmental	0.237	Land requirement	0.412	0.064	0.149	0.461	0.326
		LCA	0.588	0.127	0.310	0.247	0.316
Engineering	0.363	Resource availability	0.410	0.995	0.000	0.000	0.004
		Efficiency of the system	0.435	0.076	0.241	0.223	0.460
		Technology maturity	0.155	0.507	0.093	0.227	0.173
Economic	0.400	Capital cost	0.562	0.207	0.199	0.324	0.270
		O & M	0.438	0.242	0.268	0.178	0.311

Table 6. The final importance scores of all energy alternatives with respect to the goal of the AHP model.

Alternatives	Final Importance Score
Solar Energy	0.299
Wind Energy	0.200
Biomass Energy	0.230
Mini-hydro Energy	0.271

The final importance scores for solar, wind, biomass and mini-hydro energy were 0.299, 0.200, 0.230 and 0.271, respectively. In terms of the overall criteria under consideration, which were environmental, engineering, and economic perspectives, solar energy was found to be more sustainable than the other alternatives for the region of Tatau, Sarawak. The main reason for this was the engineering superiority of solar energy over the other alternatives, especially with regards to the resource availability of solar energy in Malaysia. The high solar irradiance of Malaysia outweighed the other resources, as the country is strategically located to have an ideal climate for solar energy [67]. Despite this, other alternatives could still operate within the minimum requirement of the electrical load in the case study area. In terms of technological maturity, the numerous successful solar energy projects in the past have proved that this type of energy system is highly reliable and established in the Malaysia region [68]. This had a significant impact on the decision as other alternatives were either still in development or had failed during previous deployment, thereby casting serious doubt on that energy system, particularly the wind energy system [84].

Despite its excellent engineering characteristics, solar energy underperformed in terms of environmental and economic perspectives, as it received a relatively low importance score in this area. Further research and development can be done in order to discover more affordable and environmentally friendly solar energy systems [84]. This can further mitigate the high energy consumption required for the manufacturing of solar energy components, which is energy intensive and may still rely on fossil fuel sources [85].

4. Conclusions

The integrated LCA-AHP approach was successfully applied to determine the best type of renewable energy for the rural area of Tatau, Sarawak. The gate-to-grave LCA was used to assess the manufacturing, construction, usage and end-of-life stages of the renewable energies under evaluation, which were solar, wind, biomass, and mini-hydro energy. Global warming potential (GWP) and acidification potential (AP) were the two environmental impacts that were evaluated. Solar energy had the greatest impact in terms of both GWP and AP, with 104 kg CO_2 eq and 528 kg SO_2 eq, respectively. The least impact among the alternatives was from mini-hydro energy, with a GWP of 10 kg CO_2 eq and an AP of 42 kg SO_2 eq, indicating that this type of renewable energy was the most sustainable environmental option. In this study, an AHP model was developed in order to determine the best renewable energy source for Tatau, Sarawak based on three criteria, namely environmental, engineering and economic. The hierarchical structure provided an easier route for evaluation, which went through every level, from the goal of the project to the criteria, then to the sub-criteria, and finally to the alternatives. The obtained AHP results differed from the LCA results in that solar energy scored the highest priority weight of 0.299, compared to 0.200, 0.230 and 0.271 for wind, biomass and mini-hydro energy, respectively. Although solar energy was not the most sustainable option from an environmental standpoint, the engineering aspect of the energy system was superior compared to the other alternatives, which heavily influenced the decision model. Prior to the actual start of the project, the decision-making process in the energy planning sector is crucial in the determination of the ideal energy system. The LCA-AHP framework in this study was proven to be robust in comparing the renewable energies that were under evaluation. With the input of coordinates for a specific area of interest into the simulation software, this framework can also applicable be for other locations, be it in Malaysia or other countries.

Author Contributions: Conceptualization, all authors; Methodology, C.A.J. and J.T.; Software, C.A.J.; Validation, L.S.T., P.L.K. and J.T.; Formal Analysis, C.A.J.; Investigation, C.A.J. and L.S.T.; Resources, L.S.T.; Data Curation, C.A.J. and J.T.; Writing—Original Draft Preparation, C.A.J., J.T. and L.S.T.; Writing—Review & Editing, J.T., P.L.K., A.M.S. and H.N.A.H.; Visualization, C.A.J. and P.L.K.; Supervision, L.S.T.; Project Administration, L.S.T.; Funding Acquisition, L.S.T., A.M.S. and H.N.A.H. All authors have read and agreed to the published version of the manuscript.

Funding: This research was jointly funded by Universiti Teknologi PETRONAS via the Joint Research Project (JRP8) funding, Universiti Teknologi Malaysia via Matching Grant (PY/2021/00347 & PY/2021/00272) and Universiti Malaysia Perlis (Grant No: 9023-00022).

Institutional Review Board Statement: Not applicable.

Informed Consent Statement: Not applicable.

Data Availability Statement: The authors confirm that all data generated or analysed to support the findings are included in the published article. The raw data that supports the findings of this study is available upon reasonable request from the corresponding author.

Conflicts of Interest: The authors declare no conflict of interest.

References

1. Energy Commission, Malaysia Energy Statistics Handbook. 2019. Available online: https://meih.st.gov.my/documents/10620/bcce78a2-5d54-49ae-b0dc-549dcacf93ae (accessed on 11 November 2020).
2. Abotah, R.; Daim, T.U. Towards building a multi perspective policy development framework for transition into renewable energy. *Sustain. Energy Technol. Assess.* **2017**, *21*, 67–88. [CrossRef]
3. Choong, J. Yeo: Malaysia Aiming for 20pc Renewable Energy Use by 2025. *Malay Mail.* 2019. Available online: https://www.malaymail.com/news/malaysia/2019/09/03/yeo-malaysia-aiming-for-20pc-renewable-energy-use-by-2025/1786768 (accessed on 11 November 2020).
4. Abdullah, W.S.W.; Osman, M.; Ab Kadir, M.Z.A.; Verayiah, R. The potential and status of renewable energy development in Malaysia. *Energies* **2019**, *12*, 2437. [CrossRef]
5. Hannan, M.A.; Begum, R.A.; Abdolrasol, M.G.; Lipu, M.H.; Mohamed, A.; Rashid, M.M. Review of baseline studies on energy policies and indicators in Malaysia for future sustainable energy development. *Renew. Sustain. Energy Rev.* **2018**, *94*, 551–564. [CrossRef]
6. Sreeraj, E.S.; Chatterjee, K.; Bandyopadhyay, S. Design of isolated renewable hybrid power systems. *Sol. Energy* **2010**, *84*, 1124–1136. [CrossRef]
7. Chen, S. Rural electrification in Sarawak, Malaysia: Potential & Challenges for Mini-Hydro & Solar Hybrid Solutions. 2016. Available online: https://www.eclareon.com/sites/default/files/Presentations/04_chen_shiun.pdf (accessed on 10 January 2021).
8. Khengwee, T.; Hoole, P.R.P.; Pirapaharan, K.; Julai, N.; Othman, A.K.H.; Anyi, M.; Haidar, A.M.A.; Hoole, S.R.H. A review of Sarawak off-grid renewable energy potential and challenges. *J. Telecommun. Electron. Comput. Eng.* **2017**, *9*, 29–33.
9. Foster, R.; Ghassemi., M.; Cota, A. *Solar Energy: Renewable Energy and the Environment*; CRC Press: Florida, FL, USA, 2009.
10. Campos-Guzmán, V.; García-Cáscales, M.S.; Espinosa, N.; Urbina, A. Life cycle analysis with multi-criteria decision making: A review of approaches for the sustainability evaluation of renewable energy technologies. *Renew. Sustain. Energy Rev.* **2019**, *104*, 343–366. [CrossRef]
11. De Marco, I.; Riemma, S.; Iannone, R. Life cycle assessment of supercritical CO_2 extraction of caffeine from coffee beans. *J. Supercrit. Fluid.* **2018**, *133*, 393–400. [CrossRef]
12. Gallucci, T.; Lagioia, G.; Piccinno, P.; Lacalamita, A.; Pontrandolfo, A.; Paiano, A. Environmental performance scenarios in the production of hollow glass containers for food packaging: An LCA approach. *Int. J. Life Cycle Assess.* **2021**, *26*, 785–798. [CrossRef]
13. Bhole, G.P.; Deshmukh, T. Multi criteria decision making (MCDM) methods and its applications. *Int. J. Appl. Sci. Eng.* **2018**, *6*, 899–915. [CrossRef]
14. Al Garni, H.; Kassem, A.; Awasthi, A.; Komljenovic, D.; Al-Haddad, K. A multicriteria decision making approach for evaluating renewable power generation sources in Saudi Arabia. *Sustain. Energy Technol. Assess.* **2016**, *16*, 137–150. [CrossRef]
15. Algarín, C.R.; Llanos, A.P.; Castro, A.O. An analytic hierarchy process based approach for evaluating renewable energy sources. *Int. J. Energy Econ. Policy* **2017**, *7*, 38–47.
16. Das, A.; Shabbiruddin. Renewable energy source selection using analytical hierarchy process and quality function deployment: A case study. In Proceedings of the 2016 Second International Conference on Science Technology Engineering and Management (ICONSTEM), Chennai, India, 30–31 March 2016; IEEE: New York, NY, USA, 2016; pp. 298–302.

17. Hilorme, T.; Tkach, K.; Dorenskyi, O.; Katerna, O.; Durmanov, A. Decision making model of introducing energy-saving technologies based on the analytic hierarchy process. *J. Manag. Inf. Decis. Sci.* **2019**, *22*, 489–494.

18. Zhang, L.Y.; Li, C.X.; Phuong, N.H. Development of biomass energy industry in Heilongjiang Province based on analytic hierarchy process. *Nat. Environ. Pollut. Technol.* **2019**, *18*, 1487–1493.

19. Ren, J.Z.; Manzardo, A.; Mazzi, A.; Zuliani, F.; Scipioni, A. Prioritization of bioethanol production pathways in China based on life cycle sustainability assessment and multicriteria decision-making. *Int. J. Life Cycle Assess.* **2015**, *20*, 842–853. [CrossRef]

20. Basri, N.A.; Ramli, A.T.; Aliyu, A.S. Malaysia energy strategy towards sustainability: A panoramic overview of the benefits and challenges. *Renew. Sustain. Energy Rev.* **2015**, *42*, 1094–1105. [CrossRef]

21. Kim, T.H.; Chae, C.U. Environmental impact analysis of acidification and eutrophication due to emissions from the production of concrete. *Sustainability* **2016**, *8*, 578. [CrossRef]

22. Ludin, N.A.; Mustafa, N.I.; Hanafiah, M.M.; Ibrahim, M.A.; Teridi, M.A.M.; Sepeai, S.; Sopian, K. Prospects of life cycle assessment of renewable energy from solar photovoltaic technologies: A review. *Renew. Sustain. Energy Rev.* **2018**, *96*, 11–28. [CrossRef]

23. Bahta, S.T. Design and Analyzing of an Off-Grid Hybrid Renewable Energy System to Supply Electricity for Rural Areas: Case Study: Atsbi District, North Ethiopia. Master's Thesis, KTH School of Industrial Engineering and Management, Stockholm, Sweden, 2016.

24. Ishikawa, N.; Soda, R. *Anthropogenic Tropical Forests: Human-Nature Interfaces on the Plantation Frontier*; Springer Nature: Singapore, 2019.

25. Jong, F.C.; Ahmed, M.M.; Aik, D.L.H. Integration of renewable energy sources optimization in Sarawak using GIS and MCDM-AHP. In Proceedings of the 2019 International UNIMAS STEM 12th Engineering Conference (EnCon), Sarawak, Malaysia, 28–29 August 2019; pp. 65–70.

26. Trowell, K.A.; Goroshin, S.; Frost, D.L.; Bergthorson, J.M. Aluminum and its role as a recyclable, sustainable carrier of renewable energy. *Appl. Energy* **2020**, *275*, 115112. [CrossRef]

27. Natural Gas Conversion Guide 2012, International Gas Union. Available online: http://agnatural.pt/documentos/ver/natural-gas-conversion-guide_cb4f0ccd80ccaf88ca5ec336a38600867db5aaf1.pdf (accessed on 21 April 2020).

28. Gómez, M.R.; Garcia, R.F.; Gómez, J.R.; Carril, J.C. Review of thermal cycles exploiting the exergy of liquefied natural gas in the regasification process. *Renew. Sustain. Energy Rev.* **2014**, *38*, 781–795. [CrossRef]

29. Huang, Y.F.; Gan, X.J.; Chiueh, P.T. Life cycle assessment and net energy analysis of offshore wind power systems. *Renew. Energy* **2017**, *102*, 98–106. [CrossRef]

30. Kaab, A.; Sharifi, M.; Mobli, H.; Nabavi-Pelesaraei, A.; Chau, K.W. Combined life cycle assessment and artificial intelligence for prediction of output energy and environmental impacts of sugarcane production. *Sci. Total Environ.* **2019**, *664*, 1005–1019. [CrossRef] [PubMed]

31. Yuguda, T.K.; Li, Y.; Xiong, W.; Zhang, W. Life cycle assessment of options for retrofitting an existing dam to generate hydro-electricity. *Int. J. Life Cycle Assess.* **2020**, *25*, 57–72. [CrossRef]

32. Hussein, I.; Raman, N. Reconnaissance studies of micro hydro potential in Malaysia. In Proceedings of the International Conference on Energy and Sustainable Development: Issues and Strategies, Chiang Mai, Thailand, 2–4 June 2010; pp. 1–10.

33. Promentilla, M.A.B.; Aviso, K.B.; Lucas, R.I.G.; Razon, L.F.; Tan, R.R. Teaching Analytic Hierarchy Process (AHP) in undergraduate chemical engineering courses. *Educ. Chem. Eng.* **2018**, *23*, 34–41. [CrossRef]

34. Saaty, T.L. The Analytic Hierarchy Process: Planning, Priority Setting, Resource Allocation. McGraw Hill International: New York, NY, USA, 1980.

35. Frischknecht, R.; Jungbluth, N.; Althaus, H.J.; Doka, G.; Dones, R.; Heck, T.; Hellweg, S.; Hischier, R.; Nemecek, T.; Rebitzer, G.; et al. The ecoinvent database: Overview and methodological framework. *Int. J. Life Cycle Assess.* **2005**, *10*, 3–9. [CrossRef]

36. Ahmad Ludin, N.; Ahmad Affandi, N.A.; Purvis-Roberts, K.; Ahmad, A.; Ibrahim, M.A.; Sopian, K.; Jusoh, S. Environmental impact and levelised cost of energy analysis of solar photovoltaic systems in selected Asia Pacific region: A cradle-to-grave approach. *Sustainability* **2021**, *13*, 396. [CrossRef]

37. Ghenai, C. Life cycle analysis of wind turbine. In *Sustainable Development–Energy, Engineering and Technologies–Manufacturing and Environment*; Ghenai, C., Ed.; InTech: Rijeka, Croatia, 2012; pp. 19–32.

38. Emissions, G.G. Comparison of Lifecycle Greenhouse Gas Emissions of Various Electricity Generation Sources. 2011. Available online: https://gssd.mit.edu/search-gssd/site/comparison-lifecycle-greenhouse-gas-61507-tue-10-31-2017-2350 (accessed on 10 January 2021).

39. Bergerson, J.; Lave, L. A Life Cycle Analysis of Electricity Generation Technologies. Health and Environmental Implications of Alternative Fuels and Technologies. 2002. Available online: https://www.cmu.edu/ceic/assets/docs/publications/working-papers/ceic-03-05.pdf (accessed on 21 April 2020).

40. Shen, X.; Kommalapati, R.R.; Huque, Z. The comparative life cycle assessment of power generation from lignocellulosic biomass. *Sustainability* **2015**, *7*, 12974–12987. [CrossRef]

41. Pang, M.; Zhang, L.; Wang, C.; Liu, G. Environmental life cycle assessment of a small hydropower plant in China. *Int. J. Life Cycle Assess.* **2015**, *20*, 796–806. [CrossRef]

42. Hanafi, J.; Riman, A. Life cycle assessment of a mini hydro power plant in Indonesia: A case study in Karai River. *Procedia Cirp.* **2015**, *29*, 444–449. [CrossRef]

43. Mulvaney, D. Solar's green dilemma. *IEEE Spectr.* **2014**, *51*, 30–33. [CrossRef]

44. Gomaa, M.R.; Rezk, H.; Mustafa, R.J.; Al-Dhaifallah, M. Evaluating the environmental impacts and energy performance of a wind farm system utilizing the life-cycle assessment method: A practical case study. *Energies* **2019**, *12*, 3263. [CrossRef]

45. Chipindula, J.; Botlaguduru, V.S.V.; Du, H.; Kommalapati, R.R.; Huque, Z. Life cycle environmental impact of onshore and offshore wind farms in Texas. *Sustainability* **2018**, *10*, 2022. [CrossRef]

46. Raadal, H.L.; Gagnon, L.; Modahl, I.S.; Hanssen, O.J. Life cycle greenhouse gas (GHG) emissions from the generation of wind and hydro power. *Renew. Sustain. Energy Rev.* **2011**, *15*, 3417–3422. [CrossRef]

47. Carrasco, L.M.; Narvarte, L.; Lorenzo, E. Operational costs of A 13,000 solar home systems rural electrification programme. *Renew. Sustain. Energy Rev.* **2013**, *20*, 1–7. [CrossRef]

48. Albani, A.; Ibrahim, M.Z. Wind energy potential and power law indexes assessment for selected near-coastal sites in Malaysia. *Energies* **2017**, *10*, 307. [CrossRef]

49. Naqvi, M.; Yan, J.; Dahlquist, E.; Naqvi, S.R. Off-grid electricity generation using mixed biomass compost: A scenario-based study with sensitivity analysis. *Appl. Energy* **2017**, *201*, 363–370. [CrossRef]

50. Dorber, M.; May, R.; Verones, F. Modeling net land occupation of hydropower reservoirs in Norway for use in life cycle assessment. *Environ. Sci. Technol.* **2018**, *52*, 2375–2384. [CrossRef] [PubMed]

51. Fthenakis, V.; Kim, H.C. Land use and electricity generation: A life-cycle analysis. *Renew. Sustain. Energy Rev.* **2009**, *13*, 1465–1474. [CrossRef]

52. Oliver, A.; Khanna, M. Demand for biomass to meet renewable energy targets in the United States: Implications for land use. *GCB Bioenergy* **2017**, *9*, 1476–1488. [CrossRef]

53. van Zalk, J.; Behrens, P. The spatial extent of renewable and non-renewable power generation: A review and meta-analysis of power densities and their application in the US. *Energy Policy* **2018**, *123*, 83–91. [CrossRef]

54. Jha, S.K.; Puppala, H. Prospects of renewable energy sources in India: Prioritization of alternative sources in terms of Energy Index. *Energy* **2017**, *127*, 116–127. [CrossRef]

55. Nonhebel, S. Renewable energy and food supply: Will there be enough land? *Renew. Sustain. Energy Rev.* **2005**, *9*, 191–201. [CrossRef]

56. Lam, H.L.; Varbanov, P.S.; Klemeš, J.J. Regional renewable energy and resource planning. *Appl. Energy* **2011**, *88*, 545–550. [CrossRef]

57. Alola, A.A.; Alola, U.V. Agricultural land usage and tourism impact on renewable energy consumption among Coastline Mediterranean Countries. *Energy Environ.* **2018**, *29*, 1438–1454. [CrossRef]

58. Fauzi, M.A.; Setyono, P.; Pranolo, S.H. Environmental assessment of a small power plant based on palm kernel shell gasification. In Proceedings of the International Conference on Science and Applied Science; AIP Publishing LLC: Surakarta, Indonesia, 2020; Volume 2296, p. 020038.

59. Vaka, M.; Walvekar, R.; Rasheed, A.K.; Khalid, M. A review on Malaysia's solar energy pathway towards carbon-neutral Malaysia beyond Covid'19 pandemic. *J. Clean. Prod.* **2020**, *273*, 122834. [CrossRef] [PubMed]

60. Petinrin, J.O.; Shaaban, M. Renewable energy for continuous energy sustainability in Malaysia. *Renew. Sustain. Energy Rev.* **2015**, *50*, 967–981. [CrossRef]

61. Sa'adi, Z.; Shahid, S.; Ismail, T.; Chung, E.S.; Wang, X.J. Distributional changes in rainfall and river flow in Sarawak, Malaysia. *Asia Pac. J. Atmos. Sci.* **2017**, *53*, 489–500. [CrossRef]

62. Solangi, K.H.; Islam, M.R.; Saidur, R.; Rahim, N.A.; Fayaz, H. A review on global solar energy policy. *Renew. Sustain. Energy Rev.* **2011**, *15*, 2149–2163. [CrossRef]

63. Chang, Y.; Phoumin, H. Harnessing wind energy potential in ASEAN: Modelling and policy implications. *Sustainability* **2021**, *13*, 4279. [CrossRef]

64. Hossain, M.; Huda, A.S.N.; Mekhilef, S.; Seyedmahmoudian, M.; Horan, B.; Stojcevski, A.; Ahmed, M. A state-of-the-art review of hydropower in Malaysia as renewable energy: Current status and future prospects. *Energy Strategy Rev.* **2018**, *22*, 426–437. [CrossRef]

65. Bong, C.P.C.; Ho, W.S.; Hashim, H.; Lim, J.S.; Ho, C.S.; Tan, W.S.P.; Lee, C.T. Review on the renewable energy and solid waste management policies towards biogas development in Malaysia. *Renew. Sustain. Energy Rev.* **2017**, *70*, 988–998. [CrossRef]

66. Umar, M.S.; Urmee, T.; Jennings, P. A policy framework and industry roadmap model for sustainable oil palm biomass electricity generation in Malaysia. *Renew. Energy* **2018**, *128*, 275–284. [CrossRef]

67. Tang, S.; Chen, J.; Sun, P.; Li, Y.; Yu, P.; Chen, E. Current and future hydropower development in Southeast Asia countries (Malaysia, Indonesia, Thailand and Myanmar). *Energy Policy* **2019**, *129*, 239–249. [CrossRef]

68. Noman, F.; Alkawsi, G.; Abbas, D.; Alkahtani, A.; Tiong, S.K.; Ekanayake, J. A Comprehensive Review of Wind Energy in Malaysia: Past, Present and Future Research Trends. *IEEE Access* **2020**, *8*, 124526–124543. [CrossRef]

69. Sampaio, P.G.V.; González, M.O.A. Photovoltaic solar energy: Conceptual framework. *Renew. Sustain. Energy Rev.* **2017**, *74*, 590–601. [CrossRef]

70. Quaranta, E.; Revelli, R. Output power and power losses estimation for an overshot water wheel. *Renew. Energy* **2015**, *83*, 979–987. [CrossRef]

71. Rawat, R.; Lamba, R.; Kaushik, S.C. Thermodynamic study of solar photovoltaic energy conversion: An overview. *Renew. Sustain. Energy Rev.* **2017**, *71*, 630–638. [CrossRef]

72. Lee, J.Y.; An, S.; Cha, K.; Hur, T. Life cycle environmental and economic analyses of a hydrogen station with wind energy. *Int. J. Hydrog. Energy* **2010**, *35*, 2213–2225. [CrossRef]
73. Nzihou, A. *Handbook on Characterization of Biomass, Biowaste and Related by-Products*; Springer Nature: Cham, Switzerland, 2020.
74. Goswami, D.Y.; Kreith, F. *Handbook of Energy Efficiency and Renewable Energy*; CRC Press: Boca Raton, FL, USA, 2007.
75. Akram, R.; Chen, F.; Khalid, F.; Ye, Z.; Majeed, M.T. Heterogeneous effects of energy efficiency and renewable energy on carbon emissions: Evidence from developing countries. *J. Clean. Prod.* **2020**, *247*, 119122. [CrossRef]
76. Leal Filho, W.; Salvia, A.L.; Do Paco, A.; Anholon, R.; Quelhas, O.L.G.; Rampasso, I.S.; Brandli, L.L. A comparative study of approaches towards energy efficiency and renewable energy use at higher education institutions. *J. Clean. Prod.* **2019**, *237*, 117728. [CrossRef]
77. Garrett-Peltier, H. Green versus brown: Comparing the employment impacts of energy efficiency, renewable energy, and fossil fuels using an input-output model. *Econ. Model.* **2017**, *61*, 439–447. [CrossRef]
78. Stavropoulos, S.; Burger, M.J. Modelling strategy and net employment effects of renewable energy and energy efficiency: A meta-regression. *Energy Policy* **2020**, *136*, 111047. [CrossRef]
79. Amer, M.; Daim, T.U. Selection of renewable energy technologies for a developing county: A case of Pakistan. *Energy Sustain. Dev.* **2011**, *15*, 420–435. [CrossRef]
80. Ahmad, S.; Tahar, R.M. Selection of renewable energy sources for sustainable development of electricity generation system using analytic hierarchy process: A case of Malaysia. *Renew. Energy* **2014**, *63*, 458–466. [CrossRef]
81. Wang, Y.; Xu, L.; Solangi, Y.A. Strategic renewable energy resources selection for Pakistan: Based on SWOT-Fuzzy AHP approach. *Sustain. Cities Soc.* **2020**, *52*, 101861. [CrossRef]
82. Abushammala, M.F.; Qazi, W.A. Evaluation of the significant renewable energy resources in Sultanate of Oman using Analytical Hierarchy Process. *Int. J. Renew. Energy Res.* **2018**, *8*, 1528–1534.
83. Ishfaq, S.; Ali, S.; Ali, Y. Selection of optimum renewable energy source for energy sector in Pakistan by using MCDM approach. *Process Integr. Optim. Sustain.* **2018**, *2*, 61–71. [CrossRef]
84. Whitacre, P. *Sustainable Materials and Manufacturing for Renewable Energy Technology Development to 2030: Proceedings of a Workshop—in Brief*; The National Academies Press: Washington, DC, USA, 2017.
85. Martinopoulos, G. Life cycle assessment of solar energy conversion systems in energetic retrofitted buildings. *J. Build. Eng.* **2018**, *20*, 256–263. [CrossRef]

*sustainability*

MDPI

Article

Is Environment a Strategic Priority of the Leading Energy Companies? Evidence from Mission Statements

Dmitry A. Ruban [1,2,*], Natalia N. Yashalova [3] and Vladimir A. Ermolaev [4]

[1] K.G. Razumovsky Moscow State University of Technologies and Management (The First Cossack University), Zemlyanoy Val Street 73, 109004 Moscow, Russia

[2] Department of Management, Higher School of Business, Southern Federal University, 23-ja Linija Street 43, 344019 Rostov-on-Don, Russia

[3] Department of Economics and Management, Business School, Cherepovets State University, Sovetskiy Avenue 10, 162600 Cherepovets, Vologda Region, Russia; natalij2005@mail.ru

[4] Department of Commodity Science and Expertise, Plekhanov Russian University of Economics, Stremyanny Lane 36, 117997 Moscow, Russia; ermolaevvla@rambler.ru

* Correspondence: ruban-d@mail.ru

Abstract: Hydrocarbon production, electricity transmission, and other energy-related activities affect the environment. It is expected that environmental issues can be among strategic priorities summarized in mission statements of energy companies. The present analysis of the mission statements of 43 leading energy companies implies that these issues are considered by 36% of the top energy companies and 37% of the fastest-growing energy companies. These considerations often co-occur with attention to a company's higher tasks and image. Most often, production ecologization is posed as a priority. The fastest-growing companies pay insufficient attention to climate changes. Conceptually, reflection of environmental issues in mission statements depends on the managerial awareness of these issues; additionally, the development of separate sustainability strategies may make environmental priorities somewhat marginal. The 'greening' of mission statements of energy companies is recommended.

Keywords: business communication; energy corporations; greening; responsibility; strategic management

 check for **updates**

Citation: Ruban, D.A.; Yashalova, N.N.; Ermolaev, V.A. Is Environment a Strategic Priority of the Leading Energy Companies? Evidence from Mission Statements. *Sustainability* **2021**, *13*, 2192. https://doi.org/10.3390/su13042192

Academic Editors: Antonio Boggia and Georgios Tsantopoulos

Received: 23 January 2021
Accepted: 12 February 2021
Published: 18 February 2021

Publisher's Note: MDPI stays neutral with regard to jurisdictional claims in published maps and institutional affiliations.

1. Introduction

Environmental issues (climate change, pollution by hydrocarbons, heavy metals, and microplastics, land degradation, biodiversity loss, deforestation, etc.) cannot be ignored by contemporary business leaders. This almost philosophical idea is developed in numerous works, including those by Bukhari et al. [1], Çop et al. [2], Lawler and Worley [3], Lenka and Kar [4], and Lozano [5]. On the one hand, the biggest corporations are often responsible for environmental damage, but they also suffer from natural resource impoverishment and an ecologically-altered workforce. On the other hand, environmental issues are on the global agenda, and showing awareness of them contributes to a positive image of corporations in the eyes of customers, business partners, and states. In other words, environmental issues are linked to both risks and opportunities. A few years after the 'Management 2.0' concept was proposed and modern managers claimed that they to aim to achieve socially-important higher tasks under this framework [6], Lawler and Worley [3] demonstrated that environmental higher tasks are also a must for business leaders caring about the sustainable growth of their corporations. Recently, Ji and Miao [7] argued that environmental responsibility is of utmost importance for achieving really successful, innovative business development.

The leading energy companies of the world are expected to be especially tied to environmental issues. A broad spectrum of relevant ideas can be found, particularly, in the works by Clerici and Gallanti [8], Dangelico and Pontrandolfo [9], Ezeonu [10], Fethi and Rahuma [11], Hashmi et al. [12], Hoffmann and Busch [13], Linn and McCormack [14], Morrow and Rondinelli [15], Raugei et al. [16], Sánchez–Ortiz et al. [17], and Smirnova and Rudenko [18]. Such companies are responsible for greenhouse gas emissions and seawater pollution, and these are also prone to developing eco-innovations and exploiting renewable energy sources. Surprisingly, the knowledge of corporate strategic treatment of environmental issues remains incomplete. Communication is vital in strategic management of organizations [19–21]. According to Steensen [22], the very type of organizational strategy chosen depends on how the strategic knowledge is communicated. Tuppen [23] and Tao and Wilson [24] emphasized the environmental aspect of corporate communication and stressed its importance, complexity, and challenges. In regard to the aforementioned factors, the documents explaining strategies of energy companies should treat the environment as a top priority.

Mission statements, which are brief, almost slogan-like summaries of strategic priorities, constitute an important channel of corporate communication. Their significance was shown by Pearce and David [25] in their already classical paper. During the three past decades, this channel has become an important research object [26,27]. Already in the pioneering works [25], attention was paid to how mission statements reflect environmental attitudes of corporations. However, the following research did not focus much on this topic. Baral and Pokharel [28] examined basic strategic documents of the largest global companies and discovered limited consideration of environmental issues. Garnett et al. [29] documented the improvement of mission statements through time in regard to how these statements take environmental sustainability into account. Molchanova et al. [30] stressed that the Russian energy corporations are more successful in posing eco-priorities. Yozgat and Karatas [31] established that less than a fifth of the leading Turkish companies have environmentally sensitive mission statements. It is also sensible to consider the work by Lenkova [32] that dealt specifically with energy companies.

The scarcity of literature on this potentially highly-important issue reveals a significant research gap: the knowledge of how mission statements of the world businesses reflect environmental issues remains incomplete, especially with regard to particular countries and industries. This gap and the relevant research question must be addressed for both theoretical and practical reasons. Theoretically, it is important to understand whether mission statements can serve as a channel for communicating the environmental priorities of energy companies. In practice, it is necessary to understand how the corporate communication policies of real companies can be improved to develop a bridge between energy leadership and environmental leadership. The present study aims at filling this gap via an examination of environmental issues in the mission statements of the world's leading energy companies. First, it is intended to document whether corporations from this important industry recognize environmental issues as a strategic priority deserving of being reflected in their mission statements and how they treat these issues. Second, the approach of a mission statement analysis for finding the environment-related notions is proposed. Third, 'greening' is conceptualized and related to corporate strategies and policy. The research question is how common and 'deep' is consideration of environmental issues in the mission statements of the leading energy companies. This study is essentially empirical, and it answers this question with analysis of the collected mission statements. This study also fill a gap by linking the understanding of business communication of environmental issues in the industry with significant environmental impacts.

2. Materials and Methods

2.1. Research Direction

A mission statement is thought to be an important tool for effective corporate communication [25,27,33–37]. On the one hand, it summarizes the very essence of a company's strategy, and, thus, it clarifies the direction(s) this company chooses to grow towards, i.e., it indicates strategic priorities. On the other hand, mission statements present the preferred business strategy in a very compact way, which is ideal for communication of the noted priorities to managerial and other staff, customers, partners, competitors, media, governments, and the general public. Such communication is especially important in the case of big corporations with significant social, political, and environmental impacts. It is also known that mission statements directly and sometimes significantly influence business performance because they permit managers to identify and to maintain priorities, employees to understand and to share these priorities, and third parties to find key points for successful collaboration. The relevant evidence is provided in numerous publications, including the works by Atrill et al. [38], Bart and Baetz [39], Bart et al. [40], Cortés–Sánchez and Rivera [41], Gharleghi et al. [42], Jovanov Marjanova and Sofijanova [43], Mersland et al. [44], and Sheaffer et al. [45]. Generally, mission statements reflect managerial opinions of their own company. Although these may only be formulated with the aim of maintaining a better company image, these statements remain 'attached' to the strategy and to managerial thoughts.

Mission statements are included in company strategic documents and official reports. These are often provided on the official web-pages and, thus, they become available for analysis [29,46–48]. In the latter case, mission statements should be distinguished from design and promotion components of web-pages because the former are official strategic statements. For the same reason, analysis of mission statements that are available on-line differs from examinations of web-page content.

One of the main directions in the study of mission statements is conducting content analysis aimed at registering the presence/absence of several standard components [26]. These components were proposed originally by Pearce and David [25] and then updated slightly by David [49]. A total of nine components are distinguished (see below). A given mission statement may include one to all nine of these standard components. The presence of some components can be registered only formally. One example of this is when a company states that it appreciates its customers. However, in the other cases, the presence of components is marked by extensive explanations. For instance, a company states that it aims to conquer the USA and Canada, and these countries are thought to be its principal markets. Generally, the content analysis of mission statements is a qualitative analytical procedure that requires deciphering of the meaning of each word and each expression in relation to the standard components.

Reflection of environmental issues in corporate mission statements was addressed by several previous researchers [25,30,31]. These issues seem to be closely related to the philosophy component, and their deeper understanding requires special ('non-standard'), more detailed component analysis [30]. Some companies stress their care towards the natural world, contribution to solutions for global environmental problems like climate change, implementation of 'green' practices like waste recycling, disclose environmental effects, etc. Energy efficiency and ecological standards are also reflected [30]. Indeed, this is often done for a better company image, although environmental priorities can really be part of the 'core' of strategies for some, if not many companies.

2.2. Sample

The present study focuses on the leading energy companies, which include hydrocarbon production companies, power generation companies, electricity transmission companies, etc. Such a broad meaning of the term 'energy company' matches its use in some highly-reputed company rankings (see below). Provisionally, the leading energy companies were compiled from two related, but essentially different rankings. The first ranking includes the top global energy companies on the basis of their business performance [50]. The second ranking comprises the fastest-growing energy companies that demonstrated the biggest growth over three years [51]. These rankings reflect the state of the world energy industry in 2019, and both lists were provided by the high-reputed 'S&P Global Platts', which is a division of the 'S&P Global' agency.

Fifty companies from each ranking are considered (a few exist in both rankings). Consideration of a larger number of companies is not sensible for two reasons. First, only the 50 fastest-growing energy companies are listed in the second ranking, and, thus, the information taken from the first ranking should be comparable in size. Alternatively, the sample would be unbalanced. The top energy companies and the fastest-growing energy companies both deserve to be judged, although they reflect different approaches to achieve success. Second, it is expected that mission statements of the less important companies differ from those of the world's leading companies [52]. Then, the official web-page of each selected company was checked in order to find its mission statement. In rare cases, clear mission statements are available only on the web-pages facilitating search of business information, i.e., outside the official web-pages. Although Pearce and David [25] originally attributed various strategic documents to mission statements, only documents specifically named as being mission statements or looking as such are considered in the present study.

Mission statements were found for half of the top energy companies and less than a half of the fastest-growing energy companies (Figure 1). These were collected for a total of 43 companies (Supplement S1). This sample seems to be appropriate for subsequent analysis, as it represents the communicated ([22]) strategic priorities of the leading companies, although they either demonstrate high performance or rapid growth (or both) (Figure 2). Only English versions of the mission statements were considered, although some companies do not represent English-speaking countries. Supposedly, corporate communication in English is not a problem for the world's leading companies, irrespective of their national affinity. It should be added that the mission statements were treated in this work anonymously to avoid occasional violation of corporate reputations.

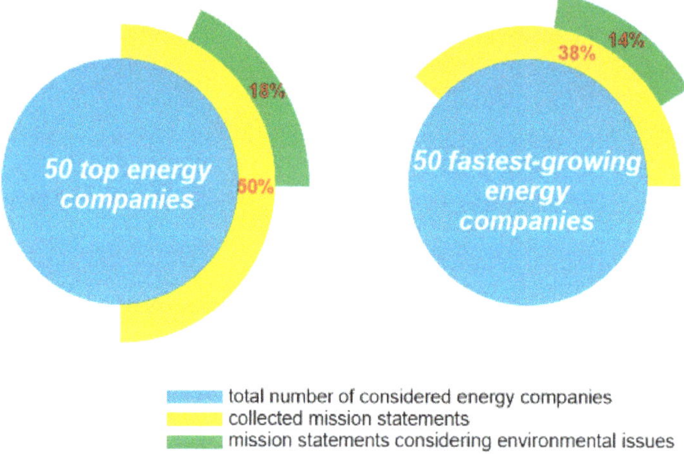

Figure 1. Relative distribution of environmental concerns in mission statements of the leading energy companies.

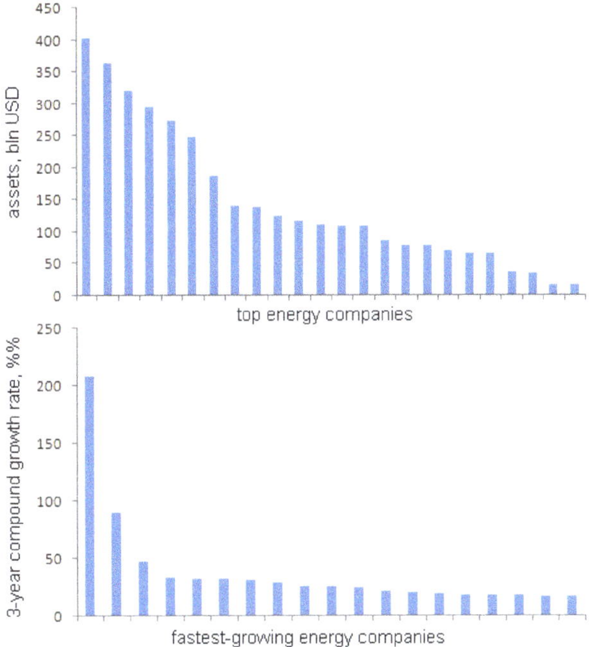

Figure 2. Distribution of the analyzed companies by the basic parameters of their selection on the basis of information from [50,51]. In the both cases, the horizontal axis indicates the considered companies ranged according to the parameter show on the vertical axis. The company names are not disclosed to avoid occasional violation of company reputations.

2.3. Analytical Procedures

The present analysis employed both qualitative and quantitative approaches. Generally, the former are linked to interpretation of text passages, and the latter are linked to calculation of component frequencies in the entirety of mission statements. The latter serves as a factual basis, gathering of which is also a part of the present study (see above).

The content of each mission statement was analyzed chiefly qualitatively, but in-depth as follows. All words and expressions relevant to environmental issues were identified (Supplement S1). As this study focuses on environment-related strategic priorities, the only relevant, 'green' mission statements were the subject of deeper analysis. The content of the mission statements considering environmental issues was analyzed in regard to the standard components of Pearce and David [25] and David [49]. The presence of the nine standard components, namely customers, markets, image, products and/or services, technology, survival, growth, and profitability, philosophy, self-understanding, and employees was checked via word-by-word interpretation. This was necessary to understand the general context of the occurrence of environmental priorities.

Then, the considered environmental issues were examined specifically. For this purpose, the classification of such issues in mission statements proposed earlier by Molchanova et al. [30] was employed, with certain modifications (with regard to the specific features of the collected mission statements of the leading energy companies). The presence of a total of five specific, environment-related components was checked, namely care for nature, production ecologization, ecological standards, climate change, and eco-responsibility. The average number of standard and specific components, and the frequency of occurrence of each component in the analyzed components were calculated.

The importance of various spatial dimensions for managers was established and conceptualized recently by Weinfurtner and Seidl [53]. Time also matters to managers, especially in regard to diversification [54], projects [55], and innovations [56,57]. Future orientation may be found in many mission statements [58]. It appears sensible to establish the spatio-temporal context of the consideration of environmental issues in the analyzed mission statements. For this purpose, we checked whether phrases bearing environment-related words and expressions deal with local, national, or global issues and treat them in past, present, or future perspectives. Indeed, spatial and temporal contexts can be indefinite in some cases. The frequencies of occurrence of each spatial and temporal context in the statements were calculated.

It appears intriguing to check how consideration of the environmental issues corresponds to consideration of the people's needs. Two reasons for such an analysis are as follows. First, it was found that social and environmental responsibilities have different importance [7]. One can even discern a company's 'competition' by their strategic priorities. Second, Pearce and David [25] did not recognize employees among their standard components, and this component appeared later, in the work by David [49]. Nonetheless, companies need to consider employees in their mission statements due to the general importance of social responsibility [59] and because employees are the target group of strategic corporate communication [60]. Moreover, Kopaneva [61] and Kopaneva and Sias [62] emphasized the efficacy of engagement of employees into company mission development. For the purposes of the present study, the consideration of employees (the employee component) and society (a part of the philosophy component) in the mission statements of all selected energy companies was established on the basis of information from Supplement S1. The distribution of these considerations was compared to those of environmental issues in order to judge their correspondence.

Finally, the mission statements were analyzed quantitatively with the 'WordItOut.com' engine. This created the so-called 'word clouds' depicting words from text passages in regard to their frequency. The 'world cloud' approach has become useful for finding text emphases [63–67]. For the present study, 'word clouds' were created for all mission statements of the top and fastest-growing energy companies, as well only for those considering environmental issues. Further comparison of these 'word clouds' sheds light on the importance of environment-related notions in the analyzed mission statements.

3. Results

3.1. General Patterns

Environmental issues were found in 36% of mission statements of the top energy companies and 37% of mission statements of the fastest-growing energy companies, which represent 18% and 15% of all leading companies, respectively (Figure 1). The former numbers were higher than the number of the world-class companies considering the environment in their basic strategic communications [28], which can be explained by the bigger relevance of energy business to environmental issues. Nonetheless, the environment seems to be a strategic priority for only about a third of all companies with mission statements. Both groups of the leading companies follow the trend of mission 'greening' comparably. Notably, the energy companies with 'green' missions represent different parts of the world (Asia, North America, and Europe) (Figure 3). Their relative concentration in the USA can be explained by the higher number of leading companies in this country. One should note the achievements of Thailand: two energy companies from this country are ranked among the leading companies, and the mission statements of the both consider the environment (Supplement S1).

The content of the mission statements of the leading mission companies differs substantially, and the combinations of the standard components also differ (Table 1). On average, both the top and fastest-growing energy companies include 5 components into their mission statements, which means these statements are not comprehensive, but are relatively well-developed. For the top energy companies, the most frequently occurring components are products/services and philosophy (100% of cases each) as well as image (67% of cases), and the least frequent are customers and employees (11% of cases each). For the fastest growing energy companies, the most frequently occurring components are philosophy (100% of cases) and image (86% of cases), and the least frequent is customers (14% of cases). The principal difference between these two groups of components is the much frequent presence of the employees component in the statements of the fastest-growing companies. All these lines of evidence imply that the general context of environment consideration in the analyzed mission statements is characterized by a moderately-diverse set of strategic priorities, with a preference for higher tasks and caring for an image. Too little attention to customers can be explained by the often-used B2B model of corporations providing energy to other enterprises. Moreover, societal needs are often posed as priorities (Supplement S1), and these are attributed to the philosophy component. This explains the low frequency of the customers component.

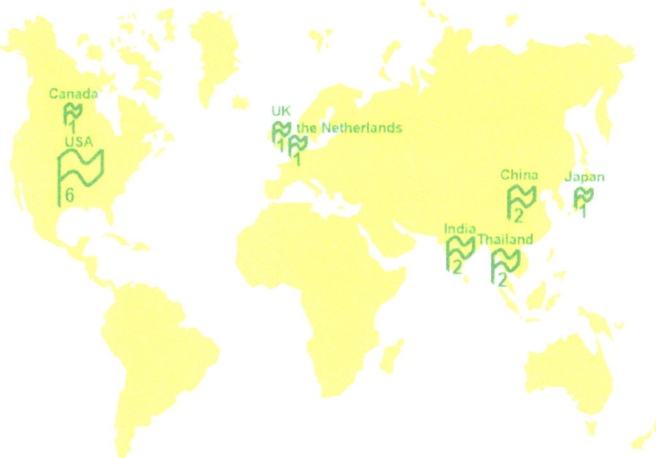

Figure 3. National affinity of the leading energy companies with mission statements considering environmental issues (based on information from Supplement S1).

Table 1. Content of the mission statements considering environmental issues.

Companies	Standard Components of Mission Statements								
	CUS	MAR	IMA	PRO	TEC	SGP	PHI	SUN	EMP
Top energy companies									
A			v	v	v	v	v	v	
B			v	v		v	v		
C			v	v			v	v	
D			v	v			v		
E	v	v	v	v		v	v	v	v
F		v		v			v		
G				v	v		v		
H		v	v	v	v		v	v	
I				v	v		v	v	
Fastest-growing energy companies									
J	v	v	v	v	v	v	v	v	v
K			v		v	v	v	v	v
L				v			v		
M			v	v		v	v		
N		v	v	v	v	v	v	v	v
O			v		v		v	v	
P			v	v			v		v

Components: CUS—customers, MAR—markets, IMA—image, PRO—products/services, TEC—technologies, SGP—survival, growth, profitability, PHI—philosophy (including environmental concerns), SUN—self-understanding, EMP—employees. Note: the order of the companies (labeled A, B, …) differs from their order in Supplement S1, i.e., the missions are treated anonymously to avoid occasional violation of company reputations.

The analyzed mission statements also differ substantially by how they reflect environmental issues (Table 2). The average number of specific components is the same for both groups of energy companies, although the statements of the fastest-growing companies are slightly less diverse: 86% of them consist of a single component in comparison to 67% of the top companies. In the mission statements, the production of ecologization (including 'green' technological processes and 'clean' energy) is the most important issue. It is established in 56% of the statements of the top energy companies and 57% of the statements of the fastest-growing companies. Care for nature is relatively frequently found, and the ecological standards and especially the eco-responsibility are the least commonly found issues. Of special interest is a specific component such as climate change. This is addressed by 44% of the mission statements of the top energy companies and by no statements for the fastest-growing energy companies. Hypothetically, the top-performing ('stable') corporations have enough inertia to address the global agenda of climate change, whereas this issue is either too ambitious or simply disinteresting for the 'accelerating' businesses.

Table 2. Environmental issues considered in the mission statements.

Companies	Environmental Issues				
	Care for Nature	**Production Ecologization**	**Ecological Standards**	**Climate Change**	**Eco-Responsibility**
Top energy companies					
A		v			
B		v			
C		v		v	
D	v				
E			v		
F	v				
G	v	v		v	
H				v	
I		v		v	
Fastest-growing energy companies					
J			v		
K		v			
L		v			
M	v				
N	v				
O		v			v
P		v			

Note: the order of the companies (labeled A, B, …) differs from their order in Supplement S1, i.e., the missions are treated anonymously to avoid occasional violation of company reputations.

Comparing the general context of environment consideration in the analyzed mission statements and the eco-content of the latter (Tables 1 and 2, see also explanations above), a kind of controversy is revealed. On the one hand, many leading companies pose higher tasks and care about their image. On the other hand, many of them do not consider the appealing and well-known issue of climate change, and do not mention a sense of eco-responsibility. Undoubtedly, focusing on production ecologization is useful for having a better image, but this seems to be a too 'narrow' way of strategic eco-thinking. The controversy is especially striking in the case of the fastest-growing companies: one would expect diversification of environment-related priorities from them to gain a competitive advantage. It cannot be tested whether the noted controversy is the result of any deficiency in strategic thinking or inaccuracies of the statement writing, but to suppose that both are true seems to be reasonable.

3.2. Spatio-Temporal Context

Consideration of environmental issues in half of the analyzed mission statements demonstrates a certain spatio-temporal context (Table 3). Interestingly, the top energy companies are more space- and less time-focused than the fastest-growing companies. This can be explained by the fact that the best-performing corporations care more about their geographical dominance and less about their time frame due to their stability. In contrast, time means more to the rapidly-growing businesses.

Table 3. Spatio-temporal context of environmental issues considered in the mission statements.

Companies	Space				Time			
	Local	National	Global	Indefinite	Past	Present	Future	Indefinite
Top energy companies								
A				v				v
B				v				v
C				v			v	
D			v					v
E		v						v
F			v					v
G		v						v
H			v			v	v	
I				v			v	
Fastest-growing energy companies								
J			v				v	
K				v		v		
L				v				v
M				v				v
N		v						v
O				v		v		
P			v			v		

Note: the order of the companies (labeled A, B, …) differs from their order in Supplement S1, i.e., the missions are treated anonymously to avoid occasional violation of company reputations.

The leading energy companies tend to adjust environmental issues to the global and national scales (Table 3). None of the mission statements communicate about local environments, which is surprising because local environmental impacts of hydrocarbon production or power station work can be significant. As for the time, both present and future perspectives of environmental issues can be found (Table 3). While the former is preferred by the fastest-growing energy companies, the top companies are more future-oriented.

3.3. Apparent People–Environment Alternative

The reflection of environmental, societal, and employee-related priorities in the analyzed mission statements is somewhat peculiar (Table 4). Although they are not opposed directly, a very uncertain correspondence between them can be observed (Table 4).

The mission statements of the top energy companies deal with society in 52% of cases, environment in 36% of cases, and employees in only 12% of cases (Table 4). As for the fastest-growing energy companies, 53% of their mission statements deal with society, 37% of the statements deal with the environment, while employees are addressed in 42% of cases (Table 4). The companies belonging to both groups tend to relate social and environmental responsibilities, but the top companies concentrate on society, whereas the fastest-growing companies balance outer (community-related) and inner (employee-related) social priorities. Moreover, the environment seems to be slightly less important than society and significantly more important than the staff for the top energy companies, whereas environment seems to be a bit less important than society and employees for the fastest-growing companies. In regard to the analyzed mission statements, it is possible to conclude that the nature 'costs' more than the workers for the top companies. Although this is a challenging and provocative proposition, it is confirmed by the findings of the present study and questions the social-versus-environment justice in the energy industry (at least, as seen in the mission statements).

Table 4. Consideration of employees, society, and the environment in all analyzed mission statements.

Employees	Society	Environment
Top energy companies		
-	-	V
-	-	-
-	-	V
-	V	-
-	V	V
-	V	-
-	-	-
-	V	V
-	-	-
-	V	-
-	-	-
-	V	-
V	-	-
V	V	V
-	-	-
-	V	V
-	-	-
-	V	-
-	-	-
-	-	-
V	V	-
-	V	V
-	V	V
Fastest-growing energy companies		
-	-	-
-	-	-
-	V	-
V	V	V
V	-	-
V	-	V
V	V	-
-	-	-
-	V	V
-	-	-
-	V	-
V	V	-
-	V	V
V	V	-
-	V	-
-	-	-
V	-	V
-	V	V
V	-	V

Note: each line corresponds to a given company in Supplement S1, the order of the companies differs from their order in Supplement S1, i.e., the missions are treated anonymously to avoid occasional violation of company reputations.

Although one would expect that caring for their corporate responsibility means equal attention towards both social and environmental responsibility, this is not supported by the findings (Table 4). Apparently, some managers responsible for mission development choose between social and environmental priorities and between 'our' people and 'them' in many cases. Such a choice is not only unexpected, but somewhat illogical. Although special investigations are necessary in order to understand its cause, it is possible to hypothesize that some companies either demonstrate a biased vision or they really face limited opportunities and need to choose which responsibility-related priorities to communicate. It is also necessary to add that caring for the environment can be understood by some energy corporations as a simultaneous caring for people, and vice versa.

3.4. 'Word Clouds'

The 'word cloud' created for all top energy companies considered in the present study implies that their mission statements emphasize energy, whereas the use of words such as the 'environment' and 'climate' is marginal (Figure 4). The situation does not change when only the environment-related statements are considered (Figure 5). The 'world cloud' created for all fastest-growing energy companies implies that their mission statements have a more diverse focus, with significant attention towards safety and value (Figure 6). Moreover, one needs to note the relatively high frequency of the word 'environment'. When the only environment-related statements of these companies are considered, the word 'environmental' becomes even more accented than 'energy' (Figure 7).

Figure 4. 'Word cloud' for the mission statements of all analyzed top energy companies.

Figure 5. 'Word cloud' for the mission statements of the analyzed top energy companies considering environmental issues.

Figure 6. 'Word cloud' for the mission statements of all analyzed fastest-growing energy companies.

Figure 7. 'Word cloud' for the mission statements of the analyzed fastest-growing energy companies considering environmental issues.

The results of the 'word cloud' analysis stress differences between the top and fastest-growing energy companies. The latter are more concerned about environmental issues, as reflected by their mission statements. Moreover, it is necessary to add that simple notion of these issues does not make statements of the top companies substantially more eco-friendly.

4. Discussion

4.1. General Interpretation

The results of the present analysis of the mission statements imply that the environment is only considered to be a strategic priority by some leading energy companies, and the environmental issues are understood rather 'narrowly'. However, the situation in this segment of the world economy is better than in other industries [28,29], similarly to what has been reported for Russia [30]. Moreover, attention to environment-related priorities raises new challenges like the apparent alternating between employee-related and environment-related priorities.

Generally, the collected evidence means the analyzed mission statements taken together are still far from the ideal corporate strategic treatment of environmental issues (of course, the situation is much better in the case of some particular companies). This is a characteristic of both the top and fastest-growing energy companies. This finding can be explained differently. First, it is possible that some leading companies are too business-focused and do not pay adequate attention to environmental responsibility. Second, it is possible that these companies prefer to focus on environmental issues in other strategic documents like sustainability reports. Third, it is also possible that their top managers responsible for strategy development and/or involved in mission statement writing are not well aware of environmental issues or do not have the necessary education to communicate about these issues properly. Although further investigations are necessary to choose between these explanations (hypothetically, all these matter), the problem with underrepresentation of environmental issues in the mission statements of the leading energy companies remains. This problem also appears to be even bigger, as these leading companies serve as examples to smaller players of the world energy industry.

Azad et al. [68] demonstrated how environmental management in energy companies can be improved, Ruka and Rashidirad [69] documented such an improvement, whereas Prechel and Istvan [70] demonstrated how internal company characteristics lead to disproportionality of environmental pollution. The present study contributes to this discourse by indicating certain weaknesses in the strategic communication of environmental corporate responsibility by the leading energy companies.

4.2. Tentative Conceptualization

In order to conceptualize the findings of this paper, a scheme of communication of environmental issues via mission statements is proposed (Figure 8). Managers responsible for strategy development may be directly aware of environmental issues (not necessarily directly related to her/his company). This is especially the case if these are new-generation managers that are always looking for higher tasks [3,6]. However, the activities of a given company may raise some environmental issues that cannot be ignored by even unaware managers, or the latter can be pressured into making changes by the broader public, media, or the state. As energy companies are closely tied to environmental issues, it would be difficult to them to avoid adding these issues to their company's agenda and reflecting them in their company's strategies. However, they need to choose the best way to do this. One option is to consider the environment as a top priority and to include this into the mission statement, which the quintessential company strategy. The results of the present study imply that just a bit more than a third of the leading energy companies pursue this option. Another option is to consider the environment somewhere in full-scale strategic documents or to limit its consideration to a separate sustainability strategy. Development of the latter is very typical for modern corporations, especially in the energy sector of the world economy [69]. The importance of these sustainability strategies is undisputable, but restriction of the environment-related priorities to only these issues makes these priorities marginal. The proposed scheme offers some other ways for corporate strategic treatment of environmental issues (Figure 8). One of them is linked to image concerns that seem to be very important in light of the findings of this study (see above). Apparently, such concerns encourage managers to reflect the environment in mission statements, irrespective of whether 'greening' is judged to be a top or marginal priority.

The proposed conceptualization is based on assumption that the leading energy companies both intend to carry out and communicate their strategies (shared strategies [22]). However, false (strategic communication, not intention), hidden (strategic intention, not communication), and especially learning (neither communication, nor intention) strategies [22] do not trigger urgent reflection of environmental issues in mission statements. The possibility cannot be excluded that some companies incline towards some of these strategies, which also explains the still insufficient 'greening' of the mission statements of the leading energy companies. Additionally, mission statements as a communication channel are linked to corporate dynamics capabilities and the relevant transformations and gaining a competitive advantage [71,72].

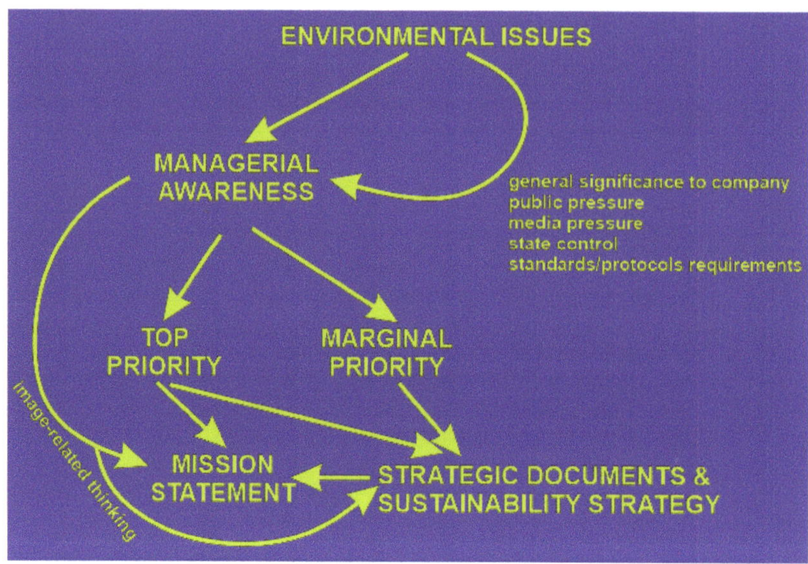

Figure 8. Complex ways for consideration of environmental issues in mission statements.

4.3. Policy Implications

The results of the present study can be implied for the improvement of corporate policies in the energy industry. In particular, these results permit making several recommendations to energy company managers. First, the energy companies need to make their mission statements more eco-friendly. This would help to increase the awareness of 'greening' among managers, employees, and the entire business community, as well as to improve a company's corporate image. Second, environmental priorities should be treated broadly. In particular, the issues of climate change, eco-responsibility, etc. deserve mentioning. This would make the environmental agenda of each given company richer and 'deeper', as well as strengthen that company's eco-image. Third, the spatio-temporal context of the consideration of environmental issues needs improvement. For instance, the local dimension of an environmental priority would make a given mission statement more appealing to society (due to its sensitivity to real problems), and consideration of corporate 'greening' in the past perspective would increase public trust towards a company (due to underlining the long-term, already-conducted pro-environmental behavior). Fourth, the 'nature or people' alternative (if it really exists) must be erased. Attention towards the environment should not diminish attention towards society and employees, and vice versa. Moreover, joint consideration of environment, society, and employees would have a synergetic positive effect on a company's image. Fifth, the development of sustainability strategies and consideration of environmental issues in lengthy strategic documents should not substitute for communication of environmental priorities via mission statements.

The leading energy companies need to care about the implementation of their mission statements. This is a highly-complex task that faces certain barriers and requires significant management commitment [73–75]. Some relevant problems in the hydrocarbon industry have been reported earlier by Sæverud and Skjærseth [76]. Broad public, media, and states should pay attention to companies' strategic statements and, particularly, to stimulating their 'greening'. As shown by the results of the present study, some companies provide excellent examples of such 'greening', and their efforts should be appreciated. The state may play a role in the improvement of corporate strategic statements, and this is especially the case for the countries launching nation-scale ecological projects [77,78]. Additionally, it is very probable that when representatives of the young generations, which demonstrate appropriate environmental awareness and readiness to act accordingly [79,80], become top managers, mission statements and relevant strategic actions will improve. Although some evidence of the young generations' environmental concerns is ambiguous [81,82], the modern education environment can really stimulate their pro-environmental thinking [83]. Indeed, 'cultivation' of such an environment in business schools developing future top managers is strongly desirable.

5. Conclusions

The analysis of the mission statements of the leading energy companies permits making three general conclusions. First, the number of the mission statements considering environmental issues remains limited, and the environmental priorities are often understood 'narrowly'. Second, the top energy companies focus on climate change more frequently than the fastest-growing energy companies. Third, there are some side effects of the consideration of environmental issues in the mission statements like the apparent alternating between protecting the environment and people. In practice, these findings indicate the remaining urgency of the improvement of the mission statements in the energy sector of the global economy, which is linked to managerial awareness of environmental issues, caring for a company's corporate image, and the role of sustainability strategies.

The main limitation of this study is the size of the dataset. Although it is representative of the leading energy companies, other companies, including some of the biggest companies in particular countries, deserve similar analysis. This clarifies important direction for further investigations. Methodologically, this study reveals the an importance of analysis not only of the biggest companies, but also of the fastest-growing companies. Both groups constitute the leading companies, and joint examination of their mission statements allows us to overcome the problem where industrial leadership is restricted to only the corporation size. It is likely that further investigations need to diversify the spectrum of leading companies. It should be underlined that the basic approach of the present study is the content analysis of the selected mission statements, i.e., finding the evidence of 'greening' in them. In-depth statistical analyses are left for the following studies, which will require more information on companies and, probably, knowledge of how mission statements changed over time.

Supplementary Materials: The following are available online at https://www.mdpi.com/2071-105 0/13/4/2192/s1, Supplement S1.

Author Contributions: Conceptualization, D.A.R. and N.N.Y.; methodology, D.A.R. and N.N.Y.; investigation, D.A.R. and V.A.E.; writing—original draft preparation, D.A.R.; writing—review and editing, D.A.R. and V.A.E. All authors have read and agreed to the published version of the manuscript.

Funding: This research received no external funding.

Institutional Review Board Statement: Not applicable.

Informed Consent Statement: Not applicable.

Data Availability Statement: Not applicable.

Acknowledgments: The authors gratefully thank G. Tsantopoulos (Greece) and E. Karasmanaki (Greece) for their editorial support, the four reviewers for their thorough consideration of this paper and important recommendations, and I. Rodella (Italy) for her helpful suggestions for the preliminary version of this work.

Conflicts of Interest: The authors declare no conflict of interest. All companies are discussed fully anonymously in the present study to avoid occasional violation of their reputations.

References

1. Bukhari, S.A.A.; Hashim, F.; Amran, A. Green Banking: A road map for adoption. *Int. J. Ethics Syst.* **2020**, *36*, 371–385. [CrossRef]
2. Çop, S.; Olorunsola, V.O.; Alola, U.V. Achieving environmental sustainability through green transformational leadership policy: Can green team resilience help? *Bus. Strategy Environ.* **2021**, *30*, 671–682. [CrossRef]

3. Lawler, E.E., III; Worley, C.G. Designing organizations for sustainable effectiveness. *Organ. Dyn.* **2012**, *41*, 265–270. [CrossRef]
4. Lenka, P.; Kar, S. Role of ethical leaders in sustainable business: An aristotelian virtue ethics perspective. *Probl. Ekorozw.* **2021**, *16*, 201–207.
5. Lozano, R. A holistic perspective on corporate sustainability drivers. *Corp. Soc. Responsib. Environ. Manag.* **2015**, *22*, 32–44. [CrossRef]
6. Hamel, G. Moon Shots for management: What great challenges must we tackle to reinvent management and make it more relevant to a volatile world? *Harv. Bus. Rev.* **2009**, *87*, 91–98.
7. Ji, H.; Miao, Z. Corporate social responsibility and collaborative innovation: The role of government support. *J. Clean. Prod.* **2020**, *260*, 121028. [CrossRef]
8. Clerici, A.; Gallanti, M. Energy efficiency: Benefits for companies, a burden for the environment. *Energ. Elettr.* **2008**, *85*, 13–21.
9. Dangelico, R.M.; Pontrandolfo, P. Being 'Green and Competitive': The Impact of Environmental Actions and Collaborations on Firm Performance. *Bus. Strategy Environ.* **2015**, *24*, 413–430. [CrossRef]
10. Ezeonu, I. Capital Accumulation, Environmental Pollution, and Public Health Challenges in the Nigerian Petroleum Industry: Lessons on Market Criminology. *Perspect. Glob. Dev. Technol.* **2020**, *19*, 181–200. [CrossRef]
11. Fethi, S.; Rahuma, A. The impact of eco-innovation on CO_2 emission reductions: Evidence from selected petroleum companies. *Struct. Chang. Econ. Dyn.* **2020**, *53*, 108–115. [CrossRef]
12. Hashmi, M.A.; Damanhouri, A.; Rana, D. Evaluation of Sustainability Practices in the United States and Large Corporations. *J. Bus. Ethics* **2015**, *127*, 673–681. [CrossRef]
13. Hoffmann, V.H.; Busch, T. Corporate carbon performance indicators: Carbon intensity, dependency, exposure, and risk. *J. Ind. Ecol.* **2008**, *12*, 505–520. [CrossRef]
14. Linn, J.; McCormack, K. The roles of energy markets and environmental regulation in reducing coal-fired plant profits and electricity sector emissions. *Rand J. Econ.* **2019**, *50*, 733–767. [CrossRef]
15. Morrow, D.; Rondinelli, D. Adopting corporate environmental management systems: Motivations and results of ISO 14001 and EMAS certification. *Eur. Manag. J.* **2002**, *20*, 159–171. [CrossRef]
16. Raugei, M.; Kamran, M.; Hutchinson, A. A prospective net energy and environmental life-cycle assessment of the UK electricity grid. *Energies* **2020**, *13*, 2207. [CrossRef]
17. Sánchez–Ortiz, J.; García-Valderrama, T.; Rodríguez-Cornejo, V.; Giner-Manso, Y. The effects of environmental regulation on the efficiency of distribution electricity companies in Spain. *Energy Environ.* **2020**, *31*, 3–20. [CrossRef]
18. Smirnova, N.V.; Rudenko, G.V. Tendencies, problems and prospects of innovative technologies implementation by Russian oil companies. *J. Ind. Pollut. Control* **2017**, *33*, 937–943.
19. Chen, Z.F.; Tao, W. Hybrid Strategy—Interference or Integration? How Corporate Communication Impacts Consumers' Memory and Company Evaluation. *Int. J. Strateg. Commun.* **2020**, *14*, 122–138. [CrossRef]
20. Popescu, C. Is environmental protection a central issue to the business strategy of high profile companies? Content analysis of website corporate communication. *J. Urban Reg. Anal.* **2017**, *9*, 87–99.
21. Valentine, S.; Hollingworth, D. Communication of Organizational Strategy and Coordinated Decision Making as Catalysts for Enhanced Perceptions of Corporate Ethical Values in a Financial Services Company. *Empl. Responsib. Rights J.* **2015**, *27*, 213–229. [CrossRef]
22. Steensen, E.F. Five types of organizational strategy. *Scand. J. Manag.* **2014**, *30*, 266–281. [CrossRef]
23. Tuppen, C. Devising a coherent company environmental communications strategy. *Eur. Environ.* **1992**, *2*, 18–21. [CrossRef]
24. Tao, W.; Wilson, C. The Impact of Corporate Communication on Company Evaluation: Examining the Message Effects of CSR, Corporate Ability, and Hybrid Strategies. *Int. J. Strateg. Commun.* **2016**, *10*, 426–444. [CrossRef]
25. Pearce, J.A.; David, F. Corporate Mission Statements: The Bottom Line. *Acad. Manag. Exec.* **1987**, *1*, 109–116. [CrossRef]
26. Alegre, I.; Berbegal-Mirabent, J.; Guerrero, A.; Mas-Machuca, M. The real mission of the mission statement: A systematic review of the literature. *J. Manag. Organ.* **2018**, *24*, 456–473. [CrossRef]
27. Desmidt, S.; Prinzie, A.; Decramer, A. Looking for the Value of Mission Statements: A Meta-Analysis of 20 Years of Research. *Manag. Decis.* **2011**, *49*, 468–483. [CrossRef]
28. Baral, N.; Pokharel, M.P. How Sustainability Is Reflected in the S&P 500 Companies' Strategic Documents. *Organ. Environ.* **2017**, *30*, 122–141.
29. Garnett, S.T.; Lawes, M.J.; James, R.; Bigland, K.; Zander, K.K. Portrayal of sustainability principles in the mission statements and on home pages of the world's largest organizations. *Conserv. Biol.* **2016**, *30*, 297–307. [CrossRef]
30. Molchanova, T.K.; Yashalova, N.N.; Ruban, D.A. Environmental Concerns of Russian Businesses: Top Company Missions and Climate Change Agenda. *Climate* **2020**, *8*, 56. [CrossRef]
31. Yozgat, U.; Karatas, N. Going Green of Mission and Vision Statements: Ethical, Social, and Environmental Concerns across Organizations. *Procedia-Soc. Behav. Sci.* **2011**, *24*, 1359–1366. [CrossRef]
32. Lenkova, O.V. The peculiarities of mission forming in Russia's oil and gas companies. *World Appl. Sci. J.* **2013**, *27*, 345–348.
33. Carmon, A.F. Is it necessary to be clear? An examination of strategic ambiguity in family business mission statements. *Qual. Res. Rep. Commun.* **2013**, *14*, 87–96. [CrossRef]
34. King, D.L.; Case, C.J.; Premo, K.M. 2012 mission statements: A ten country global analysis. *Acad. Strateg. Manag. J.* **2013**, *12*, 77–94.
35. Lin, Y.; Ryan, C. From mission statement to airline branding. *J. Air Transp. Manag.* **2016**, *53*, 150–160. [CrossRef]

36. Tenca, E. Remediating corporate communication through the web: The case of about us sections in companies' global websites. *ESP Today* **2018**, *6*, 84–106. [CrossRef]
37. Verboven, H. Communicating CSR and Business Identity in the Chemical Industry through Mission Slogans. *Bus. Commun. Q.* **2011**, *74*, 415–431. [CrossRef]
38. Atrill, P.; Omran, M.; Pointon, J. Company mission statements and financial performance. *Corp. Ownersh. Control* **2005**, *2*, 28–35. [CrossRef]
39. Bart, C.K.; Baetz, M.C. The relationship between mission statements and firm performance: An exploratory study. *J. Manag. Stud.* **1998**, *35*, 823–853. [CrossRef]
40. Bart, C.K.; Bontis, N.; Taggar, S. A model of the impact of mission statements on firm performance. *Manag. Decis.* **2001**, *39*, 19–35. [CrossRef]
41. Cortés–Sánchez, J.D.; Rivera, L. Mission statements and financial performance in Latin-American firms. *Bus. Theory Pract.* **2019**, *20*, 270–283. [CrossRef]
42. Gharleghi, E.; Nikbakht, F.; Bahar, G. A survey of relationship between the characteristics of mission statement and organizational performance. *Res. J. Bus. Manag.* **2011**, *5*, 117–124. [CrossRef]
43. Jovanov Marjanova, T.; Sofijanova, E. Corporate mission statement and business performance: Through the prism of Macedonian companies. *Balk. Soc. Sci. Rev.* **2014**, *3*, 179–198.
44. Mersland, R.; Nyarko, S.A.; Szafarz, A. Do social enterprises walk the talk? Assessing microfinance performances with mission statements. *J. Bus. Ventur. Insights* **2019**, *11*, e00117. [CrossRef]
45. Sheaffer, Z.; Landau, D.; Drori, I. Mission statement and performance: An evidence of "Coming of Age". *Organ. Dev. J.* **2008**, *26*, 49–62.
46. Bartkus, B.; Glassman, M.; McAfee, B. Do large European, US and Japanese firms use their web sites to communicate their mission? *Eur. Manag. J.* **2002**, *20*, 423–429. [CrossRef]
47. Lee, M.-Y.; Fairhurst, A.; Wesley, S. Corporate social responsibility: A review of the top 100 US retailers. *Corp. Reput. Rev.* **2009**, *12*, 140–158. [CrossRef]
48. Sones, M.; Grantham, S.; Vieira, E.T. Communicating CSR via pharmaceutical company web sites: Evaluating message frameworks for external and internal stakeholders. *Corp. Commun. Int. J.* **2009**, *14*, 144–157. [CrossRef]
49. David, F.R. How companies define their mission. *Long Range Plan.* **1989**, *22*, 90–97. [CrossRef]
50. The Platts. Top 250 Global Energy Company Rankings. Available online: https://www.spglobal.com/platts/top250/rankings (accessed on 15 March 2020).
51. Platts. Top 250 Fastest Growing Companies. Available online: https://www.spglobal.com/platts/top250/fastest-growing (accessed on 15 March 2020).
52. King, D.L.; Case, C.J.; Premo, K.M. Does company size affect mission statement content? *Acad. Strateg. Manag. J.* **2014**, *13*, 21–34.
53. Weinfurtner, T.; Seidl, M. Towards a spatial perspective: An integrative review of research on organisational space. *Scand. J. Manag.* **2019**, *35*, 101009. [CrossRef]
54. Oberfield, Z.W. Accounting for time: Comparing temporal and atemporal analyses of the business case for diversity management. *Public Adm. Rev.* **2014**, *74*, 777–789. [CrossRef]
55. Delisle, J. Uncovering temporal underpinnings of project management standards. *Int. J. Proj. Manag.* **2019**, *37*, 968–978. [CrossRef]
56. Ellwood, P.; Horner, S. In search of lost time: The temporal construction of innovation management. *R D Manag.* **2020**, *50*, 364–379. [CrossRef]
57. Wrede, M.; Dauth, T. A temporal perspective on the relationship between top management team internationalization and firms' innovativeness. *Manag. Decis. Econ.* **2020**, *41*, 542–561. [CrossRef]
58. Bartkus, B.; Glassman, M.; McAfee, R.B. A comparison of the quality of European, Japanese and U.S. mission statements: A content analysis. *Eur. Manag. J.* **2004**, *22*, 393–401. [CrossRef]
59. Low, M.P.; Siegel, D. A bibliometric analysis of employee-centred corporate social responsibility research in the 2000s. *Soc. Responsib. J.* **2019**, *16*, 691–717. [CrossRef]
60. Klemm, M.; Sanderson, S.; Luffman, G. Mission statements: Selling corporate values to employees. *Long Range Plan.* **1991**, *24*, 73–78. [CrossRef]
61. Kopaneva, I.M. Left in the Dust: Employee Constructions of Mission and Vision Ownership. *Int. J. Bus. Commun.* **2019**, *56*, 122–145. [CrossRef]
62. Kopaneva, I.; Sias, P.M. Lost in translation: Employee and organizational constructions of mission and vision. *Manag. Commun. Q.* **2015**, *29*, 358–384. [CrossRef]
63. DePaolo, C.A.; Wilkinson, K. Get Your Head into the Clouds: Using Word Clouds for Analyzing Qualitative Assessment Data. *TechTrends* **2014**, *58*, 38–44. [CrossRef]
64. Filatova, O. More Than a Word Cloud. *Tesol J.* **2016**, *7*, 438–448. [CrossRef]
65. Fu, Y.; Xiong, K.; Zhang, Z. Ecosystem services and ecological compensation of world heritage: A literature review. *J. Nat. Conserv.* **2021**, *60*, 125968. [CrossRef]
66. Mulay, P.; Joshi, R.; Chaudhari, A. Distributed Incremental Clustering Algorithms: A Bibliometric and Word-Cloud Review Analysis. *Sci. Technol. Libr.* **2020**, *39*, 289–306. [CrossRef]

67. Ruban, D.A.; Yashalova, N.N. Lost in Missions? Employees as a Top Strategic Priority of the World's Biggest Banks. *J. Risk Financ. Manag.* **2021**, *14*, 46. [CrossRef]
68. Azad, A.K.; Rasul, M.G.; Islam, R.; Ahmed, S.F. A study on energy and environmental management techniques used in petroleum process industries. In *Exergy for a Better Environment and Improved Sustainability 2*; Springer: Cham, Switzerland; pp. 219–230.
69. Ruka, A.; Rashidirad, M. Exploring the environmental strategy of big energy companies to drive sustainability. *Strateg. Chang.* **2019**, *28*, 435–443. [CrossRef]
70. Prechel, H.; Istvan, A. Disproportionality of Corporations Environmental Pollution in the Electrical Energy Industry. *Sociol. Perspect.* **2016**, *59*, 505–527. [CrossRef]
71. Best, B.; Miller, K.; McAdam, R.; Moffett, S. Mission or margin? Using dynamic capabilities to manage tensions in social purpose organisations' business model innovation. *J. Bus. Res.* **2021**, *125*, 643–657. [CrossRef]
72. Shevchenko, I.K.; Razvadovskaya, Y.V.; Kaplyuk, E.V.; Rudneva, K.S. Developing indicators for assessing the dynamic capabilities of industrial enterprises. *Terra Econ.* **2020**, *18*, 121–139. [CrossRef]
73. Bey, N.; Hauschild, M.Z.; McAloone, T.C. Drivers and barriers for implementation of environmental strategies in manufacturing companies. *Cirp Ann. Manuf. Technol.* **2013**, *62*, 43–46. [CrossRef]
74. Cater, B.; Cater, T.; Prašnikar, J.; Ivaškovic, I. Environmental strategy and its implementation: What's in it for companies and does it pay off in a posttransition context? *J. East Eur. Manag. Stud.* **2018**, *23*, 55–88. [CrossRef]
75. Minkov, I. Organizational culture—A key factor for the creation and implementation of the company strategy. *Ikon. Izsled.* **2009**, *18*, 91–123.
76. Sæverud, I.A.; Skjærseth, J.B. Oil companies and climate change: Inconsistencies between strategy formulation and implementation? *Glob. Environ. Politics* **2017**, *7*, 42–62. [CrossRef]
77. Nosachevskiy, K.; Nosachevskaya, E.; Afanasjeva, L. Implementation of National Projects in Russia: Approaches, Organization, Management Control. Vision 2025: Education Excellence and Management of Innovations through Sustainable Economic Competitive Advantage. In Proceedings of the 34th International-Business-Information-Management-Association (IBIMA) Conference, Madrid, Spain, 13–14 November 2019; IBIMA: Madrid, Spain; pp. 150–155.
78. Potravny, I.M.; Novoselov, A.L.; Gangut, I.B. Formalization of the general model of the green economy at the regional level. *Econ. Reg.* **2016**, *2*, 438–450.
79. Karasmanaki, E.; Galatsidas, S.; Tsantopoulos, G. An investigation of factors affecting the willingness to invest in renewables among environmental students: A logistic regression approach. *Sustainability* **2019**, *11*, 5012. [CrossRef]
80. Zerinou, I.; Karasmanaki, E.; Ioannou, K.; Andrea, V.; Tsantopoulos, G. Energy saving: Views and attitudes among primary school students and their parents. *Sustainability* **2020**, *12*, 6206. [CrossRef]
81. Gómez-Román, C.; Lima, M.L.; Seoane, G.; Alzate, M.; Dono, M.; Sabucedo, J.-M. Testing common knowledge: Are northern europeans and millennials more concerned about the environment? *Sustainability* **2021**, *13*, 45. [CrossRef]
82. Lehnert, M.; Fiedor, D.; Frajer, J.; Hercik, J.; Jurek, M. Czech students and mitigation of global warming: Beliefs and willingness to take action. *Environ. Educ. Res.* **2020**, *26*, 864–889. [CrossRef]
83. Janmaimool, P.; Khajohnmanee, S. Roles of environmental system knowledge in promoting university students' environmental attitudes and pro-environmental behaviors. *Sustainability* **2019**, *11*, 4270. [CrossRef]

 *sustainability*

Article

The Citizens' Views on Adaptation to Bioclimatic Housing Design: Case Study from Greece

Veronika Andrea [1], Stilianos Tampakis [2,*], Paraskevi Karanikola [1] and Maria Georgopoulou [1]

[1] Department of Forestry and Management of the Environment and Natural Resources, Democritus University of Thrace, 68200 Orestiada, Greece; vandrea@fmenr.duth.gr (V.A.); pkaranik@fmenr.duth.gr (P.K.); mar.georgopoulou@gmail.com (M.G.)

[2] School of Forestry and Natural Environment, Faculty of Agriculture, Forestry and Natural Environment, Aristotle University of Thessaloniki, 54124 Thessaloniki, Greece

* Correspondence: stampaki@for.auth.gr; Tel.: +30-231-099-2756

Received: 2 May 2020; Accepted: 12 June 2020; Published: 18 June 2020

Abstract: Bioclimatic housing design is regarded as an important pillar towards energy policies. Additionally, it is closely affiliated with the performance of energy efficiency of buildings. The citizens' views and their adaptation to energy saving practices can be utilized as an important data base in order to design, improve and properly manage urbanization and environmental challenges in the residential sector. For the capitalization of the citizens' views in Orestiada, the newest city in Greece, simple random sampling was applied on data that were collected via personal interviews and with the use of a structured questionnaire. Reliability and factor analyses were applied for the data processing along with hierarchical log-linear analysis. The latter was utilized for the statistical clustering of citizens into given distinct groups—clusters, arising by factor analysis. The main findings revealed that the citizens are merely aware of bioclimatic principles, while only a small percentage of 28.8% adopts some primary bioclimatic disciplines. Conclusively, it should be noted that there is a need for effective planning towards empowerment on energy efficiency in the residential sector of the city. Notwithstanding, it should not be disregarded the need for the incorporation of conceptual frameworks in urban planning. This is an approach that prerequisites public awareness and the stakeholders' participation in decision making processes.

Keywords: bioclimatic buildings; sustainable buildings; bioclimatic housing design; energy efficiency; citizens' views; raising-awareness; adaptation; decision making

1. Introduction

The current allocation of urbanization indicates that some 55% of the population worldwide live in cities. It is true that if this trend continues by 2050, the urban population will exceed the double of its current size [1]. Intense urbanization could be explained as an indicator of economic and social advance. However, it constitutes an important means that puts pressure on the infrastructure and resilience of our planet [2]. At the same time, cities have to deal with great technological advances, which inevitably cause constant environmental degradation and exhaustion of the natural resources. In particular, most of the countries worldwide are called to handle high urbanization, which in turn leads to an increase of their ecological footprint. There is also a record indicating the deterioration of the ecosystem's capacity. Cities are the main pillars of both economic motion and energy consumption [3]. Certainly, urban areas constitute complex ecosystems with special features. In fact, they operate as integrated units and extensive entities of multilevel and critical interactions that demand significant amounts of energy for their operational needs [4]. Intrinsically, this is how they should be considered, as integrated systems.

However, a crucial sustainability challenge still remains on how to enhance environmental quality and simultaneously accomplish the humans' development goals. Therefore, a challenging target in urban planning stands in the development of cities that will provide the prerequisites for better standards in the quality of life. Some of the most important ones are quality, health and safety and sustainability in all aspects of societal cohesion. Though, in setting a conceptual basis for effective urban planning, it presupposes the balance of natural ecosystems and the viability of cities [5].

The geographical allocation of the urban population is considered as an important characteristic for the better understanding of the citizens' environmental awareness and consciousness [6]. Moreover, the identification and efficient management of vital developments in the performance of ecosystem services should be used for sustainability purposes and in order to ensure a better quality of life for citizens. To this end, the citizens' adaptation to energy efficiency measures is considered as a priority. Special attention has been shown over the last decade, by the European policies on operational energy use. In fact, all the European Commission (EC) strategies and new measures are moving towards the design of new smart buildings. These buildings are supposed to have extremely low energy demands. Additionally, the EC policies include embodied energy of buildings and the development of sustainable buildings. More specifically, EC policy decisions aim to sustainable use of resources throughout their life cycle in achieving significant reduction on their environmental impacts [7]. In addition, the concept of smart cities is listed as a substantial goal for EC. Under this scheme, European cities are expected to transform into units that will provide a better quality of life for their citizens, taking into consideration both technical and political processes. Thus, the exploitation of smart citizens' typology is critical in order to design the required infrastructures. The latter should be determined through a procedure that has the citizens' participation in decision making as a major prerequisite [8].

Moreover, it should be noted that houses represent the major part of the existing buildings in most urban areas. For this particular purpose, bioclimatic housing design should be regarded as a field of special attention. Bioclimatic architecture principles are used to improve the housing energy efficiency through the proper capitalization of the existing environmental and location conditions, such as temperature, humidity, solar radiation and prevailing winds [9]. An expected solution among several habitability optimum conditions, is the thermal comfort achieved by least energy consumption, so that a residence satisfies the citizens' needs for health and wellness. Thus, the citizens' views, as building users, are important sources of information towards effective indoor environmental quality performance and adaptation level on bioclimatic measures [10]. Moreover, recent research has indicated that bioclimatic design should not be focused exclusively on indoor environments. In fact, outdoor assets such as balconies and the outdoor surroundings should be regarded as an integrated unit in efficient bioclimatic design, while it is of utmost importance for future research to shed light into the associations among certain community practices, features and their adaptation to bioclimatic housing design [11].

The Energy Efficiency Performance Status for Buildings in the Study Area

The building sector is very significant for energy and environmental goals of EC. Not to mention, that smart and sustainable buildings are proven to effectively contribute to the social and financial level of cities, formatting the standards for a better quality of life [12]. Energy efficiency is considered as a major matter for the building sector in Europe. Therefore, EC has issued two directives for all its members to comply with in the energy sector. Namely, Directive 2010/31/EU (EPBD), attributed to Energy Performance of Buildings, and Directive 2012/27/EU, addressing the Energy Efficiency scheme. In Greece, EPBD transposition was enacted by the national law N.3661/2008. The "Hellenic Regulation on the Energy Efficiency of Buildings—KENAK" outlines the general calculation approach in line with the European principles. Indeed, the Technical Chamber of Greece has prepared and elaboratively explained all the technical requirements included in KENAK.

In 2010, a Presidential Decree (PD 100/2010) was issued including: the qualifications required from the construction field professionals; new capacity building and vocational training framework;

monitoring systems; the specific regulations needed by experts in order to perform the energy audits of buildings; and inspections of heating and air-conditioning systems. Consequently, from January 2011, the energy performance assessment was mandatory for new buildings including energy performance certificates [13].

According to the 2011 Census of Buildings by the Hellenic Statistical Authority, 55% of the country's residential buildings were built before 1980, which means that they are thermally unprotected. Due to the economic crisis, the number of buildings built after 2010 with KENAK's minimum requirements were estimated as being only 1.5% in the building stock of Greece. In the Region of Eastern Macedonia and Thrace, that the Municipality of Orestiada administratively belongs, the vast majority of buildings are residential ones. Taking for granted that KENAK came into force in 2010, it is clear that the building stock of the Municipality of Orestiada does not meet the energy efficiency standards according to the European directives.

The age of the buildings can be divided into three main periods. These are grouped according to the existing legal framework. The legislation was initially adopted in 1980 with the introduction of Hellenic Building Thermal Insulation Regulation and updated in 2010 with the implementation of KENAK. Consequently, the age classes that affect energy efficiency in the buildings of Orestiada are the following four:

(1) First class: Buildings constructed before 1980. There is no thermal insulation regulation and these buildings are thermally unprotected.
(2) Second class: Buildings constructed from 1981 to 2000, when the Building Thermal Insulation Regulation was put in force.
(3) Third class: Buildings constructed from 2001 to 2010. These buildings meet more construction standards, namely the standards of Hellenic Building Thermal Insulation Regulation; Greek seismic code; and the Greek Regulation for Reinforced Concrete that was put in force in 2000.
(4) Fourth class: Buildings constructed from 2010 until today, in which KENAK is applied.

The aim of the study was to investigate the citizens' views on the basic principles of bioclimatic housing design and further analyzes their adaptation level to certain patterns and measures implemented in their houses. The study is expected to serve as indicative for areas with similar characteristics as Orestiada, while a point to underline is that the oldest buildings are less than 100 years old in the total area of the Municipality. The specific objectives focus on the citizens' facilitation in adapting energy efficiency attitudes in the residential sector, serving the need for a transition to a low-carbon society. The main findings underscore low adaptation on bioclimatic housing design principles mainly restricted on passive solar systems. Furthermore, light was shed on the institutional framework that should be developed, including the development of decision making processes towards multilevel cooperation of key stakeholders in the energy performance of buildings. It is therefore apprised the imperative for environmental raising awareness strategies devoted to the citizens.

2. Materials and Methods

2.1. Study Area

The Municipality of Orestiada (Figure 1) as an administrative unit belongs to the Prefecture of Eastern Macedonia and Thrace, situated in the Regional Union of Evros in Greece. This municipality was established in 2011. Before 2011 the former Municipalities were four including Orestiada, Vissa, Trigono, and Kiprinos. Orestiada is situated in the northernmost part of Greece, bordering with Bulgaria to the north and Turkey to the east. The total holding of the Municipality is 955.6 km^2 and the population is estimated as being 37,695 inhabitants according to the 2011 census. The seat of the municipality is the city of Orestiada, with a population of 18,426 citizens according to the 2011 census. Orestiada has an altitude of 40 m above sea level. It is the most newly established city in Greece that was constructed after the Lausanne Treaty in 1923. The first citizens were refugees from Karagats and

Andrianoupolis. In terms of urbanity, it is mainly an urban municipality with an important primary development sector. Regarding its architecture, it is characterized by its modern street plan, while the topographic plan of the city aged in 1923, is characterized by the exemplary road construction, wide one-way streets and large squares (Figure 2) that most of the modern Greek cities lack.

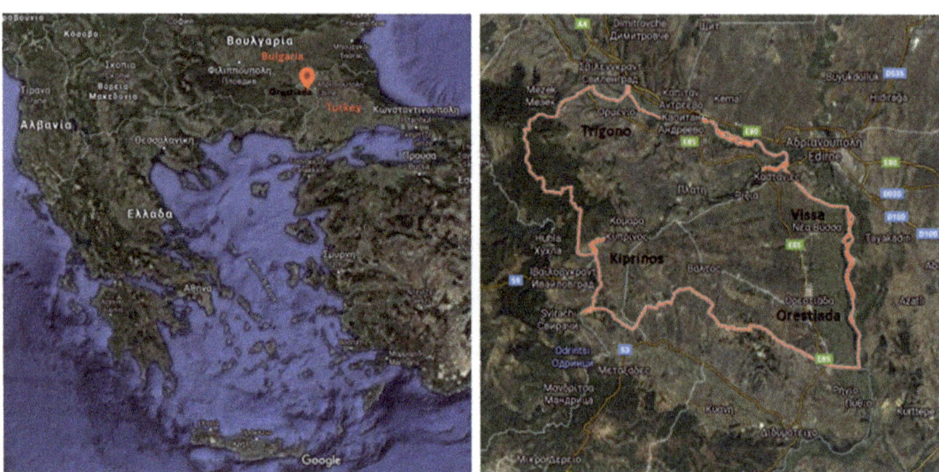

Figure 1. The location and map of the Municipality of Orestiada, Greece (source: Google Maps).

Figure 2. The city of Orestiada, Greece (source: Municipality of Orestiada).

2.2. The Survey

The population that took part in the study included the households in the Municipality of Orestiada. As sampling procedure framework, the domestic electricity consumers' list was chosen. The utilization of households for the sampling procedure constitutes a common example of using teams rather than sample units. In fact, it is considered as an efficient method which is also less pricey [14]. Moreover, interviews took place with a face-to-face method and simple random sampling was applied [15,16]. The average time for the interview was estimated as lasting approximately 25 min. The survey was divided into four sections:

(1)   The demographic profile of the locals;
(2)   Bioclimatic housing design and the house surroundings;
(3)   Adaptation to environmentally friendly and energy efficiency practices;
(4)   The implementation of Hierarchical log-linear analysis on the citizens' awareness on bioclimatic housing design.

The data collection was conducted in 2018 from March to May. The interviews were assigned to a skilled student on the interviewing method, and were conducted during 17:00–21:00 on both weekdays and weekends. The households were selected in a random way by the use of random number tables. The personal interviews took place with the contribution of one family member per household. The rate of response was estimated as being very high (98%) for the survey. The participants ought to have been 18 years old or older in order to comply with the restrictions in Greek law. In case a member of a household could not be found or denied to fulfil the questionnaire, new sample units were chosen.

In order to collect the data, there were organized face-to-face interviews by the use of a structured questionnaire. The questions involved various topics and they were formatted by Likert-scaled and/or close-ended questions. The questionnaire aimed to investigate the citizens' views on bioclimatic housing design principles, energy efficiency practices they implement, the establishment and utilization of green spaces in the buildings, the means of heating and cooling installations they use, the utilization of renewable energy, and their attitudes addressing participation in environmental decision making and awareness. The questionnaire used in the survey is provided in the Supplementary Materials.

2.3. Research Method

The research method used simple random sampling and according to its formulas, there were estimated the numerical mean (y), the standard error (s) for the quantitative variables, and for the standard error for population proportion (Sp) addressing the qualitative variables [14]. The survey was conducted through personal interviews and with the use of a structured questionnaire. The population that took part in the study was the total of the citizens in the households of Orestiada, Greece.

To estimate the sample size, pre-sampling was implemented. The sample size was determined as being 50 citizens. Thus, for each quantitative variable, its variation or standard deviation was determined, and for each qualitative variable, the proportion was calculated [14]. The finite population correction should be disregarded due to the fact that n the size of the sample is high compared with the population N 30 size [16]. The utilization of a questionnaire is not restricted in the estimation of a single variable of the population, yet is addressed to more variables. Thus, the calculation provided a sample size of 400 citizens (for probability $(1 - \alpha)100 = 95\%$, e = 0.049 and without the correction of the finite population). The collection of the research data was conducted in 2018 with the use of SPSS statistical package for the data processing.

Two analyses—reliability and factor were used for the multivariable processing [17]. In order, to examine the questionnaire's internal reliability [18], namely if the data had the tendency to evaluate the same thing, the Cronbach's reliability coefficient was used. A coefficient is regarded satisfactory when its value is equal or higher than 0.70 [19]. Moreover, coefficients with values higher than 0.80 are regarded very satisfactory. Practically, values lower than 0.60 for reliability coefficients are acceptable in various studies [14].

In order for the tests to be useful, reliability must be ensured. In particular, reliability must also be valid, which is controlled via the implementation of factor analysis [16]. The latter constitutes a statistical method that is used in order to examine the existent common factors inside a group of variables [17]. Especially, another analysis was implemented, termed principal component analysis. This analysis is using a spectral analysis of the variance (correlation) matrix. The option of the specific number of factors is a process characterized by special dynamics. It also requires the assessment of the model in a repeated performance. The solution of two factors was applied in the case study. Furthermore, the rotation of the matrix principal components was implemented with the aim of the maximum variance rotation method, suggested by Kaiser [20,21]. It should be taken into consideration that when it should be determined which of the variables belong to a specific factor, the loads in this factor play a critical role. This means that after their rotation, the variables belong to that factor where their loads appear higher than 0.5 [18].

Hierarchical log-linear analysis was utilized for the statistical clustering of citizens into given distinct groups—clusters arising by the factor analysis. This analysis was used to examine two groups of variables. Namely, the adaptation to environmentally friendly practices for energy efficiency (continuous variables) and the citizens' views on the bioclimatic housing design (categorical variables). This analysis serves as an exploratory tool that aims to identify clusters of similar objects in a large number of observations. The condition is then accepted that the variables are independent, which allows the manipulation of categorical and continuous variables simultaneously, with the former following a polynomial distribution and the latter a normal distribution. Before the implementation of hierarchical log-linear analysis, the anticipated frequencies in the contingency table were examined. The classes were practically grouped in order to meet the criteria described by Tabachick and Fidell [22].

In addition, with the implementation of Pearson's X^2 check, the relationship between other variables and each cluster was investigated separately. Therefore, for each cluster, it was possible to recognize its identity in a more accurate way.

3. Results

3.1. The Demographic Profile of the Citizens

The investigated population included the citizens of the Municipality of Orestiada, Greece. The establishment of the municipality is the most recent one in Greece, and accordingly, the housing took place from 1923 and beyond. It is characterized by modern street and topographic plans, while the oldest house is built after 1923. The locals' demographic features were collected during the interviews. From the representations of Table 1, it is apparent that the majority are men (57.2%), aged between 18–30 years old (39.0%); married (74.5%) with two children (41.0%), having completed high school education (29.5%). Regarding their occupation, they were mostly public servants (28%).

Table 1. Demographic characteristics of the locals.

Gender	Male	Female		
	57.2% (s_p = 0.0247)	42.8% (s_p = 0.0247)		
Age	18–30	31–40	41–50	>50
	39.0% (s_p = 0.0244)	23.5% (s_p = 0.0212)	36.2% (s_p = 0.0240)	30.5% (s_p = 0.0230)
Marital Status	Unmarried	Married	Divorced or Widowed	
	19.2% (s_p = 0.0197)	74.5% (s_p = 0.0218)	6.2% (s_p = 0.0121)	
Childhood without Children	One Child	Two Children	Three Children	More than Three
22.2% (s_p = 0.0208)	22.0% (s_p = 0.0207)	41.0% (s_p = 0.0246)	14.5% (s_p = 0.0176)	0.2% (s_p = 0.0025)
Educational Level	Primary School	Secondary School	Technical School	
	4.2% (s_p = 0.0101)	8.2% (s_p = 0.0138)	15.0% (s_p = 0.0179)	
	High School	Technological Ed.	University	
	29.5% (s_p = 0.0228)	16.8% (s_p = 0.0187)	25.2% (s_p = 0.0217)	
Profession			Farmers	
Laborer	Private Employee	Public Servants	Livestock farmers	Pensioners
0.8% (s_p = 0.0043)	18.2% (s_p = 0.0193)	28.0% (s_p = 0.0224)	7.2% (s_p = 0.0130)	14.5% (s_p = 0.0176)
Freelancers	Students	Housewives	Unemployed	
21.2% (s_p = 0.0205)	1.8% (s_p = 0.0066)	5.0% (s_p = 0.0109)	3.2% (s_p = 0.0089)	

3.2. Bioclimatic Housing Design and the House Surroundings

The majority of the citizens agreed that there is need to embrace bioclimatic housing design by the design phase of the buildings that will passively meet their needs for heating and cooling. In particular, more than half of the residents (55.8%, sp = 0.0248) totally agree to its implementation in houses, while an important percentage of 35.5% (sp = 0.0239) also agrees. 7.5% (sp = 0.0132) of the citizens have no opinion as they stated neither agree nor disagree. Finally, a small percentage of 1.2% (sp = 0.0056) disagree and consider that bioclimatic design is not necessary in homes.

However, their knowledge on passive solar systems appears to be limited as merely the 29.80% are aware that the south side of the building is important for the efficient exploitation of solar radiation (Table 2). On the contrary, more than half (55.0%) seem to recognize the importance of the northern side for proper wind protection and heat retention; yet, even in this case, the results indicate that citizens' awareness on bioclimatic housing design remains at low levels.

Table 2. Building side—exploitation of solar radiation and protection from winds and removal of the heat.

Building Side	Exploitation of Solar Radiation	Protection from Winds/Removal of the Heat
to the north	17.2%, s_p = 0.0189	55.0%, s_p = 0.0249
to the east	27.5%, s_p = 0.0223	10.0%, s_p = 0.0150
to the south	29.8%, s_p = 0.0229	15.2%, s_p = 0.0180
to the west	17.0%, s_p = 0.0188	8.2%, s_p = 0.0138
not aware	8.5%, s_p = 0.0139	11.5%, s_p = 0.0153

More specifically, 37.8% of respondents state that there are openings (doors and/or windows) of average size to the north side of the building, while only 34.0% state that there are openings of average size with a south orientation (Table 3) in their residence. The openings to the south serve as a solar collector, so there should be double glazed, insulated window coverings and proper placement of the door frames [23].

Table 3. The openings' size (doors and windows) on the north-south axis in the citizens' residences.

Openings' Size	Wide	Average	Narrow	Do Not Exist
to the north	8.8%, s_p = 0.0141	37.8%, s_p = 0.0242	25.0%, s_p = 0.0217	28.5%, s_p = 0.0226
to the east	39.5%, s_p = 0.0244	38.0%, s_p = 0.0243	6.2%, s_p = 0.0121	16.2%, s_p = 0.0184
to the south	32.2%, s_p = 0.0234	34.0%, s_p = 0.0237	6.5%, s_p = 0.0123	27.2%, s_p = 0.0223
to the west	21.0%, s_p = 0.0204	48.5%, s_p = 0.0250	7.5%, s_p = 0.0132	23.0%, s_p = 0.021

Thus, for the northern latitudes (including the research area), large openings are suggested to be designed to the south including single or double glazing. Moreover, openings of average dimensions on the eastern and western walls are also ideal for this geographical area, while narrow openings with double glazing should be considered as efficient for the northern side of the buildings [24]. The findings reveal that in Orestiada, the suggested principles of bioclimatic housing design are not widely adapted. Namely, the openings oriented to the south are recorded as average size by 34% of the citizens and as wide by 32.2%. As regards to the openings to the east, 39.5% claim to have wide openings and 38% have average ones. The north openings are of average size for 37.8% of the houses and there are narrow openings in 25% of the homes with no openings to the north for 28.5% homes (Table 3). The data reflects a neutral conceptualization by the citizens regarding passive solar systems applied as part of bioclimatic housing design that is focused on the placement of openings (windows, doors, etc.). Indeed, the design of openings took place during the design phase of the building. However, it could be regarded as positive that primary steps towards bioclimatic housing design are adopted by the construction field and professionals of Orestiada for the orientation of openings.

Concerning the interior space installations on the north-south axis, it is evident that for the dining room and living room, there is preference for installation in the middle of the house (35.8% and 32.8% respectively). The kitchen and the toilet are located to the north (35.5% and 34.8% respectively); and the bedrooms to the south (39.2%) and to the north (35%) (Table 4). During the design phase of interior spaces, these rooms should be organized and grouped in a more efficient way. Design should be based on the fact that spaces which will be used the most should be placed to the south side of the building in order to ensure the desired thermal comfort where high levels are usually preferred. In contrast, for tspaces that are not used as often, there are restricted demands for thermal comfort conditions such as high temperatures. These spaces should be placed in the intermediate thermal zone. The rest of the spaces, usually the auxiliary ones, should be placed on the north side of the building, in order to provide a kind of "protection zone" and insulate the other two space categories. By following these basic installation principles for interior spaces, it is possible to separate the external environment from the internal one, in which mainly higher temperatures prevail. This is a convenient way to reduce energy losses from the most used spaces of the house [25]. According to the findings, interior space installations on the north-south axis are randomly set in the houses of Orestiada. An explanation could be that the designers did not prioritize this principle. In other cases, we would expect, at least for the living room, to be situated mainly to the south. Construction experts might list, as more important, other features such as the better utilization of the plot and the premises of the building, or the view from the rooms.

Table 4. Interior space installations on the north-south axis.

Interior Space Installations	North	In the Middle of the House	South	Do Not Exist
Bedrooms	35.0%, s_p = 0.0238	18.5%, s_p = 0.0194	39.2%, s_p = 0.0244	7.2%, s_p = 0.0130
Living room	29.2%, s_p = 0.0227	32.8%, s_p = 0.0235	27.8%, s_p = 0.0224	10.2%, s_p = 0.0152
Kitchen	35.5%, s_p = 0.0239	27.8%, s_p = 0.0224	25.2%, s_p = 0.0217	11.5%, s_p = 0.0160
Dining room	22.0%, s_p = 0.0207	35.8%, s_p = 0.0240	27.2%, s_p = 0.0223	15.0%, s_p = 0.0179
Toilet	34.8%, s_p = 0.0238	25.2%, s_p = 0.0217	19.0%, s_p = 0.0196	21.0%, s_p = 0.0204

Only a low percentage (25%) of buildings have a southern orientation. The urban planning and layout of the main road network along the east-west or north-south axis is what predetermines the orientation of the facades of the buildings. This causes certain constraints to the designers as it is not possible to take advantage of the available environmental and thermal benefits. Inevitably, they are led to design buildings with many problems, such as overheating of interiors (in buildings with east or west orientation), as well as insolation from solar radiation concerning buildings with north orientation. Added to that, it should be noted that although southern orientation may be possible for the construction of a building, shading conditions that are established by the surrounding buildings could provoke an unpleasant thermal situation for residents. Thus, interactions between the height of the buildings and the width of the streets should be also taken into account [26].

The citizens were asked about the type of their residence. Respectively, 32.2% (sp = 0.0234) answered that they live in a detached house, 25.5% (sp = 0.0218) on the first floor, 26% (sp = 0.0219) on the second and 16.3% (sp = 0.0184) reside in an apartment on a higher floor. Another issue to investigate for the type of residence is the surroundings and the outdoor environment. Trees and high neighboring buildings could affect homes from a bioclimatic point of view. Thus, the citizens were also asked about the house orientation on the north-south axis of the land upon which the building is situated. 34.5% (sp = 0.0238) are said to have a house in the middle of the plot, 34% (sp = 0.0237) south of the plot and 30.2% (sp = 0.0230) north of the plot, while 1.2% (sp = 0.0056) of the residents did not respond.

Furthermore, a high percentage of 64.2% (sp = 0.0240) answered that there is a garden in their residence while 35.8% (sp = 0.0240) claimed not to have one. The citizens that stated they had a garden were then asked to provide more information. Namely, regarding the size of the garden, 3.5% (sp = 0.0092) said that it is over 300 m^2; 7% (sp = 0.0128) have a garden between 150 and 300 m^2; 19.8% (sp = 0.0199) between 50 and 150 m^2; 24% (sp = 0.0214) have a garden sized from 20 and 50 m^2 and 10% (sp = 0.0150) have one smaller than 20 m^2. Additionally, 17% (sp = 0.0188) of the citizens stated that the garden is located around the plot, 16.8% (sp = 0.0187) to the north, 15.5% (sp = 0.0181) to the south and 14.8% (sp = 0.0177) in the middle of the plot.

The number of evergreen and deciduous trees found in home gardens is given in Table 5, while the type of trees depended on the planting location as shown in Table 6. Trees and shrubs, which are porous barriers, allow crossing a part of the wind, reduce turbulence and create a wider protection zone. They are superior in terms of fencing in and reducing wind speed, as they can reduce wind speed by 50% at a distance equal to five times their height [27]. However, it seems that the citizens are not able to conceive that evergreen trees should be planted in northern locations to serve insulation means for cold northern winds in winter. In fact, they associate garden trees with the shade they offer during seasons of high temperatures. Thus, 36.8% (sp = 0.0241) of the citizens state that the best season to relax in their garden is spring, 19.8% (sp = 0.0199) in summer, 7.2% (sp = 0.0130) in autumn and 0.5% (sp = 0.0035) in winter. It should also be noted that 35.8% do not have a garden.

Table 5. Number of trees in the residence garden.

No. of Trees in the Garden	Evergreen Trees	Deciduous Trees
No tree	47.0%, $s_p = 0.0250$	51.0%, $s_p = 0.0250$
1	10.0%, $s_p = 0.0150$	13.5%, $s_p = 0.0171$
2	14.8%, $s_p = 0.0177$	12.0%, $s_p = 0.0162$
3–5	18.0%, $s_p = 0.0192$	16.5%, $s_p = 0.0186$
>5	10.2%, $s_p = 0.0152$	7.0%, $s_p = 0.0128$

Table 6. Tree species depending on the planting location.

Location in the Garden	More Evergreen Trees	More Deciduous Trees
North side	20.5%, $s_p = 0.0202$	15.0%, $s_p = 0.0179$
East side	13.5%, $s_p = 0.0171$	14.5%, $s_p = 0.0176$
South side	15.0%, $s_p = 0.0179$	11.8%, $s_p = 0.0161$
West side	5.2%, $s_p = 0.0112$	8.8%, $s_p = 0.0141$
Did not answer	45.8%, $s_p = 0.0249$	50.0%, $s_p = 0.0250$

In order to make a forward to bioclimatic design, interior spaces do not suffice. Surroundings are also significant spaces that should be designed in line with bioclimatic principles. For the successful incorporation of green practices, it is crucial for citizens to adopt them in their residence. Basic guidelines include the proper landscape design, ideally with the introduction of low demanding native species [12], while the construction of green roofs could also be used for the balance of the microclimate in many cases [28].

3.3. Adaptation to Environmentally Friendly and Energy Efficiency Practices

The level of the citizens' adaptation to environmentally friendly and energy efficiency practices was examined and analyzed with the aim of descriptive statistics and the use of the statistical mean. In particular, citizens were asked to evaluate the capitalization of certain practices aiming to prevent energy losses in their houses including insolation preservation during winter; means and installations for the protection from strong winter winds; ways for the minimization of heating losses during winter; the protection from the summer sun; taking advantage of the summer breezes; and as regards to the removal of the heat which accumulates in summer in their residence. In Table 7, the results are interpreted, showing that the most important practices that have been adopted by the citizens of Orestiada are the minimization of the heating losses during winter, the protection from the summer sun and the protection from strong winter winds. It should be noted that the climate in the broader region is characterized by severe winters and hot summers.

Reliability analysis was performed to examine the consistency of the equivalent questions of the above multivariable. There is a significant ranking of the reliability coefficient alpha with a value of 0.808. This is strong evidence for the research data in evaluating the same thing. Prior to the implementation of factor analysis, all the necessary checks were conducted. More specifically, the data appropriateness was checked; while it was also examined for the appropriateness of all the variables used in the model. The results of the factor analysis are shown in Table 8. The representations of the loads have shown that there is partial correlation in two factors for the six variables. There is a positive correlation between the variables' loads and the factor that is affecting the total degree fluctuation for this variable. The variables "belonging" to a specific factor are the ones with loads that overcome 0.5 value.

Table 7. The citizens' adaptation on environmentally friendly and energy efficiency practices in their residence.

Variable	Statistical Mean	Standard Error
Insolation preservation during winter	5.77	2.576
Protection from strong winter winds	5.79	2.464
Minimization of the heating losses during winter	6.34	2.193
Protection from the summer sun	6.00	2.156
Take advantage of the summer breezes	5.29	2.210
Removal of the heat which accumulates in summer	5.11	2.121

Table 8. Factor analysis loadings after rotation (numbers represented in bold indicate the factor that belongs to each variable).

Variables	Factor Loadings	
	1	2
Insolation preservation during winter	0.384	**0.673**
Protection from strong winter winds	**0.854**	0.125
Minimization of the heating losses during winter	**0.840**	0.146
Protection from the summer sun	**0.724**	0.328
Take advantage of the summer breezes	0.075	**0.905**
Removal of the heat which accumulates in summer	0.198	**0.820**

The burdens in bold represent the variables that belong to the specific factor.

The first factor termed as *Energy efficiency practices of major acceptance* consists of the variables "Protection from strong winter winds", "Minimization of the heating losses during winter" and "Protection from the summer sun". Accordingly, the second factor named as *Energy efficiency practices of minor acceptance* includes the variables "Insolation preservation during winter", "Take advantage of the summer breezes" and "Removal of the heat which accumulates in summer". The extraction of the two factors reveals that this grouping addresses the basic interventions which are closely affiliated with primary construction issues and are proven to be more acceptable by the citizens (listed in the first factor). On the other hand, the adaptation to some particular practices that prerequisite further investments, installation of a special equipment for the saving of energy or further construction works seem to have lower acceptance by the citizens.

3.4. The Implementation of Hierarchical Log-Linear Analysis on the Citizens' Awareness on the Bioclimatic Housing Design

In the next stage of data processing, the hierarchical log-linear analysis was conducted, in order to further examine the citizens' awareness on the bioclimatic housing design in line with the energy efficiency practices they adopt (Figure 3). Hence, with the application of two-step cluster analysis, the observations were listed into three clusters comprising the optimum solution.

Particularly, from the 385 citizens, 28.8% are listed in the first cluster, 33.2% in the second cluster and 37.9% in the third one. As regards to the variables' (continuous and categorical) relative significance towards the clusters' formation, the diagrammatic representations of Figure 3 indicate the tests with statistical significance. Variables are important in creating the cluster when the statistical value exceeds the critical value. In particular, for the continuous variables, it was observed that the variable "Energy efficiency practices of major acceptance" is in a close proximity with the critical value limit of the first cluster. Furthermore, the variable "Energy efficiency practices of minor acceptance" tends to hold an important role for the second cluster formation. Concerning the third cluster, both variables of "Energy efficiency practices of major acceptance" and "Energy efficiency practices of minor acceptance" are the reason for this formulation (Figure 3a,c,e). Whereas, concerning the categorical variables, the value of the statistical X^2 addressing both variables in the three clusters was higher compared with the limits of the critical value. This was explained by the fact that all the categorical variables in the analysis strongly affected the shaping of the three clusters (Figure 3b,d,f).

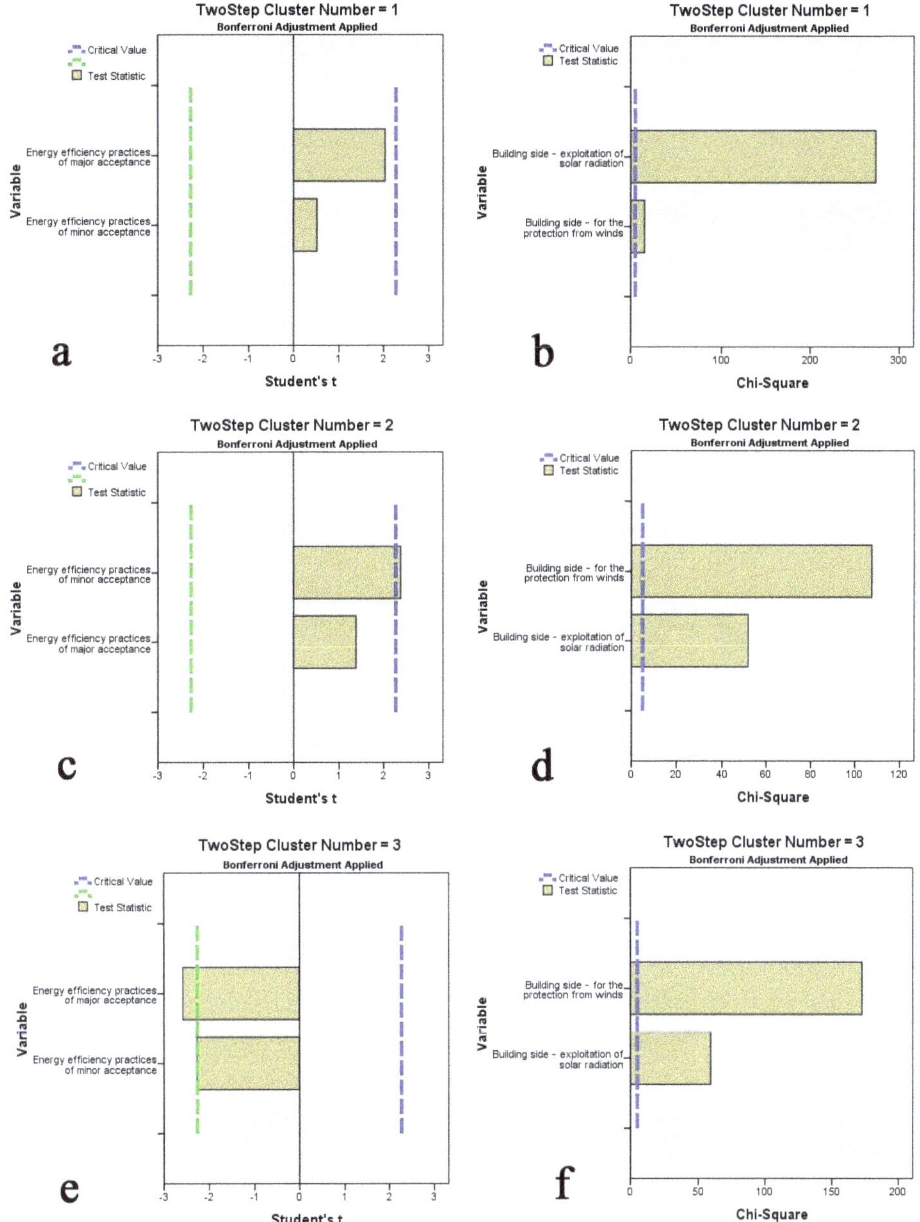

Figure 3. Diagrammatic representations of statistical tests of variables per cluster, with quantitative variables (**a**,**c**,**e**), respectively categorical variables (**b**,**d**,**f**).

The characteristics of the three clusters are represented in Table 9. In particular, the practices adopted in the houses, with the highest acceptance by the citizens, have a positive value in the first and second cluster and a negative value in the third cluster. In addition, the practices adopted in houses with the lowest acceptance have a positive value in the second cluster, neutral to positive in the first cluster and negative in the third cluster. According to these findings, it seems that the citizens of the first and second cluster consider the adoption of environmentally friendly and energy efficiency practices under a more positive light. In addition, their positive concept is mainly estimated on practices engaging the efficient management of solar and wind energy.

Table 9. Interpretation of the cluster observations.

Variables	Cluster 1	Cluster 2	Cluster 3
Energy Efficiency Practices of Major Acceptance	positive	positive	negative
Energy Efficiency Practices of Minor Acceptance	neutral to positive	positive	negative
Building Side—Exploitation of Solar Radiation	south	something else	something else
Building Side— for the Protection from Winds	north	north	something else
With the use of Person X^2 check			
Need for Bioclimatic Housing Design	disagree or totally agree	agree	neither agree nor disagree
Openings' Size (Windows and Doors to the North)	narrow	average or narrow	do not exist or wide
Openings' Size (Windows and Doors to the South)	wide or average	wide	do not exist or average
Interior Space Installations on The North-South Axis—Living Room	south	in the middle of the house or do not exist	in the middle of the house or do not exist
Interior Space Installations on the North-South Axis—Dining Room	south or north	in the middle of the house	in the middle of the house or do not exist
Type of Residence	detached house or on a higher floor	detached house or on the first floor	first and second floor
House Orientation on the North-South Axis	north of the land upon which the building is situated	south of the land upon which the building is situated	in the middle of the land upon which the building is situated
Educational Level	university	technological ed. and high school	lower level of education

The characteristics of each cluster of citizens are supplemented by the results of the analysis of the categorical variables. More specifically, it was observed that the citizens of the first cluster are aware and seem to comprehend that the side of the building that is important for the insolation is the south part. Moreover, they also acknowledge that the north part of the building is important for the protection of winds.

On the other side, as regards to the citizens of the second cluster, they only understood that the side of the building that should be given importance for the protection from the winds is the north part. Eventually, the citizens of the third cluster show the least knowledge on bioclimatic housing design patterns and disciplines.

Indeed, with the aid of Pearson's X^2 ($\alpha < 0.005$) and in the lower part of Table 9, it is provided the correlation among the three clusters with other variables about the citizens' characteristics.

The representations are the following:

- The citizens of the first cluster disagree or totally agree with the need for bioclimatic housing design. They also claim that as regards to the construction and architecture of their houses, there are narrow windows and door openings to the north and wide or average to the south.

Furthermore, as regards to the interior space installations that are most occupied, the living room is in the south and the dining room is located in the south or north part of the house. Finally, the citizens of this cluster own a detached house or an apartment on a higher floor of a building. Their residence is in the north part of the land upon which the building is situated. The correlation with the education level shows that these citizens are of an advanced educational status as they mainly hold a university degree.

- Citizens that belong to the second cluster merely agree with the need for integrating bioclimatic housing design. The same also state that they reside in a house with average or narrow openings of windows and doors to the north and that these openings are wide to the south axis. Added to that, they declare that their living room is situated in the middle of the house. The citizens of the second cluster are owners of a detached house or an apartment on the first floor and their house is built on the south part of the land upon which the building is situated. Finally, their education level is intermediate in comparison with the citizens of the first and the third cluster.

- The third cluster citizens neither agree nor disagree with the need for establishing bioclimatic housing design. Concerning the existing status of their residence in terms of basic bioclimatic housing design, they describe the window and door openings as being wide to the north or not existing at all. As regards to the south axis, they claim to have average window and door openings or none at all. Moreover, interior space installations in their residence involve a living room in the middle of the house or these do not exist. As for the dining room, they indicate the same positioning. Concerning the type of residence, they have an apartment on the first or second floor, in the middle of the land upon which the building is situated. In this cluster, citizens are of a lower educational level.

4. Discussion

It is evident that the existing situation of the building stock in the Municipality of Orestiada is a challenging issue as part of an effort to harmonize elements of bioclimatic housing design and their adaptation in the existing of future buildings. In fact, the embracing of this point of view coincides with the fact that only 1.5% of the buildings in Orestiada were constructed after 2010 and accordingly meet the minimum energy performance requirements of the Hellenic Regulation on the Energy Assessment of Buildings—KENAK. Additionally, it should be taken into consideration that the majority of the citizens are not aware of the need and benefits arising with the establishment of bioclimatic systems; although, according to Manzano-Agugliaro et al. [29], such systems will serve as a reduction of energy consumption along with the enhancement of the climate comfort level in their houses.

In fact, the hierarchical log-linear analysis revealed that despite the fact that the citizens of the first and second cluster hold a positive view on the adoption of environmentally friendly and energy efficiency practices, this mainly addresses energy efficiency on solar and wind systems that are established with the basic architecture principles. These interventions mostly took place during the construction phase of their residence, such as the determination of the side of the building that can take advantage of the insolation or, respectively, provide protection from winter winds. However, the majority of citizens that belong in the third cluster seem to hold poor knowledge on bioclimatic housing design patterns and disciplines.

Moreover, it is of outmost importance for the citizens to understand and adopt new green practices and efficiently use energy in their residence [15]. The basic principles of this effort are related with sustainability goals, namely the sustainable exploitation of the natural environment and its resources. Thus, the citizens' adaptation to primary and, in turn, to advanced bioclimatic patterns aims to preserve energy and establish a holistic approach towards environmentally friendly attitudes at both passive and active levels of construction solutions.

Furthermore, bioclimatic housing design implies the construction of houses designed in a way that will meet the citizens' energy needs, through the optimal utilization of the building comfortably and the existing local and microclimatic environment [30]. The adaptation to a bioclimatic residence presupposes that the main openings and the main face of the house should be oriented to the south, while, in the north, there should be small openings and solid walls. Additionally, the openings should be designed in a way to provide transparent ventilation [31,32]. Nonetheless, most of the citizens of Orestiada seem not to be aware of energy and/or bioclimatic contemporary architecture interventions. In fact, only the citizens of the first cluster are the advocates of adapting bioclimatic housing design. The adaptation of specific patterns and systems plays a crucial role in achieving improved interior environments by the conservation of energy and the incorporation of practices that have proven to be more efficient in the building construction sector; such as the passive thermal systems for the improvement of thermal comfort in heating and cooling of interior spaces [12]. In order to achieve efficient energy management on a city level, it is of great importance to make a forward in utilization of green energies. Self-efficiency in energy production on site could be an innovative solution the housing sector; namely by the use of renewable energy [33] such as solar or wind energy.

The citizens of the third cluster, who are apparently the most reluctant in adapting bioclimatic housing design, argue that their windows and door openings are wide and oriented to the north or that they do not exist. Moreover, as regards to the south axis, they claim to have average window and door openings or none at all. However, there is a different view concerning the citizens of the first cluster; they stated to have, at least, narrow windows and door openings to the north in their houses and wide or average ones to the south. The citizens of this cluster have proven to adapt more efficiently to passive solar strategies that are based on cooling or heating strategies, which passively absorb or protect from solar radiation and are using slightly or not at all mechanical devices [34]. According to Bughio et al. [10], cross-ventilation via the proper placement of openings, could improve external shading devices, and create more comfortable indoor environmental quality.

As regards to the positioning of the most used rooms of the house, namely the living room and the dining room, only the citizens of the first cluster have stated to have the living room in the south part of the house. The citizens of the second and third cluster claim to have placed the living room in the middle of the house, while, for some citizens of the third cluster, the living room does not exist at all. Similar responses were recorded for the dining room positioning. From the first phase of designing a building, designers should consider the internal existing conditions required to control variation of room temperature and humidity, and the positioning of specific rooms such as the kitchen, laundry and bathroom. The design and recommendation for the position of the rooms should be in combination with the external climatic local conditions [35]. Indeed, the room design should fulfill the standards for optimum thermal comfort with low energy consumption and sustain a good indoor air quality [36]. Concerning the positioning of the living room, this should provide good indoor conditions during daytime, when it is mainly used [37].

Concerning the orientation of the house in the plot and the ownership status of the house, the citizens of the first cluster that belong to the highest educational level seem to be the most privileged and in accordance with basic bioclimatic housing design principles. This arises by the fact that they own a detached house or an apartment on a higher floor of a building, which is situated in the north part of the land upon which the building is situated. In line with Akadiri et al. [12], the orientation of the building is part of bioclimatic design that aims to passive energy utilization. Thus, it should be taken into consideration and conceptualized in a macro and microclimate framework from the beginning of the design procedure by the architects. In a similar study that was conducted in Finland, a country with a different climatic condition, it seems that there is a strong association between the citizens' income and amount of energy they consume to fulfill their needs. The same study also revealed that carbon emissions are similar for detached houses in suburban regions and for dense building blocks in rural areas [38].

A gradual relegation of the ownership conditions and bioclimatic benefits is observed respectively in the second and third cluster. Particularly, the citizens of the second cluster are of an intermediate educational level, owners of a detached house or an apartment on the first floor, and their house is situated at the south of the property land. The citizens of the third cluster have the lowest educational level and they reside in an apartment on the lowest floors, that is situated in the middle of the land upon which the building is situated. It was proven that the proper orientation of the building contributes to natural ventilation, to lighting and to the reduction of energy consumption and emissions by devices for artificial light and air-conditioning systems [39]. In addition, other findings from areas with similar climate conditions and Mediterranean climate, such as Italy, have shown that except for orientation, openings, exposure to weather conditions and employment of the local natural resources, that the building shape plays a critical role in the thermal behavior of buildings [40].

5. Conclusions

The majority of the respondents are aware of the term 'bioclimatic housing design' yet, their knowledge on passive solar systems is limited. The citizens of Orestiada recognize the importance of the northern house orientation that is foreseen in order to ensure protection from winds as well as heat retention. However, some of the basic principles of bioclimatic design, such as the southern location of the building, seems not to have a significant adaptation in the housing of Orestiada.

On the other hand, the majority of the citizens believe that the basic principles of bioclimatic design are applied in their houses. Indeed, for an important amount of the citizens, it seems that there is a partial application of bioclimatic housing design. Notwithstanding, for the minority of the citizens, there is no adaptation, even as regards to the basic principles of bioclimatic design.

In line with the findings highlighted by the application of hierarchical log-linear analysis, it could be underscored that the citizens of the first cluster seem to adapt at least to the basic principles of bioclimatic housing design; while for the ones that belong to the second cluster, there is a mere adaption to the primary principles. Nonetheless, in the last group of citizens, none of the basic principles of bioclimatic design seems to be adopted.

Reasonably, there is a gap of awareness on innovative energy saving systems. It is evident that there was low acceptance by the citizens on the need to install advanced systems of energy saving purposes such as active solar energy systems. Possibly, this was regarded as a questioning issue as these systems demand further investments and direct costs that are included at the first stage of their installation. In Greece, there is an ongoing financial program for the transformation of the building sector into the sustainable sector. The program "Saving at home" was introduced in 2007 and it is supported by the European Regional Development Fund and national funds. Unfortunately, there is still low absorption of structural funds. An answer for this situation could be the low dissemination of the program and its potential. Another important issue is that citizens regard it as questionable. In fact, there is a lot of mistrust due to the fact that most of the candidates for implementing the program and improving their residence in line with KENAK standards, have applied in the past and had no feedback for years. Added to that, a lot of bureaucratic and demanding procedures constitute another reason for the citizens' abstention.

Another point to consider is the microclimate of the area. Particularly, the significant humid climate of Orestiada, the moderate sunshine and the restricted but intensely hot summer period, comprise some of the key factors for which bioclimatic housing design should be adapted in the major part of the buildings in the broader area. This means that there is an indisputable need for the designing of buildings with the engagement of proper and rational principles in terms of location and orientation of the building; size—orientation and allocation of the door and window openings; the protection of the building shell (thermal insulation, wind protection and solar radiation protection)—which is the most important feature that designers of housing in Greece, such as engineers and architects, should take into account. It is also essential for the homeowners to be aware of the bioclimatic framework and set performance indicators as a prerequisite for a contract assignment to a housing designer or to a technician.

Taking everything into consideration, it would be argued that sustainable and bioclimatic buildings do not suffice to protect the natural environment. Raising awareness and creating consciousness in environmental, sustainability and energy efficiency issues stands as a necessary condition for the citizens to be able to recognize the value of embracing bioclimatic housing design practices. The citizens Orestiada are quite aware and environmentally conscious. Therefore, with the aim of raising awareness programs, they will have access to higher levels of information, which, in turn, will serve as a means of adaptation to bioclimatic practices.

Nevertheless, the measures taken within a policy framework of sustainable development ought not to be considered as fragmented and isolated. In fact, efficient city management should regard cities as integrated units. Nowadays, an incorporated strategy is the first and foremost demand in order to address the intense problems of the natural and urban environment under a holistic point of view. If we are willing to establish strategic planning for the integrated management of smart cities, aiming both at sustainability and the improvement of citizens' quality of life, this implies the effective participation, cooperation and interaction of key stakeholders in decision making. In conclusion, it appears that the props of this procedure are the public services for construction and technical schemes, as well as the enforcement of certain policies, measures and regulations in the field of transport, renovation, energy, economic development, social cohesion and other parameters that set up the picture of a city.

Finally, due to research limitations, it was not possible to investigate some specific characteristics of the housing sector in the Municipality of Orestiada in order to attribute a more elaborative view of the existing establishments. To this end, it is highly recommended, for future research, a deeper examination of specific features that comprise the city housing typology such as building heights, floor numbers, structural materials, orientation and household size. Additionally, a better analysis is suggested for the outdoor spaces such as access to gardens and watering mechanisms used. Other points that need to be investigated in the future is if renewable energy systems can be established and used to cover the household energy needs, the introduction of innovation and smart systems for energy efficiency and also the citizens' performance in taking up funding programs such as the periodical European structural and investment funds, for the transition to an energy efficient community.

Supplementary Materials: The questionnaire is available online at http://www.mdpi.com/2071-1050/12/12/4984/s1.

Author Contributions: Conceptualization, V.A.; Formal analysis, S.T. and M.G.; Methodology, S.T. and M.G.; Project administration, P.K.; Resources, V.A. and P.K.; Supervision, S.T.; Writing—original draft, V.A.; Writing—review & editing, V.A. and P.K. All authors have read and agreed to the published version of the manuscript.

Funding: This research received no external funding.

Conflicts of Interest: The authors declare no conflict of interest.

References

1. United Nations, Department of Economic and Social Affairs. 68% of the World Population Projected to Live in Urban Areas by 2050, Says UN. Available online: https://www.un.org/development/desa/en/news/population/2018-revision-of-world-urbanization-prospects.html (accessed on 24 April 2020).

2. Romero-Lankao, P.; Gnatz, D.M.; Wilhelmi, O.; Hayden, M. Urban Sustainability and Resilience: From Theory to Practice. *Sustainability* **2016**, *8*, 1224. [CrossRef]

3. Ahmed, Z.; Zafar, M.W.; Ali, S. Danish, Linking urbanization, human capital, and the ecological footprint in G7 countries: An empirical analysis. *Sustain. Cities Soc.* **2020**, *55*, 102064. [CrossRef]

4. Bajcinovci, B.; Jerliu, F. Achieving Energy Efficiency in Accordance with Bioclimatic Architecture Principles. *Environ. Clim. Technol.* **2016**, *18*, 54–63. [CrossRef]

5. Ji, Z.; Xu, Y.; Wei, H. Identifying Dynamic Changes in Ecosystem Services Supply and Demand for Urban Sustainability: Insights from a Rapidly Urbanizing City in Central China. *Sustainability* **2020**, *12*, 3428. [CrossRef]

6. Zoraghein, H.; O'Neill, B.C. U.S. State-level Projections of the Spatial Distribution of Population Consistent with Shared Socioeconomic Pathways. *Sustainability* **2020**, *12*, 3374. [CrossRef]

7. European Commission. Energy Performance of Buildings Directive. Available online: https://ec.europa.eu/energy/topics/energy-efficiency/energy-efficient-buildings/energy-performance-buildings-directive_en (accessed on 23 April 2020).

8. Calzada, I. (Smart) Citizens from Data Providers to Decision-Makers? The Case Study of Barcelona. *Sustainability* **2018**, *10*, 3252. [CrossRef]

9. Tovar Alcázar, M.R.; García Chávez, J.R. Educational Program for Promoting the Application of Bioclimatic and Sustainable Architecture in Elementary Schools. *Energy Procedia* **2014**, *57*, 999–1004. [CrossRef]

10. Bughio, M.; Schuetze, T.; Mahar, W.A. Comparative Analysis of Indoor Environmental Quality of Architectural Campus Buildings' Lecture Halls and its' Perception by Building Users, in Karachi, Pakistan. *Sustainability* **2020**, *12*, 2995. [CrossRef]

11. Oliveira, S.; Marco, E. Role of 'Community Spaces' in Residents' Adaptation to Energy-Efficient Heating Technologies—Insights from a UK Low-Energy Housing Development. *Sustainability* **2018**, *10*, 934. [CrossRef]

12. Akadiri, P.O.; Chinyio, E.A.; Olomolaiye, P.O. Design of A Sustainable Building: A Conceptual Framework for Implementing Sustainability in the Building Sector. *Buildings* **2012**, *2*, 126–152. [CrossRef]

13. KENAK—Greek Regulation for the Energy Efficiency of Buildings. Available online: https://www.buildup. eu/en/practices/publications/kenak-greek-regulation-energy-efficiency-buildings (accessed on 25 April 2020).

14. Matis, K. *Forest Sampling*; Democritus University of Thrace: Xanthi, Greece, 2001.

15. Hoyos, D. The state of the art of environmental valuation with discrete choice experiments. *Ecol. Econ.* **2010**, *69*, 1595–1603. [CrossRef]

16. Pagano, M.; Gauvreau, K. *Elements of Biostatistics*; Ellin Publications: Athens, Greece, 2000.

17. Siardos, G.K. Multivariate Statistical Analysis Methods. In *Part I: Exploring the Relations between Variables*; Zitis Publications: Thessaloniki, Greece, 1999.

18. Frangos, C.K. *Methodology of Market Research and Data Analysis with the Use of the Statistical Package SPSS for Windows*; Interbooks Publications: Athens, Greece, 2004.

19. Howitt, D.; Gramer, D. *Statistics with the SPSS 11 for Windows*; Kleidarithmos Publications: Athens, Greece, 2003.

20. Djoufras, I.; Karlis, D. *Elements of Multivariate Data Analysis*; University of the Aegean: Chios, Greece, 1997.

21. Harman, H.H. *Modern Factor Analysis*; The University of Chicago Press: Chicago, IL, USA, 1976.

22. Tabachick, B.G.; Fidell, L.S. *Using Multivariate Statistics*, 2nd ed.; Harper and Row: New York, NY, USA, 1989.

23. Antonopoulou, S. Bioclimatic Architecture and Sustainable Development—Methods and Examples in Specific Buildings. Bachelor's Thesis, School of Environment, Geography and Applied Economics, Department of Home Economics and Ecology, Harokopio University, Athens, Greece, 2009.

24. Kaniadaki, M. Energy Efficiency Housing. Energy Efficiency Methods and Systems in Modern Greek Housing. Bachelor's Thesis, Department of Mechanical Engineering, Technological Educational Institute of Crete, Heraklion, Greece, 2011.

25. Zoumbourlis, G.; Etmektzoglou, S. Bioclimatic Design in the Building Construction Sector. Bachelor's Thesis, Department of Civil Engineering, Technological Educational Institute of Piraeus, Aigaleo, Greece, 2014.

26. Hegazi, K. Bioclimatic Construction and Sustainable Development. Bachelor's Thesis, School of Mechanical Engineering, National Technical University of Athens, Athens, Greece, 2009.

27. Rizou, S. Bioclimatic and Energy Efficiency Building Management. The Case of a Multi-Use Space of Shopping Mall and Restaurants in Volos Greece. Master's Thesis, Interdepartmental Master Programme Law and Energy Engineering, Aristotle University of Thessaloniki Faculty of Law, School of Law, Thessaloniki, Greece, 2017.

28. Salkini, H.; Greco, L.; Lucente, R. Towards Adaptive Residential Buildings Traditional and Contemporary Scenarios in Bioclimatic Design (the Case of Aleppo). *Procedia Eng.* **2017**, *180*, 1083–1092. [CrossRef]

29. Manzano-Agugliaro, F.; Montoya, F.G.; Sabio-Ortega, A.; García-Cruz, A. Review of bioclimatic architecture strategies for achieving thermal comfort. *Renew. Sustain. Energy Rev.* **2015**, *49*, 736–755. [CrossRef]

30. Mohammed, U.A.; Alibaba, H.Z. Application of Bioclimatic Design Strategies to Solve Thermal Discomfort in Maiduguri Residences, Borno State Nigeria. *Imp. J. Interdiscip. Res. (IJIR) Peer Rev. Int. J.* **2018**, *4*, 227–233.

31. Center for Renewable Energy Sources. Bioclimatic Design and Passive Solar Systems. Available online: http://www.cres.gr/kape/energeia_politis/energeia_politis_bioclimatic_eng.htm (accessed on 20 April 2020).

32. Bodach, S.; Lang, W.; Hamhaber, J. Climate responsive building design strategies of vernacular architecture in Nepal. *Energy Build.* **2014**, *81*, 227–242. [CrossRef]

33. Heinonen, J.; Kyrö, R.; Junnila, S. Dense downtown living more carbon intense due to higher consumption: A case study of Helsinki. *Res. Lett.* **2011**, *6*, 034034. [CrossRef]

34. Duraković, B. Passive Solar Heating/Cooling Strategies. In *PCM-Based Building Envelope Systems. Green Energy and Technology*; Springer: Cham, Switzerland, 2020; pp. 39–40. [CrossRef]

35. Sada, G.K.A.; Salih, T.W.M. Enhancing Indoor Air Quality for Residential Building in Iraq as a Typical Case of Hot Arid Regions. In Proceedings of the Workshop on Indoor Air Quality in Hot Arid Climate, Kuwait City, Kuwait, 3–4 April 2017; Yassin, M.F., Ed.; Kuwait Institute for Scientific Research: Kuwait City, Kuwait, 2017; pp. 113–121.

36. Pamonpol, K.; Areerob, T.; Prueksakorn, K. Indoor Air Quality Improvement by Simple Ventilated Practice and Sansevieria Trifasciata. *Atmosphere* **2020**, *11*, 271. [CrossRef]

37. Abdallah, A.S.H. New passive cooling as a technique for occupant comfort and indoor air quality in hot arid climate, Egypt. In Proceedings of the Workshop on Indoor Air Quality in Hot Arid Climate, Kuwait City, Kuwait, 3–4 April 2017; Yassin, M.F., Ed.; Kuwait Institute for Scientific Research: Kuwait City, Kuwait, 2017; pp. 162–168.

38. Heinonen, J.; Junnila, S. Implications of urban structure on carbon consumption in metropolitan areas. *Res. Lett.* **2011**, *6*, 014018. [CrossRef]

39. Chel, A.; Kaushik, G. Renewable energy technologies for sustainable development of energy efficient building. *Alex. Eng. J.* **2018**, *57*, 655–669. [CrossRef]

40. Albatici, R.; Passerini, F. Bioclimatic design of buildings considering heating requirements in Italian climatic conditions. A simplified approach. *Build. Environ.* **2011**, *46*, 1624–1631. [CrossRef]

 *sustainability*

Article

Towards a Territorially Just Climate Transition—Assessing the Swedish EU Territorial Just Transition Plan Development Process

John Moodie *, **Carlos Tapia**, **Linnea Löfving, Nora Sánchez Gassen and Elin Cedergren**

Nordregio, SE-111 86 Stockholm, Sweden; carlos.tapia@nordregio.org (C.T.); linnea.lofving@nordregio.org (L.L.); nora.sanchezgassen@nordregio.org (N.S.G.); elin.cedergren@nordregio.org (E.C.)
* Correspondence: john.moodie@nordregio.org

Abstract: The move towards a climate neutral economy and society requires policymakers and practitioners to carefully consider the core technical, social, and spatial dimensions of a just transition. This paper closely examines the processes undertaken during the development of EU Territorial Just Transition Plans (TJTPs) for the three Swedish regions of Gotland, Norrbotten, and Västra Götaland. The aim is to establish whether the content and actions outlined in the TJTPs were driven by the technical, social, or spatial dimensions of a just transition. The analysis is primarily based on a socio-economic and governance impact assessment conducted in each region as part of the TJTP formulation process. These data are also supported by observations of the TJTP development process by the article authors who were part of the team put together by DG Reform to work with the preparation of the TJTPs. The paper finds that the TJTPs development process was largely driven by technical considerations, rather than spatial and socio-economic issues. This indicates that a more open and inclusive place-based territorial approach to climate transition policy formulation and implementation is required. A balance between the technical, social, and spatial elements of a just transition is needed if policies are going to meet the requirements of local and regional citizens and provide sustainable socio-economic growth and environmental protection, without risks of delocalizing energy-intensive processes to other regions.

Keywords: climate transitions; climate justice; regional development; territorial plans; place-based policy

Citation: Moodie, J.; Tapia, C.; Löfving, L.; Gassen, N.S.; Cedergren, E. Towards a Territorially Just Climate Transition—Assessing the Swedish EU Territorial Just Transition Plan Development Process. *Sustainability* **2021**, *13*, 7505. https://doi.org/10.3390/su13137505

Academic Editors: Georgios Tsantopoulos and Evangelia Karasmanaki

Received: 27 May 2021
Accepted: 1 July 2021
Published: 5 July 2021

Publisher's Note: MDPI stays neutral with regard to jurisdictional claims in published maps and institutional affiliations.

1. Introduction

Climate transitions will challenge our economies and societies during the years to come. These transformations provide a great opportunity to avert climate risks and define long-term sustainable socio-economic and environmental development pathways; however, before the benefits of systemic climate transition can be fully realized, societies will have to deal with the short-term socio-economic and governance-related impacts caused by these changes. The impacts of climate transformations will not be neutral from a territorial perspective. On the contrary, due to agglomeration economies and regional specialization trajectories over the past decades, the impacts of climate transitions will be highly concentrated in specific countries and regional areas across Europe and the world [1].

The concept of the 'just transition' is starting to gain momentum as it lays at the heart of the European Union's (EU) Green Deal [2] and the United Nations (UN) Agenda 2030 and Sustainable Development Goals [3]. While the just transition is an amorphous concept defined and interpreted differently by diverse actors, an overview of the literature on this topic highlights three core conceptual dimensions. Firstly, the technical dimension relates to the shift towards climate neutral carbon free technologies; secondly, the social justice dimension focuses on citizen involvement in the transition process, preserving jobs and protecting the most vulnerable in society from the potentially damaging socio-economic impacts of climate policies; and thirdly, the spatial dimension aims to ensure that transition policies are based on territorial specificities that meet the needs of local and regional citizens [4].

The EU fully recognizes that all three dimensions need to be assessed in relation to the development and implementation of just transition climate processes and policies [1]. The EU's Just Transition Mechanism, under the European Green Deal, is committed to identifying and supporting Europe's most affected regions to cope with the technical, social, and territorial specific impacts of the transition to a low carbon economy, focusing on those regions and sectors highly dependent on fossil fuels and energy-intensive processes. A core instrument of the Just Transition Mechanism is the Just Transition Fund, which aims to facilitate socio-economic diversification in regions most affected by climate transition. EU member state access to the Just Transition Fund, implemented within the remit of EU Cohesion Policy, is conditional on the development of Territorial Just Transition Plans (TJTPs) [1].

TJTPs outline regional climate transition processes up until 2030. They identify which EU member state territories should be supported, the main climate-related challenges they face, and the targeted socio-economic and environmental actions and governance mechanisms needed to help meet the threats and opportunities posed by the transition. Central for the European Commission is that the TJTPs are developed and implemented through open and inclusive processes, based on local and regional knowledge and expertise to ensure that climate transition policies meet the needs of citizens and leave no one behind [1].

In Sweden, the draft TJTPs have been developed in close collaboration between national government representatives, national agencies, regional and municipal authorities, and key sectoral actors. The European Commission and Swedish government identified four counties eligible for the Just Transition Fund based on their dependence on carbon intensive industries, namely, Gotland county, Norrbotten county, Västerbotten county, and Västra Götaland county [5,6]. Based on the EU Commission's and the Swedish Government's assessment, the implementation of the fund in Sweden is to be concentrated on those regions where the most greenhouse emissions-heavy industries are located [7]. The TJTP development process took place between October 2020 and January 2021 and the regional TJTP drafts developed are now in the process of being reviewed and validated by the Swedish government and EU.

Nordregio's research team was tasked by DG Reform to directly assist and support national and regional actors in the preparation of the Swedish TJTPs. As part of this process, Nordregio conducted a territorial socio-economic and governance impact analysis within three of the four eligible Swedish regions, Gotland, Norrbotten, and Västra Götaland. The Swedish Agency for Economic and Regional Growth (Tillväxtverket) conducted the analysis for the fourth eligible region, Västerbotten, and the authors of this paper were not involved in this process. Therefore, Västerbotten is not part of this article.

The impact analysis of the three Swedish regions was based on a mixed quantitative and qualitative research approach, including: (1) a socio-economic impact analysis using a combination of socio-demographic and regional economic assessment methods, including a classic economic base and input-output analyses; (2) a desk-based review of primary national and regional climate strategy documents and transition roadmaps developed by regional industries; and (3) semi-structured interviews conducted with national and regional policymakers, public officials, and key sectoral actors. The empirical results presented in Section 6 are based on these analyses.

This paper aims to identify the main technical, social, and spatial challenges each region faces in relation to the green transition and establishing which of the core just transition dimensions informed the overall direction of the contents of the TJTPs. The analysis is primarily based on the socio-economic and governance impact assessment conducted in each region as part of the TJTP formulation process. This information is supported by direct observations made by the authors, who were directly involved in the TJTP formulation process.

The paper is structured as follows. The first section examines the evolution and key features of the just transition concept, closely examining their central technical, social, and spatial dimensions. This is followed by a discussion of how these different dimensions are interpreted and reflected on within key EU and Swedish climate transition policy documents. The next section provides an empirical overview of the territorial analysis conducted to inform the development of TJTPs for the three Swedish regions of Gotland, Norrbotten, and Västra Götaland, focusing on the main economic, social, and governance challenges presented by the transition. The discussion section reflects on these empirical findings, particularly to what extent the final TJTPs were driven by technical, social, and spatial dimensions of a just transition. We conclude by providing some reflections on the future direction of climate transition policies, in particular the need to take an open and inclusive place-based territorial approach to transition policy formulation and implementation, and the importance of policies striking a balance between the technical and spatial and social justice elements embedded in the notion of just transition towards climate neutrality.

2. Defining the Just Transition

The origins of the term 'just transition' can be traced back to the United States during the 1970s when the leader of the American Oil, Chemical, and Atomic Workers Union, Tony Mazzacchi, encouraged the labor unions and national government to engage in peacetime planning to support wartime workers at risk of losing their jobs due to disarmament. Later, in the 1980s, Mazzacchi argued that a "super fund for workers" was required to provide financial and education support for blue-collar workers at risk of unemployment in industries threatened by new environmental legislations [8]. In the 1990s, the 'super fund' was retroactively described as a 'just transition fund' and the term was officially endorsed by various North American labor organizations. In the 21st century, the concept of just transition has become synonymous with the global policy discourse around the need to combat climate change and the shift towards a climate resilient low-carbon economy [9].

The term 'just transition' has been defined and applied differently depending on the context in which it is being used in and who is using it. The just transition concept has been used as a framework by the International Trade Union Confederation (ITUC) to strike a balance between meeting climate and environmental objectives, while protecting and equipping workers whose jobs, livelihoods, and communities are most at risk from climate change or climate interventions [10]. This is confirmed by Rosemberg [11] who notes that the just transition "can be understood as the conceptual framework in which the labour movement captures the complexities of the transition towards a low-carbon and climate-resilient economy, highlighting public policy needs and aiming to maximize benefits and minimize hardships for workers and their communities in this transformation".

In a report by the 'Just transition initiative', it is stated that the methods for achieving just transitions are unclear and that the outcome is dependent on how the transition is implemented in terms of scale, context, and time [9]. The report also addresses the importance of understanding the core principles of the just transition, while at the same time accepting the range of definitions held among stakeholders. While different actors define the just transition differently, including international organizations, unions, environmental lobbyists, and civil society groups, there is broad agreement that its implementation depends on more equitable and inclusive governance processes, policies, and investments.

The concept 'just transition' builds on two related concepts, namely social justice and socio-technical transitions. The term 'transition' was applied and developed further by Frank Geels in his 2002 paper on innovation and climate adaptation [12]. In this paper, Geels introduced the notion of 'sociotechnical transitions' and explained why and how these occur. Geels argues that sociotechnical transitions are an outcome of recurrent innovation and technology substitution processes, such as those required to move from a fossil dependency to climate neutrality. Geels explains that technical transitions occur as the outcome of linkages between developments at multiple levels (multilevel perspective). Here, he distinguishes between 'sociotechnical regimes' and 'sociotechnical landscape', the former being a group of key stakeholders and the latter an asset of heterogeneous exogeneous and endogenous factors at play. These include determinants like oil prices, economic growth, conflicts, migration, broad political coalitions, cultural and normative values, environmental problems, or climate change. According to Geels, radical innovations are developed when ongoing processes at the levels of regime and landscape create a 'window of opportunity' [12]. These windows may be created by tensions in the sociotechnical regimes or by shifts in the landscape which put pressure on the regime. In the case of the just transition, we can interpret climate change as the sociotechnical landscape, which has put pressure on businesses, stakeholders, and institutions (the sociotechnical regimes) to make changes.

The term 'just' in just transitions relates to the concept of social justice developed by philosophers and thinkers, including Rawls, Locke, Rousseau, and Kant [13]. This literature identifies a tension between two important paradigms of social justice, namely the distributional and procedural components of justice. These two paradigms are not considered as opposites but are both required to uphold justice. The distributional paradigm relates to the equal distribution of goods, services, and opportunities, as well as burdens [14], while the procedural paradigm revolves around just institutions and procedures, focusing on the extent to which individual and organizational actors are able to meaningfully participate in the decision making process [15]. Just procedures are necessary, but not sufficient for the fairness of the outcome, while attention to the outcome may mask the injustices of the process. Social justice is, therefore, a normative concept which implies that society and the state should strive towards achieving social justice for all citizens. Recognitional considerations form an important part of the procedural dimension with an emphasis on the role of disadvantaged and vulnerable minority groups whose voices are often left unheard in decision making processes, including policy making [4].

Both the 'socio-economic' and 'just' elements are clearly reflected in the just transition guidelines developed by the ITUC and International Labor Organization (ILO) [16]. These outline the need for high levels of investment in low-carbon technologies; a social dialogue and consultations between policymakers and affected groups; proactive labor market policies based on social protection and worker rights, including financial support and retraining opportunities for the unemployed; early-stage research into the potential socio-economic impacts of climate policies; and the development of local economic diversification plans [10]. The guidelines also make a distinction between social investment policies (e.g., active labor market policies, training and re-skilling policies to increase workers' employability in a greener economy) and social protection policies (e.g., unemployment and minimum income benefits) and mean that both are needed in a just transition framework.

The local and regional implications of the climate transition also highlight the need for a spatial dimension to justice. Spatial justice emphasizes how macro forces can cause local injustices, which resonates with how some regions or municipalities dependent on industries with high emissions will be unevenly impacted by the climate transition. The spatial element of the just transition has been a focal point for international policymakers who have highlighted the importance of regional context and place-based responses in the development and implementation of spatially just transition plans. The different dimensions of the just transition in relation to EU policy are discussed more in the following section.

3. The EU Just Transition—A Territorial Specific Approach

The EU has a long history of supporting and guiding regions in industrial transition. The original European Coal and Steel Community provided technical and economic assistance for different regions to undergo transitions or reconstructions of industries. More recently, the 2017 Initiative for Coal Regions in Transition focused on the transition of European coal regions, which included the development of the EU Just Transition Platform to facilitate the exchange of best practices and discuss strategies and projects with the potential to kick-start the transition process in declining coal regions [17].

In 2019, the European Commission adopted the European Green Deal [2]. The strategy states that the European Union should have no net emissions of greenhouse gases in 2050 to reach the Paris Agreement's goal of holding warming "well below" 2 °C and pursuing efforts to keep warming below 1.5 °C. To achieve this goal and ensure that 'no person or place is left behind' as a consequence of climate transitions, the Just Transition Mechanism (JTM) was introduced [18]. Its main objective is to ensure that the transition takes place in an effective and fair manner. The JTM is split into three main pillars. Pillar one, the Just Transition Fund, provides financial support to those regions and sectors most dependent on fossil fuels, and thereby also more affected by the transition. Pillar two, the InvestEU Just Transition Scheme, provides budgetary support for private sector investments that support transition. Finally, pillar three, the Public Loan Facility, supports public sector transition investments.

The Commission notes that support will be available to all member states, focused on regions that are the most carbon-intensive and people and citizens most vulnerable to the transition. The JTM will protect them by:

- supporting the transition to low-carbon and climate-resilient activities;
- creating new jobs in the green economy;
- offering re-skilling opportunities;
- investing in public and sustainable transport;
- providing technical assistance;
- investing in renewable energy sources;
- improving digital connectivity;
- providing affordable loans to local public authorities; and
- improving energy infrastructure, district heating, and transportation networks [1].

The aims and objectives outlined by the JTM indicate an attempt to balance the socio-technical and social justice related elements of the just transition. The socio-technical elements are reflected in the EU's commitment to investing in carbon-neutral and environmentally friendly production technologies, transport, renewable energies, and digitalization, whereas the social elements are highlighted in the worker protection rights and reskilling opportunities offered. In a key passage within a communication document on the EU Green Deal, the Commission further outlines the core social elements of the just transition:

"This transition must be just and inclusive. It must put people first, and pay attention to the regions, industries and workers who will face the greatest challenges. Since it will bring substantial change, active public participation and confidence in the transition is paramount if policies are to work and be accepted. A new pact is needed to bring together citizens in all their diversity, with national, regional, local authorities, civil society and industry working closely with the EU's institutions and consultative bodies." [2]

In this excerpt, there is a strong emphasis on the social justice element of the transition, particularly in relation to the need for open and inclusive social dialogue with citizens to promote the trust and acceptance of policies.

Sabato and Fronteddu [19] argue that the understanding of the just transition in the European Green Deal promotes social investment policies (e.g., active labor market policies, training and re-skilling policies to increase workers' employability in a greener economy) in favor of the need to ensure the protection of citizens through traditional social protection policies (e.g., unemployment and minimum income benefits). [1] They argue that EU's policy approach to the transition is not consistent with the Guidelines of the ILO [16] which emphasize the need to place territorial/sectorial policies and social investment policies within a strong social protection system guaranteeing social rights to all citizens. As will be discussed further, the Swedish welfare system is relatively strong in comparison to many other countries which means that social investment policies to some extent will be placed in a social protection system, as guided by the ILO. However, this will not be the case for all member states. That said, the Communication of the European Green Deal refers to the European Pillar of Social Rights principles as a reference framework to promote equal opportunities and access to the labor market; fair working conditions; and social protection and inclusion [20], which emphasize a broader objective of the promotion of social rights, even though the practical implications of this are unclear.

Other priorities are implicit and embedded in the concept of just transition, as formulated in the JTM. The spatial and territorial dimensions of the just transition is one of those embedded principles. This priority is regularly advocated within the EU transition policy documents, with the Commission highlighting the important role of regions and cities in guiding just transition processes: "Citizens, depending on their social and geographic circumstances, will be affected in different ways. Not all Member States, regions and cities start the transition from the same point or have the same capacity to respond. These challenges require a strong policy response at all levels." [2]. The territorial focus of the EU's just transition plans is in keeping with the Commission's recent emphasis on territorial governance and place-based policymaking, such as the development of regional smart specialization strategies, in addition to applying the concept of 'active subsidiarity' which advocates a central role of regions and cities in EU policy formulation and implementation [21].

The region-specific focus also links climate transition to the concept of 'regional resilience' of local communities and regions [22]. Regional resilience refers to a set of regional and local economic, social, and institutional traits that characterize the ability of regions to respond to a shock and maintain system stability and durability, as well as adapt to structural changes and move to new development pathways [23]. Although the carbon transition is often closely linked to more or less 'spontaneous' technological changes and innovations, it can be considered a policy-induced shock to regions, with new rules putting pressure on policymakers, industries, businesses, workers, and citizens alike (e.g., new tax regimes, shift on investments, trade deals, revised regulations and laws, or as in the case of the JTM, supportive investments and incentives). Many variables are identified as important to endure an economic shock, such as the existing economic path of a region, regional economic structures, resources, capabilities, and competences. It can also be business cultures and any supportive measures implemented by different institutions at national and subnational levels (e.g., welfare policies and programs). The OECD has identified four areas that drive regional resilience, including clear leadership and management; strategic and integrated approaches; public sector skills; and open and transparent governments [24]. This highlights the importance of regional governance processes and actors in formulating and implementing a just transition.

The significant role of regional structures and actors is most visible in the Commission proposals for member states to develop TJTPs in regions most affected by climate issues. The TJTPs are to be developed through a dialogue between the European Commission, national government representatives, and regional actors. As the European Commission notes, "these plans set out the challenges in each territory, as well as the development needs and objectives to be met by 2030" [1]. They identify the types of operations envisaged and specify governance mechanisms. The approval of the TJTPs opens the doors to dedicated financing under the pillars of the JTM. EU member states are currently in the process of developing and ratifying their TJTP draft proposals. The following sections examine and analyze the process of developing TJTPs in three Swedish regions, focusing on the extent to which the TJTPs developed were driven by technical, social, and territorial place-based dimensions which meet the needs and requirements of local stakeholders and citizens.

4. Research Methods

The following analysis of the processes undertaken to develop TJTPs in three Swedish regions is primarily based on an economic, social, and governance impact assessment conducted by Nordregio researchers at the request of DG Reform and the Swedish government. This assessment was conducted using a combination of qualitative and quantitative research methods. The three analytical strands presented below were performed on each of the three regions in scope.

A socio-economic analysis included a regional economic base analysis and an impact assessment. The economic base analysis was conducted to characterize regional economies and their evolution prior to the adoption of the TJTP. This included the calculation of location quotients for selected sectors and a shift-share analysis focusing on employment patterns in the industries under investigation. Such empirical analyses were performed using regional statistics on economic production (in value added) and employment (in full-time equivalents) provided by Statistics Sweden and the regional statistical offices. The analysis of potential socio-economic impacts linked to the decarbonization of the industries under scrutiny were evaluated by means of an input-output analysis. This assessment was performed on the symmetric input-output tables (SIOT) for year 2018 provided by Eurostat. The regional employment and value-added effects were estimated through regionalized input-output coefficients. The potential impacts were modeled under worst-case scenarios that assumed a discontinuation of the activities of the major industrial emitters in each region.

A social analysis looked at demographic dynamics, including aging processes, migration trends, population projections, and statistics related to educational outcomes in the different counties. Particular attention was put on the capacity of regional economies to attract and retain trained workforce. Gender aspects were considered both from the demographic balance as well as from labor segregation perspectives.

A governance assessment was based on a desk-based examination of primary EU, national, and regional level climate strategy documents, including climate roadmaps developed by key sectoral actors within the three Swedish regions. The data and information were supplemented by authors' observations of the TJTP development process. Authors formed part of the team assembled by DG Reform and Swedish government representatives to produce first drafts of the TJTPs. As part of the process, the Nordregio research team had first-hand access and involvement in meetings in which they could observe discussions between key stakeholders including government representatives, national agencies, and regional and local public authorities. Finally, semi-structured interviews with national and regional level actors were also performed which allowed for more detailed discussion in relation to the main economic-, social-, and governance-related issues posed by climate transition policies.

5. The Swedish Climate Policy Framework

In 2017, the Swedish Parliament adopted a climate policy framework outlining Sweden's approach for complying with the Paris Agreement. The framework sets ambitious targets for climate and energy and goes further than the EU's 2050 climate neutrality foci and current energy and climate objectives for 2030. Sweden has committed to reducing all net emissions of greenhouse gas (GHG) into the atmosphere to zero by 2045 and using 100 percent renewable energy in 2040. The Climate Policy Framework also includes a Climate Act which regulates the government's climate policy work, including its overall aims and how they should be implemented. The Framework is based on the development of a climate policy action plan every four years [5]. This plan should demonstrate how the government's overall policies in all relevant spending areas contribute to achieving the 2030 and 2040 milestones and the long-term emissions target by 2045. A Climate Policy Council has also been established as an authority in the form of an independent interdisciplinary expert body that is tasked with evaluating how the government's overall policy is compatible with the climate objectives decided by the parliament and government.

The term 'just transition' is rarely used within Swedish national and regional climate and energy strategies, except for documents connected to the European Just Transition Mechanism. The terms 'climate transition' and 'green transition' are instead commonly used, often linked to sustainable development and the UN's Agenda 2030. There are very few public policies and policy documents exploring the socio-economic impacts of the green transition in Sweden. However, many of the goals and objectives outlined within the climate framework and other initiatives focus on the technical elements required to support the transition. This includes financial support to improve energy efficiency, financial support for industrial companies to reduce their emissions through technical advances, a digitalization strategy, and a strategy for a circular economy. The Swedish Trade Union Confederation (LO) has argued that the just transition in Sweden has mainly focused on technological questions, and it is important that transition plans reflect the needs of both industry and workers affected by climate policies [25].

In response, the undersecretary to the Minster for Climate and Environment in Sweden highlighted in 2020 that the ambitious Swedish climate policies need to develop in parallel with social justice, social security, equity, and gender equality [25]. In June 2020, the Ministry for Foreign Affairs also published a report that operationalized the "leave no one behind" principle from the UNs Agenda 2030 in Sweden. The report stresses seven main messages: realizing human rights and gender equality; strengthening empowerment and participation; advancing the transition towards resource-efficient, resilient, and climate-neutral economies; promoting multidimensional poverty reduction; promoting social dialogue and decent work; progressively realizing universal social protection; and improving data and monitoring. The report states that "special attention must be paid to the social and gender dimension of the transition, in order to ensure that no one, particularly people living in poverty, is left behind when society implements measures to become climate-neutral" [26].

The work to reach the environmental objectives has been designed as a concerted effort across the whole of society, including public agencies, business communities, stakeholder organizations, and, not least, individual citizens. Many regions, County Administrative Boards, and municipalities have also developed climate and energy strategies [27–29]. As part of Fossil-Free Sweden, 22 sectors have developed their own road maps for fossil-free competitiveness, including the sectors in focus for the JTF, the cement industry, the mining and minerals industry, the steel industry, and the petroleum industry. Regional and local authorities play a central role in the implementation of climate and energy transition plans in Sweden [30]. The implementation of the objectives outlined in the Swedish Climate Framework draws on multi-level governance processes with a central role for regional public authorities and industries with territorial knowledge and expertise. Table 1 provides an overview of the key administrative levels involved in the development and realization of the TJTPs in Sweden, along with their formal tasks.

Table 1. Overview of the key administrative levels in the process outlining the TJTPs.

European	European Commission **Support for the development of the TJTPs is provided by DG REFORM.**

National	**National authorities** The national authorities are responsible for implementing the decisions made by the Parliament and the Government. **Role in TJTP process:** An authority group has been established to lead and coordinate the process, align priorities as well as to assist with expertise and data. These consist of: • Swedish Agency for Economic and Regional Growth (managing authority) • Swedish Public Employment Service • Swedish Energy Agency • Swedish Environmental Protection Agency

County Administrative Boards
National authorities operating at county level, responsible for the state administration in the county (in those areas where no other authority is responsible for special administrative tasks). The County Administrative Boards shall work to ensure that national goals have an impact in the county, while also taking into account regional conditions.

Role in TJTP process: The county administrative boards have the mission to promote, coordinate, and lead the regional work in the implementation of the government's policy regarding energy conversion and reduced climate impact with a long-term perspective. Together with the regions, they have important roles in the implementation of the TJTP. These have participated in meetings at regional level and provided input during the work.

County Administrative board of Norrbotten	County Administrative board of Västra Götaland	County Administrative board of Gotland

Regional/ County	**Regional Authorities** Led by political assemblies with responsibilities for, e.g., health care and social care, public transportation, and regional development. Counties and regions cover the same geographical area. **Role in TJTP process:** The regions have the responsibility for regional development and thus a key role in the implementation of the plan. The region has an active role in moving the regional economy from fossil dependence to a sustainable society and an especially important role in linking sectors as well as the regional and national level in the innovation system to create the necessary triple-helix collaborations.

Region Norrbotten	Region Västra Götaland	Region Gotland

Local	**Local authorities** Local government responsible for, e.g., primary and secondary school, preschool activities, elderly care, physical and comprehensive planning, roads, water and sewage issues, and energy issues. They also issue different types of permits, such as building permits. **Role in TJTP process:** Municipalities play an important role in Sweden's climate work. Due to the proximity to the citizens, their roles for spatial planning and as large employers are the significant climate actors in the work towards set climate goals. The municipalities drive local development in collaboration with companies, organizations, and residents, and will therefore play a significant role in the implementation of the TJTP. The municipalities participated in the drafting of the TJTPs through a written consultation process on the plans and related outputs.

• Gällivare municipality • Öxelösund municipality • Luleå municipality	• Lysekil municipality • Stenugnssund municipality	• Gotland Municipality

Source: Government Offices of Sweden, 2020; TJTP Gotland; TJTP Norrbotten; TJTP Västra Götaland.

6. Developing EU Territorial Just Transition Plans in Sweden

The industries and counties that are proposed to be covered by the EU's new climate fund are the steel industry in Norrbotten, the cement industry on Gotland, refineries and the petro-chemical industry in Västra Götaland, and the metal industry in Västerbotten. These counties and industries have been selected as they contribute to the largest shares of carbon emissions in Sweden. In absolute terms, Gotland, Norrbotten, and Västra Götaland are accountable for around 19.8 million metric tons of carbon dioxide equivalents (MMTCDE). This represents a third (34.7 percent) of total fossil-based greenhouse gas (GHG) emissions in Sweden [31]. The largest GHG contributor is Västra Götaland (11.9 MMTCDE; 20.8 percent of total GHG emissions), followed by Norrbotten (5.2 MMTCDE; 9.2 percent of total GHG emissions) and Gotland (2.7 MMTCDE; 4.8 percent of total GHG emissions). In 2018, the two Swedish counties with the highest emission intensities in terms of fossil-based GHG emissions per unit of Gross Regional Domestic Product (GRDP) were Gotland and Norrbotten. Västra Götaland ranked sixth in the list. Even if Gotland in economic and employment terms represents a small share of Sweden's economy (0.5 percent and 0.6 percent, respectively), it contributes to almost 5 percent of total GHG emissions at national level. As a result, Gotland is the most carbon-intensive county in Sweden [31].

The diagnostic work for developing the TJTPs in the four selected Swedish regions is taking place in dialogue with different actors within existing collaborative structures across multiple levels of governance. This work is being conducted in dialogue with different actors within existing collaborative structures across multiple levels of governance. The Swedish Agency for Economic and Regional Growth (Tillväxtverket) is tasked by the government to ensure preparations for the Fund. The government has identified the Swedish Public Employment Service, the Swedish Energy Agency, and the Swedish Environmental Protection Agency as especially involved agencies in the preparations. These government agencies have worked in close collaboration with the County Administrative Boards, the regions, and municipalities in the preparation of the TJTPs [7]. Local public authorities have in turn engaged in dialogue with the public and private sector, academia, and other interested parties. The plans are also drafted in dialogue with key industry stakeholders such as Cementa AB and the HeidelbergCement Group representing the cement industry on Gotland; SSAB, Lulekraft AB, and LKAB Kiruna from the HYBRIT initiative in the Norrbotten county; as well as the chemical and refining industries, including Preem AB, Borealis AB, and ST1 in the Västra Götaland county. Other key sectoral agencies, stakeholders, and actors have been an important part of formulating the plans. These include Vattenfall and Svenska Kraftnät in the energy sector, higher education institutions, academia, and other interested civil society and labor organizations.

The TJTPs for each transition region have the same structure, including the following seven core elements; one, an overview of the national climate policy framework, highlighting key national climate targets; two, a national level assessment of the territorial impacts of climate change, identifying which regions within the country contribute the most to gas and carbon emissions; three, an assessment of the potential economic, social, and environmental impacts of the transition on the selected region; four, an analysis of the development needs required in each region to meet climate targets; five, an outline of the main type of actions planned to deliver the regional transition; six, performance-related output and results indicators are highlighted for measuring the impact of the proposed actions; and seven, the governance structured and key stakeholders needed to develop and implement the plans are outlined. The proposed regional just transition fund support actions are shown in Table 2. The outlined actions must contribute to a positive development regarding gender equality, integration and diversity, the environment, and the living conditions of young people. These horizontal criteria form the basis for the Just Transition Fund support efforts, as outlined in the plans.

Table 2. Proposed actions in the Swedish Territorial Just Transition Plans.

Type of Action	Gotland	Norrbotten	Västra Götaland
Investments for the use of clean energy technologies and infrastructure, reduction of greenhouse gas emissions, energy efficiency, and renewable energy.	Support the cement industry with fuels substitution to waste-based and bio-based fuels as well as new cement grades and materials in the production. Investments in improving infrastructures for a flexible and robust energy system in the island and the connections to the mainland.	Transition to carbon neutral steel production (support to EU ETS steel industry facilities in Norrbotten).	Transition to the production and use of green hydrogen in production processes in the refinery and petro-chemical industry; increasing production capacity for biofuels to reduce CO_2 emissions.
Investments in research and innovation and promotion of advanced technology transfer.	RD&I for developing efficient and commercially available technology for CCS, alternative fuels and electrification, as well as new cement grades and materials in the production	Innovation for the production of innovation-critical raw materials and materials necessary for a transition to a fossil-free society; RD&I for large-scale energy storage and development and implementation of fossil-free technologies and other alternative energy carriers and raw materials.	Support for mapping Västra Götaland's ability to form a hydrogen cluster; RD&I for new raw materials and secondary materials, including waste streams, for biofuel production as well as separation, use and storage of carbon dioxide (CCS/CCU).
Skills upgrading and retraining of employees.	-	Mapping of the steel industry's skills needs; support for networks and clusters for skills-enhancing initiatives in the steel industry and its value chain, retraining and skills development of existing and new workforce, skills validation measures and strengthen the companies' strategic work with skills issues.	-
Investments to promote the circular economy, including through measures to prevent and reduce waste, resource efficiency, reuse and recycling.	Minimize waste and increase recycling in the cement production.	Efforts to promote the use of recycled materials as a raw material; support for environmentally friendly production processes and resource efficiency.	Improve the use of resources and recycling in refineries and the chemical industry; re-use of waste oils (refineries) and chemical recycling of plastics (polyethylene) for chemical industry.

Source: Swedish TJTPs for Gotland, Norrbotten, and Västra Götaland.

Draft versions of the TJTPs were sent out between November 2020 (Gotland and Norbotten) and February 2021 (Västra Götaland and Västerbotten) for collection of comments from relevant authorities, companies, organizations, and municipalities. After compiling the comments from the consultation, revision, and completion of the program proposals from the Agency for Regional and Economic Growth and the Ministry of Enterprise and Industry, the final proposals were presented to the Government Offices for validation in March 2021. The following sub-sections provide an overview of the key findings from economic, social, and governance analysis within the TJTPs, focusing on the main commonalities and differences from across the three selected Swedish transition regions, Gotland, Norbotten, and Västra Götaland.

6.1. Economic Analysis

Discussions surrounding the development of the regional EU TJTPs revolved largely around the technical dimension of the shift towards climate neutrality. The focus is on the degree of maturity and feasibility of specific carbon-free technologies, vis-a-vis the socio-economic costs of the transformation. The Swedish national government is committed to devoting the EU Just Transition Fund to the development of climate-neutral processes by existing regional fossil-dependent industries. This vision emerges in the proposed actions within each TJTP, which are skewed towards the technical elements of climate transitions (Table 2). Most of the interventions target specific industrial actors and some have been even designed at plant level. These priorities are also reflected in regional energy and climate policy documents. The primary focus on the technical dimension of the transition is driven by the view that EU transition funding should cover the economic costs of technological change. This might derive from the perception that the Swedish welfare state and other funding sources have the capacity to address territorial socio-economic challenges caused by the transition.

Large fossil fuel-based industries make up an important part of the economic profile of the selected Swedish just transition regions. While these primary industries represent a comparatively modest share of employment and gross value added (GVA), their relevance in specific localities and their contribution to overall regional economic growth is substantial. Moreover, the affected industries provide products with high strategic value, including cement, steel, and basic chemicals. Most of these commodities still have limited substitutability by fossil-free alternatives. This explains why national and regional policymakers are committed to achieving the technical dimension of transition policies. A failure to support these industries in the shift to climate neutral technologies would have major economic consequences at both national and regional levels.

This situation holds particularly in Norrbotten, where the manufacture of basic metals and fabricated metal products sector (grouped in categories 24–25, according to Eurostat's Statistical classification of economic activities, NACE Rev. 2) is the sixth largest employer [32]. In total, 2610 persons worked in this sector in 2018. Still, since the iron ore used by local steel plants is sourced locally, steel processing is closely connected to local mining. Combined, the mining and steel sectors clearly dominate the regional economy. In 2018, 6900 persons were directly employed in these industries, which corresponds to 10.7 percent of the total number of jobs in Norrbotten. Employment in both sectors is often concentrated in a small number of large mines and plants. In Kiruna, for instance, around 2175 workers are hired by the state-owned company LKAB, a producer of iron ore, corresponding to 17.5 percent of local jobs [33]. In Gällivare, the same company employs 1175 people (13.5 percent of all employees in the municipality) [34]. Between 2010 and 2018, the manufacture of basic metals and fabricated metals sectors in Norrbotten suffered a decline in employment (436 jobs lost) that contrasts with a period of relatively high economic dynamism in the region and in Sweden as a whole. Still, in comparative terms, the sector performed better in Norrbotten than in other Swedish regions. This suggests that this county provides good conditions for this sector that still account for a very relevant share of jobs in some localities. In Luleå, for example, the SSAB EMEA AB steel plant employs 1325 persons, corresponding to 1.2 percent of all jobs in the region [35]. Based on the employment multipliers from our input-output analysis (Table 3), if the activity of this plant was discontinued as a response to stringent climate policies and regulations, a total of 1965 full-time equivalent (FTE) jobs might be lost in Norrbotten. This includes the initial jobs lost in the SSAB EMEA AB plant itself, as well as 640 indirect jobs lost in the region because of backward linkages in the steel value chain.

Table 3. Employment indicators and multipliers for selected economic sectors in Gotland, Norrbotten, and Västra Götaland (2018).

Indicator	Gotland: Other Non-Metallic Mineral Products (CPA_23)	Västra Götaland: Coke and Refined Petroleum Products (CPA_C19), Chemicals and Chemical Products (CPA_C20), and Basic Pharmaceutical Products and Pharmaceutical Preparations (CPA_C21)	Norrbotten: Basic Metals (CPA_C24) and Fabricated Metal Products, Except Machinery and Equipment (CPA_C25)
Initial employment effect	0.329	0.119	0.308
First-round employment effect	0.051	0.057	0.104
Industrial support employment effect	0.017	0.210	0.045
Production-induced employment effect	0.067	0.034	0.149
Simple employment multiplier	0.396	0.092	0.457
Type 1A employment multiplier	1.154	1.481	1.339
Type 1B employment multiplier	**1.204**	**1.769**	**1.483**

Note: Employment multipliers were calculated by means of input-output analyses on the symmetric input-output tables provided by Eurostat (naio_10_cp1700). National input-output coefficients were calculated by direct allocation of competing imports. Input-output coefficients were regionalized by applying Flegg's location quotient (FLQ), method proposed by Flegg, Webber, and Elliott (1995). Even if several methods can be used to regionalize the input-output coefficients, it is now well established that the FLQ method can give more precise results than alternative approaches, like SLQ or the CILQ [36,37]. Employment multipliers highlighted in bold represent the extra number of persons employed in all industries in the economy for one extra person employed in the industries under investigation. These multipliers hence summarize the aggregated direct and indirect employment effects for every new job created in the relevant sectors, including initial, first round and industrial support induced output effect. Type 1B employment multipliers were hence used to calculate the employment effects described on the text.

In Gotland, the limestone and non-metallic mineral manufacturing industries are small but important contributors to the island's economy. Even if the number of direct jobs provided by these sectors is rather limited (398 workers in 2018), its relevance for Gotland's economy is considerably greater than for other regions in Sweden. In 2018, these sectors represented 3.2 percent of region's total employment [32]. Based on the location quotients, in employment terms, the industry for non-metallic mineral products is in fact the single most overrepresented economic activity in Gotland's economy. Cement and limestone industries are particularly important for rural Gotland, and particularly the northern part of the island, where the main extractive and processing plants are located. The shift-share analysis performed for 2010–2018 data shows that Gotland's economy has comparatively lower competitiveness levels than other Swedish regions. These conditions also affect the non-metallic manufacturing sector. Even if the sector is performing comparatively better in Gotland than its peers at national level, it is still a very traditional industry with a weaker competitive position in comparison to other sectors and high dependence on national economic dynamics (essentially, its level of output is driven by the demand in the building construction sector). Based on the worst-case scenario that all the 230 FTE jobs in the Cementa AB plant at Slite could be lost as a response to climate-driven decisions, by applying the regionalized employment multipliers for 2018 presented in Table 3, we find that the hypothetical number of direct and indirect FTE jobs lost in the island would be around 277.

Västra Götaland is one of the three largest regional economies in Sweden. The region has an expansive economy and a strong and export-oriented industrial sector. The manufacture of refined petroleum and chemical industry, (NACE sectors 19–20), excluding pharma (NACE 21), represents a small share of the economy in the region. In 2018, these sectors employed 5669 persons (1.5 percent of the total workforce in the region) and, together with the pharmaceutical industry, generated 15,272 million SEK in value added (3.2 percent of region's total GVA). [32] From a territorial perspective, the largest dependence on economic activity in the petrochemical sector concentrates in a restricted number of municipalities in Västra Götaland. Dependence is higher in smaller communities, particularly Lysekil and Stenungsund, where chemical and refining sectors occupy a large proportion of the working population. In the Lysekil municipality, Preem is by far the largest private employer, with 600 people directly engaged by the company [38]. In Stenungsund, it is estimated that the chemical industry cluster provides at least 2350 direct jobs. [39] In the 2010–2018 period, both the regional economy in general and the petrochemical sector showed positive economic trajectories. The refining sector (NACE 19) maintained a stable workforce over the last decade (from 1430 in 2007 to 1611 in 2018). The chemical sector (NACE 20) oscillated between 4500 employees in 2010 and 4058 in 2018 [40]. According to our shift-share analysis, regional competitive factors seem to be driving the overall good employment performance, particularly in the chemical sector. By applying the Type 1B multiplier on employment data for the refining industry provided by the Västra Götaland region [40], we estimated that the direct and indirect employment effects caused by a hypothetical discontinuation of the activities of the local oil refineries might lead to a potential loss of 2850 FTE jobs in the county. This figure represents around 0.6 percent of total employment in Västra Götaland.

6.2. Social Analysis

Contrary to our worst-case scenarios, national and regional policymakers and stakeholders interviewed as part of the TJTP development process stressed that no job losses are anticipated or considered likely as a result of the climate policies introduced in any of the Swedish just transition regions [41]. On the contrary, the transition to a sustainable industrial system could even lead to investments and job creation. This growth potential could present challenges caused by existing socio-economic development trends occurring within the three regions.

Regional labor shortages could pose a significant challenge to climate transition processes due to negative demographic development trends. Population aging is creating a shift in the labor force with the large number of people reaching retirement age leading to higher recruitment needs. The population in Norrbotten is declining; the population size on the county level decreased by around three percent between 2000 and 2020 due to negative population growth and the outmigration of young skilled people [42]. Similarly, in Gotland, the working age population fell from 32,700 persons in 2009 to around 32,100 persons in 2019 [43]. Consequently, both Norrbotten and Gotland today have a smaller labor supply from which companies can recruit employees than ten years ago. These regions are reliant on attracting skilled immigrants to overcome this shortfall, so regional authorities have focused on making these areas more attractive places to live by providing affordable quality housing and better transport accessibility links.

In order to reap the opportunities and benefits presented by the climate transition, access to the right competencies and skills is required. Plans to increase energy efficiency, to electrify industries, to increase the use of biogas, and to develop options for carbon capture and storage require skills that are not always readily available. In Norrbotten, a relatively high proportion of young people, particularly boys, leave school without qualifying for higher secondary education (gymnasiebehörighet). In the school year 2018/2019, around 14 percent of pupils belonged to this group. The proportion of non-qualifying pupils was higher in some smaller rural municipalities such as Pajala and Övertorneå than in the regional centers Luleå and Piteå [44]. Low educational attainments are a challenge for young people who may experience difficulties in finding employment, but they are also a challenge for businesses. Already in 2017, one out of four businesses in Norrbotten stated that they experienced challenges in recruiting people with the right competencies [45], and that this was a major obstacle to development and growth. A positive development in this context is that educational attainment levels in Norrbotten have increased during the last decade. In 2019, 31.6 percent of the adult population had attained post-secondary education.

This proportion is lower than for the Swedish population at national level (37.9 percent in 2019). Nonetheless, it is a substantial change in comparison to 2010, when only 27.3 percent of people in Norrbotten had attended post-secondary education or training [43].

Similarly, in Gotland, 25 percent of employers report difficulties in finding staff, particularly with highly specialized competencies, as educational attainment levels of the population are comparatively low [46]. In Sweden as a whole, more than 38 percent of the population over 16 years have taken part in or concluded post-secondary education. In Gotland, it is only 31 percent [43]. These trends are also reflected in Västra Götaland, where several population groups struggle to enter the labor market due to low levels of education, particularly younger people. Almost one out of five students in the county leave comprehensive schools (grundskola) without qualifying for upper secondary education (gymansieskola). This negatively influences their employment prospects and career options [39]. The proportion of students who do not qualify for higher secondary education has fluctuated during the last years in Västra Götaland and in Sweden as a whole. Västra Götaland has closely followed the Swedish trend, with proportions of non-qualifying students increasing especially rapidly between the years 2015/2016 and 2016/2017. As a result of these trends, all Swedish transition regions need to provide local citizens with access to life-long learning opportunities, vocational training, and reskilling. This requires establishing close links between industries and businesses to ensure that student education courses match the needs of employers.

Women are another target group for sectors affected by the climate transition. The labor markets within core transition industries in Gotland, Norrbotten, and Västra Götaland are gender-segregated to a relatively strong degree. For example, in Gotland, around half of all women work in the public sector, in particular in health care and social care, while around 80 percent of men work in the sectors that are important in the climate transition—including the cement and limestone industry, energy, transport, and the building sector [46]. In Gotland, there are strong educational attainment differences, with men less likely to have obtained higher education than women. This is also the case in Västra Götaland and Norrbotten, where the risk of leaving comprehensive schools without moving on to higher education is more pronounced among young men than among young women. Subsequently, attracting more women to work in these traditional sectors could be one solution to potential labor shortages in these regions [39].

One social group that is particularly vulnerable to climate change and the climate transition in Norrbotten is the Sámi people. Climate change, as well as policies and measures to support the climate transition, affect the context in which the Sámi preserve their unique culture and traditional livelihoods. Changes in temperatures and precipitation influence the conditions for reindeer herding, for instance through changes in food supply for reindeers as well as water and flooding conditions. At the same time, laws and policies to accelerate the phasing out of fossil fuels, such as the increased use of biomaterial and installations of wind turbines or hydropower infrastructure, also have implications for the land and water use of the Sámi people [47]. Due to these potential negative impacts of climate change and climate transition measures on the Sámi's traditional way of life, it is important that their right to be consulted is always respected. In terms of implementing the TJTP for Norrbotten, public authorities have begun the process of presenting and discussing proposals in meetings with the Council of the European Social Fund (ESF Council), Vinnova, and the Geological Survey of Sweden (SGU), and the Sámi Parliament. Dialogue with the Sámi Parliament is particularly significant as climate change has a major impact on the conditions for Sámi culture and land use. Nonetheless, Sámi representatives have argued that policies and legislation to support the climate transition have so far not adequately reflected their interests and that their traditional rights have not been respected. It is increasingly considered that, in order to increase the legitimacy of the climate transition process in Norrbotten, the Sámi community needs to be more strongly involved in decision making processes in the region, particularly in those related to land management issues [48].

6.3. Governance Analysis

Sweden has a decentralized system of governance in which regional and local public authorities play a central role in delivering public policies and services due to their close proximity to citizens. The institutional stakeholders at regional and local level in the three counties include the County Administrative Boards (CAB)—a national government authority operating in 21 counties with a responsibility for coordinating the climate transition work at regional and local level and ensuring that decisions from parliament and the Government are implemented in the counties. The CABs are also tasked with coordinating the work on regional climate and energy strategies, but the responsibility is shared. Regions and municipalities (self-governing local authorities) have a central role in the climate transition, not least in relation to strategic and physical planning, regional development work, education, stakeholder involvement, and advisory aspects. There are also local climate and environmental action plans at municipal level with localized climate policy targets [49]. Local authorities also play an important role in facilitating stakeholder coordination and promoting the conditions for collaboration between actors in the region, such as industries, businesses, associations, and other NGOs in relation to regional development. Local authorities in Gotland, Norrbotten, and Västra Götaland have facilitated dialogue between these key stakeholders in the development of regional and climate energy strategies [7].

Joint efforts between public authorities, regional and local companies, professional associations, and industry-relevant players are required to ensure that the climate transition takes place while maintaining the competitiveness and value of regions [31]. In all three Swedish just transition regions analyzed, high levels of cooperation and social capital have been built up through various EU and national innovation projects. In Norrbotten, the Hybrit Initiative, Reemap Project, and Sustainable Underground Mining Project have brought together key industries including SSAB, LKAB, and Vattenfall to develop carbon free production processes and circular economy initiatives. In Gotland, Cementa AB engages in various partnerships and R&D initiatives for the transition to fossil-free industrial processes. The industrial climate transition initiative Fossil-Free Sweden has set out a roadmap for a climate neutral cement industry in Sweden. The initiative has been led by Cementa AB [30]. The roadmap also notes that the construction sector and mining industries are closely linked to the cement industry's transition. Furthermore, Cementa AB, SMA Mineral AB, and Vattenfall, among others, collaborate under the umbrella initiative CemZero, conducting pilot studies and investigating the preconditions for climate neutral production processes. CemZero involves three research projects carried out in collaboration across the public and private sector as well as academia (Umeå University), partly funded by the Swedish Energy Agency [50]. The business collaboration Tillväxt Gotland also enables industrial actors to collaborate via the Industrigruppen Gotland in matters related to business growth, skills, and competence supply issues and transition.

In Västra Götaland, several science parks gather research and academia, public actors, businesses, and NGOs for the energy and climate transition. For example, Johanneberg Science Park hosts the West Swedish Chemical and Materials Cluster. High levels of cooperation and social capital are demonstrated through numerous collaboration projects in the region. The long-term project Climate Leading Process Industry supports the transition in a region with intense production of chemicals and materials. Here, the large chemical companies in Stenungsund (Adesso Bioproducts, Borealis, INOVYN, Nouryon, Perstorp) and Västra Götalandsregionen work together with a joint vision on Sustainable Chemistry by 2030. The West Sweden Chemicals and Materials Cluster includes research and innovation actors besides the chemical industry companies themselves, such as Chalmers University of Technology, SP Technical Research Institute of Sweden, as well as companies from other industry sectors, such as Renova and Göteborg Energi.

In Sweden, stable energy systems and electricity networks are key factors for the transition. For instance, it is estimated that the new HYBRIT plant planned in Gällivare and the H2 Green Steel plant in Boden together can increase the yearly electricity consumption in Sweden by around 45–50 percent: this will require significant expansion of the renewable energy production [51]. This process is parallel to similar developments in the other Swedish regions. In Västra Götaland, the capacity of the electricity networks risks being insufficient in the short term [7]. In Gotland, developing the infrastructure that supports the energy transition and electrification of the cement production is also a key challenge. Cementa's need for electricity is expected to increase tenfold by 2030, as a result of the transition to carbon dioxide neutrality [7]. Improved transmission capacity of electricity between Gotland and the mainland is a prerequisite for large-scale electrification of the industry. A new cable connection needs to be in place within the next 10 years to align with Cementa's transition timeline. The relatively short time frame is emphasized as a key challenge in terms of planning, permits, and establishment. The design of the cable and investments is being investigated by Svenska Kraftnät, Vattenfall, Eldistribution, and Gotland Energy (GEAB).

Public authorities in all three regions work closely with higher education institutes. Gothenburg University, Chalmers University of Technology, Luleå Univeristy of Technology, and Uppsala University Campus Gotland carry out multidisciplinary research projects and have provided knowledge and expertise in the drafting of the regional energy and climate strategies. Municipalities, lower education providers, companies, and universities also play an important role in ensuring that there is a good match between access to a skilled workforce and employers' demand for skills. Public authorities in Gotland work with Uppsala University Campus Gotland and Teknik College Gotland to ensure that higher education and vocational training programs meet the needs of local industries and businesses. Norrbotten's Regional Competence Council coordinates actions on skills development and governance in the region. The task force is made up of municipalities, Swedish public employment service, Region Norrbotten (the County Council), the County Administrative Board, Luleå University of Technology (LTU), and vocational education schools. LTU also has an extensive range of training aimed at the steel industry's value chain and gathers research in mining and process technology. The Mining and Steel Industry Research Institute, Swerim, with process engineering and equipment in Luleå, is an important link between academia and industry. In Västra Götaland, there is a strong foundation for municipal collaboration projects to ensure skills supply in different sectors and industries in the region. These include for example Validation West (Validering Väst) that supports structure in the region to strengthen the skills validation process. Throughout the municipal associations branch, specific Competence Councils are also set up as a collaboration arena, ensuring coordinated efforts between labor market actors, academy, sectoral organizations, and other regional actors. The Gothenburg Region also has a competence hub to ensure coordinated efforts to support individuals and business life in transition.

7. Discussion

The core findings from the analyses presented above have informed and guided the preparation of the EU TJTPs for the Swedish transition regions of Gotland, Norrbotten, and Västra Götaland. These regional assessments highlight that the just transition has an important spatial dimension as each Swedish region is different in terms size, governance structure, and socio-economic dynamics. Gotland is a small rural and tourism-dependent economy that has very specific economic environment conditions determined by its insularity. Norrbotten is an extractive and increasingly technology and innovation-driven economy that is performing rather well and will probably keep doing so in a global context characterized by a transition from oil-based to metal-based energy carriers. Västra Götaland is a leading economy in the Nordics, with a broad industrial basis and a long-lasting tradition in innovation and technology-oriented productions. The diverse nature of Swedish regions makes it important to note that there is no one-size-fits-all model for just transition planning. This is largely reflected in the TJTPs that have been tailored to meet regional and territorial specificities and needs requirements.

Trade unions and civil society groups have voiced concern that the Swedish just transition is too focused on the economic and technical dimensions of climate policies, while neglecting the potential regional social impacts of transition. An assessment of the main transition actions outlined within the draft regional TJTPs plans suggest that this criticism is not unjustified, as they largely focus on developing the technical infrastructure required for reducing emissions in fossil fuel-dependent industries. The economic and technical imperative at the heart of the regional action plans has been driven by the national government during the TJTP development process.

During TJTP development discussions, this proved a source of tension with the EU who were more inclined to consider the social elements of the transition. The national government relented on its position and accepted the introduction of more social-orientated actions into the TJTPs when the regional territorial analysis highlighted that reskilling and retraining was essential in all Swedish transition regions. While some more socially targeted actions have been introduced within the plans, most of the actions remain focused on the technological enablers of the transition. This confirms the thesis that, also in Sweden, the national government tends to dominate multi-level discussions related to EU Cohesion Policy funding, with sub-national regional and local actors being constrained by political asymmetries of power [21,52].

Even though no job losses are anticipated within key regional industries, our analyses have highlighted some important social development trends within each region that might impact on the effective implementation of just transition plans. These issues need to be more clearly addressed within the regional TJTPs, including recommendations and solutions for overcoming these obstacles. Specific regions are more vulnerable to the effects of transition not only because of the regions' dependence on industries with high emissions, but also because of intrinsic conditions. Regions outside larger cities are likely to be more affected by industrial decline because of social development stressors such as the shrinking population caused by low birth rates and outmigration to urban areas, which make them more vulnerable to external shocks and policy reforms such as those brought about by the climate transition. Similarly, low-skilled workers are usually more vulnerable since it is more difficult to adapt to other occupations and because many affected industries are located in regions where the economy is not diversified.

Indeed, ensuring that companies in the cement and limestone, petro-chemical, and steel industries have access to the right skills and competencies is of fundamental importance for green just transition in Sweden. The TJTPs outline specific actions for skills development, but they are vague. Regional and local policy makers and stakeholders in Gotland, Västra Götaland, and Norrbotten can add further specification by focusing on four broad strategies to ensure future competence supply for the green just transition. Possible actions include: (1) providing access to higher education, re-training, and upskilling courses expanded and offered to all population groups; (2) encouraging people who are currently inactive or unemployed to upgrade their skills, facilitating their integration into the labor market; (3) developing tailored programs to attract skilled staff from other parts of Sweden and from abroad to fill competence gaps within regional labor markets; and (4) in addition to the already existing industry road maps, social road maps could be created to highlight the social dimension of the transition and to form collective social action.

The key governance structures and stakeholders needed for the development and implementation of climate transition policies in Sweden are highlighted within the TJTPs. This can be considered a key enabler for the technical solutions. For instance, in the preparation of the new national electrification strategy it has been emphasized that ensuring the electrification process to support the climate transition requires improved collaboration between governance levels as well as streamlining and expansion of the network [53]. However, the plans do not provide any detailed specification on the role and responsibilities of these governance levels and actors at different stages of the transition policy cycle. According to Swedish climate policies, climate transition processes should be open and inclusive of all stakeholders and citizens, particularly underrepresented minority groups [5].

There is little information within the TJTPs regarding public consultation on the transition proposals. There is, therefore, a need for further analysis on how best to include the public in the transition process from both a policy input and policy dissemination perspective. This is particularly the case in the Norrbotten region where there are still tensions with the indigenous Sámi community regarding transition proposals and discussions with the Sámi Parliament remain ongoing. In order to meet the goal of 'leaving no one behind' during the climate transition, but also to avoid breaching the legal rights and prerogatives of the Sámi people, their involvement should be enhanced.

Our territorial analysis reveals that each region has the governance structures and stakeholders needed to develop and implement transition policies. Existing governance structures are based on strong links between national government, national agencies and regional public authorities, and high levels of social capital and collaboration between key stakeholders, including industries, businesses, high education institutions, labor unions, and civil society organizations. Governance challenges remain, however, including a lack of dialogue and co-operation between the regional and local electricity grid-owners regarding conditions for energy supply in the comprehensive or physical planning that the municipalities carry out. As the structures for coordination are weak, there is a risk that electricity users, community planners, and network companies will not be made aware of the limitations that exist in the electricity network. Lengthy permitting procedures may also affect many sectors concerned by climate transitions. Current procedures have been designed to ensure compliance with highly conservative technical and safety regulations in sectors such as energy production, storage, and distribution. Permitting procedures should be simplified in a way that safety and stability are maintained while ensuring that the processes are significantly shortened. This example further reinforces the idea that the Swedish just transition process should pay more attention to the governance dimension of transitions to also enable its technical components. Further cross-sectoral collaboration and communication will be required to overcome these challenges.

8. Conclusions

With the introduction of the EU's Green Deal and Just Transition Mechanism, the concept of the 'just transition' is starting to grow and gain momentum. There are many lessons to be drawn for both Sweden and other EU member states from the Swedish experience of developing TJTPs. Most significantly, it is important that EU action plans, and other climate and energy related policies, are developed based on a well-defined just transition concept. Because of the broad nature of the just transition notion, it is used and interpreted differently in different contexts. It is important to balance between acknowledging the rights of the EU, nation states, and regions to interpret and implement the concept in different ways, while also attempting to provide some coherence, so that plans can be implemented smoothly in a way that meets the needs of local citizens.

Swedish policymakers and the TJTPs they are developing need to fully recognize the different technical, social, and spatial elements of the shift towards climate neutrality. In the Swedish case, there has been a tendency for national policymakers to prioritize the more technical elements of the transition, driven by the economic cost of helping industries transfer to climate neutral, carbon-free technologies. It is not surprising that the actions outlined in the TJTPs emphasize the technical and economic imperatives of transition as the TJTP development process was driven by representatives of the Swedish national government, regional public authorities, and sectoral actors. Indeed, the actions outlined in the TJTPs are entirely consistent with their climate policies and roadmaps which are focused on the technical aspects of shifting to carbon neutral technologies and processes and neglect the potential social impacts of the transition. This position was exacerbated by the processes used in the formulation of TJTPs, as societal groups, NGOs, and citizens were largely excluded from this process, with the notable exception of the Sámi Parliament in Norrbotten. The European Commission note that the transition can be successful only if policies are designed with the involvement of citizens and accepted by them [2]. More specifically, an 'active social dialogue' is recognized as an essential element to ensure that the transition is successful and accepted by workers and companies [2]. In Sweden, greater consultation of societal groups and citizens in the TJTP process might have acted as a counterweight to the political and technical priorities of the Swedish government, regional authorities, and sectors, creating a balance between the technical and social actions outlined in the TJTPs. It is, therefore, vital that social groups and local citizens be consulted in any evaluation and amendments of the plans if they are to be fully accepted by society.

Transition plans need to strike a balance between these important economic and technical imperatives, while ensuring that the social justice and spatial dimensions of the transition are not ignored. To make sure that the social and spatial elements are not neglected, policymakers and practitioners could, one, develop social roadmaps that highlight the social challenges posed by climate policies within each transition region, and two, ensure that climate transition policies are based on areas of regional resilience strengths and opportunities. These measures would make policies responsive to local needs and objectives, and more accessible to citizens in helping overcome the challenges posed by regional socio-economic trends, including an aging work force, lower education levels, outmigration, and gender imbalances.

The social impacts of the transition should be minimal in the Swedish transition regions if, as noted by interviewees, there are no resultant job losses in the industries most affected by the shift to carbon neutral technologies. Furthermore, the Swedish welfare system is considered to be well equipped to deal with the possibility of job losses and the resultant social challenges presented by increasing unemployment. In this regard, the strength of the Swedish welfare system has played a significant role in driving the political decision of the Swedish national government to focus TJTP actions on the more technical and economic aspects of transition, including covering the costs of the development and implementation of carbon neutral technologies and processes in key industries. The structural limitations of the Just Transition Fund identified by Sabato and Fronteddu [19] have not played such a significant role in influencing the content of the TJTPs in Sweden. The narrow focus of the Just Transition Funds on skills and retraining they outline is potentially beneficial to Swedish transition regions and industries, but has largely been a secondary consideration within the actions outlined in the TJTPs, despite regional impact assessments highlighting the need for both youth and elderly reskilling within Swedish transition regions. In countries with less generous welfare systems, the narrow focus on the Just Transition Funds on skills and retraining will not help overcome the more short-term challenges presented by unemployment. There is, therefore, a need to ensure that the focus on the Just Transition Funds on skills and retraining is not viewed as an alternative to traditional social welfare measures and that the Just Transition Funds effectively complement and fill gaps in national and regional level welfare policies [19].

An effective climate transition process requires smooth collaboration across multiple levels of governance. Ensuring that EU, national, and regional level goals and objectives are represented within transition plans requires decision-making based on equality and reciprocal persuasion in which actors try to understand each other's incentives and underlying assumptions. Strong regional and local leadership is particularly important for facilitating open and inclusive dialogue between key regional and local levels stakeholders and citizens. The Swedish case regions have shown that effective collaboration is often based on the social capital built up through existing regional networks, mainly involving industries, businesses, and higher education institutes. However, it is also vital that transition processes are open and transparent to facilitate participatory governance throughout the policy cycle which involves labor unions, civil society organizations, and minority groups. Moving forward, the development of national and regional climate transition platforms might be one way to maintain smooth and effective collaborations across governance levels. Such platforms could help establish sectoral synergies and business value chains, coordinate public and private financing, identify existing and newly emerging socio-economic challenges and opportunities, and monitor the impacts of the transition. Collaboration in relation to these issues will be essential if the technical, social, and territorial dimensions of the just transition are to be addressed in a balanced and equal manner.

An important criterion for just transitions is that 'no one is left behind', but the transition plans should also ensure that environmental burdens and social impacts are not transferred to other regions. The Swedish examples presented here show how that, even if the socio-economic impacts of the transitions might be large in some settings, in most cases, these impacts are highly localized and could be totally or partially overcome by the strong economic inertia and innovation capacity of the Swedish economy. In some cases, it might be even easier to replace jobs in certain activities by occupations in other sectors with smaller climate footprints rather than investing in the transformation of the problematic energy-intensive industries. Still, the products provided by these industries, including cement, steel, and chemical compounds, will continue to be demanded and consumed by the Swedish economy. Delocalizing the production of these materials to other regions would not reduce the global environmental burden and would not contribute to tackling the climate emergency either. In this sense, the just transition concept acquires an ethical dimension that goes beyond the social justice principle that lays at its core. It is essential that the just transition plans in Sweden and elsewhere in Europe take up this challenge and contribute to ensuring that the climate transitions are performed in situ, without alienating the employment basis, production capacity, and sectoral specialization of the affected regions.

Author Contributions: Conceptualization: all authors; methodology: all authors; software: not applicable; validation: all authors; formal analysis: all authors; investigation: all authors; resources: not applicable; data curation: all authors; writing—original draft preparation: all authors; writing—review and editing: all authors; visualization: not applicable; supervision: not applicable; project administration: not applicable; funding acquisition: not applicable. All authors have read and agreed to the published version of the manuscript.

Funding: This research received no external funding.

Institutional Review Board Statement: Not applicable.

Informed Consent Statement: Not applicable.

Data Availability Statement: Not applicable.

Acknowledgments: We would like to thank the journal editors and three reviewers for their helpful and constructive feedback that helped to improve the structure and focus of the paper immeasurably. We would also like to thank our colleagues who helped in the preparation of the Swedish Territorial Just Transition Plan drafts, particularly those at Tillväxtverket, Trinomics, and regional authorities of Gotland, Norrbotten and Västra Götaland.

Conflicts of Interest: The authors declare no conflict of interest.

References and Note

1. European Commission. The Just Transition Mechanism: Making Sure No One Is Left Behind. Available online: https://ec.europa.eu/info/strategy/priorities-2019-2024/european-green-deal/actions-being-taken-eu/just-transition-mechanism_en (accessed on 23 June 2021).
2. European Commission. Communication from the Commission—The European Green deal, COM, 2019, 640 Final. Available online: https://eur-lex.europa.eu/legal-content/EN/TXT/?uri=COM%3A2019%3A640%3AFIN (accessed on 23 June 2021).
3. United Nations. Transforming Our World: The 2030 Agenda for Sustainable Development, A/RES/70/1, 21st October 2015. Available online: https://sdgs.un.org/2030agenda (accessed on 1 July 2021).
4. Preston, C.; Carr, W. Recognitional Justice, Climate Engineering, and the Care Approach. *Ethics Policy Environ.* **2018**, *3*, 308–323. [CrossRef]
5. Government Offices of Sweden. The Swedish Climate Policy Framework. 2017. Available online: https://www.government.se/495f60/contentassets/883ae8e123bc4e42aa8d59296ebe0478/the-swedish-climate-policy-framework.pdf (accessed on 2 July 2021).
6. European Commission. Overview of Investment Guidance on the Just Transition Fund 2021–2027 per Member State (Annex D). 2020. Available online: https://ec.europa.eu/info/sites/default/files/annex_d_crs_2020_en.pdf (accessed on 1 July 2021).

7. Swedish Agency for Economic and Regional Growth. Operativt Program: Fonden för en Rättvis Omställning. 2020. Available online: https://tillvaxtverket.se/download/18.29ac13dc177db549d2a3d2eb/1615384939727/Operativt%20program%20Fro.pdf (accessed on 1 July 2021).

8. Labour Network for Sustainability, Just Transition—Just What Is It? 2021. Available online: https://www.labor4sustainability.org/uncategorized/just-transition-just-what-is-it/ (accessed on 23 June 2021).

9. CSIS; Climate Investment Funds. *Just Transition Concepts and Relevance for Climate Action—A Preliminary Framework, a Study by the Just Transition Initiative*; CSIS and CIF: Washington, DC, USA, 2020.

10. ITUC. *Climate Justice: There Are No Jobs on a Dead Planet, Frontlines Briefing*; ITUC: Brussels, Belgium, 2015.

11. Rosemberg, A. *Strengthening Just Transition Policies in International Climate Governance*; Policy Analysis Brief, The Stanley Foundation: Muscatine, IA, USA, April 2017.

12. Geels, F.W. Technological Transitions as Evolutionary Reconfiguration Processes: A Multi-Level Perspective and a Case-Study. *Res. Policy* **2002**, *31*, 1257–1274. [CrossRef]

13. Rawls, J. *A Theory of Justice, Revised Edition*; Belknap/Harvard University Press: Cambridge, UK, 1999. [CrossRef]

14. Liljenfeldt, J.; Pettersson, Ö. Distributional Justice in Swedish Wind Power Development—An Odds Ratio Analysis of Windmill Localization and Local Resident's 'Socio-Economic Characteristics'. *Energy Policy* **2017**, *105*, 648–657. [CrossRef]

15. Ryder, S.S. Developing an Intersectionally-Informed, Multi-Sited, Critical Policy Ethnography to Examine Power and Procedural Justice in Multis-calar Energy and Climate Change Decision-making Processes. *Energy Res. Soc. Sci.* **2018**, *45*, 266–275. [CrossRef]

16. ILO. *Guidelines for a Just Transition Towards Environmentally Sustainable Economies and Societies for All*; ILO Publications: Geneva, Switzerland, 2015.

17. European Commission, Just Transition Platform: Accompanying Member States and Regions to Achieve a Just Transition. 2021. Available online: https://ec.europa.eu/info/strategy/priorities-2019-2024/european-green-deal/actions-being-taken-eu/just-transition-mechanism/just-transition-platform_en (accessed on 24 June 2021).

18. European Commission. *Proposal for a Regulation of the European Parliament and of the Council—Establishing the Just Transition Fund, COM/2020/22*; European Commission: Brussels, Belgium, 14 January 2020. Available online: https://eur-lex.europa.eu/legal-content/EN/TXT/?uri=CELEX%3A52020PC0022 (accessed on 1 July 2021).

19. Sabato, S.; Fronteddu, B. *A Socially Just Transition through the European Green Deal*; ETUI Research Paper, Working Paper; The European Trade Union Institute: Brussels, Belgium, 2020.

20. Inclusion Europe, The European Pillar of Social Rights. Available online: https://www.inclusion-europe.eu/social-pillar/ (accessed on 24 June 2021).

21. Moodie, J.R.; Meijer, M.W.; Salenius, V.; Kull, M. Territorial governance and Smart Specialisation: Empowering the sub-national level in EU regional policy. *Territ. Polit. Gov.* **2021**, 1–21. [CrossRef]

22. Giacometti, A.; Teräs, T. *Regional Economic and Social Resilience: An Exploratory in Depth Study in the Nordic Countries*; Nordregio Report; Nordregio: Stockholm, Sweden, 2019; Volume 2. [CrossRef]

23. Di Caro, P. Recessions, recoveries and regional resilience: Evidence on Italy. *Camb. J. Reg. Econ. Soc.* **2014**, *8*, 273–291. [CrossRef]

24. OECD. Resilient Cities—Preliminary Version. 2016. Available online: https://www.oecd.org/fr/gov/politique-regionale/resilient-cities.htm (accessed on 24 June 2021).

25. Swedish Society for Nature Conservation & Nature and Youth Sweden. Webinar How Do We Ensure a Just Climate Transition (Webbinarium: Hur Säkrar vi en Rättvis Klimatomställning?), 2 September 2020. Available online: https://www.youtube.com/watch?v=pYrcEJd3DKU&ab_channel=Naturskyddsf%C3%B6reningen (accessed on 24 June 2021).

26. Government Offices of Sweden. Sweden and the Leaving No One behind Principle—Nationally and Globally. 2020. Available online: https://sustainabledevelopment.un.org/content/documents/26672Sweden_and_LNOB_2020.pdf (accessed on 1 July 2021).

27. County Administrative Board of Norrbotten. *Climate and Energy Strategy for the County of Norrbotten 2020–2024. Objectives for 2045*; County Administrative Board of Norrbotten: Luleå, Sweden, 2019.

28. County Administrative Board of Gotland. *Together Towards 2030—An Energy and Climate Strategy for Gotland (Tillsammans mot 2030-En Energi och Klimatstrategi för Gotland)*; County Administrative Board Gotland: Visby, Sweden, 2019.

29. County Administrative Board of Västerbotten. *Together for the Climate: Climate and Energy strategy for Västerbotten (Tillsammans för Klimatet-Klimat och Energistrategi för Västerbottens)*; Grejja Kommunikation AB: Umeå, Sweden, 2020.

30. Fossil Free Sweden. *Roadmaps for Fossil Free Competitiveness—Fossil-Free Development of Swedish Industry*; Fossil Free Sweden: Stockholm, Sweden, 2021.

31. Bonde, I.; Kuylenstierna, J.; Bäckstrand, K.; Eckerberg, K.; Kåberger, T.; Löfgren, Å.; Rummukainen, M.; Sörlin, S. *Report of the Swedish Climate Policy Council*; Report Number 3; Swedish Climate Policy Council: Stockholm, Sweden, 2020; ISBN 978-91-984671-4-7.

32. Statistics Sweden. Local Kind of Activity Unit (Table RegionalBasf07). 2021. Available online: https://www.statistikdatabasen.scb.se/pxweb/sv/ssd/ (accessed on 23 June 2021).

33. Regionfakta, Största Arbetsgivare i Kiruna Kommun 2019. 2021. Available online: https://www.regionfakta.com/norrbottens-lan/norrbottens-lan/kiruna/arbete1/kommunens-15-storsta-arbetsgivare/ (accessed on 23 June 2021).

34. Regionfakta, Största Arbetsgivare i Gällivare Kommun 2019. 2021. Available online: https://www.regionfakta.com/norrbottens-lan/norrbottens-lan/gallivare/arbete1/kommunens-15-storsta-arbetsgivare/ (accessed on 23 June 2021).

35. Regionfakta, Största Arbetsgivare i Luleå Kommun 2019. 2021. Available online: https://www.regionfakta.com/norrbottens-lan/norrbottens-lan/lulea/arbete1/kommunens-15-storsta-arbetsgivare/ (accessed on 23 June 2021).
36. Flegg, A.T.; Webber, C.D.; Elliott, M.V. On the Appropriate Use of Location Quotients in Generating Regional Input–Output Tables. *Reg. Stud.* **1995**, *29*, 547–561. [CrossRef]
37. Jahn, M.; Flegg, A.T.; Tohmo, T. Testing and Implementing a New Approach to Estimating Interregional Output Multipliers Using Input–Output Data for South Korean Regions. *Spat. Econ. Anal.* **2020**, *15*, 165–185. [CrossRef]
38. Preem, A.B. Preemraff Lysekil. Available online: https://www.preem.se/om-preem/om-oss/vad-vi-gor/raff/preemraff-lysekil/ (accessed on 2 July 2021).
39. Region Västra Götaland. *Var det Bättre Förr? Regionrapport 2020—Hållbar Utveckling i Västra Götaland*; VGR Analys; Västra Götalandsregionen Regionen: Vänersborg, Sweden, 2020.
40. Region Västra Götaland. Arbetsmarknad. 2021. Available online: https://www.vgregion.se/ov/data-och-analys/statistik-analysportalen/samhalleochbefolkning/arbetsmarknad/ (accessed on 2 July 2021).
41. Stakeholder Interviews Performed by Nordregio and Trinomics during 2020–2021
42. Statistics Sweden. Population by Region, Age and Year. 2021. Available online: https://www.scb.se/en/finding-statistics/statistics-by-subject-area/population/population-composition/population-statistics/ (accessed on 23 June 2021).
43. Statistics Sweden. Population by Level of Education, Region, Year and Age. 2020. Available online: https://www.scb.se/en/finding-statistics/statistics-by-subject-area/education-and-research/education-of-the-population/educational-attainment-of-the-population/ (accessed on 23 June 2021).
44. Regionfakta. Utbildning—Grundskola—Gymnasiebehörighet (Data Available for Norrbotten, Västra Götaland and Gotland). 2021. Available online: https://www.regionfakta.com/ (accessed on 24 June 2021).
45. Ejdemo, T.; Johansson, J.; Söderholm, P. *Möjligheter Och Hinder för en Grön Energiomställning: Erfarenheter Från Andra Regioner Med Lärdomar för Norrbotten*; Luleå Tekniska Universitet, Gröna Energiinvesteringar, Energikontor Norr: Luleå, Sweden, 2017.
46. Region Gotland. *Vårt Gotland 2040—Regional Utvecklingsstrategi för Gotland*; RS 2018/1237; Region Gotland: Visby, Sweden, 2018.
47. Cambou, D. Uncovering injustices in the Green Transition: Sámi rights in the development of wind energy in Sweden. In *Arctic Review on Law and Politics*; NOASP: Oslo, Norway, 2020; Volume 10, pp. 310–333. [CrossRef]
48. OECD. *Mining Regions and Cities Case of Västerbotten and Norrbotten, Sweden, OECD Rural Studies Series 2021*; OECD: Paris, France, 2021. [CrossRef]
49. Government Offices of Sweden. Sweden's Long-Term Strategy for Reducing Greenhouse Gas Emissions. 2020. Available online: https://unfccc.int/sites/default/files/resource/LTS1_Sweden.pdf (accessed on 23 June 2021).
50. Wilhelmsson, B.; Kollberg, C.; Larsson, J.; Eriksson, J.; Eriksson, M. *CemZero: A Feasibility Study Evaluating Ways to Reach Sustainable Cement Production via the Use of Electricity*; Vattenfall and Cementa: Stockholm, Sweden, 2012.
51. Sveriges Television. Ekonomibyrån—Stålbadet. Video. 2021. Available online: https://www.svtplay.se/video/30586078/ekonomibyran/ekonomibyran-sasong-3-avsnitt-10 (accessed on 24 June 2021).
52. Bache, I. The extended gatekeeper: Central government and the implementation of EC regional policy in the UK. *J. Eur. Public Policy* **1999**, *6*, 28–45. [CrossRef]
53. Government Offices of Sweden. Nationell Elektrifieringsstrategi. Intressentmöte. 30 March 2021. Available online: https://www.energiforetagen.se/globalassets/dokument/elektrifiering/presentation-intressentmote-elektrifieringsstrategin-30-mars.pdf (accessed on 1 July 2021).

 *sustainability*

Article

Optimizing Treatment of Cesspool Wastewater at an Activated Sludge Plant

Piotr Bugajski , **Agnieszka Operacz** * , **Dariusz Młyński**, **Andrzej Wałęga** and **Karolina Kurek**

Department of Sanitary Engineering and Water Management, Faculty of Environmental Engineering and Land Surveying, University of Agriculture in Krakow, Al. Mickiewicza 21, 30-059 Kraków, Poland; p.bugajski@urk.edu.pl (P.B.); d.mlynski@urk.edu.pl (D.M.); a.walega@urk.edu.pl (A.W.); karolina.kurek@urk.edu.pl (K.K.)
* Correspondence: a.operacz@urk.edu.pl; Tel.: +48-126-624-045

Received: 26 October 2020; Accepted: 1 December 2020; Published: 7 December 2020

Abstract: The purpose of this work was to determine the optimal percentage of wastewater from cesspool in the mixture of wastes subjected to treatment processes, which will not have a negative impact on the functioning of the collective treatment plant. The study was carried out over a period of two years, with 48 samples of wastewater flowing in from the sewage network and delivered with the slurry tanker collected and subjected to physical and chemical analysis. The analysis included: Biochemical Oxygen Demand (BOD_5), Chemical Oxygen Demand (COD), and Total Nitrogen (TN). In addition, the study defined the daily balance of the amount of inflowing and transported wastewater. Based on the analysis carried out, it was found that the unit loads of BOD_5, COD and TN in the mixture of wastewater subjected to the treatment process will be at the level of loads assumed in the project, when the share of supplied wastewater, i.e., from cesspool, will be at the level of 5% of the total amount of wastewater. Considering that in the analysed period the total average daily amount of wastewater subjected to the treatment process was 253.5 $m^3 \cdot d^{-1}$, the optimal amount of wastewater delivered should be 12.7 m^3 in each day of the week.

Keywords: wastewater; sewerage; liquid waste tanks (cesspool); partial correlation; organic and biogenic pollution

---

1. Introduction

Along with the growing expansion of housing construction in rural areas in Poland, the volume of water used by residents increases, thus also the volume of generated sewage increases. Unfortunately, in many cases, professional water supply and sewage systems are not provided along with the expansion of housing. In Poland, in areas where there are no collective sewage systems, residents use the so-called septic tanks (tanks for liquid waste) in which sewage is collected for disposal to a collective treatment plant. The numbers show how big this problem is in Poland.

In rural areas, according to current Central Statistical Office (CSO) data [1], in 2018 there were 21,775.50 thousand tanks for liquid waste, commonly known as cesspool. It still is the most common way of wastewater disposal in Poland, in rural areas where there are no collective or individual sewage systems. The correct operation of such tanks consists in regular emptying of wastewater, which should then be transported by a professional slurry tanker to a collective wastewater treatment plant, whose technological line is suitable to treat this type of wastewater. Unfortunately, according to national literature reports, many such tanks are improperly used by residents, because the wastewater coming from them goes illegally to the environment, i.e., land or flowing water [2–4]. Such practices lead to pollution of land, surface and underground waters [5]. The main reason for this is the intention of users to reduce the costs of exporting and utilizing wastewater from cesspits in a collective treatment plant. To prevent this type of practice, it is necessary to control the leak-tightness of this type of tanks and the regularity of their emptying by a professional company with the appropriate slurry tanker.

Another problem posed by wastewater from cesspool is their utilization in a collective wastewater treatment plant, because the concentrations of pollutants contained in them are often several times, and sometimes several dozen times, higher than typical wastewater flowing into the sewage system [6–8]. High concentrations of pollutants in wastewater from cesspool result from the saving habits of residents to save and thus use a small amount of water, which results in an increase in the concentration of pollutants [9,10]. In addition, the long-term storage of wastewater in tanks creates the conditions for anaerobic decomposition of organic pollutants and the occurrence of sewage rotting, which causes the release of an unpleasant odour of hydrogen sulfide [11–13]. The volume of liquid waste tanks should be designed to be emptied with a 3–4 week time interval. Too long intervals between emptying these tanks result in the wastewater being rotten and similar in composition to sewage sludge with very high hydration [14]. In practice, wastewater from cesspools is transported to the area of the collective wastewater treatment plant on an irregular basis on each day of the week [8,15], and this type of practice is conducive to disruption of biological wastewater treatment processes, as these processes are sensitive to major changes in both the quantity and quality of treated wastewater. Therefore, the amount of wastewater from cesspool should be dosed (batched) with great caution to the total amount of wastewater subjected to treatment [16–20]. Adding (mixing) in a short time, e.g., directly from the slurry tanker sewage from cesspool to sewage flowing into the sewage network, will result in a sudden increase in the amount of treated sewage, as well as a sudden increase in the so-called "impact" of the pollutant load in the wastewater mixture subjected to the treatment process [21].

As it has been shown in the publications concerning the problem of neutralization of sewage from septic tanks, there are alternative treatment systems to transport to collective treatment plants, which is an expensive process and may have a negative impact on biological processes in collective treatment plants. As Forbis-Stokes et al. [21] stated, it is possible to treat sewage from septic tanks with a mobile installation at the place where they are generated. Another solution in this aspect is the treatment of sewage from septic tanks in wetland sewage treatment plants. As Jong and Tang [22] stated, wetland wastewater treatment plants, while maintaining an appropriate operating regime, demonstrate high efficiency in septic tank waste material treatment. In addition, as Mancl and Rosencrans indicated, it is possible to drain sewage from septic tanks into properly prepared fields, which allows water to be retained in the soil, thus increasing water retention, which is crucial nowadays [23].

However, in Poland, now and in the following years, the basis for the disposal of sewage from septic tanks is their selection and transport to a collective treatment plant. This applies to the majority of collective treatment plants in rural communes in Poland. Therefore, there is an urgent need to indicate the optimal proportion (volume) of sewage delivered by the slurry rolling stock, so that other sewage does not adversely affect the treatment of all sewage.

The overall aim of the analysis was to calculation an optimal organic and biogenic load into the wastewater treatment plant (WWTP) by blending strong cesspool wastewater with weak sanitary wastewater.

The detailed aim of the study was to determine the optimal percentage (amount) of sewage from cesspool in the mixture of wastes subjected to treatment processes, at which the unit load (Ul) of organic and biogenic pollutants will be at the level of the unit load assumed in the project for the wastewater treatment plant. The unit load of the analysed indicators assumed in the project is: 60 $g \cdot I^{-1} \cdot d^{-1}$ for Biochemical Oxygen Demand (BOD_5), 120 $g \cdot I^{-1} \cdot d^{-1}$ for Chemical Oxygen Demand (COD) and 13 $g \cdot I^{-1} \cdot d^{-1}$ for Total Nitrogen (TN). The study results have an important practical aspect, as many small, collective wastewater treatment plants in rural areas in Poland have problems with setting the daily limit of sewage delivered by the slurry tanker, which will not adversely affect the functioning of the facility. The novelty of this work is the answer to an important question asked by the operators of wastewater treatment plants in Poland: "How much sewage from cesspool that is transported by slurry tanker can be introduced into the technological system of the wastewater treatment plant, so as not to disturb the purification processes?" The test results presented in this publication indicate the need to build or modernize technological systems for the collection and uniform dosing of collective wastewater from cesspools, so that this wastewater does not adversely affect biological treatment processes in wastewater treatment plants with bioreactors with activated sludge.

2. Materials and Methods

2.1. Characteristics of the Sewerage System

The analysed sewage system includes a sewage network with a total length of 6900 m and a diameter of collectors DN = 0.2 m together with a collective sewage treatment plant, whose designed daily average capacity is 500 m^3·d^{-1}. During the study period, 360 residential buildings were connected to the sewer network. The technological system of the sewage treatment plant consists of a dense grate, a sand pit with a grease and oil separator and a radial bioreactor with a central integrated secondary settling tank. The diameter of the bioreactor is DN = 11 m, with a depth of H = 5.9 m, while the diameter of the secondary settling tank is DN = 3.5 m, with a depth of H = 5.9 m. After mechanical treatment, the sewage flows to the bioreactor, where its biological treatment takes place. The wastewater flows into the nitrification zone and then into the denitrification zone. The sewage then flows to the secondary settling tank from where, after the sedimentation of the deposits, it flows to the river (stream without proper name). The wastewater treatment plant station also has a sewage catchment station transported by a slurry tanker from cesspool. In the commune, where there is no sewage system, most residential buildings have tanks for liquid impurities, cesspool from which wastewater is transported by means of slurry tanker to the treatment plant on weekdays. The scheme of the technological layout of the wastewater treatment plant is presented in Figure 1.

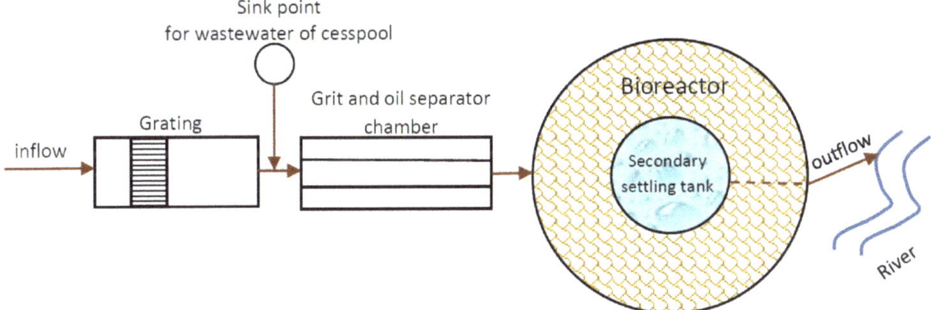

Figure 1. Scheme of the technological layout of the wastewater treatment plant (WWTP).

2.2. Analytical Methods

The study was carried out in the period of two years, 2013 and 2014. During this period, 48 samples of wastewater flowing in from the sewage network and delivered with the slurry tanker were collected and analysed. Incoming wastewater samples were taken from the inflow channel using an autosampler type wastewater sampling device that was programmed for the wastewater flow rate. On the other hand, samples of wastewater from cesspool were taken from the drainage station ("Sink point" on Figure 1). Impurity indicators, BOD$_5$, COD, and TN, were analysed in both types of wastewater. Samples of wastewater were subjected to the physical–chemical analysis in accordance with reference methods set out in the applicable legal acts.

- BOD$_5$—measurement of oxygen after 5 days of incubation at 20 °C in OXI TOP—197 WTW
- COD$_{cr}$—the bichromate method according to PN-ISO 6060: 2006
- Total Nitrogen (Kjeldahl) according to PN-EN 25663: 200

In the days on which the wastewater samples were taken, the amount of wastewater of inflowing Q_1 and the amount supplied of wastewater from cesspools Q_2 were also determined. The amount of flowing wastewater was measured using measuring systems consisting of a probe of the level of the wastewater mirror above the triangular overflow located in the drainage channel. The amount of sewage delivered was determined on the basis of entries in the operating log regarding the amount of sewage delivered with the slurry tanker.

3. Results and Discussion

During the tests, the average daily sewage inflow from the sewage network amounted to $Q_1 = 238.5 \text{ m}^3 \cdot \text{d}^{-1}$ to the said sewage treatment plant, while the average daily sewage delivered by the slurry tanker was $Q_2 = 15.1 \text{ m}^3 \cdot \text{d}^{-1}$, which constituted 5.7% of their share in the total amount of wastewater subjected to the treatment process. However, the amount and frequency of wastewater from cesspool and transported by slurry tanker to the treatment plant was irregular on each day of the week. Wastewater from cesspools was delivered only on weekdays, i.e., from Monday to Friday. In the examined period, on weekdays, the amount of sewage delivered ranged from 5 to 28 $\text{m}^3 \cdot \text{d}^{-1}$. This represented from 2.1% to 12.4% (median 13.5 $\text{m}^3 \cdot \text{d}^{-1}$) of their share in the total amount of treated wastewater. This type of irregular delivery of wastewater from cesspools to the treatment plant is undesirable, as the irregularity of the amount of sewage flowing in and the load of pollutants contained in it causes disruption of wastewater treatment processes [7,16–18].

In the introductory part of the analysis on the quality of treated wastewater, the study presents pollutant indicators in wastewater flowing in from the sewage network and in wastewater from cesspool supplied by slurry tanker. In inflowing wastewater, the BOD_5 median value was 236.0 $\text{mg} \cdot \text{dm}^{-3}$, the COD median value was 390.0 mg·dm-3, while the median of TN concentration was 57.5 $\text{mg} \cdot \text{dm}^{-3}$. In inflowing sewage, the coefficient of variation for BOD_5 was Cv = 18%, for COD it was Cv = 14%, and for Total Nitrogen it was Cv = 13%, which indicates in all cases a small differentiation of this indicator according to the scale proposed by Mucha [24]. Based on the results of analyses regarding the size of pollutant indicators in inflowing wastewater, it was found that their values corresponded to typical domestic sewage described in the literature [25–28]. The median value of the analysed indicators in sewage from cesspool was as follows: for BOD_5—3825.0 $\text{mg} \cdot \text{dm}^{-3}$, for COD—7750.0 $\text{mg} \cdot \text{dm}^{-3}$, and for TN—585.0 $\text{mg} \cdot \text{dm}^{-3}$. As demonstrated, the values of indicators in wastewater from non-drainage tanks are much higher than the inflowing wastewater, which is confirmed by literature reports on the quality of wastewater from cesspool [6,8]. The variability of the values of the analysed indicators in wastewater was delivered at the level of mean variability according to the Mucha scale [24]. The coefficient of variation Cv oscillated between 20% and 25%. Characteristic values of the analysed indicators in inflowing and delivered wastewater are presented in Table 1.

Table 1. Statistical characteristics of concentration indicators of contamination in raw wastewater from sewer system and from cesspool.

Parameters	Types of Wastewater	Statistics					
		Average mg·dm^{-3}	Median mg·dm^{-3}	Min. mg·dm^{-3}	Max. mg·dm^{-3}	Standard Deviation mg·dm^{-3}	Coefficient of Variation %
Biochemical Oxygen Demand (BOD$_5$)	sewers	242.8	236.0	178.0	340.0	42.7	18
	cesspool	4005.6	3825.0	1560.0	6530.0	1003.9	25
Chemical Oxygen Demand (COD)	sewers	387.1	390.0	253.0	493.0	52.4	14
	cesspool	7595.3	7750.0	4390.0	9870.0	1489.5	20
Total Nitrogen (TN)	sewers	57.2	57.5	39.0	73.0	7.2	13
	cesspool	592.7	585.0	380.0	810.0	124.9	21

To determine the optimal, i.e., design unit load of the analysed indicators, i.e., BOD_5, COD and TN in the wastewater mixture subjected to the treatment process, an analysis was carried out in the following stages:

−   Determination of the concentration of indicators in the mixture of inflowing sewage;

- Determination of the unit load of indicators in inflowing sewage;
- Determination of the unit load of indicators in the mixture of inflowing and delivered wastewater;
- Determining the optimal share of the amount of wastewater delivered in the total wastewater mixture so as to obtain the concentrations assumed in the project.

Based on the values of the analysed indicators, i.e., BOD_5, COD and TN in inflowing sewage and delivered sewage from cesspools as well as taking into account the quantitative balance of both types of sewage, the value of these indicators was calculated in the mixture of sewage subjected to the treatment process. In order to calculate the values of the analysed indicators in the wastewater mixture, the weighted average formula was used (1):

$$W_a = \frac{W_1 \cdot Q_1 + W_2 \cdot Q_2}{Q_1 + Q_2} \tag{1}$$

where:

W_a—value of the indicator in the wastewater mixture $(g \cdot m^{-3})$;
W_1—value of the indicator in inflowing wastewater $(g \cdot m^{-3})$;
W_2—value of the indicator in delivered wastewater $(g \cdot m^{-3})$;
Q_1—amount of inflowing sewage $(m^3 \cdot d^{-1})$;
Q_2—amount of sewage delivered $(m^3 \cdot d^{-1})$.

Based on the results of the calculated weighted average (1), it was found that the median BOD_5 value in the wastewater mixture was 438.5 $mg \cdot dm^{-3}$, the median COD was 773.1 $mg \cdot dm^{-3}$ and the median TN was 89.3 $mg \cdot dm^{-3}$. With regard to organic indicators in the wastewater mixture, the range of values for BOD_5 ranged from 247.0 to 815.6 $mg \cdot dm^{-3}$, while for COD it was from 415.6 to 1495.8 $mg \cdot dm^{-3}$. In both cases, the variability of the values of these indicators expressed by the coefficient of variation Cv was 28%, which indicates their average differentiation. The range of TN concentration in the wastewater mixture oscillated from 57.4 to 126.0 $mg \cdot dm^{-3}$ and was characterized by a small variation at the level of Cv = 18%.

The next stage of the analysis was to determine in the wastewater flowing from the sewage network a unit load of organic pollutants expressed as BOD_5 and COD and a unit load of biogenic pollutants expressed as TN. In order to calculate the unit load (per one inhabitant) of individual pollution indicators, Formula (2) was used:

$$UL_i = \frac{W_x \cdot Q_1}{In} \tag{2}$$

where:

UL_i—unit load of the indicator in inflowing wastewater $(g \cdot I^{-1} \cdot d^{-1})$;
W_x—concentration of indicator in inflowing wastewater $(g \cdot m^{-3})$;
Q_1—amount of inflowing wastewater $(m^3 \cdot d^{-1})$;
In—total number of inhabitants connected to the sewage network (In = 1200).

In wastewater flowing from the sewage network, the median BOD_5 unit load was 46.1 $g \cdot I^{-1} \cdot d^{-1}$ and it was a 23.2% lower BOD_5 unit load than assumed in the project. The median COD load was 75.7 $g \cdot I^{-1} \cdot d^{-1}$ and was lower by 36.9% for the load assumed in the project. In the case of unit load of TN in inflowing wastewater, it was found that the median of this parameter was 11.4 $g \cdot I^{-1} \cdot d^{-1}$ and it was smaller than assumed by 18.6%. As can be seen in all 3 cases, the unit load of the analysed indicators was lower as a result of design assumptions. Characteristic unit loads of the analysed indicators in the inflow wastewater are presented in Table 2. Based on the calculated coefficients of variation Cv for the unit loads of the analysed indicators in the inflow wastewater, it was found that they were at the level of average differentiation during the test period.

Table 2. Statistical characteristics of unit load of analysed indicators in inflow wastewater.

Parameters	Statistics					
	Average $g \cdot I^{-1} \cdot d^{-1}$	Median $g \cdot I^{-1} \cdot d^{-1}$	Min. $g \cdot I^{-1} \cdot d^{-1}$	Max. $g \cdot I^{-1} \cdot d^{-1}$	Standard Deviation $g \cdot I^{-1} \cdot d^{-1}$	Coefficient of Variation %
BOD_5	48.3	46.1	29.4	81.3	11.5	24
COD	77.1	75.7	41.7	123.7	15.9	21
TN	11.4	11.4	6.9	17.5	2.3	20

The reason for the unit load lower than assumed in the design of the organic and biogenic indicators tested is the inflow of accidental (rainfall) and infiltration waters to the sewage system. According to the interview with the operator of the sewage system in question, rainwater is discharged into the sewage network from illegally connected roof gutters from residential buildings. As Kaczor et al. [29] and Nowobilska-Majewska and Bugajski [30] indicated, the inflow of rainwater to the sewage network significantly reduces the concentration of organic and biogenic pollutants in the wastewater subject to treatment. Rainwater entering the sewage system intended for the disposal of only domestic sewage affects periodic disruptions in the operation of sewage treatment plants and causes higher costs of wastewater treatment [29,31]. In addition, the analysed sewage network is partly located under the groundwater occurrence level, which causes groundwater to flow into the sewer collectors through leaks in their connections. The foundation of sewer collectors below the groundwater level causes that these waters infiltrate the collectors through all kinds of leaks and causes an increase in the amount of wastewater flowing into the treatment plant [32]. As Madryas et al. described in their research [33], the intensity of the infiltration water inflow to sewage channels is directly proportional to the height of the groundwater table above the pipe.

Because the treatment process is subjected to a mixture of inflowing and delivered wastewater from cesspool, the size of unit loads for BOD_5, COD and TN in the wastewater mixture is analysed using Formula (3). To calculate the unit load in the wastewater mixture, the sum of Q_1 inflows and Q_2 supplied wastewater was adopted, and its weighted average was used as the value of the given indicator.

$$\text{ULmix.} = \frac{W_a \cdot (Q_1 + Q_2)}{\text{In} + \text{Ic}} \tag{3}$$

where:

ULmix.—unit indicator load in the wastewater mixture $(g \cdot I^{-1} \cdot d^{-1})$;
W_a—weighted average indicator in the wastewater mixture $(g \cdot m^{-3})$;
Q_1—amount of inflowing wastewater $(m^3 \cdot d^{-1})$;
Q_2—amount of wastewater delivered $(m^3 \cdot d^{-1})$;
In—total number of inhabitants connected to the sewage network (In = 1200);
Ic—number of inhabitants served by the slurry tanker-average per day (Ic = 10).

In the mixture of inflowing and delivered wastewater, which were subjected to the treatment process, the median unit load of BOD_5 was 90.5 $g \cdot I^{-1} \cdot d^{-1}$, and it was higher than the designed value (60 $g \cdot I^{-1} \cdot d^{-1}$) by 50.8%. The median COD in mixed wastewater was 155.2 $g \cdot I^{-1} \cdot d^{-1}$. Compared to the designed (120 $g \cdot I^{-1} \cdot d^{-1}$) unit load of COD, it was higher by 29.3%. The median TN in mixed wastewater was 18.5 $g \cdot I^{-1} \cdot d^{-1}$ and it was higher than the value assumed in the design (13 $g \cdot I^{-1} \cdot d^{-1}$) by 42.3%. As stated in all 3 cases, the unit load volumes of the analysed indicators in the wastewater mixture were significantly higher than the load assumed in the project. At the same time, it should be noted that this type of situation only took place on days when wastewater was delivered by means of slurry tanker. In relation to the unit load of the analysed indicators in the wastewater mixture, a greater unevenness was observed compared to the unit load of these indicators in the inflowing sewage. The unevenness of the unit load in the wastewater mixture expressed as the coefficient of unevenness Cv was for BOD$_5$—35%, for COD—35% and for TN—26%. Characteristics of unit load of analysed indicators in mixed wastewater are presented in Table 3.

Table 3. Statistical characteristics of unit load of analysed indicators in mixed wastewater.

Parameters	Statistics					
	Average $g{\cdot}I^{-1}{\cdot}d^{-1}$	Median $g{\cdot}I^{-1}{\cdot}d^{-1}$	Min. $g{\cdot}I^{-1}{\cdot}d^{-1}$	Max. $g{\cdot}I^{-1}{\cdot}d^{-1}$	Standard Deviation $g{\cdot}I^{-1}{\cdot}d^{-1}$	Coefficient of Variation %
BOD$_5$	99.5	90.5	39.8	185.7	35.1	35
COD	172.6	155.2	67.0	297.7	59.7	35
TN	18.7	18.5	10.3	30.5	4.8	26

Based on the analysis of the unit load of individual indicators in the inflow sewage and in the mixture of inflowing and delivered sewage, it was found that the unit load of the indicators in the inflowing sewage is lower than assumed in the project, while the unit load in the sewage mixture is too high in relation to the specified load in the sewage treatment plant design.

Because it was found that in the wastewater mixture, the unit load of the examined indicators increases with the increasing amount (percentage share) of added wastewater delivered by means of Pearson's linear correlation analysis, the following was determined:

- impact of the percentage (%) of wastewater delivered in the wastewater mixture on the unit load of the examined indicators;
- impact of the concentration of the analysed indicators in the delivered sewage on the unit load of these indicators in the sewage mixture.

Using the data covering the percentage (%) of the amount of wastewater delivered in the total wastewater mixture (independent variable) and the unit load data of the examined indicators in the wastewater mixture (dependent variable), the strength of the relationship of these two variables was determined in Figures 2–4.

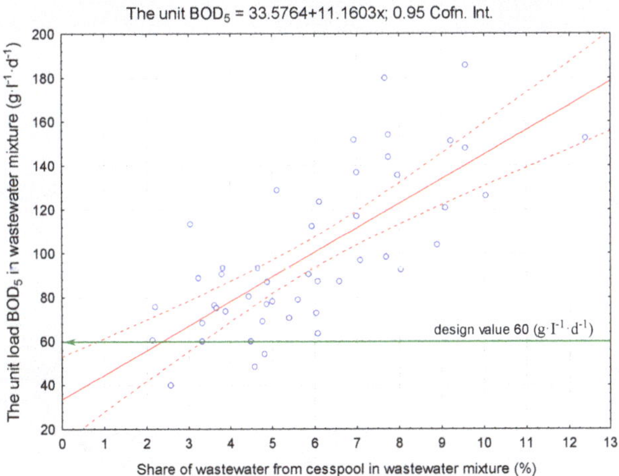

Figure 2. Connection share of wastewater from cesspool in wastewater mix (%) with unit load BOD$_5$ in wastewater mix and results of linear regression analysis.

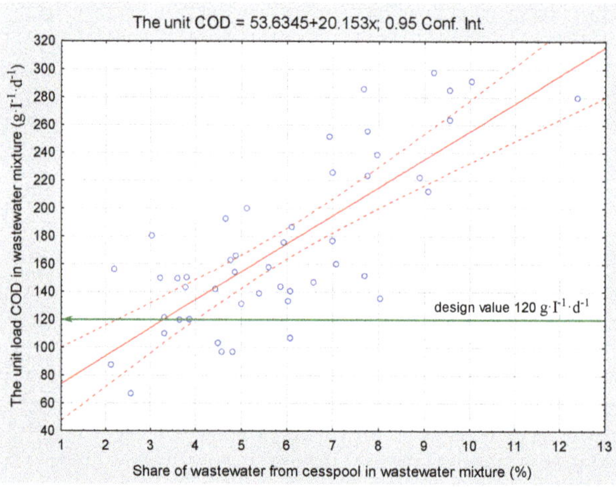

Figure 3. Connection share of wastewater from cesspool in wastewater mix (%) with unit load COD in wastewater mix and results of linear regression analysis.

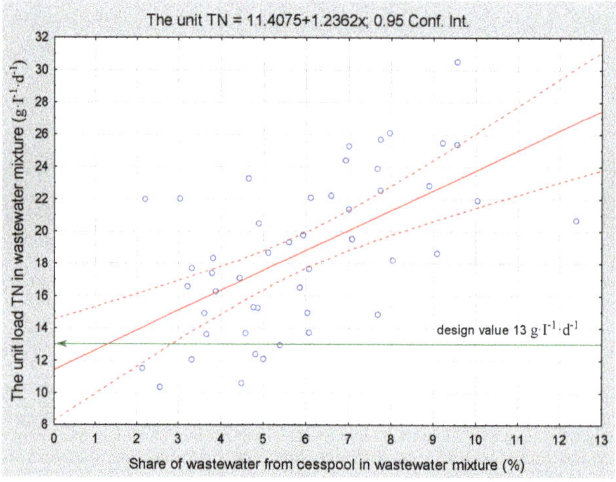

Figure 4. Connection share of wastewater from cesspool in wastewater mix (%) with unit load TN in wastewater mix and results of linear regression analysis.

Based on the analysis of the impact of the percentage of sewage delivered to the BOD_5 unit load in the sewage mixture, a correlation of $r_{xy} = 0.73$ was found, which in the scale proposed by Stanisz [34] defines this level of correlation as very high. In the analysed case, the correlation is statistically significant at the level of $\alpha = 0.05$. From the equation describing the regression line presented in Figure 2, it may be stated that a change (%) in the sewage supplied in the total sewage mixture by 1% causes a change in the BOD_5 unit load by 11.1 $g{\cdot}I^{-1}{\cdot}d^{-1}$. The dependence of the influence of the percentage of sewage delivered to the COD unit load in the sewage mixture was determined at the level of $r_{xy} = 0.78$, which also indicates the relationship of these variables at a very high level. In the case of the COD load from the equation describing the regression line shown in Figure 3, it is stated that a change in the proportion (%) of sewage delivered in the total sewage mixture by 1% causes a change in the COD unit load by 20.1 $g{\cdot}I^{-1}{\cdot}d^{-1}$. In the case of the analysed biogenic indicator, i.e., TN, the impact of the share (%) of sewage delivered on the unit load of Total Nitrogen in the wastewater mixture was at a high level, as indicated by the calculated correlation coefficient of $r_{xy} = 0.59$. From the equation describing the regression line presented in Figure 4, it can be stated that a change in the share (%) of sewage delivered in the total sewage mixture by 1% causes a change in the unit load of TN by $1.2\ g\cdot I^{-1}\cdot d^{-1}$. In all analysed cases of studied relationships, the correlation is statistically significant at the level of $\alpha = 0.05$.

In relation to the analysis of the correlation relationship between the concentration of organic and biogenic impurities (BOD_5, COD and TN) in the supplied wastewater and the size of the unit load of these parameters in the wastewater mixture, it was found that the correlation between the BOD_5 size of the delivered wastewater and the unit load of this parameter in the wastewater mixture is $r_{xy} = 0.61$. As follows from the equation describing the regression line in Figure 5, a change in BOD_5 value in the supplied sewage by $100\ g\cdot m^{-3}$ causes a change in the BOD_5 unit load in the sewage mixture by $2.1\ g\cdot I^{-1}\cdot d^{-1}$. The correlation of the COD value in the supplied sewage and the COD unit load in the sewage mixture was $r_{xy} = 0.52$. The equation describing the regression line in Figure 6 indicates that with a change in COD value of $100\ g\cdot m^{-3}$ in delivered sewage, there is a change in the COD unit load in the sewage mixture by $2.1\ g\cdot I^{-1}\cdot d^{-1}$. Whereas the correlation of the Total Nitrogen concentration in the supplied sewage and the unit load of Total Nitrogen in mixed sewage was $r_{xy} = 0.50$, and as the equation describing the regression line in Figure 7 shows, along with the change in the Total Nitrogen concentration in the sewage delivered by $100\ g\cdot m^{-3}$, the unit load changes TN in sewage mixed by $1.8\ g\cdot I^{-1}\cdot d^{-1}$. The level of correlation of the analysed variables in all cases in the scale proposed by Stanisz [34] was high. In the analysed cases, the correlation is statistically significant at the level of $\alpha = 0.05$.

Because the variability of the unit load BOD_5, COD and Total Nitrogen in the wastewater mixture depends on the percentage (%) in them of wastewater from cesspits delivered to the sewage treatment plant by the slurry tanker and on the size of these parameters in the supplied sewage, a partial correlation analysis was performed. Partial correlation analysis will allow to determine simultaneously the strength (relationship) of the relationship of two dependent variables to one independent variable.

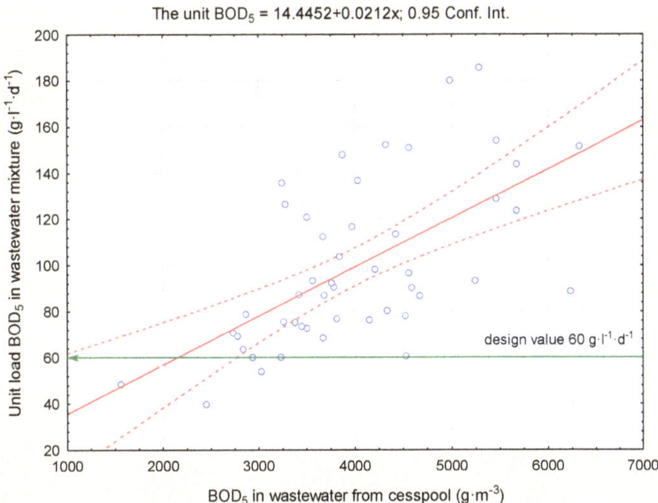

Figure 5. Connection BOD_5 in wastewater from cesspool with unit load BOD_5 in wastewater mix and results of linear regression analysis.

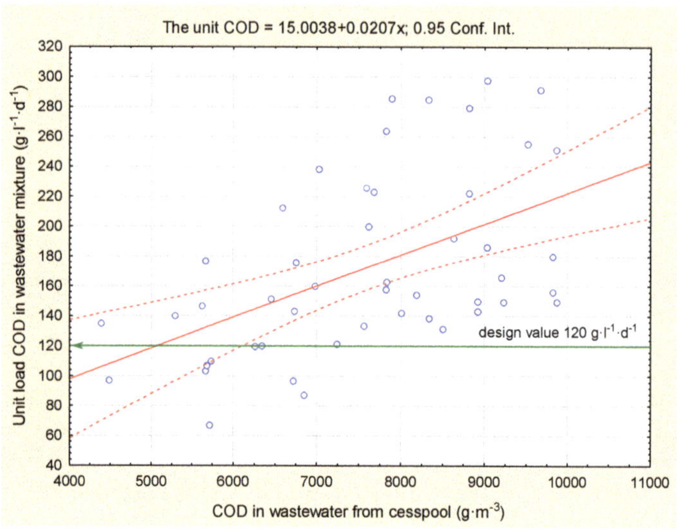

Figure 6. Connection COD in wastewater from cesspool with unit load COD in wastewater mix and results of linear regression analysis.

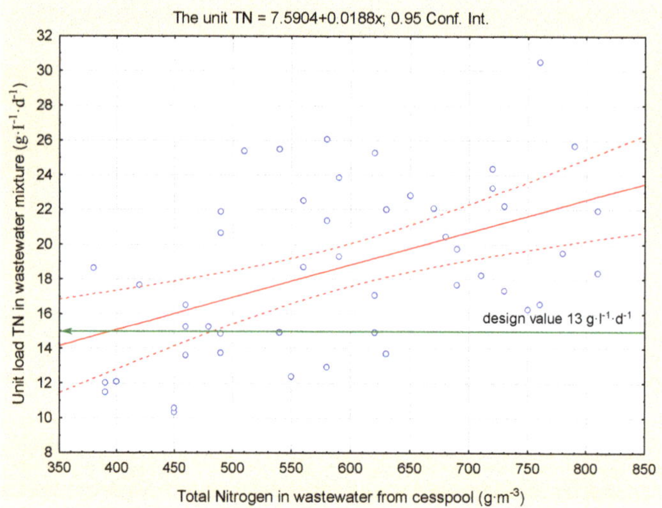

Figure 7. Connection TN in wastewater from cesspool with unit load TN in wastewater mix and results of linear regression analysis.

Based on the partial correlation analysis regarding the BOD_5 unit load in the wastewater mixture, it was found that the unit load of this parameter in the wastewater mixture depends on the percentage (%) of sewage delivered to them, as well as the value of this parameter in the supplied wastewater. However, the results of the partial correlation analysis indicate that the unit BOD_5 load in the wastewater mixture is more dependent on the percentage share of wastewater delivered than on the value of this indicator contained in it. The impact of the percentage of delivered wastewater on the BOD_5 unit load in the wastewater mixture was determined at the correlation level $R_c = 0.80$, while the impact of the BOD_5 value in the wastewater on the BOD_5 unit load in the total wastewater mixture was determined at the level of $R_c = 0.72$. On the scale provided by Stanisz [34], in both cases the relationship is at a very high level. The significance of the calculated correlation coefficients was tested by the Student's *t*-test at the significance level of $\alpha = 0.05$. In both cases, the significance of the studied relationships was found. In the case of partial correlation analysis regarding COD, it was found that the share of supplied sewage has a greater impact on the COD unit load in the wastewater mixture than the value of this parameter contained therein. The impact of the percentage share of sewage delivered to the COD unit load in the wastewater mixture was determined at the correlation level of $R_c = 0.87$, while the effect of the value of COD in the wastewater on the COD unit load in the total wastewater mixture was determined at the level of $R_c = 0.75$. In both cases, the correlation relationship is at a very high level, and the examined relationships are statistically significant at the level of $\alpha = 0.05$. With reference to the unit load of Total Nitrogen in the wastewater mixture, it was found that its concentration in the wastewater mixture at a very high level has a percentage (%) of the supplied wastewater, where $R_c = 0.72$, while at a high level, the effect of the concentration of this parameter in the wastewater was noted as delivered, where $R_c = 0.66$.

In order to indicate the optimal amount of supplied sewage, which was added to the inflowing sewage, so that the unit load of the analysed indicators was at the assumed level in the project, the nomograms are presented in Figures 8–10. Nomograms for individual indicators were developed based on the results of partial correlations. From the developed nomograms in Figures 8–10, it is possible to forecast (predict) the unit load of a given indicator in the wastewater mixture depending on the percentage (%) of wastewater delivered in the wastewater mixture and on the value $(g \cdot m^{-3})$ of this indicator in the supplied wastewater and the share percentage (%) of sewage delivered. The optimal load of the analysed indicators can be described by the formulas below:

- $ULmix_{BOD5}$ $(g \cdot l^{-1} \cdot d^{-1})$ = $-28.7907 + 9.8888 \cdot BOD_5$ in wastewater form cesspool + 0.0174% share of delivered wastewater
- $ULmix_{COD}$ $(g \cdot l^{-1} \cdot d^{-1})$ = $-86.1077 + 19.4221 \cdot COD$ in wastewater form cesspool + 0.0190% share of delivered wastewater
- $ULmix_{TN}$ $(g \cdot l^{-1} \cdot d^{-1})$ = $-1.2271 + 1.3171 \cdot TN$ in wastewater form cesspool + 0.0205% share of delivered wastewater

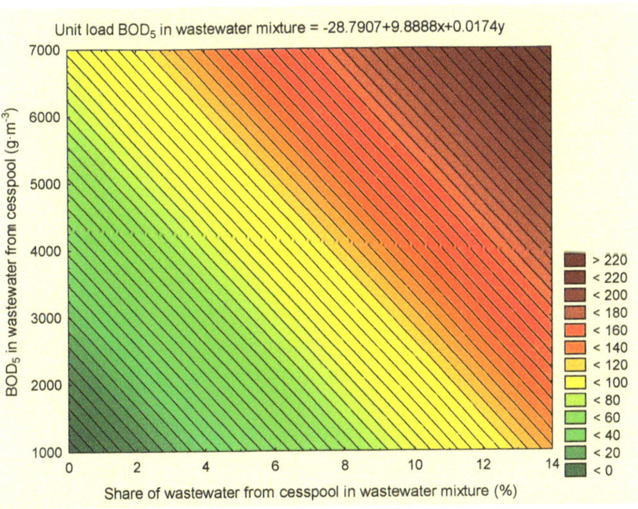

Figure 8. Nomogram to forecast unit load BOD₅ in the wastewater mix on the basis of percentage share of the inflow wastewater in the wastewater mix and the value BOD₅ in inflowing wastewater.

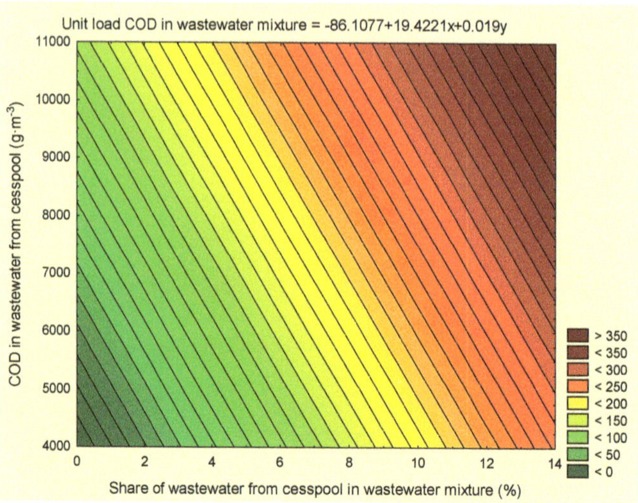

Figure 9. Nomogram to forecast unit load COD in the wastewater mix on the basis of percentage share of the inflow wastewater in the wastewater mix and the value COD in inflowing wastewater.

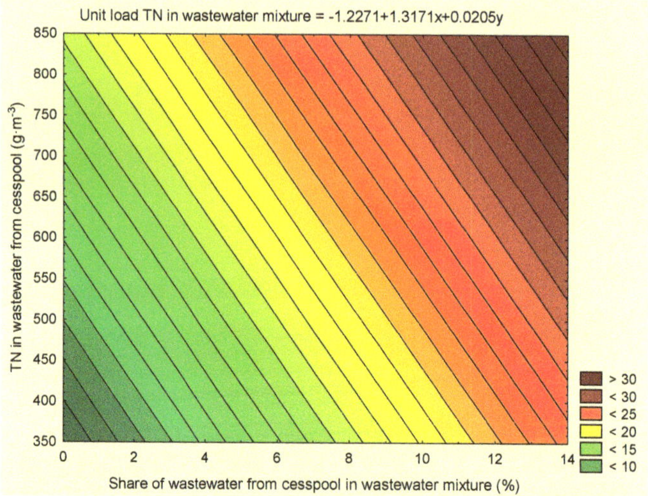

Figure 10. Nomogram to forecast unit load TN in the wastewater mix on the basis of percentage share of the inflow wastewater in the wastewater mix and the value TN in inflowing wastewater.

That the unit load in the wastewater subjected to the treatment process oscillated within the limits of the designed load, assuming that in delivered wastewater the median BOD_5 value is 3825.0 $g·m^{-3}$, the COD value is 7750.0 $g·m^{-3}$ and the Total Nitrogen concentration is 585.0 $g·m^{-3}$, and the percentage of wastewater transported in the mixture should be between 4% and 6% (average 5%). Assuming the average daily amount of treated wastewater, which during the study period was 253.5 $m^3·d^{-1}$, the amount of wastewater transported by the slurry tanker should be from 10.2 $m^3·d^{-1}$ to 15.2 $m^3·d^{-1}$ (average 12.7 $m^3·d^{-1}$).

4. Conclusions

Based on the analysis carried out, it was found that the unit loads of BOD$_5$, COD and TN in the mixture of wastewater subjected to the treatment process will be at the level of loads assumed in the project, when the share of sewage delivered from cesspool will be at the level of 5% in the total amount of wastewater. Bearing in mind that the period of conducted research, where the total average daily amount of sewage was 253.5 m^3·d^{-1}, the amount of wastewater from cesspool delivered should be 12.7 m^3·d^{-1}. An important aspect and practical guideline for wastewater plants operators is the need for sewage-transported wastewater from cesspool to be dosing evenly every day of the week, including Saturday and Sunday, to the sewage flowing in from the sewage system. Because sewage from cesspool is delivered and mixed with sewage flowing only on weekdays, i.e., from Monday to Friday, it is advisable to build a tank with the right volume to collect sewage delivered so that it is possible to collect these wastes and then their even-dosing in an appropriate proportion on every day of the week. In the analysed case, the volume of the retention reservoir for the supplied sewage should provide a two-day volume resulting from the guidelines indicated in the analysis, i.e., 25.4 m^3 (2 × 12.7 m^3·d^{-1}). Moreover, it is very important that the wastewater from this reservoir is dosed evenly to the treatment system over the weekend, e.g., with an interval of 0.5 m^3·h^{-1}. The technological system of the wastewater treatment plant for the reception and dosage of sewage from septic tanks should be rebuilt. All supplied sewage should go to the collection point, then flow to the retention tank and then to the treatment technological system. The retention tank should be equipped with a properly programmed system (pump + controller) for even dosing of sewage to the process line of the treatment plant. Personnel servicing the sewage treatment plant should constantly monitor the amount of inflowing wastewater and its quality (concentration of pollution indicators) in order to determine the possibility of increasing the amount of sewage from cesspool, which may be an admixture of sewage subjected to the treatment process. Along with the extension of the sewage network in the commune, which will contribute to an increase in the amount of sewage flowing into the treatment plant, it is possible to increase the number of farms from which sewage from cesspool will be transported to the WWTP.

Author Contributions: Conceptualization, P.B.; methodology, P.B. and A.O.; software, P.B.; validation, P.B. and K.K.; formal analysis, P.B. and A.O.; investigation, P.B. and A.O.; resources, P.B. and D.M.; data curation, P.B., A.W. and K.K.; writing—original draft preparation, P.B. and A.O.; writing—review and editing, P.B. and D.M.; visualization, P.B.; supervision, P.B., A.O. and A.W. funding acquisition, P.B., K.K., and A.O. All authors have read and agreed to the published version of the manuscript.

Funding: Publication is funded by the Polish National Agency for Academic Exchange under the International Academic Partnerships Programme from the project "Organization of the 9th International Scientific and Technical Conference entitled Environmental Engineering, Photogrammetry, Geoinformatics—Modern Technologies and Development Perspectives".

Conflicts of Interest: The authors declare no conflict of interest.

References

1. Central Statistical Office. *Gospodarka Mieszkaniowa i Infrastruktura Komunalna w 2018 r*; Główny Urząd Statystyczny: Warszawa, Poland, 2019. (In Polish)
2. Błażejewski, R.; Nawrot, T. Jak uszczelnić system gromadzenia i dowożenia nieczystości ciekłych? *GazWoda I Tech. Sanit.* **2009**, *9*, 2–3. (In Polish)
3. Heinonen-Tanski, H.; Savolainen, R. Disinfection of Septic Tank and Cesspool Wastewater with Peracetic Acid. *J. Hum. Environ.* **2003**, *32*, 358–361. [CrossRef] [PubMed]
4. Palarz, H. *Nielegalny Pobór Wody i Nielegalne Odprowadzenie Ścieków—Aspekty Prawne*; Wolters Kluwer, S.A.: Warszawa, Poland, 2015. (In Polish)
5. Jóźwiakowski, K.; Listosz, A.; Gizińska-Górna, M.; Pytka, A.; Marzec, M.; Sosnowska, B.; Kowalczyk-Juśko, A.; Grzywna, A.; Mazur, A.; Obroślak, R. Effect of anthropogenic pollutants on the quality of surface waters and groundwaters in the catchment basin of lake Bialskie. *J. Ecol. Eng.* **2016**, *17*, 154–162. [CrossRef]

6. Dymaczewski, Z. *Poradnik Eksploatatora Oczyszczalni Ścieków*; PZITS o/Wielkopolski: Poznań, Poland, 2011. (In Polish)

7. Zdebik, D.; Głodniok, M.; Zawartka, P. Badania symulacyjne procesu fermentacji w układzie komory psychrofilnej i komory mezofilnej w odniesieniu do ilości wytwarzanego biogazu. *Inżynieria Ekol.* **2015**, *42*, 63–67. (In Polish) [CrossRef]

8. Jeleń, U.; Wyrwik, S. Wpływ ścieków dowożonych beczkowozami na prawidłową pracę małej oczyszczalni ścieków na podstawie eksploatacji oczyszczalni w Trzebini-Sierszy. *Forum Eksploatatora* **2003**, *3*, 5–8. (In Polish)

9. Bergel, T. Practical implication of tap water consumtpion structure in rural households. *J. Ecol. Eng.* **2017**, *18*, 231–237. [CrossRef]

10. Bugajski, P.; Kurek, K.; Młyński, D.; Operacz, A. Designed and real hydraulic load of household wastewater treatment plants. *J. Water Land Dev.* **2019**, *40*, 155–160. [CrossRef]

11. Ingallinella, A.M.; Sanguinetti, G.G.; Vazquez, H.P.; Fernández, R.G. Treatment of wastewater transported by vacuum trucks. *Water Sci. Technol.* **1996**, *33*, 239–246. [CrossRef]

12. Borchardt, M.A.; Chyou, P.-H.; DeVries, E.O.; Belongia, E.A. Septic System Density and Infectious Diarrhea in a Defined Population of Children. *Environ. Health Perspect.* **2003**, *111*, 742–748. [CrossRef]

13. Sobsey, M.D.; Wallis, C.; Melnick, J.L. Chemical disinfection of holding tank sewage. *J. Appl. Microbiol.* **1974**, *28*, 861–866. [CrossRef]

14. Bugajski, P.; Chmielowski, K.; Cupak, A.; Wąsik, E. Influence of sewage from septic tanks on the variability concentration of pollutants in sewage undergoing purification processes. *Infrastruct. Ecol. Rural Areas* **2016**, *2*, 517–526.

15. Bugajski, P.; Satora, S. The balance of wastewater inflowing and brought to the treatment plant based on example of the chosen object. *Infrastruct. Ecol. Rural Areas* **2009**, *5*, 73–82.

16. Elmitwalli, T.A.; Ralf, O. Anaerobic biodegradability and treatment of grey water in upflow anaerobic sludge blanket (UASB) reactor. *Water Res.* **2007**, *41*, 1379–1387. [CrossRef] [PubMed]

17. Krzanowski, S.; Wałęga, A. Effectiveness of organic substance removal in household conventional activated sludge and hybrid treatment plants. *Environ. Prot. Eng.* **2008**, *34*, 5–12.

18. Ladu, J.-L.C.; Lü, X. Effects of hydraulic retention time, temperature, and effluent recycling on efficiency of anaerobic filter in treating rural domestic wastewater. *Water Sci. Eng.* **2014**, *7*, 168–182.

19. Lu, S.; Pei, L.; Bai, X. Study on method of domestic wastewater treatment through new-type multi-layer artificial wetland. *Int. J. Hydrog. Energy* **2015**, *40*, 11207–11214. [CrossRef]

20. Nowobilska–Majewska, E.; Bugajski, P. The Impact of Selected Parameters on the Condition of Activated Sludge in a Biologic Reactor in the Treatment Plant in Nowy Targ, Poland. *Water* **2020**, *12*, 2657. [CrossRef]

21. Forbis-Stokes, A.A.; Arumugam, A.; Ravindran, J.; Deshusses, M.A. Technical evaluation and optimization of a mobile septage treatment unit. *J. Environ. Manag.* **2020**, *277*, 111361. [CrossRef]

22. Jong, V.S.W.; Tang, F.E. Septage Treatment Using Pilot Vertical-flow Engineered Wetlands System. *Pertanika J. Sci. Technol.* **2014**, *22*, 613–625.

23. Mancl, K.; Rosencrans, R. *Water* augmentation through onsite wastewater management. In Proceedings of the 9th National Symposium on Individual and Small Community Sewage Systems, ASAE, St Joseph, MI, USA, 11–14 March 2001; pp. 358–364.

24. Mucha, J. *Geostatistical Methods in Documenting Deposits. Script, Department of Mine Geology*; Wydawnictwo AGH: Kraków, Poland, 1994; p. 155. (In Polish)

25. Abbassi, B.E.; Abuharb, R.; Ammary, B.; Almanaseer, N.; Kinsley, C. Modified Septic Tank: Innovative Onsite Wastewater Treatment System. *Water* **2018**, *10*, 578. [CrossRef]

26. Gajewska, M. Influence of composition of raw wastewater on removal of nitrogen compounds in multistage treatment wetlands. *Environ. Prot. Eng.* **2015**, *41*, 19–30. [CrossRef]

27. Koutsou, O.P.; Gatidou, G.; Stasinakis, A.S. Domestic wastewater management in Greece: Greenhouse gas emissions estimation at country scale. *J. Clean. Prod.* **2018**, *188*, 851–859. [CrossRef]

28. Nowobilska–Majewska, E.; Bugajski, P. The analysis of the amount of pollutants in wastewater after mechanical treatment in the aspect of their susceptibility to biodegradation in the treatment plant in Nowy Targ. *J. Ecol. Eng.* **2019**, *20*, 135–143. [CrossRef]

29. Kaczor, G.; Chmielowski, K.; Bugajski, P. Wpływ sumy rocznej opadów atmosferycznych na objętość wód przypadkowych dopływających do kanalizacji sanitarnej. *Rocz. Ochr. Środowiska* **2017**, *19*, 668–681. (In Polish)

30. Nowobilska–Majewska, E.; Bugajski, P. Influence of the amount of inflowing wastewater on concentrations of pollutions contained in the wastewater in the Nowy Targ sewerage system. *E3s Web Conf.* **2019**, *86*, 24. [CrossRef]

31. Kaczor, G.; Chmielowski, K.; Bugajski, P. Influence of extraneous waters on the quality and loads of pollutants in wastewater discharged into the treatment plant. *J. Water Land Dev.* **2017**, *33*, 73–78. [CrossRef]

32. Cieślak, O.; Pawełek, J. Dopływ wód obcych do kanalizacji sanitarnej na przykładzie gminy Mézos we Francji. *Instal* **2014**, *7–8*, 90–95. (In Polish)

33. Madryas, C.; Przybyła, B.; Wysocki, L. *Badania i Ocena Stanu Technicznego Przewodów Kanalizacyjnych*; Dolnośląskie Wydawnictwo Edukacyjne: Wrocław, Poland, 2010. (In Polish)

34. Stanisz, A. *Przystępny Kurs Statystyki, Tom 1*; Wydawnictwo StatSoft Polska Sp. z o.o.: Kraków, Poland, 1998.

Publisher's Note: MDPI stays neutral with regard to jurisdictional claims in published maps and institutional affiliations.

 *sustainability*

Review

Heat-Mitigation Strategies to Improve Pedestrian Thermal Comfort in Urban Environments: A Review

Nazanin Nasrollahi [1], Amir Ghosouri [1], Jamal Khodakarami [1] and Mohammad Taleghani [2,*]

[1]   Department of Architecture, Faculty of Technology and Engineering, Ilam University, 69134 Ilam, Iran; n.nasrollahi@ilam.ac.ir (N.N.); amirghosouri72@gmail.com (A.G.); j.khodakarami@ilam.ac.ir (J.K.)

[2]   Leeds School of Architecture, Leeds Beckett University, Leeds LS1 3HE, UK

*   Correspondence: m.taleghani@leedsbeckett.ac.uk

Received: 4 November 2020; Accepted: 27 November 2020; Published: 30 November 2020

Abstract: Thermal comfort is one of the main factors affecting pedestrian health, and improving thermal comfort enhances walkability. In this paper, the impact of various strategies on thermal-comfort improvement for pedestrians is thoroughly evaluated and compared. Review studies cover both fieldwork and simulation results. These strategies consist of shading (trees, buildings), the orientation and geometry of urban forms, vegetation, solar-reflective materials, and water bodies, which were investigated as the most effective ways to improve outdoor thermal comfort. Results showed that the most important climatic factors affecting outdoor thermal comfort are mean radiant temperature, wind speed, and wind direction in a microclimate. The best heat-mitigation strategy for improving thermal comfort was found to be vegetation and specifically trees because of their shading effect. The effect of height-to-width (H/W) ratio in canyons is another important factor. By increasing H/W ratio, the thermal-comfort level also increases. Deploying highly reflective materials in urban canyons is not recommended, as several studies showed that they could reflect solar radiation onto pedestrians. Results also showed that, in order to achieve a satisfactory level of thermal comfort, physiological and psychological factors should be considered together.

Keywords: thermal comfort; outdoor environments; pedestrians; heat mitigation; microclimates

---

1. Introduction

A significant part of the global rural population has migrated to cities, and urban populations are rapidly growing [1]. In 2003, the United Nations predicted that about 61% of the global population will live in cities by 2030 [2]; this has already happened in many developed and developing countries. Population growth in cities is aligned with the increase in construction and urban densification [3], which ultimately result in thermal discomfort in cities.

The outdoor environment is of high importance in cities, since it includes various pedestrian activities. The comfort level of pedestrians in such spaces has direct impact on the presence of people in outdoor environments [4–7].

Thermal comfort is one of the most important factors affecting the quality of outdoor environments for pedestrians [8]. Better thermal comfort leads to the presence of more people in open spaces [9]. Urban open spaces such as town squares, green spaces, or parks bring different environmental, social, and economic benefits [8]. Thermal discomfort, on the other hand, reduces the power of thinking and concentration of pedestrians [10]. Therefore, thermal comfort in hot and cold seasons is considered a necessity for users of outdoor environments. Thermal comfort in open spaces is crucial and must be thoroughly considered when designing an open space since it is affected by a wide range of variables [11].

1.1. Research Method

These steps were followed to prepare the review paper:

(1) Data collection: research began with peer-reviewed papers published in English within the ScienceDirect, Scopus, Wiley, and Springer databases.
(2) Postprocessing of collected data: papers were categorised on the basis of their general topics, forming the four main sections of the paper.
(3) Classification: papers in each section (each heat-mitigation strategy) were studied, and the findings of each study were recorded.
(4) Writing up the body: each section was comprehensively written, including studies from different climates for each chapter (heat-mitigation strategy).
(5) Conclusion and final review: the conclusion was written considering that it should respond to the reviewed sources in terms of the widely used research methods, used software, studied climates, etc.

Regarding the structure of this paper, the chronology of outdoor thermal-comfort studies is first introduced. Our research keywords were "outdoor thermal comfort", "heat-mitigation strategies", "thermal-comfort indices", and "urban-canyon geometries". Studies were covered that had been published since the 1970s. Second, different research methods used in outdoor studies are presented, and the frequency of using different methods is shown with a graph. Third, the different simulation software used in modelling outdoor comfort studies is presented. Different indices used for measuring outdoor thermal comfort are also addressed. Lastly, different heat-mitigation strategies within urban environments are reviewed. The main contribution (and novelty) of this paper to the current body of the literature is that, on the basis of different research methods and indices of thermal-comfort studies, heat-mitigation strategies are comprehensively presented. In contrast to previous review studies that focused on nature-based solutions, canyon geometries, or green/reflective materials, all these strategies are reviewed here, considering their research method(s), geography, comfort index, and effectiveness in improving pedestrian thermal comfort. Results of this review paper help to better understand different outdoor-thermal-comfort approaches that are practised in different climates and countries, and the selection of suitable strategies in practice (see Figure 1).

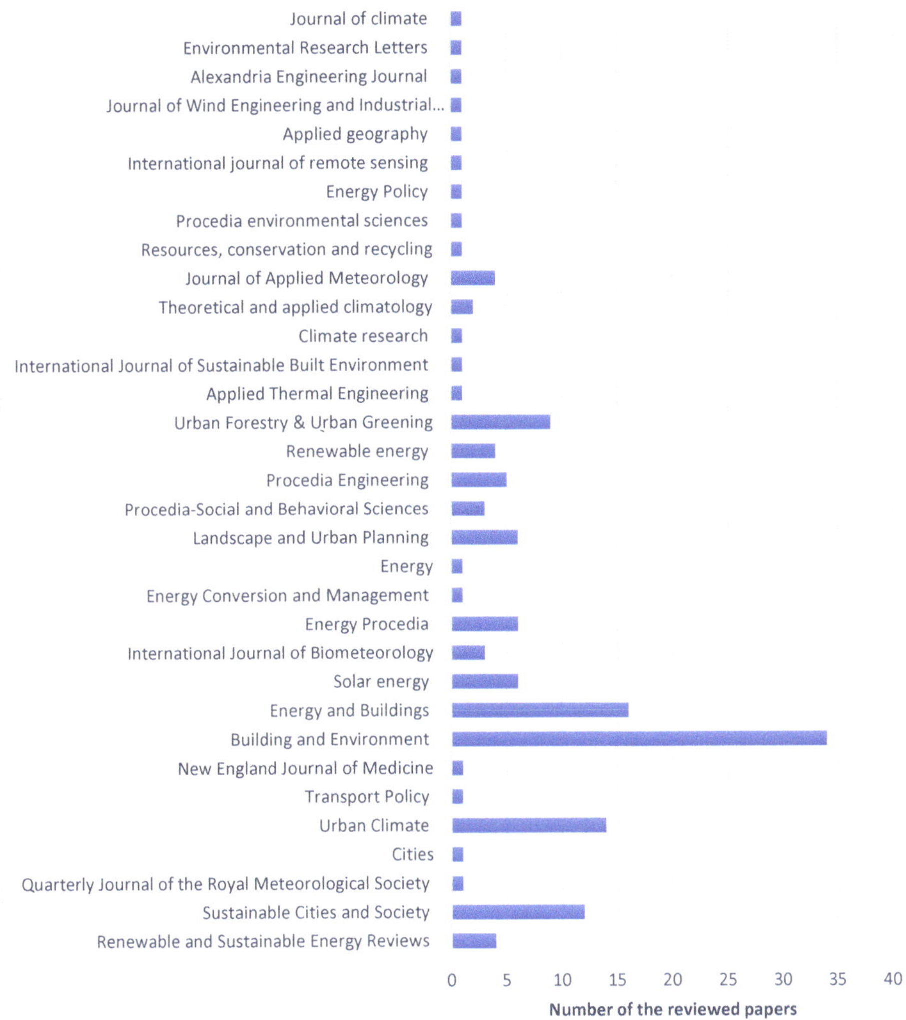

Figure 1. Journal specification and frequency of sources used in this review article.

Due to the complexity of outdoor environments (compared to indoor environments), thermal comfort in open spaces is less studied. The beginning of such studies dates back to the last few decades of the 20th century. Figure 2 shows the annual record of publications in this topic. There were few studies in the 20th century regarding outdoor thermal comfort. In Figure 3, the most common research methods on thermal comfort are shown with fieldwork, simulations, and their combination. However, in recent years, the development of simulation software has led to a rapid growth in the number of simulation-based studies in combination with fieldwork.

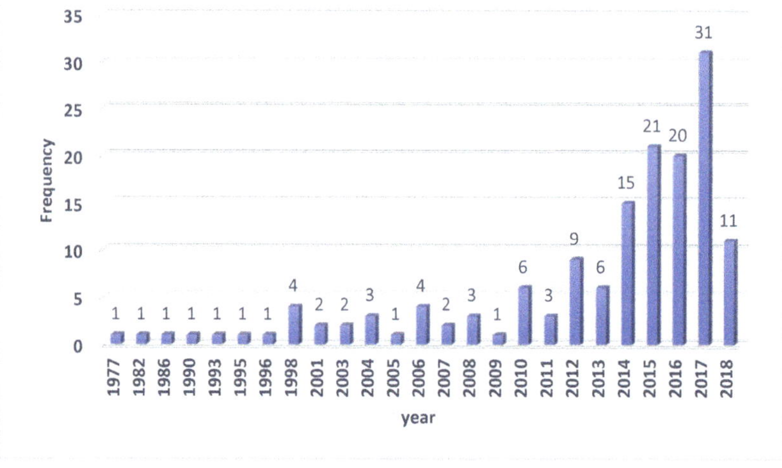

Figure 2. Year-dispersion graph of related studies.

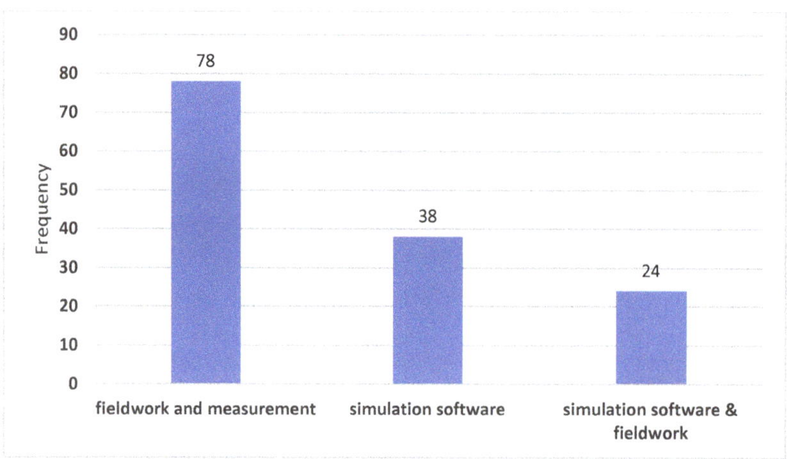

Figure 3. Research methodology regarding outdoor thermal comfort.

Regarding the novelty of this paper:

(a) Outdoor-thermal-comfort papers published from 1977 onwards are comprehensively examined. Papers from journals with high-impact factors were specifically considered. In total, 153 studies were reviewed.

(b) Most previous studies focused on specific climates. In this paper, various heat-mitigation strategies in different global climates, from Canada to Australia, were reviewed.

(c) Previous papers reviewed either nature-based solutions (green strategies such as living walls) or urban design solutions (e.g., canyon effects). This study utilises a holistic approach to include all aspects of heat-mitigation strategies for urban designers and planners.

1.2. Background of Outdoor-Thermal-Comfort Studies

In 1971, the first studies were carried out regarding the impact of microclimates on outdoor activities [4]. Using the number of people sitting on shaded and unshaded benches showed that sunny or shady conditions affected people's willingness to stay or leave. It could be concluded that the physical conditions of a location affect thermal comfort.

In 1982, Fanger [4] suggested and presented the predicted mean vote (PMV), which predicts the average heat response of people on a 7-point scale to assess their thermal comfort. In 1987, Mayer and Hoppe [4] presented the physiologically equivalent temperature (PET) for the assessment of thermal comfort in external environments (see Appendix A for PET ranges).

In 2001, one of the very first studies in the field of outdoor thermal comfort was based on people's behaviour. In this study, Nikolopoulou et al. [12] examined the thermal-comfort conditions within open spaces in Cambridge, United Kingdom. They evaluated the sensory perception of every individual on a scale of 1–5. In this study, only 35% of the participants experienced the desired thermal comfort. It was concluded that a physiological approach for the assessment of outdoor thermal conditions is not sufficient, while the health history and expectations of individuals play a significant role.

In 2004, Thorsson et al. [13] investigated the impact of biological conditions on people's behavioural patterns via 280 questionnaires in a park as a resting place in Gothenburg, Sweden. A comparison of the results showed that thermal expectations had significant impact on the mental assessment of individuals regarding the thermal comfort of their surrounding environment.

In 2010, Lin et al. [14] studied the effect of shadowing on thermal comfort in outdoor environments. They conducted 12 field tests at a university campus in central Taiwan with a tropical climate. They evaluated the thermal conditions of the campus using RayMan software to calculate the PET index. It was concluded that in the very hot summers and mild winters of Taiwan, a thermally comfortable microclimate is possible with the shading impact of trees and buildings.

In 2012, Makaremi et al. [11] used PET to assess the outdoor environment of the Malaysian Putra University campus (tropical climate). They found out that shaded places have a longer period of acceptable temperature range. Furthermore, while studying the temperature tolerance of native and non-native students, they found out that native students could tolerate a higher temperature rate in comparison with non-native students due to their thermal adaptation to Malaysia's climate.

Huang et al. [15] investigated temperature differences within a university in northwestern China considering different scenarios with increased green spaces, water elements, and highly reflective surfaces using ENVI-met. It was concluded that increasing green spaces led to a maximal reduction of temperature by 0.3 °C, as well as a decrease in maximal mean radiant temperature by 32.1 °C.

Taleghani [16] concluded that, among different climatic factors, mean radiant temperature has the greatest impact on thermal comfort in outdoor environments. He also found that using vegetation in urban environments is better than using highly reflective surfaces.

Salata et al. [17] measured air temperature within the campus of Sapienza University of Rome, Italy (Mediterranean climate). They found that concrete pavements had higher albedo and lower thermal capacity than those of asphalt, and this could improve thermal conditions.

Studies in the past few decades were mainly based on measurements, field observations, and questionnaires. These studies further examined the causes and effects that affect human thermal comfort in outdoor environments. In recent decades, simulation tools for outdoor environments have revolutionised the development of these studies. These simulation programmes evaluate the outdoor thermal environment using various indices. Figures 4 and 5 show the extent of using the software and indices used in the reviewed studies in this paper.

On the basis of Figures 4 and 5, it is evident that ENVI-met and RayMan, respectively, are the most popular simulation tools. In addition, PET and PMV indices are widely used to evaluate thermal environments in various studies.

In many studies conducted in recent years, outdoor-thermal-comfort assessment was performed on the basis of PET index. In these studies, there is a table defining the relationship between thermal perception and PET index or acceptable temperature in the climate. Table 1 shows the neutral or acceptable temperature in a number of studied climates, and the acceptable temperature range or neutral PET for different climates in order to obtain a desirable thermal condition.

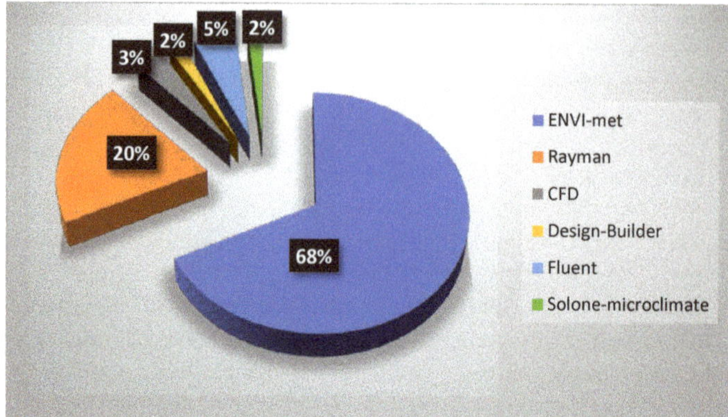

Figure 4. Usage percentage of various simulation tools regarding outdoor thermal comfort.

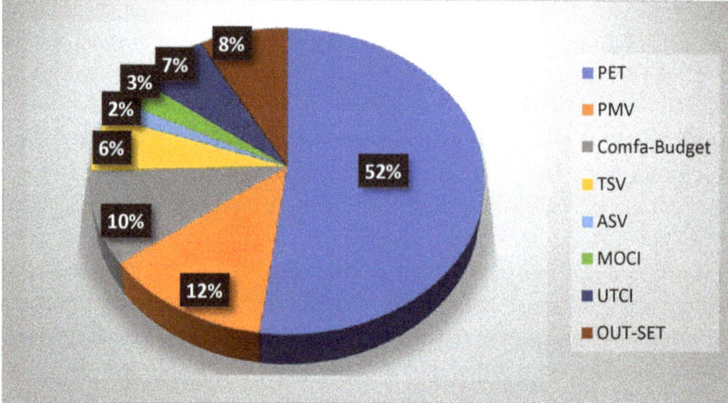

Figure 5. Usage percentage of various indices to assess outdoor-thermal-comfort conditions.

Table 1. Neutral physiologically equivalent temperature (PET; acceptable temperatures) in different climates.

Geographical Region	Climate	Temperature Range (°C)	References
Malaysia	Temperate	18–23	[11]
Malaysia	Subtropical	26–30	[11]
Isfahan, Iran	Hot and dry	23.06–29.73	[18]
Central and western Europe	Temperate	18–23	[19]
Taiwan	Tropical	26–30	[19]
Crete, Greece	Mediterranean	20–25	[20]
Athens, Greece	Mediterranean	18–23	[21]
Hong Kong	Hot and humid	28	[22]
Nis, Serbia	Temperate	18–23	[23]
Sao Paulo, Brasilia	Hot and humid	27.2	[24]
Hong Kong	Tropical	25–29	[25]
Sydney, Australia	Subtropical	26.2	[26]
Belo Horizonte, Brasilia	Tropical	19–27	[27]
Belo Horizonte, Brasilia	Tropical	16–30	[28]
Freiburg, Germany	Continental	18–28	[28]
Ibadan, Nigeria	Tropical	23–27	[29]
Dhaka, Bangladesh	Tropical	28.5–32.8	[30]
Singapore	Tropical	26–31.7	[31,32]
Guangzhou, China	Subtropical	28.54–31	[32]

In this review, the impact of some of the most important and influential variables on the thermal comfort of outdoor environments is investigated. Table 2 illustrates and categorises research themes in the field of outdoor thermal comfort.

Table 2. Themes in field of outdoor comfort. Note: H/W, height to width; SVF, sky-view factor.

	Subject	Number	References
1-	Climatic parameters affecting outdoor comfort	17	[8,10,16,18–20,31,33–42]
2-	Effect of shading on outdoor comfort	15	[21–23,34,42–52]
3-	Effect of H/W and SVF on outdoor comfort	25	[14,21,24,30,37,38,53–71]
4-	Effect of trees (shading and morphology) on outdoor comfort	26	[10,21,24,25,33,35,53,55,57,72–88]
5-	Effect of orientation and geometric form of urban canyons on outdoor thermal comfort	12	[24,37,39,63,89–96]
6-	Effect of green, blue, and white surfaces on outdoor comfort	12	[15,16,97–106]
7-	Effect of vegetation on outdoor comfort	25	[17,48,56,67,102,103,107–125]
8-	Effect of ceiling and green walls in urban canyons on outdoor comfort	10	[62,68,82,126–132]
9-	Impact of modern materials in urban canyons on outdoor comfort	10	[133–142]
10-	Effect of water elements on outdoor comfort	6	[36,143–147]
11-	Impact of psychological factors on outdoor thermal comfort	7	[9,12,26,148–153]

Figure 6 shows the type of urban spaces used in the reviewed thermal-comfort studies. Most studies were performed in the field of outdoor thermal comfort in public spaces. The study of urban canyons, university campuses, and urban parks follows. The fewest studies regarding this topic are about historical sites, residential complexes, and urban squares.

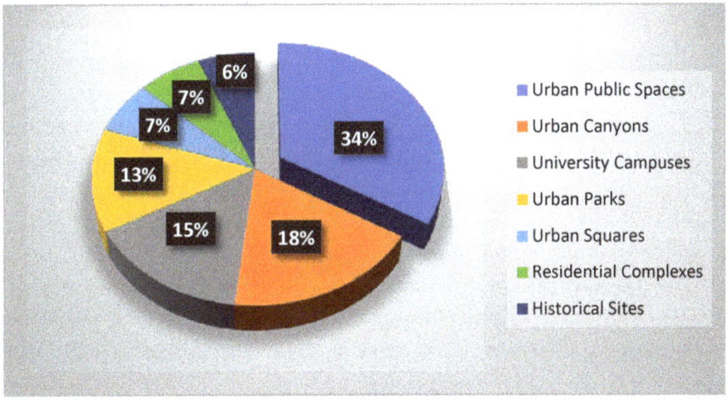

Figure 6. Usage percentage of studies in field of outdoor comfort.

The distribution map of these studies in different climates is shown in Figure 7, showing that these studies focused more on Europe and East Asia.

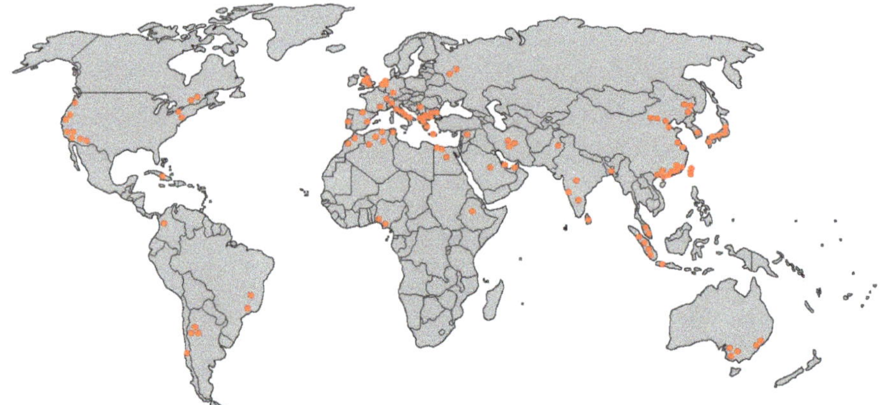

Figure 7. Climate-dispersion map of outdoor-thermal-comfort studies.

2. Climatic Parameters Affecting Thermal Comfort

Climatic parameters that affect thermal comfort are air and mean radiation temperature, relative humidity, and wind speed. Among climatic parameters stated in several studies, mean radiation temperature (mostly derived from solar radiation) is known as the most influential factor affecting thermal comfort in outdoor environments [16,18,31,33–38].

Taleghani et al. [39] researched outdoor external thermal comfort by examining five different urban forms in the Netherlands. In another study on a university campus in Hong Kong, it was also observed that radiation temperature and wind speed play major roles in creating thermal conditions in outdoor environments [19]. Mahmoud [40] also investigated the thermal-comfort level in an urban park in Cairo, Egypt with a hot and dry climate. They found out that the most important factors affecting outdoor thermal comfort are mean radiant temperature and wind speed. In the analysis of outdoor thermal comfort at the Guangzhou Higher-Education Megacentre, Li and Lixiu [41] found that air temperature and mean radiant temperature are the most influential factors on outdoor thermal comfort. In another study, Tsitoura et al. [20] obtained similar results with those of Li and Lixiu while investigating thermal comfort in the island of Crete, Greece.

Yoshida et al. [10] showed that radiant and air temperature are the most important factors with regard to outdoor thermal comfort while examining the effects of tree canopies on the thermal environment of the University of Osaka, Japan. In a study of thermal comfort in Harbin, China, Jin et al. [42] found that, in warm seasons, radiant temperature has the greatest effect on thermal comfort in outdoor environments, followed by wind speed and air temperature.

Chen et al. [8] examined a city square/park during the cold seasons of Shanghai, and considered air temperature and mean radiant temperature as the most important factors regarding outdoor thermal comfort in winter. They also found that people's presence in outdoor environments during winter is directly related to available solar radiation; the longer the sun is available, the longer the amount of time that people spend in outdoor environments.

Reviewing thermal-comfort studies in outdoor environments, it is evident that, in order to obtain thermal comfort, it is important to address strategies such as shading and the use of green–white–blue surfaces while considering the form and geometry of city canyons, and the psychological factors of individuals. In most thermal-comfort studies, the main focus is on only one or two climatic factors and the strategies to improve them. However, the strategies used for other climatic factors may have an adverse effect on other factors. Therefore, comprehensive attention on all climatic factors is recommended.

3. Shading Effect

Controlling solar radiation is the most important factor affecting outdoor thermal comfort, especially in hot seasons [43–45]. Kariminia et al. [34] investigated outdoor thermal comfort within the two urban squares of Naghshe-Jahan and Jolfa in the hot arid climate of Isfahan, Iran. Using field measurements and questionnaires, they found out that Jolfa has more comfortable hours mainly due to the shading effect of its walls.

Hwang et al. [50] explored the effect of urban canopies on thermal comfort in different seasons. By using RayMan software, they showed that, in summer, spring, and autumn, shading is recommended, whereas in winter, there is minimal need for shadowing. Therefore, they proposed deciduous trees.

Ng and Cheng [22] investigated thermal comfort in the hot and humid climate of Hong Kong using field measurements. They concluded that, at a temperature of 29.7 °C, surfaces exposed to direct radiation experienced a surface temperature in the range of 50–60 °C, while this value for shaded surfaces was in the range of 30–34 °C.

Watanabe et al. [51] studied thermal comfort within the campus of the University of Nagoya, Japan in both shaded and unshaded areas. They found that, under solar-radiation intensity of 800 W/m^2, the universal-effective-temperature index was reduced by 18.4 °C by the shading of buildings, and by 16.2 °C by the pergola.

Morakinyo et al. [52], in an experiment on the outdoor thermal conditions of two buildings (one with and one without tree shading) at Akure University in Nigeria (tropical climate), found that the air temperature around the building without tree shading was always higher than that of the building with shading.

Djekic et al. [23] studied the impact of sidewalk materials on raising the local temperature in summer in Nis (Serbia; temperate climate). They showed that temperature differences between the shaded and unshaded surfaces were up to 20 °C.

Most studies on the effect of urban canopy shading on outdoor thermal comfort were conducted in warm seasons. Some canopies reduce the amount of sunlight in winter and increase thermal discomfort. Therefore, the effect of shading on thermal comfort should be examined in both warm and cold seasons.

3.1. H/W and SVF

Shading by buildings is an important strategy for creating thermally comfortable conditions for pedestrians in urban canyons. In addition, a low sky-view factor (SVF) or less openness to the sky in urban canyons caused by tall buildings and trees improves thermal comfort during warm seasons [14,53–56]. In several studies, various proportions of building height to street width (H/W) were investigated with respect to thermal comfort [53–60].

Yang et al. [61] investigated thermal comfort in high-rise-building areas of Singapore using ENVI-met. They concluded that, in a warm and humid climate, a height-to-width ratio of 3 and above can provide outdoor thermal comfort for pedestrians.

Jamei and Rajagopalan [62] used ENVI-met and examined a microclimate in Melbourne, Australia. They concluded that, by increasing the height of buildings, temperature drops by 1–4 °C.

Achour-Younsi and Kharrat [63] studied the H/W ratio of three urban streets in Tunisia with subtropical Mediterranean climates using ENVI-met. They found that the universal-thermal-climate-index (UTCI) difference between H/W of 0.25 and 4 was 8.48 °C. Furthermore, as H/W increases, thermal comfort improves.

Johansson [64] investigated the effect of urban geometry on outdoor comfort by comparing a shallow street (low H/W) with a deep street (high H/W) in Fez (Morocco; hot and dry climate). It was concluded that, during warm summer days, comfortable hours at the deep streets were more than those in the shallow one.

Kariminia et al. [38], using ENVI-met, looked at the role of geometry on the thermal comfort of visitors from a historical site in Isfahan, Iran (hot and dry climate). Their results suggested that, by increasing the H/W ratio from 0.1 to 0.3 in a historic square, PET was decreased by 1.6 °C. This decreased the discomfort period by 3 h.

Rodríguez-Algeciras et al. [65] studied 4 different H/W ratios within the central courtyards at Camagüey in Cuba (warm and humid climate). They showed that a H/W ratio of 3 in comparison with 0.5 reduced mean radiant temperature by up to 20 °C.

It was concluded that increasing H/W and the consequent shading effect improve thermal conditions in urban canyons [21,24,30,37,66–71]. Many studies were conducted in order to increase thermal comfort by increasing H/W during warm seasons. However, increasing the H/W does not improve thermal comfort in winter, so this solution is not recommended in cold climates. Further studies are needed to determine how this solution could work in different climates.

3.2. Trees

Trees are considered as a strategy to enhance thermal comfort in outdoor environments for different reasons, including their shading effect [72–76]. Several studies were used to reduce air temperature and radiant temperature, control wind speed and moisture, and generally improve thermal-comfort conditions [33,53,77–79]. In some studies, trees were identified as the most effective strategy for thermal comfort in outdoor environments among various other approaches [21,57,80–82].

Ruiz et al. [57] investigated 12 different urban streets in Mendoza, Argentina and concluded that there was a 60% improvement in thermal comfort in streets with trees compared to bare ones.

Johansson et al. [24] studied thermal conditions in 6 different urban environments (). The study was performed in the warm and humid climate of Sao Paulo, Brazil using the BRAMS and ENVI-met simulation packages. Results showed that the vegetated area had the highest thermal comfort.

Stocco et al. [55] studied 3 different areas in Mendoza, Argentina and found that areas with the highest tree density had the lowest air temperature.

Tree-growth scenarios are being investigated using tree forecast prediction models in a span of 30 years ranging from 2002 to 2032 in Milan, Italy. It was observed that, with the growth of trees and the increase in their umbrellas, a decrease in radiant temperature, and thermal-comfort improvement were estimated [83].

Tree Morphology

Trees are of great importance in outdoor environments due to their effective shading effect and for improving thermal-comfort conditions in urban streets. In some studies, a tree species with its specific morphology and climatic conditions in the study area was discussed.

Kong et al. [84] looked at the impact of tree types on outdoor thermal environments in Hong Kong. They concluded that trees with larger crowns, such as *Macaranga tanarius*, *Ficus microcarpa*, and *Acacia confusa* are more recommended over those with small crowns such as *Melaleuca*, *Leucadendron*, and *Livistona chinensis*.

Hanafi and Alkama [85] investigated the role of vegetation in outdoor environments in the warm and dry climate of Biskra, Algeria. They observed that *Ficus* trees (as a group of large crowns) were the most suitable.

Correa et al. [35] studied 3 different streets with widths of 16, 20, and 30 m in Mendoza, Argentina. They concluded that *Platanus* × *acerifolia* had the best performance among 3 common tree species in that area.

Yoshida [10], examining the effect of tree shadows on the thermal environment within the University of Osaka, Japan, concluded that trees with smaller leaves perform better than those with larger leaves in terms of thermal comfort.

In another study in Hong Kong (tropical climate), Morakinyo et al. [25] examined the influence of 8 tree species on outdoor thermal environments and concluded that leaf-area index is the most important physiological factor of trees. They recommended trees with narrower crown width, less density, and greater trunk height in high-density canyons.

Zhao et al. [86] analysed the biological properties of five different tree species in Harbin, China. They found that the species of *Populus* × *berolinensis*, *Populus alba*, and *Acer saccharum* had greater impact on climatic and thermal comfort during summer, with maximal PET reductions of 4.7 to 15.9 °C, respectively. It was also concluded that tree umbrella width and density were the most important biological factors in creating thermal comfort.

In another study, the impact of different tree-planting scenarios on the thermal comfort of outdoor environments was examined in Phoenix, Arizona, United States (hot and dry climate). The most appropriate scenario was the planting of trees with a distance of two trees, followed by a cluster model without overlapping the canopy [87].

Morakinyo et al. [88] studied the influence of common tree species in Hong Kong regarding thermal comfort of outdoor environments. They found that dense and medium-sized trees are suitable for shallow streets, while low-density trees were suitable for deep canyons. They also found that the most important features of trees for improving the thermal comfort of outdoor environments were leaf-area index, trunk height, tree height, and crown diameter.

Despite the important role of trees in shading and improving thermal comfort, few studies considered them in the context of various climates. This can help urban planners include the most suitable types of trees in terms of thermal comfort before designing urban canyons.

3.3. Urban-Canyon Orientation

The orientation of urban canyons with regard to the direction of sunlight and the prevailing wind in each climate is an important factor for creating the desired thermal comfort in outdoor environments [89].

Targhi and Van Dessel [90] studied different points in north–south and east–west streets by using ENVI-met and RayMan on 2 July 2014 (the hottest day of the year) in Winchester, USA. They found that the north–south street was more comfortable due to the solar radiation.

Achour-Younsi and Kharrat [63] investigated the H/W of three streets in Tunis (subtropical Mediterranean climate). The obtained results showed that, in all streets with a fixed H/W ratio, the best orientation was north–south, while the worst orientation was east–west.

Johansson et al. [24] found that, in the warm and humid climate of Brazil's Sao Paulo, streets with northwest–southeast and southwest–northeast orientations performed better in terms of thermal conditions in comparison with north–south and east–west orientations.

In another study, four different orientation scenarios were simulated at a university campus in Dubai (hot and dry climate). Using ENVI-met(Essen, Germany), it was found out that 2 scenarios with low-rise buildings that were facing the wind flow enhanced overall thermal comfort [91].

Ali-Toudert and Mayer [37] simulated a comparison between urban canyons with a similar H/W and different orientations using ENVI-met in Ghardaia, Algeria (hot and dry climate). They found that northeast–southwest and northwest–southeast streets thermally perform better than those with north–south and east–west orientations. In Concepción, Chile (mild climate), diagonal urban canyons were found to have the best performance in terms of both physical and psychological thermal comfort [92].

Cao et al. [93] investigated the effect of street orientation on local thermal comfort in Guangzhou, China using Fluent. Their final results showed that having the same wind flow and street direction increases average wind speed while decreasing mean radiant temperature.

3.4. Geometrical Forms

Different urban forms can create various microclimatological conditions in regard to the pedestrian comfort in cities.

Taleghani et al. [39] studied five different urban geometrical forms (singular north–south and east–west, linear north–south and east–west, and central courtyard) in the Netherlands. The central courtyards (that received low solar radiation) were the best form, while singular (with a high amount of received direct sunlight) were considered to be the worst. In another study, conducted by Taleghani et al. [94] in the Netherlands, the central courtyard was considered to be the best and the singular form the worst for heating and cooling energy consumption.

Xi et al. [95] researched various geometric forms at the University of Guangzhou (subtropical climate) and concluded that the pilot form had the best thermal performance, and could reduce the air temperature during the summer by 2 to 3 °C.

Field measurements of the building materials, pavements, and urban geometry of 4 residential neighbourhoods in Rome during 2015 and 2016 showed that suburban areas were more thermally comfortable in comparison with those downtown [96]. It was concluded that the denser urban geometry in the Mediterranean region causes thermal dissatisfaction in outdoor environments [96]. However, this conclusion cannot be considered valid for other climates, especially hot climates where shading can improve thermal comfort.

4. Green, Blue, and White Surfaces

Among different available strategies to improve thermal comfort in outdoor environments, it is important to consider surfaces covering urban streets such as pavements, building facades, and roofs. Several studies addressed the role of green surfaces (vegetation), water surfaces (blue elements), and white surfaces (high-reflection surfaces).

Taleghani [16] studied the role of vegetation and highly reflective materials as the most common solutions for improving thermal comfort in urban canyons. It was concluded that vegetation and reflective surfaces significantly decrease ambient air temperature. However, high albedo surfaces reflect the sun's rays and cause thermal discomfort, which is why the use of vegetation is recommended more.

Martins et al. [97] used ENVI-met to simulate thermal-comfort conditions within a new urban area in Toulouse, France. It was concluded that increasing vegetation and water surfaces led to a 7 and 2 °C reduction in PET, respectively.

Morille and Musy [98] assessed three strategies of green, water, and reflective surfaces using the SOLENE-Microclimate simulation tool in Lyon, France. They found that green surfaces had the best thermal performance. However, highly reflective materials reflected solar radiation back to the pedestrians, and poorer thermal performance was observed in comparison with that of green and water surfaces.

Huang et al. [15] studied the impact of increasing green, water, and white surfaces at a university campus in northwest China. Results showed that green surfaces decreased air temperature and mean radiant temperature by 0.3 and 32.1 °C, respectively. However, the reduction in air temperature for white surfaces was 1.1 °C, while mean radiant temperature was increased by 5.4 °C. Water surfaces had minimal effect on reducing air temperature and mean radiant temperature.

The current climate in Athens, Greece (Mediterranean climate) was compared with scenarios where green, water, and white surfaces were added to the land cover [99]. It was concluded that the second scenario (more water bodies) provided better improvement for outdoor thermal comfort.

In another study, green, water, and reflective surfaces were analysed as the main strategies for either preventing or decreasing the effects of heat islands in a university campus in Hong Kong, and the authors concluded that green and water surfaces significantly reduced heat, while reflective surfaces had an adverse effect [100].

Taleghani and Berardi studied a crowded area during the hottest days of 2015 in Toronto, Canada. The increase in albedo level from 0.1 to 0.3 and 0.5 resulted in a decrease in air temperature by 0.5 and 1 °C, respectively. However, this increase in albedo led to an increase in the reflection of sun rays, and subsequently an increase in the thermal discomfort of pedestrians [101].

Numerous studies indicated that highly reflective surfaces, despite reducing air temperature, have an adverse effect in terms of a greater reflection of sun rays in urban canyons, which ultimately increases the thermal discomfort of pedestrians [101–103]. In general, it is recommended to use surfaces with high reflectivity on the roofs of buildings in order to reduce the energy consumption of the buildings [104–106]. However, the use of such materials in horizontal and vertical surfaces within the urban canyons is not recommended.

4.1. Vegetation

Using vegetation in urban areas such as parks improves overall thermal conditions by decreasing air temperature and mean radiant temperature, and increasing the humidity of the surrounding environment [107–114].

In several studies, the role of vegetation in various urban areas was considered, and the need to use vegetation to enhance thermal comfort was proven [17,48,56,67,103,115–120].

Radhi et al. [121] studied the effect of artificial islands in Bahrain on climatic parameters using computational-fluid-dynamics (CFD) analysis. It was concluded that the mean radiant temperature difference between a vegetated area and concrete surfaces was up to 5 °C.

Barakat et al. [122] assessed 3 different microclimates by using ENVI-met in Alexandria, Egypt (hot and dry climate). They suggested that thermal conditions can be improved by reducing the pavement areas, and increasing greenery surfaces, water bodies, and the number of trees. It is well documented that these changes reduce air temperature and average radiation temperature while increasing humidity.

Georgi and Dimitriou [123] analysed the effect of vegetation on improving thermal conditions of Chania in the island of Crete (Mediterranean climate). In an area of 100 m^2, they assessed 3 different vegetation strategies, namely, the planting of 8 trees, using 4 cooling fans, and implementing a cladding canopy. It was concluded that tree planting was financially the best strategy for improving thermal conditions.

Jeong et al. [124] studied and compared the satisfaction degree of people in a forest–urban district in the central area of Seoul. They found that 79.3% of the people in the forest–urban area experienced a comfortable situation, while this value within the central region of the city was 31.1%.

Klemm et al. [125] evaluated 9 streets with a similar geometry in Utrecht, aiming to study the physical and psychological impacts of green spaces on thermal comfort. Results showed that mean radiant temperature in streets with trees was 39% lower than the bare streets.

In a study performed with ENVI-met on the thermal environment of a historic site in Rome, researchers concluded that vegetation improved overall thermal conditions while decreasing the PMV index by 1.5 °C at the middle of a hot summer day [102].

4.2. Green Roof and Wall

Roofs account for about 20%–25% of urban surfaces [126]. Green roofs and walls are two examples of adding vegetation to urban canyons. In a study on Melbourne's urban design with the use of vegetation, it was found that green roofs did not improve the PET index for pedestrians [62].

Perini and Magliocco [68] analysed the cooling impact of green roofs and surfaces on the ground level in the Mediterranean climate of three Italian cities: Genoa, Rome, and Milan. They concluded that green surfaces on the ground performed more efficiently than green roofs did. This was because they reduce air temperature and mean radiant temperature (and consequently PMV) at the height of 1.6 m (pedestrian level). However, green roofs were effective in reducing the cooling load of the buildings.

Taleghani et al. [127] assessed two central courtyards at Portland State University (Portland, OR, USA) during summer 2013. One wall was built using red bricks, while the other was a vegetated green wall. They continuously measured thermal conditions within the centre of the two yards bounded by both walls and found that air temperature in the centre of the yard with the green walls at 16:30 was 4.7 °C lower than that of the bare courtyard.

Alexandri and Jones [128] found that green roofs and walls in hot and dry climates had the greatest impact on the improvement of thermal comfort in 9 different cities. This conclusion regarding green walls can be extended to all climates. They also stated that green walls have a greater effect than that of green roofs on reducing air temperature in urban canyons.

Zhang et al. [82] studied the thermal impact of three strategies, namely, green roofs, green facades, and cool roofs in a school in Tianjin, China. They concluded that green facades had the least effect on improving the thermal environment.

Morakinyo et al. [129] examined the thermal benefits of green facades in Hong Kong. They found that, by greening 30–50% of the facades within the city, up to 1 °C air temperature could be reduced in Hong Kong. This situation led to the improvement in thermal conditions by at least one unit.

In general, it can be concluded that green roofs reduce the energy consumption in buildings [130–132]. However, they have little impact on outdoor thermal comfort, especially on the pedestrian level [62,68,128]. On the other hand, green walls can be effective in improving the overall thermal comfort of urban canyons [82,127,129].

4.3. Highly Reflective Materials

Reflective (white) and cold surfaces in urban canyons (especially in dense areas) can reflect the solar radiation, allow for heat to flow back to the atmosphere, and mitigate urban heat islands [133–138].

Castaldo et al. [139] used two types of cold-red and cold-grey concrete composites to study a dense historical site in Rome. They could observe reductions in the Mediterranean outdoor comfort index (MOCI) of 15% and 30% for cold-red and cold-grey surfaces, respectively.

In another study on a historical site in Italy, Rosso et al. [140] used ENVI-met to simulate three concrete types (red, white, and grey), and a specific type of marble in 7 different scenarios. They found that using red concrete on the walls, and grey concrete or marble in horizontal surfaces on the ground level of the urban canyons helped to decrease heat islands while increasing thermal comfort in urban canyons.

Rossi et al. [141] combined thermal, visual, and acoustical comfort using an acoustic white velvet fabric in an effort to study a historical site in Italy.

Lin and Ichinose [142] compared a type of travertine with concrete blocks commonly used on sidewalks in Japan during summer and autumn. The travertine stone was recommended to be replaced by the concrete blocks that have high thermal capacity, high reflectivity, and low thermal conductivity.

4.4. Water Bodies

Water bodies are considered to be a strategy to improve thermal comfort in outdoor environments due to their ability to increase humidity and reduce air temperature in urban areas [143–145].

Xu et al. [146] explored the most suitable place for an exhibition in terms of thermal comfort in Shanghai. They found areas with 10 to 20 m distance from the Huansha artificial water body as the most comfortable places.

A study on thermal comfort around bodies of water in Japan, with the main focus the direction and size of the ponds, suggested that larger ponds prone to prevailing wind can more significantly improve thermal comfort [147].

Mazhar et al. [36] investigated thermal comfort in the outside environments of Lahore (warm and dry climate). They compared Shalimar Gardens with a central courtyard in the Alhambra Arts Council, and concluded that Shalimar Gardens provided a higher level of thermal comfort because of their vegetation and huge water ponds.

5. Psychological Factors Affecting Thermal Comfort

In numerous studies, in addition to microclimatic parameters affecting thermal comfort in the outdoor environment, the effects of psychological factors were investigated [148]. Many studies noted that individuals can consciously or unconsciously adapt to their thermal conditions and achieve thermal comfort [149].

Spagnolo and de Dear [26] rated the thermal comfort of 1018 people in open and semiopen areas in a field study in Sydney, Australia. They found that the number of individuals feeling thermally comfortable in indoor environments was far less than that of those comfortable in outdoor environments.

This shows that people in outdoor environments have higher thermal expectations, and they accept a wider range of temperature levels as comfortable.

Nikolopoulou et al. [12] investigated four major touristic sites in Cambridge during spring, summer, and winter. According to their 1431 questionnaires, the thermal-comfort expectations of each individual and their understanding of their thermal environment had significant effects on their thermal perceptions.

In a study in Barranquilla, Colombia (tropical weather), the thermal conditions of pedestrians in five points of the city were investigated. It was determined that people tend to find neutral thermal conditions and cooler environments as ideal thermal conditions. However, in similar studies in tropical climates, it was reported that warm and very hot conditions are desired. This reflects the impact of psychological factors on the perception of individuals of the thermal environment [150].

Lin [151], in a study aimed at discovering the relationship between psychological factors and the use of an urban square in a tropical region, found that, when people decided to stay within the square (sitting in the square etc.), they had high thermal satisfaction, and this satisfaction level dropped when they were forced to cross the square.

In general, thermal comfort relies on both physical and psychological factors. Therefore, people's thermal expectations are an effective factor that determines their level of thermal comfort [9,148,152].

6. Conclusions

The main purpose of this study was to identify parameters affecting the thermal comfort of pedestrians in outdoor environments. More than 150 studies were reviewed. These studies were chosen because they addressed the improvement of outdoor thermal comfort in different climates, used different indices of human thermal comfort, and implemented different research methods (field measurements, computer simulations, or a combination).

The reviewed heat-mitigation strategies were divided into four sections: climatic parameters; shading effects; green, blue and white surfaces; and psychological parameters. Here, the most important lessons learnt from these studies are summarised:

- The conducted studies used different data-collection methods. Field measurements and computer simulations are frequently used in order to analyse thermal environments. The most frequently used strategies to improve pedestrians' thermal comfort include climatic design solutions that are related to the physical properties of urban spaces. These include shading (by buildings and trees), street orientation, and geometrical forms of urban canyons, vegetation, water surfaces, and highly reflective materials. However, many studies confirmed that, in addition to the aforementioned strategies, the psychological factors (perception of people of their thermal environment) and thermal expectations of individuals are also necessary in order to create favourable thermal conditions.

- Among climatic factors affecting outdoor thermal comfort, mean radiant temperature has the greatest effect, followed by wind speed and direction. As a result, strategies that reduce mean radiant temperature (such as shading) are more effective. Results showed that increasing H/W in urban canyons improves outdoor thermal comfort.

- Many studies concluded that deploying different types of vegetation is the best strategy to improve thermal comfort, with an emphasis on trees. By comparing green roofs and walls, it could generally be concluded that green roofs reduce energy consumption for buildings. However, they have little impact on pedestrian thermal comfort. Green walls are more effective than green roofs in improving thermal conditions in urban canyons.

- The use of highly reflective materials in urban canyons tends to increase thermal discomfort due to the reflection of solar radiation (despite the fact that they reduce air temperature). It is not recommended to use such materials in surfaces of urban canyons. Nevertheless, they could be deployed on roofs in order to reduce air temperature in buildings to reduce their overall energy consumption.

- Regarding the orientation and geometries of urban canyons, orientation with respect to the direction of sunlight and the prevailing wind is an important factor for creating favourable thermal conditions. In addition, the various forms and geometries of urban canyons can affect microclimatic and thermal conditions for pedestrians. There is a direct relationship between area density and thermal comfort. In dense areas, heat is trapped due to less ventilation. Comparing different urban forms, courtyards were found to be the best form in terms of thermal comfort and energy consumption.

7. Recommendations for Future Studies

This paper recommends three topics for further research:

- Considering the significant role of trees in shading and improving outdoor thermal comfort, studies are required on the geometry of trees in each climate. In this way, the most suitable tree species could be used (planted) in order to more efficiently improve outdoor thermal conditions.
- Despite evidence showing that water bodies reduce air temperature and increase humidity, there are not many extensive studies regarding the number, depth, form, and location of water surfaces in urban areas with different climates.
- Using highly reflective surfaces in urban canyons is not recommended, as they reradiate solar radiation back to pedestrians. Further studies can provide scientific solutions in order to regulate and optimally use these materials.

Author Contributions: All authors contributed equally to the research and writing. All authors have read and agreed to the published version of the manuscript.

Funding: This research received no external funding.

Conflicts of Interest: The authors declare no conflict of interest.

Appendix A

Table A1. Different ranges of physiologically equivalent temperature (PET) index for different grades of human thermal perceptions [153].

Thermal Perception Grades	PET (°C)
Extreme cold stress	Below 4
Strong cold stress	4.1 to 8
Moderate cold stress	8.1 to 13
Slight cold stress	13.1 to 18
No thermal stress	18.1 to 23
Slight heat stress	23.1 to 29
Moderate heat stress	29.1 to 35
Strong heat stress	35.1 to 41
Extreme heat stress	Above 41

References

1. Jamei, E.; Rajagopalan, P.; Seyedmahmoudian, M.; Jamei, Y. Review on the impact of urban geometry and pedestrian level greening on outdoor thermal comfort. *Ren. Sustain. Energy Rev.* **2016**, *54*, 1002–1017. [CrossRef]
2. Ignatius, M.; Nyuk Hien, W.; Steve Kardinal, J. Urban microclimate analysis with consideration of local ambient temperature, external heat gain, urban ventilation, and outdoor thermal comfort in the tropics. *Sustain. Cities Soc.* **2015**, *19*, 121–135. [CrossRef]

3. Oke, T.R. The energetic basis of the urban heat island. *Q. J. Royal Meteorol. Soc.* **1982**, *108*, 1–24. [CrossRef]
4. Chen, L.; Edward, N.G. Outdoor thermal comfort and outdoor activities: A review of research in the past decade. *Cities* **2012**, *29*, 118–125. [CrossRef]
5. Coccolo, S.; Kampf, J.; Scartezzini, J.-L.; Pearlmutter, D. Outdoor human comfort and thermal stress: A comprehensive review on models and standards. *Urban Clim.* **2016**, *18*, 33–57. [CrossRef]
6. Hass-Klau, C. A review of the evidence from Germany and the UK. *Transp. Policy* **1993**, *1*, 21–31. [CrossRef]
7. Hakim, A.A.; Petrovitch, H.; Burchfiel, C.M.; Ross, W.; Rodriguez, B.L.; White, L.R.; Yano, K.; Curb, J.D.; Abbott, R.D. Effects of walking on mortality among nonsmoking retired men. *N. Eng. J. Med.* **1998**, *338*, 94–99. [CrossRef]
8. Chen, L.; Wen, Y.; Zhang, L.; Xiang, W.-N. Studies of thermal comfort and space use in an urban park square in cool and cold seasons in Shanghai. *Build. Environ.* **2015**, *94*, 644–653. [CrossRef]
9. Shooshtarian, S.; Priyadarsini, R. Study of thermal satisfaction in an Australian educational precinct. *Build. Environ.* **2017**, *123*, 119–132. [CrossRef]
10. Yoshida, A.; Hisabayashi, T.; Kashihara, K.; Kinoshita, S.; Hashida, S. Evaluation of effect of tree canopy on thermal environment, thermal sensation, and mental state. *Urban Clim.* **2015**, *14*, 240–250. [CrossRef]
11. Makaremi, N.; Salleh, E.; Jaafar, M.Z.; Ghaffarian Hoesini, A. Thermal comfort conditions of shaded outdoor spaces in hot and humid climate of Malaysia. *Build. Environ.* **2012**, *48*, 7–14. [CrossRef]
12. Nikolopoulou, M.; Nick, B.; Koen, S. Thermal comfort in outdoor urban spaces: Understanding the human parameter. *Solar Energy* **2001**, *70.3*, 227–235. [CrossRef]
13. Thorsson, S.; Maria, L.; Sven, L. Thermal bioclimatic conditions and patterns of behaviour in an urban park in Göteborg, Sweden. *Int. J. Biometeorol.* **2004**, *48.3*, 149–156. [CrossRef] [PubMed]
14. Tzu-Ping, L.; Matzarakis, A.; Hwang, R.-L. Shading effect on long-term outdoor thermal comfort. *Build. Environ.* **2010**, *45.1*, 213–221.
15. Huang, Q.; Meng, X.; Yang, X.; Jin, L.; Liu, X.; Hu, W. The ecological city: Considering outdoor thermal environment. *Energy Procedia* **2016**, *104*, 177–182. [CrossRef]
16. Taleghani, M. Outdoor thermal comfort by different heat mitigation strategies-A review. *Ren. Sustain. Energy Rev.* **2018**, *81*, 2011–2018. [CrossRef]
17. Salata, F.; Golasi, I.; Petiti, D.; de Lieto Vollaro, E.; Coppi, M.; de Lieto Vollaro, A. Relating microclimate, human thermal comfort and health during heat waves: An analysis of heat island mitigation strategies through a case study in an urban outdoor environment. *Sustain. Cities Soc.* **2017**, *30*, 79–96. [CrossRef]
18. Nasrollahi, N.; Hatami, Z.; Taleghani, M. Development of outdoor thermal comfort model for tourists in urban historical areas; A case study in Isfahan. *Build. Environ.* **2017**, *125*, 356–372. [CrossRef]
19. Niu, J.; Liu, J.; Lin, Z.; Mak, C.; Tse, K.T.; Tang, B.S.; Kwok, K.C.S. A new method to assess spatial variations of outdoor thermal comfort: Onsite monitoring results and implications for precinct planning. *Build. Environ.* **2015**, *91*, 263–270. [CrossRef]
20. Tsitoura, M.; Theocharis, T.; Tryfon, D. Evaluation of comfort conditions in urban open spaces. Application in the island of Crete. *Energy Convers. Manag.* **2014**, *86*, 250–258. [CrossRef]
21. Shashua-Bar, L.; Ioannis, X.T.; Milo, H. Passive cooling design options to ameliorate thermal comfort in urban streets of a Mediterranean climate (Athens) under hot summer conditions. *Build. Environ.* **2012**, *57*, 110–119. [CrossRef]
22. Ng, E.; Vicky, C. Urban human thermal comfort in hot and humid Hong Kong. *Energy Build.* **2012**, *55*, 51–65. [CrossRef]
23. Djekic, J.; Djukic, A.; Vukmirovic, M.; Djekic, P.; Dinic Brankovic, M. Thermal comfort of pedestrian spaces and the influence of pavement materials on warming up during summer. *Energy Build.* **2018**, *159*, 474–485. [CrossRef]
24. Johansson, E.; Spangenberg, J.; Gouvea, M.L.; Freitas, E.D. Scale-integrated atmospheric simulations to assess thermal comfort in different urban tissues in the warm humid summer of São Paulo, Brazil. *Urban Clim.* **2013**, *6*, 24–43. [CrossRef]
25. Morakinyo, T.E.; Kong, L.; Lun Lau, K.K.; Yuan, C.; Ng, E. A study on the impact of shadow-cast and tree species on in-canyon and neighborhood's thermal comfort. *Build. Environ.* **2017**, *115*, 1–17. [CrossRef]
26. Spagnolo, J.; De Dear, R. A field study of thermal comfort in outdoor and semi-outdoor environments in subtropical Sydney Australia. *Build. Environ.* **2003**, *38.5*, 721–738. [CrossRef]

27. da Silveira Hirashima, S.Q.; Sad de Assis, E.; Nikolopoulou, M. Daytime thermal comfort in urban spaces: A field study in Brazil. *Build. Environ.* **2016**, *107*, 245–253. [CrossRef]
28. da Silveira Hirashima, S.Q.; Katzschner, A.; Ferreira, D.G.; Sad de Asis, E.; Katzscher, L. Thermal comfort comparison and evaluation in different climates. *Urban Clim.* **2016**, *23*, 219–230. [CrossRef]
29. Omonijo, A.G. Assessing seasonal variations in urban thermal comfort and potential health risks using physiologically equivalent temperature: A case of Ibadan, Nigeria. *Urban Clim.* **2017**, *21*, 87–105. [CrossRef]
30. Ahmed, K.S. Comfort in urban spaces: Defining the boundaries of outdoor thermal comfort for the tropical urban environments. *Energy Build.* **2003**, *35*, 103–110. [CrossRef]
31. Yang, W.; Nyuk, H.W.; Jusuf, S.K. Thermal comfort in outdoor urban spaces in Singapore. *Build. Environ.* **2013**, *59*, 426–435. [CrossRef]
32. Zhao, L.; Zhou, X.; Li, L.; He, S.; Chen, R. Study on outdoor thermal comfort on a campus in a subtropical urban area in summer. *Sustain. Cities Soc.* **2016**, *22*, 164–170. [CrossRef]
33. Tong, S.; Wong, N.H.; Tan, C.L.; Jusuf, S.K.; Ignatius, M.; Tan, E. Impact of urban morphology on microclimate and thermal comfort in northern China. *Solar Energy* **2017**, *155*, 212–223. [CrossRef]
34. Kariminia, S.; Shamshibrand, S.; Hashim, R.; Saberi, A.; Petkovic, D.; Roy, C.; Motamedi, S. A simulation model for visitors' thermal comfort at urban public squares using non-probabilistic binary-linear classifier through soft-computing methodologies. *Energy* **2016**, *101*, 568–580. [CrossRef]
35. Correa, E.; Ruiz, M.A.; Canton, A.; Lesino, G. Thermal comfort in forested urban canyons of low building density. An assessment for the city of Mendoza, Argentina. *Build. Environ.* **2012**, *58*, 219–230. [CrossRef]
36. Mazhar, N.; Brown, R.D.; Kenny, N.; Lenzholzer, S. Thermal comfort of outdoor spaces in Lahore, Pakistan: Lessons for bioclimatic urban design in the context of global climate change. *Landsc. Urban Plan.* **2006**, *138*, 110–117. [CrossRef]
37. Ali-Toudert, F.; Helmut, M. Numerical study on the effects of aspect ratio and orientation of an urban street canyon on outdoor thermal comfort in hot and dry climate. *Build. Environ.* **2006**, *41.2*, 94–108. [CrossRef]
38. Kariminia, S.; Sabarinah, S.A.; Ahmadreza, S. Microclimatic conditions of an urban square: Role of built environment and geometry. *Procedia-Soc. Behav. Sci.* **2015**, *170*, 718–727. [CrossRef]
39. Taleghani, M.; Kleerekoper, L.; Tenpierik, M.; van den Dobbelsteen, A. Outdoor thermal comfort within five different urban forms in the Netherlands. *Build. Environ.* **2015**, *83*, 65–78. [CrossRef]
40. Mahmoud, A.H.A. Analysis of the microclimatic and human comfort conditions in an urban park in hot and arid regions. *Build. Environ.* **2011**, *46*, 2641–2656. [CrossRef]
41. Li, L.; XiaoQing, Z.; Lixiu, Y. The analysis of outdoor thermal comfort in Guangzhou during summer. *Proc. Eng.* **2017**, *205*, 1996–2002. [CrossRef]
42. Jin, H.; Siqi, L.; Jian, K. The thermal comfort of urban pedestrian street in the severe cold area of northeast china. *Energy Proc.* **2017**, *134*, 741–748. [CrossRef]
43. Cheung, P.K.; Jim, C.Y. Comparing the cooling effects of a tree and a concrete shelter using PET and UTCI. *Build. Environ.* **2018**, *130*, 49–61. [CrossRef]
44. Martinelli, L.; Tzu-Ping, L.; Matzarakis, A. Assessment of the influence of daily shadings pattern on human thermal comfort and attendance in Rome during summer period. *Build. Environ.* **2015**, *92*, 30–38. [CrossRef]
45. Middel, A.; Selover, N.; Hagen, B.; Chhetri, N. Impact of shade on outdoor thermal comfort—A seasonal field study in Tempe, Arizona. *Int. J. Biometeorol.* **2016**, *60*, 1849–1861. [CrossRef] [PubMed]
46. Jin, H.; Liang, Q.; Bo, W. Field research and study of campus thermal environment in winter in severe cold areas. *Energy Proc.* **2017**, *134*, 607–615. [CrossRef]
47. Paolini, R.; Mainini, A.G.; Poli, T.; Vercesi, L. Assessment of thermal stress in a street canyon in pedestrian area with or without canopy shading. *Energy Proc.* **2014**, *48*, 1570–1575. [CrossRef]
48. Yahia, M.W.; Johansson, E. Landscape interventions in improving thermal comfort in the hot dry city of Damascus, Syria—The example of residential spaces with detached buildings. *Land. Urban Plan.* **2004**, *125*, 1–16. [CrossRef]
49. Ali, S.B.; Suprava, P. Thermal comfort in urban open spaces: Objective assessment and subjective perception study in tropical city of Bhopal, India. *Urban Clim.* **2018**, *24*, 954–967. [CrossRef]
50. Hwang, R.-L.; Tzu-Ping, L.; Matzarakis, A. Seasonal effects of urban street shading on long-term outdoor thermal comfort. *Build. Environ.* **2011**, *46*, 863–870. [CrossRef]

51. Watanabe, S.; Nagano, K.; Ishii, J.; Horikoshi, T. Evaluation of outdoor thermal comfort in sunlight, building shade, and pergola shade during summer in a humid subtropical region. *Build. Environ.* **2014**, *82*, 556–565. [CrossRef]

52. Morakinyo, T.E.; Ahmed, A.B.; Olumuyiwa, B.A. Comparing the effect of trees on thermal conditions of two typical urban buildings. *Urban Clim.* **2013**, *3*, 76–93. [CrossRef]

53. Andreou, E. Thermal comfort in outdoor spaces and urban canyon microclimate. *Ren. Energy* **2013**, *55*, 182–188. [CrossRef]

54. Boukhelkhal, I.; Bourbia, P.R.F. Thermal comfort conditions in outdoor urban spaces: Hot dry climate-ghardaia-algeria. *Proc. Eng.* **2016**, *169*, 207–215. [CrossRef]

55. Stocco, S.; Cantón, M.A.; Correa, E.N. Design of urban green square in dry areas: Thermal performance and comfort. *Urban Forest. Urban Green.* **2015**, *14*, 323–335. [CrossRef]

56. Bourbia, F.; Boucheriba, F. Impact of street design on urban microclimate for semi arid climate (Constantine). *Ren. Energy* **2010**, *35*, 343–347. [CrossRef]

57. Ruiz, M.A.; Sosa, B.; Correa, E.N.; Canton, A. Design tool to improve daytime thermal comfort and nighttime cooling of urban canyons. *Landsc. Urban Plan.* **2017**, *167*, 249–256. [CrossRef]

58. Jihad, A.S.; Mohamed, T. Modeling the urban geometry influence on outdoor thermal comfort in the case of Moroccan microclimate. *Urban Clim.* **2016**, *16*, 25–42. [CrossRef]

59. Lobaccaro, G.; Juan, A.A. Comparative analysis of green actions to improve outdoor thermal comfort inside typical urban street canyons. *Urban Clim.* **2015**, *14*, 251–267. [CrossRef]

60. Hosseini, S.H.; Ghobadi, P.; Ahmadi, T.; Calutit, J.C. Numerical investigation of roof heating impacts on thermal comfort and air quality in urban canyons. *Appl. Therm. Eng.* **2017**, *123*, 310–326. [CrossRef]

61. Yang, W.; Nyuk, H.W.; Yaolin, L. Thermal comfort in high-rise urban environments in Singapore. *Proc. Eng.* **2015**, *121*, 2125–2131. [CrossRef]

62. Jamei, E.; Priyadarsini, R. Urban development and pedestrian thermal comfort in Melbourne. *Solar Energy* **2017**, *144*, 681–698. [CrossRef]

63. Achour-Younsi, S.; Fakher, K. Outdoor thermal comfort: Impact of the geometry of an urban street canyon in a Mediterranean subtropical climate–Case study Tunis, Tunisia. *Proc. Soc. Behav. Sci.* **2016**, *216*, 689–700. [CrossRef]

64. Johansson, E. Influence of urban geometry on outdoor thermal comfort in a hot dry climate: A study in Fez, Morocco. *Build. Environ.* **2006**, *41*, 1326–1338. [CrossRef]

65. Rodríguez-Algeciras, J.; Tablada, A.; Chaos-Years, M.; la Paz, G.D.; Matzarakis, A. Influence of aspect ratio and orientation on large courtyard thermal conditions in the historical centre of Camagüey-Cuba. *Ren. Energy* **2018**, *125*, 840–856. [CrossRef]

66. Qaid, A.; Lamit, H.B.; Ossen, D.R.; Shahminan, R.N.R. Urban heat island and thermal comfort conditions at micro-climate scale in a tropical planned city. *Energy Build.* **2016**, *133*, 577–595. [CrossRef]

67. Ragheb, A.A.; Ingy, I.E.; Sherif, A. Microclimate and human comfort considerations in planning a historic urban quarter. *Int. J. Sustain. Built Environ.* **2016**, *5*, 156–167. [CrossRef]

68. Perini, K.; Magliocco, A. Effects of vegetation, urban density, building height, and atmospheric conditions on local temperatures and thermal comfort. *Urban Forest. Urban Green.* **2014**, *13*, 495–506. [CrossRef]

69. Chatzidimitriou, A.; Yannas, S. Microclimate design for open spaces: Ranking urban design effects on pedestrian thermal comfort in summer. *Sustain. Cities Soc.* **2016**, *26*, 27–47. [CrossRef]

70. Martinelli, L.; Matzarakis, A. Influence of height/width proportions on the thermal comfort of courtyard typology for Italian climate zones. *Sustain. Cities Soc.* **2017**, *29*, 97–106. [CrossRef]

71. Emmanuel, R.; Johansson, E. Influence of urban morphology and sea breeze on hot humid microclimate: The case of Colombo, Sri Lanka. *Clim. Res.* **2006**, *30*, 189–200. [CrossRef]

72. Heisler, G.M. Effects of individual trees on the solar radiation climate of small buildings. *Urban Ecol.* **1986**, *9*, 337–359. [CrossRef]

73. Coutts, A.M.; White, E.C.; Tapper, N.J.; Beringer, J.; Livesley, S.J. Temperature and human thermal comfort effects of street trees across three contrasting street canyon environments. *Theor. Appl. Climatol.* **2016**, *124*, 55–68. [CrossRef]

74. Saito, I.; Osamu, I.; Tadahisa, K. Study of the effect of green areas on the thermal environment in an urban area. *Energy Build.* **1990**, *15*, 493–498. [CrossRef]

75. Nichol, J.E. High-resolution surface temperature patterns related to urban morphology in a tropical city: A satellite-based study. *J. Appl. Meteorol.* **1996**, *35*, 135–146. [CrossRef]
76. Vailshery, L.L.; Madhumitha, J.; Harini, N. Effect of street trees on microclimate and air pollution in a tropical city. *Urban Forest. Urban Green.* **2013**, *12*, 408–415. [CrossRef]
77. Lenzholzer, S. Research and design for thermal comfort in Dutch urban squares. *Res. Conserv. Rec.* **2012**, *64*, 39–48. [CrossRef]
78. Nasir, R.A.; Ahmad, S.S.; Ahmed, A.Z.; Ibrahim, N. Adapting human comfort in an urban area: The role of tree shades towards urban regeneration. *Proc. Soc. Behav. Sci.* **2015**, *170*, 369–380. [CrossRef]
79. Tan, Z.; Ka-Lun Lau, K.; Ng, E. Planning strategies for roadside tree planting and outdoor comfort enhancement in subtropical high-density urban areas. *Build. Environ.* **2017**, *120*, 93–109. [CrossRef]
80. Sun, S.; Xu, X.; Lao, Z.; Liu, W.; Garcia, E.H.; He, L.; Zhu, J. Evaluating the impact of urban green space and landscape design parameters on thermal comfort in hot summer by numerical simulation. *Build. Environ.* **2017**, *123*, 277–288. [CrossRef]
81. Yang, Y.; Zhou, D.; Gao, W.; Chen, W.; Peng, W. Simulation on the impacts of the street tree pattern on built summer thermal comfort in cold region of China. *Sustain. Cities Soc.* **2018**, *37*, 563–580. [CrossRef]
82. Zhang, A.; Bokel, R.; van den Dobbelsteen, A.; Sun, Y.; Huang, Q.; Zhang, Q. An integrated school and schoolyard design method for summer thermal comfort and energy efficiency in Northern China. *Build. Environ.* **2017**, *124*, 369–387. [CrossRef]
83. Picot, X. Thermal comfort in urban spaces: Impact of vegetation growth: Case study: Piazza della Scienza, Milan, Italy. *Energy Build.* **2004**, *36*, 329–334. [CrossRef]
84. Kong, L.; Ka Lu Lau, K.; Yuan, C.; Chen, Y.; Xu, Y.; Ren, C.; Ng, E. Regulation of outdoor thermal comfort by trees in Hong Kong. *Sustain. Cities Soc.* **2017**, *31*, 12–25. [CrossRef]
85. Hanafi, A.; Djamel, A. Role of the urban vegetal in improving the thermal comfort of a public place of a contemporary Saharan city. *Energy Proc.* **2017**, *119*, 139–152. [CrossRef]
86. Zhao, X.; Guojie, L.; Tianyu, G. Research on optimization and biological characteristics of Harbin trees based on thermal comfort in summer. *Proc. Eng.* **2017**, *180*, 550–561. [CrossRef]
87. Zhao, Q.; Sailor, D.J.; Wentz, E.A. Impact of tree locations and arrangements on outdoor microclimates and human thermal comfort in an urban residential environment. *Urban Forest. Urban Green.* **2018**, *32*, 81–91. [CrossRef]
88. Morakinyo, T.E.; Ka Lun Lau, K.; Ren, C.; Ng, E. Performance of Hong Kong's common trees species for outdoor temperature regulation, thermal comfort and energy saving. *Build. Environ.* **2018**, *137*, 157–170. [CrossRef]
89. Nunez, M.; Timothy, R.O. The energy balance of an urban canyon. *J. Appl. Meteorol.* **1977**, *16*, 11–19. [CrossRef]
90. Targhi, M.Z.; Van Dessel, S. Potential contribution of urban developments to outdoor thermal comfort conditions: The influence of urban geometry and form in Worcester, Massachusetts, USA. *Proc. Eng.* **2015**, *118*, 115–1161. [CrossRef]
91. Taleb, H.; Dana, T. Enhancing the thermal comfort on urban level in a desert area: Case study of Dubai, United Arab Emirates. *Urban Forest. Urban Green.* **2014**, *13*, 253–260. [CrossRef]
92. Lamarca, C.; Jorge, Q.; Cristián, H. Thermal comfort and urban canyons morphology in coastal temperate climate, Concepción, Chile. *Urban Clim.* **2018**, *23*, 159–172. [CrossRef]
93. Cao, A.; Qiong, L.; Qinglin, M. Effects of orientation of urban roads on the local thermal environment in Guangzhou City. *Proc. Eng.* **2015**, *121*, 2075–2082. [CrossRef]
94. Taleghani, M.; Tenpierk, M.; van den Dobbelsteen, A.; de Dear, R. Energy use impact of and thermal comfort in different urban block types in the Netherlands. *Energy Build.* **2013**, *67*, 166–175. [CrossRef]
95. Xi, T.; Li, Q.; Mochida, A.; Meng, A. Study on the outdoor thermal environment and thermal comfort around campus clusters in subtropical urban areas. *Build. Environ.* **2012**, *52*, 162–170. [CrossRef]
96. Zinzi, M.; Carnielo, E. Impact of urban temperatures on energy performance and thermal comfort in residential buildings. The case of Rome, Italy. *Energy Build.* **2017**, *157*, 20–29. [CrossRef]
97. Martins, T.A.L.; Adolphe, L.; Bonhomme, M.; Bonneaud, F.; Faraut, S.; Ginestat, S.; Michel, C.; Guyard, W. Impact of urban cool island measures on outdoor climate and pedestrian comfort: Simulations for a new district of Toulouse, France. *Sustain. Cities Soc.* **2016**, *26*, 9–26. [CrossRef]
98. Morille, B.; Marjorie, M. Comparison of the impact of three climate adaptation strategies on summer thermal comfort–Case study in Lyon, France. *Proc. Environ. Sci.* **2017**, *38*, 619–626. [CrossRef]

99. Gaitani, N.; Mihalakakou, G.; Santamouris, M. On the use of bioclimatic architecture principles in order to improve thermal comfort conditions in outdoor spaces. *Build. Environ.* **2007**, *42*, 317–324. [CrossRef]

100. Zhao, T.F.; Fong, K.F. Characterization of different heat mitigation strategies in landscape to fight against heat island and improve thermal comfort in hot–humid climate (Part I): Measurement and modelling. *Sustain. Cities Soc.* **2017**, *32*, 523–531. [CrossRef]

101. Taleghani, M.; Berardi, U. The effect of pavement characteristics on pedestrians' thermal comfort in Toronto. *Urban Clim.* **2018**, *24*, 449–459. [CrossRef]

102. Salata, F.; Golasi, I.; de Lieto Vollaro, A. How high albedo and traditional buildings' materials and vegetation affect the quality of urban microclimate. A case study. *Energy Build.* **2015**, *99*, 32–49. [CrossRef]

103. Taleghani, M. The impact of increasing urban surface albedo on outdoor summer thermal comfort within a university campus. *Urban Clim.* **2018**, *24*, 175–184. [CrossRef]

104. Akbari, H.; Damon Matthews, H. Global cooling updates: Reflective roofs and pavements. *Energy Build.* **2012**, *55*, 2–6. [CrossRef]

105. Akbari, H.; Konopacki, S. Calculating energy-saving potentials of heat-island reduction strategies. *Energy Policy* **2005**, *33*, 721–756. [CrossRef]

106. Akbari, H.; Melvin, P.; Haider, T. Cool surfaces and shade trees to reduce energy use and improve air quality in urban areas. *Solar Energy* **2001**, *70*, 295–310. [CrossRef]

107. Hwang, Y.H.; Qin Jie, G.L.; Yeow Kwang, D.C. Micro-scale thermal performance of tropical urban parks in Singapore. *Build. Environ.* **2015**, *94*, 467–476. [CrossRef]

108. Skoulika, F.; Santamorius, M.; Koloktsa, D.; Boemi, N. On the thermal characteristics and the mitigation potential of a medium size urban park in Athens, Greece. *Land. Urban Plan.* **2014**, *123*, 73–86. [CrossRef]

109. Chang, C.R.; Huang Li, M. Effects of urban parks on the local urban thermal environment. *Urban Forest. Urban Green.* **2014**, *13*, 672–681. [CrossRef]

110. Chang, C.R.; Ming-Huang, L.; Shyh-Dean, C. A preliminary study on the local cool-island intensity of Taipei city parks. *Landscape Urban Plan.* **2007**, *80*, 386–395. [CrossRef]

111. Oliveira, S.; Andrade, H.; Vaz, T. The cooling effect of green spaces as a contribution to the mitigation of urban heat: A case study in Lisbon. *Build. Environ.* **2011**, *46*, 2186–2194. [CrossRef]

112. Feyisa, G.L.; Klaus, D.; Meilby, H. Efficiency of parks in mitigating urban heat island effect: An example from Addis Ababa. *Landsc. Urban Plan.* **2014**, *123*, 87–95. [CrossRef]

113. Yu, C.; Wong Nyuk, H. Thermal benefits of city parks. *Energy Build.* **2006**, *38*, 105–120. [CrossRef]

114. Spronken-Smith, R.A.; Oke, T.R. The thermal regime of urban parks in two cities with different summer climates. *Int. J. Remote Sens.* **1998**, *19*, 2085–2104. [CrossRef]

115. Morris, K.I.; Chan, A.; Morris, K.J.K.; Ooi, M.C.G.; Oozeer, M.Y.; Abakar, Y.A.; Nadzir, M.S.M.; Mohammed, I.Y.; Al-Qrimli, H.F. Impact of urbanization level on the interactions of urban area, the urban climate, and human thermal comfort. *Appl. Geogr.* **2017**, *79*, 50–72. [CrossRef]

116. Gómez, F.; Luisa, G.; Jabaloyes, J. Experimental investigation on the thermal comfort in the city: Relationship with the green areas, interaction with the urban microclimate. *Build. Environ.* **2004**, 1077–1086. [CrossRef]

117. Lin, B.; Li, X.; Zhu, Y.; Qin, Y. Numerical simulation studies of the different vegetation patterns' effects on outdoor pedestrian thermal comfort. *J. Wind Eng. Ind. Aerod.* **2008**, *96*, 1707–1718. [CrossRef]

118. Klemm, W.; Heusinkveld, B.G.; Lenzholzer, S.; Jacobs, M.H.; van Hove, B. Psychological and physical impact of urban green spaces on outdoor thermal comfort during summertime in The Netherlands. *Build. Environ.* **2015**, *83*, 120–128. [CrossRef]

119. Amani-Beni, M.; Biao, Z.; Jie, X. Impact of urban park's tree, grass and waterbody on microclimate in hot summer days: A case study of Olympic Park in Beijing, China. *Urban Forest. Urban Green.* **2018**, *37*, 1–6. [CrossRef]

120. Karakounos, I.; Dimoudi, A.; Zoras, S. The influence of bioclimatic urban redevelopment on outdoor thermal comfort. *Energy Build.* **2018**, *158*, 1266–1274. [CrossRef]

121. Radhi, H.; Stephen, S.; Essam, A. Impact of urban heat islands on the thermal comfort and cooling energy demand of artificial islands—A case study of AMWAJ Islands in Bahrain. *Sustain. Cities Soc.* **2015**, *19*, 310–318. [CrossRef]

122. Barakat, A.; Hany, A.; Zeyad, E.S. Urban design in favor of human thermal comfort for hot arid climate using advanced simulation methods. *Alex. Eng. J.* **2017**, *9*, 533–543. [CrossRef]

123. Georgi, J.N.; Dimos, D. The contribution of urban green spaces to the improvement of environment in cities: Case study of Chania, Greece. *Build. Environ.* **2010**, *45*, 1401–1414. [CrossRef]

124. Jeong, M.A.; Sujin, P.; Song, G.S. Comparison of human thermal responses between the urban forest area and the central building district in Seoul, Korea. *Urban Forest. Urban Green.* **2016**, *15*, 133–148. [CrossRef]

125. Klemm, W.; Heusinkveld, B.G.; Lenzholzer, S.; van Hove, B. Street greenery and its physical and psychological impact on thermal comfort. *Land. Urban Plan.* **2015**, *138*, 87–98. [CrossRef]

126. Besir, A.B.; Cuce, E. Green roofs and facades: A comprehensive review. *Ren. Sustain. Energy Rev.* **2018**, *82*, 915–939. [CrossRef]

127. Taleghani, M.; Sailor, D.J.; Tenpierk, M.; van den Dobbelsteen, A. Thermal assessment of heat mitigation strategies: The case of Portland State University, Oregon, USA. *Build. Environ.* **2014**, *73*, 138–150. [CrossRef]

128. Alexandri, E.; Phil, J. Temperature decreases in an urban canyon due to green walls and green roofs in diverse climates. *Build. Environ.* **2008**, *43*, 480–493. [CrossRef]

129. Morakinyo, T.E.; Lai, A.; Lau, K.K.L.; Ng, E. Thermal benefits of vertical greening in a high-density city: Case study of Hong Kong. *Urban Forest. Urban Green.* **2017**, *37*, 42–55. [CrossRef]

130. Castleton, H.F.; Stovin, V.; Beck, S.B.M.; Davisnon, J.B. Green roofs; building energy savings and the potential for retrofit. *Energy Build.* **2010**, *42*, 1582–1591. [CrossRef]

131. Bevilacqua, P.; Mazzeo, D.; Brumo, R.; Arcuri, N. Experimental investigation of the thermal performances of an extensive green roof in the Mediterranean area. *Energy Build.* **2016**, *122*, 63–79. [CrossRef]

132. Vijayaraghavan, K. Green roofs: A critical review on the role of components, benefits, limitations and trends. *Ren. Sustain. Energy Rev.* **2016**, *57*, 740–752. [CrossRef]

133. Santamouris, M. Cooling the cities–a review of reflective and green roof mitigation technologies to fight heat island and improve comfort in urban environments. *Solar Energy* **2014**, *103*, 682–703. [CrossRef]

134. Sailor, D.J. Simulated urban climate response to modifications in surface albedo and vegetative cover. *J. Appl. Meteorol.* **1995**, *34*, 1694–1704. [CrossRef]

135. Synnefa, A.; Danodu, A.; Santamorius, M.; Tomboru, M.; Soulakellis, N. On the use of cool materials as a heat island mitigation strategy. *J. Appl. Meteorol. Climatol.* **2008**, 2846–2856–2856. [CrossRef]

136. Rosenfeld, A.H.; Akbari, H.; Romm, J.J.; Pomerantz, M. Cool communities: Strategies for heat island mitigation and smog reduction. *Energy Build.* **1998**, *28*, 51–62. [CrossRef]

137. Jacobson, M.Z.; John, E.T.H. Effects of urban surfaces and white roofs on global and regional climate. *J. Clim.* **2012**, *25*, 1028–1044. [CrossRef]

138. Menon, S.; Akbari, H.; Mahanama, S.; Sednev, I.; Levinson, R. Radiative forcing and temperature response to changes in urban albedos and associated CO_2 offsets. *Environ. Res. Lett.* **2010**, *5*, 014005. [CrossRef]

139. Castaldo, V.L.; Rosso, F.; Golasi, I.; Piselli, C.; Salata, F.; Pisello, A.L.; Ferrero, M.; Cotana, F.; Vollaro, A.L. Thermal comfort in the historical urban canyon: The effect of innovative materials. *Energy Procedia* **2017**, *134*, 151–160. [CrossRef]

140. Rosso, F.; Golassi, I.; Castaldo, V.L.; Piselli, C.; Salata, F.; Ferrero, M.; Cotana, F.; de Lieto Vollaro, A. On the impact of innovative materials on outdoor thermal comfort of pedestrians in historical urban canyons. *Ren. Energy* **2018**, *118*, 825–839. [CrossRef]

141. Rossi, F.; Anderini, E.; Castellani, B.; Nicolini, A.; Morini, E. Integrated improvement of occupants' comfort in urban areas during outdoor events. *Build. Environ.* **2015**, *93*, 285–292. [CrossRef]

142. Ye, L.; Ichinose, T. Experimental evaluation of mitigation of thermal effects by "Katsuren travertine" paving material. *Energy Build.* **2014**, *81*, 253–261.

143. Nishimura, N.; Nomura, T.; Iyota, H.; Kimoto, S. Novel water facilities for creation of comfortable urban micrometeorology. *Solar Energy* **1998**, *64*, 197–207. [CrossRef]

144. Broadbent, A.M.; Coutts, A.M. The microscale cooling effects of water sensitive urban design and irrigation in a suburban environment. *Theor. Appl. Climatol.* **2017**, *134*, 1–23. [CrossRef]

145. Taleghani, M.; Tenpierk, M.; van den Dobbelsteen, A.; Sailor, D. Heat mitigation strategies in winter and summer: Field measurements in temperate climates. *Build. Environ.* **2014**, *81*, 309–319. [CrossRef]

146. Xu, J.; Wei, Q.; Huang, X.; Zhu, Z.; Li, G. Evaluation of human thermal comfort near urban waterbody during summer. *Build. Environ.* **2010**, *45*, 1072–1080. [CrossRef]

147. Syafii, N.I. Thermal environment assessment around bodies of water in urban canyons: A scale model study. *Sustain. Cities Soc.* **2017**, *34*, 79–89. [CrossRef]

148. Rutty, M.; Scott, D. Bioclimatic comfort and the thermal perceptions and preferences of beach tourists. *Int. J. Biometorol.* **2015**, *59*, 37–45. [CrossRef]

149. Shooshtarian, S.; Priyadarsini, R.; Amrit, S. A comprehensive review of thermal adaptive strategies in outdoor spaces. *Sustain. Cities Soc.* **2018**, *41*, 647–665. [CrossRef]
150. Villadiego, K.; Velay-Dabat, M.A. Outdoor thermal comfort in a hot and humid climate of Colombia: A field study in Barranquilla. *Build. Environ.* **2014**, *75*, 142–152. [CrossRef]
151. Lin, T.P. Thermal perception, adaptation and attendance in a public square in hot and humid regions. *Build. Environ.* **2009**, *44*, 2017–2026. [CrossRef]
152. Taleghani, M. Dwelling on Courtyards: Exploring the Energy Efficiency and Comfort Potential of Courtyards for Dwellings in The Netherlands. PhD Thesis, Architecture and the Built Environment, Delft University of Technology, Delft, The Netherlands, 2014; pp. 1–354.
153. Matzarakis, A.; Mayer, H.; Iziomon, M.G. Applications of a universal thermal index: Physiological equivalent temperature. *Int. J. Biometeorol.* **1999**, *43*, 76–84. [CrossRef] [PubMed]

Publisher's Note: MDPI stays neutral with regard to jurisdictional claims in published maps and institutional affiliations.

 *sustainability*

Article

Are Global Companies Better in Environmental Efficiency in India? Based on Metafrontier Malmquist CO_2 Performance

Yongrok Choi [1],* , Hyoungsuk Lee [2] and Jahira Debbarma [1],*

[1]   Program in Industrial Security Governance, Inha University; 100 Inha-ro, Incheon 22212, Korea
[2]   East Asia Environment Research Center, Inha University, Inharo 100, Nam-gu, Incheon 22212, Korea; zard2303@naver.com
*   Correspondence: yrchoi@inha.ac.kr (Y.C.); jahira@inha.edu (J.D.)

Received: 18 August 2020; Accepted: 9 October 2020; Published: 12 October 2020

Abstract: There is a rapid increase in inflows of foreign direct investment (FDI) into developing countries such as India. Some researchers argue that FDI has a positive impact on sustainable development in terms of environmental efficiency and brings innovative green technology to the host country. In contrast, others claim that FDI brings considerable pollution to the host country, and their motive is only to yield profit. To address this issue, this paper analyzes environmental efficiency between FDI and domestic firms in India for seven years between 2012 and 2018. The research aims to evaluate the performance of FDI firms in terms of environmental efficiency in India after implementing certain policy regulations, nationally and globally. In this analysis, we use the non-radial metafrontier Malmquist CO_2 performance index (NMMCPI) with three decomposition indices: efficiency change index, best practice gap index, and technological gap change index. Our empirical results indicate that domestic firms have performed well in terms of better catch-up and innovation performance. On the other hand, FDI firms only demonstrated higher technology leadership performance, indicating weaker catch-up performance and weaker innovation performance. From the results, we proposed that policymakers should harmonize between the FDI promotion and regulation in its sustainable performance because global companies are not sensitive to the local regulations, and not very proactive in implementing the global standard of eco-friendliness.

Keywords: foreign direct investment; environmental efficiency; non-radial metafrontier Malmquist index; partial-factor CO_2 emission performance; India

---

1. Introduction

At present, carbon dioxide (CO_2) emissions from human activities have increased rapidly due to the rapid industrialization, urbanization, population explosion, and exploitation of natural resources, etc. According to the Mauna Loa Atmospheric Reference Observatory in Hawaii in 2018, CO_2 levels reached 411 parts per million, the highest monthly average ever recorded in history [1]. A recent report indicates that countries like China, United States, and India are the major emitters of CO_2 in the world. In terms of CO_2 output, India has recorded more than half of the increase in global CO_2 since 2013 [2].

1.1. Socio-Managerial Contexts

Foreign direct investment (FDI): In recent decades, we have also witnessed a sharp increase in inflows of foreign direct investment (FDI) to developing countries. FDI in India began in 1991 under the Foreign Exchange Management Act (FEMA) with a baseline of USD 1 billion in 1990. Since then, India has been one of the most popular destinations preferred for FDI. India ranks the 9th to receive

FDI inflows as in 2019, against 12th in the previous year, among the largest recipients of FDI in the world reported by UNCTAD, 2020. Indian government proactively induced FDI due to its quantitative contribution in the local economy as well as the qualitative contribution for innovative technology transfer and advanced know-how for the sustainable development of India. Unfortunately, it does not always seem true in the performance of FDI firms. As shown in Figure 1, the trend in FDI inflows from 2001 to March 2016 corresponds to CO_2 emissions in India [3,4]. This means that there is a direct relationship between CO_2 emissions and FDI inflows into India. The Indian government may be concerned about this phenomenon as it expects qualitative contributions from FDI companies, including sustainable and environmentally friendly management of these global firms.

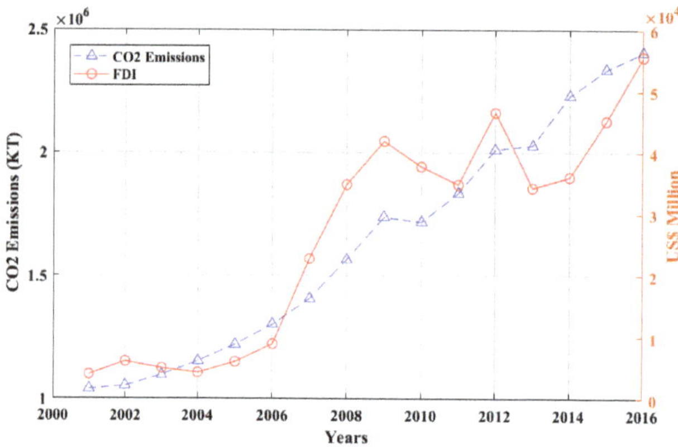

Figure 1. CO_2 emissions and foreign direct investment (FDI) inflows in India.

Gross Domestic Products (GDP): FDI is one of the important factors that promote economic growth in India. Thus, the Indian government has gradually relaxed restrictions on FDI to achieve rapid growth. For example, on 25 September 2014, the Prime Minister of India published an international marketing strategy called "Make in India" to encourage foreign firms to invest in India to strengthen the country's manufacturing sector [5]. Figure 2 provides some details on FDI equity inflows by sector in India as of 2018 [4]. As shown in Figure 2, the main sectors of FDI inflows to India are services sectors, computer software and hardware, telecommunications, construction development, trading, automobile industry, chemicals, infrastructure, drugs and pharmaceuticals, etc.

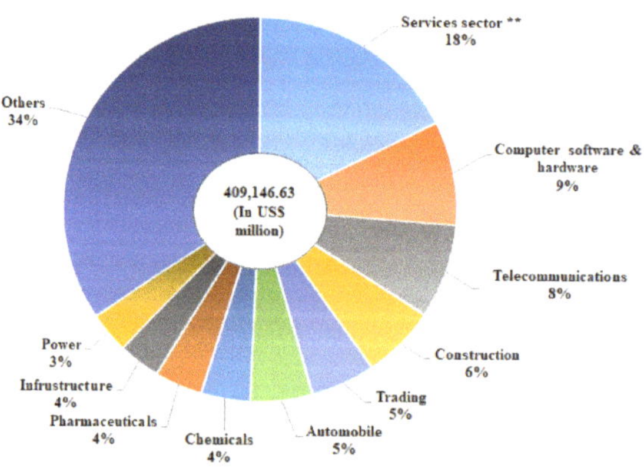

Figure 2. FDI inflows by sectors in India (as of 2018).

Corporate Social Responsibility (CSR): Due to strict environmental regulations in developed countries, investors have been discouraged from investing in pathways that produce much natural pollution and thus create a heavier response on its corporate social responsibility (CSR) for the community, resulting in overseas investments toward developing countries, where the environmental regulations are not very severe, and the global company has the invisible privileges on the local regulations. For these reasons, FDI investors still prefer countries where legal, environmental regulations have not yet been determined, ignoring the negative consequences of the investment they make, which in turn leads to environmental degradation. Due to the need for all kinds of industrial activities such as low-income level, insufficient property rights, and lack of development of environmental awareness, developing countries do not give much importance to environmental regulations [6]. As a result, the increased transfer from developed to developing countries by high CO_2-emitting industries such as energy-intensive industries has started to create an environmental hazard. Consequently, to avoid this type of negative impact situation and to increase its rigor, developing countries have also begun to improvise their environmental regulations. For example, the Reserve Bank of India has said that political action is needed to put in place an enabling environment for fostering green investment as well as advancing the net-zero carbon delivery program in India [7].

Energy-related policies in India: Some of the highlighted energy-related policy Acts and plans proposed and adopted by the Indian government are Energy and carbon taxes supported by the United Nations Framework Convention on Climate Change. A carbon tax is a step to help India reduce the amount of CO_2 released per unit of gross domestic product (GDP) and reach its voluntary target of 25% over 2005 levels by 2020. India has already introduced a national carbon tax, generally in the transport and energy sector of 50 rupees (Rs) per ton of coal consumed by any economic sector on July 1, 2010. The carbon tax has been further increased, and currently, the carbon tax amounts to Rs 400 per ton in India in all sectors emitting CO_2. Nationally Determined Contributions (NDCs) under the Paris Agreement, 2016. India is one of the few developing countries on the way to achieving the goals of the Paris Agreement by the "nationally determined contributions" or NDCs. On 2 October 2016, India ratified the Paris Agreement with a 4.1% reduction goal by 2030 in global CO_2 emissions. Based on NDCs, the three main objectives have been set by the Indian government. The first objective is to increase the share of non-fossil fuels to 40% of total electricity production capacity. The second objective is to reduce emission intensity in the economy from 33% to 35% by 2030 compared to the 2005 level. The third objective is to create an additional carbon sink of 2.5–3 billion tons of CO_2 equivalent for the period 2020–2030.

Now, the question is whether this rosy pathway does affect the sustainable management of FDI companies in India. The Indian government gave easier access on the market to the FDI firms with the expectation of these FDI firms' leadership coming from the global standard on the qualitative contribution to the local economy, including the environmentally friendly economic activities in India.

1.2. Research Motivation and Objective

In the literature, however, many studies have argued on the potential impacts of FDI inflows and environmental sustainability, such as pollution from CO_2 emissions, energy consumption, etc. The influence of environmental regulations on FDI firm performance for decades has been an increasingly important subject for researchers around the world [8]. In one view, FDI can bring advanced technologies to the host country and improve the country's sustainable growth because FDI raises its level of technical progress through "learning by doing" in sustainable ways [9,10]. Another extreme point of view is that developed countries are shifting polluting industries to developing countries. It has brought much pollution to developing countries by transferring energy-intensive and much pollution-oriented plants. Therefore, the increase in the inflow of FDI could have adverse effects on the host countries, causing severe environmental pollution and degradation of the environment [11–15]. It is a widely debated subject whether India benefits from FDI or not in environmental perspectives.

Based on the above argument, this research aims to evaluate the environmental efficiency between FDI and domestic firms in India and determine the performance of FDI in the period between 2012 and 2018. Since the evaluation should be based on the dynamic effect over time with multi-inputs/outputs model, we will use the non-radial metafrontier Malmquist index to analyze our data.

1.3. Research Contribution

There are two types of firms in India; FDI and domestic firms. The FDI firm implies that a company takes controlling ownership in a business entity in another country. To take advantage of cheaper wages, FDI firms invest directly in the fast-growing Indian market and change the business environment in India. A domestic firm is defined as a local investor who can conduct business in his home country. Since this research aims to compare all FDI firms with domestic ones, it will be essential to also focus on the average variation in the productivity of each firm over time.

This study contributes to understanding how vital FDI is to build a sustainable ecosystem in India, taking into account the environmental impact. How will regulatory policies affect these firms to reduce CO_2 emissions and to improve environmental performance in India? To the best of our knowledge, no one has explored this research area in India so far. Since there is no comprehensive comparison between FDI firms and domestic firms to determine the environmental efficiency of FDI in India, this paper attempts to compare FDI and domestic firms by measuring their environmental efficiency. Therefore, the purpose of this research is empirically to analyze whether FDI inflows to India increase its CO_2 emissions or vice-versa.

The structure of the paper is as follows; in Section 2, we present a literature review in the related areas to find out our strategic variables and methodologies; in Section 3, we develop our empirical models to evaluate the environmental efficiency of FDI and domestic firms in India; we discuss our empirical result and its implications in Section 4; in Section 5, we conclude our study by providing some policy implications.

2. Literature Review

Many policymakers and academics professionals have paid considerable attention to the relationship between FDI inflows and the effectiveness of environmental sustainability in the host country. However, the impact of FDI on the environment is still unclear. Some studies argue that FDI can improve the environmental efficiency of the host country [16–18], while others find that it can damage the environment of the host country [19].

Most of the literature in energy and environment (E&E) analysis uses the multi-inputs and outputs to handle the coupling and decoupling between the potentially conflicting variables of desirable outputs such as profits or GDP, and undesirable outputs such as CO_2 emission, as shown in Table 1. Since the data envelope analysis (DEA) is very popular for handling multi-inputs/outputs models, it is also possible to utilize this approach for our research in the comparison between FDI and domestic firms as well [20,21]. The DEA approach does not need any specific form of the production function for this multi-inputs/outputs model. Thus, it is very popular to evaluate the relative efficiency for all decision-making units (DMUs), FDI, and domestic firms in our model. Thus, we first examine research papers that are related to the environmental efficiency of FDI using DEA. As shown in Table 1, we classify the previous research into three categories. First, FDI has a positive effect on the environmental efficiency of the host country [16]. Second, FDI harms (negative) the environmental efficiency of the host country [19]. Third, FDI may have a mixed effect with positive and negative impacts on the environmental efficiency of the host country [22–24].

Table 1. Comparison of the research on the environmental efficiency of FDI.

Reference(s)	Field of Research	Measurement	Method	Input/Output	Conclusions
Pan et al. (2019) [25]	FDI Quality in China	Energy efficiency	SBM-DEA	Input—energy, capital, and labor. Desirable output-GDP of each province. Undesirable output—CO_2 emissions.	Positive
Wang (2017) [26]	FDI on energy efficiency in China	Energy efficiency	Sequential DEA	Input—capital, labor, and energy. Desirable output—GDP. Undesirable output—SO_2 and CO_2.	Positive
Mastromarco et al. (2017) [27]	FDI and time	Efficiency	DEA	Input—labor and capital. Output—GDP.	Positive
Yue et al. (2016) [28]	FDI from China's Experience	Efficiency	SBM-DEA	Input—capital, labor, and energy. Desirable output—GDP. Undesirable output—SO_2.	Positive
Song et al. (2015) [29]	FDI in China	Efficiency	DEA	Input—variables passed the test. Desirable output—the number of patents.	Positive
Yang et al. (2019) [30]	FDI and export in China	Environmental efficiency	SBM-DEA	Input—labor, capital stock, energy consumption, and water consumption. Desirable output—industrial added value. Undesirable output—CO_2 emissions, SO_2 emissions, and wastewater.	Negative
Lei et al. (2013) [31]	FDI attractiveness from Chinese provinces	Bottleneck	DEA	Input—material capital, human capital, energy, and degree of openness. Output—FDI Performance Index and FDI Potential Index.	Negative
Zang et al. (2012) [32]	FDI in developing countries	Energy efficiency	Super-efficiency DEA		Negative
Monaheng et al. (2019) [33]	FDI and economic performance (BRICS)	Economic performance	DEA (managerial disposability)	Input—oil, labor force, and capital. Desirable output—GDP. Undesirable output—economic production.	Mixed
Guo et al. (2013) [11]	Regional Influence of FDI in China	Energy efficiency	DEA	Input—capital, labor and energy. Desirable output—GDP. Undesirable output—environmental pollution	Mixed
Luo et al. (2013) [34]	FDI of China	Environmental performance	DEA	Input—capital, labor, and energy. Output—GDP.	Mixed

Pan et al. [25] used the slacks-based measure (SBM)-Data Envelopment Analysis (DEA) model to measure more accurately the energy efficiency of the quality of FDI in China. They found that the quality of FDI had a significant positive effect on energy efficiency in China. Wang [26] measured the energy efficiency of FDI on energy efficiency in China and concluded that FDI significantly improves energy efficiency. Mastromarco et al. [27] analyzed FDI and time on catching up and found that FDI influences productivity by increasing efficiency and improving technological change. Yue et al. [28] measured the effectiveness of FDI from China using the SBM-DEA. They found that FDI significantly improved energy efficiency. Song et al. [29] examined FDI in China using the DEA model to measure efficiency. They found that FDI inflows can play a positive role in local economic and technological development in China, especially in some rapidly economically developing areas.

However, Yang et al. [30] compared the environmental efficiency of FDI and exports from China. They concluded that FDI reduces industrial environmental efficiency in China. Likewise, export from China also has a significant negative impact on industrial environmental efficiency. Lei et al. [31] have researched the attractiveness of FDI in Chinese provinces and found that only eastern provinces have great development potential. Zang et al. [32] have used a super-efficiency DEA model to evaluate energy efficiency from FDI in developing countries and found that FDI has a negative impact in developing countries.

Monaheng et al. [33] examined FDI and economic performance in BRICS countries. They found that FDI had a positive impact on the economic performance of the BRICS countries except for China. Guo et al. [11] analyzed the regional influence of FDI in China by measuring energy efficiency using the DEA model. They recognized that FDI inflows improve local energy efficiency in the central and eastern regions but reduce energy efficiency in the western region. Finally, Luo et al. [34] have studied FDI from China to measure environmental performance. They have argued that FDI has a positive impact on the environment as well as a negative impact on the country's environmental efficiency. For example, FDI has improved local energy efficiency in the eastern and central regions due to the aggregate result of the technological effect, the scale effect, and the structure effect of FDI. However, FDI has reduced energy efficiency in the western zone due to the lower level of economic development.

As shown in Table 1, the authors used diverse DEA models to measure economic performance, environmental efficiency, energy efficiency, CO_2 emission performance, and efficiency of FDI in the host country. Existing literature has extensively studied the effects of FDI on a country's economic development and environmental pollution simultaneously. However, there is no comprehensive comparison between FDI firms and domestic firms to evaluate the environmental performance of FDI. Therefore, this paper attempts to measure the environmental efficiency of FDI in India by comparing the environmental efficiency of FDI and domestic firms in India.

3. Model Design and Specification

3.1. Production Technology Set

In E&E studies, the term "environmental production technology" is defined as the basic directional distance function of all the interrelated variables without any specific production frontier constraints a priori, because environmental production technology unambiguously encompasses multiple outputs, differently from other parts of traditional production theory in economics. This environmental production technology encompasses two classical inputs of the capital (x^1) employees (x^2) and the specialized input of energy (x^3), with the sales turnover variables (B) as desirable output and CO_2 (C) as an undesirable output. Therefore, we selected the three inputs and two outputs in this study, based on the traditional approach to the production function. We collected data from the annual reports of each FDI and domestic firms from different sectors that emit CO_2. Environmental production technology is defined as the causal relationship between the firm's capital (x^1), employee (x^2), and energy (x^3) as an input, and the desirable as well as undesirable outputs of firms' sales turnover (B) and CO_2 (C), respectively. According to Lee and Choi (2018), this can be expressed in the mathematical form [35]:

$$T = \left\{ \left(x^1, x^2, x^3, B, C : x \text{ can produce } (B, C) \right) \right\} \tag{1}$$

where the set of production possibilities T is assumed to satisfy the standard axioms of the production theory [36]. For example, finite amounts of input can only produce limited amounts of output because inactivity is always possible [37]. Additionally, inputs and desirable outputs are often assumed with undesirable output freely disposable (weak disposability), implying that a reduction of the undesirable output such as CO_2 should match with less desirable outputs in the production process. The elimination of CO_2 can be possible if and only if by stopping production, on T concerning regulated environmental technologies (null-jointness) [38]. These two hypotheses can be expressed in the mathematical expression as follows:

i.   If $\left(x^1, x^2, x^3, B, C \right) \in T$ and $0 \leq \theta \leq 1$, then $\left(x^1, x^2, x^3, \Theta B, \Theta C \right) \epsilon T$

ii.   If $\left(x^1, x^2, x^3, B, C \right) \in T$ and $B = 0$, then $C = 0$

Once the environmental production technology (T) is defined, we can specify the frontier. Consequently, we can formulate T for N FDI and domestic firms under constant returns to scale as follows [39]:

$$T = \{ \left(x^1, x^2, x^3, B, C \right) : \sum_{n-1}^{N} Z_n x_{in} \leq x_i \text{ , where } i = x^1, x^2, x^3$$
$$\sum_{n-1}^{N} Z_n B_n \geq B, \quad \sum_{n-1}^{N} Z_n C_n = C, \tag{2}$$
$$Z_n \geq 0, \, n = 1, 2, \ldots, N \}$$

where Zn is an intensity variable. By using a convex combination, it can build environmental production technology. Although the assumptions of variable returns to scale are widely adopted in the various literature, we selected the constant returns to scale (CRS) for T in this study because the CRS method is easily generalized to multiple types of firms.

3.2. CRS Non-Radial Directional Distance Function

There are two types of distance functions for E&E studies, which are widely adopted Shephard distance function and the directional distance function (DDF) [40,41]. Shephard distance function is known for minimizing the inputs and undesirable output, at the same rate, maximizes the desirable output simultaneously. However, this function has several limitations. The most outstanding difficulty in Shephard distance function may come from the radial approach for its coupling issues between the desirable and undesirable outputs. To solve this coupling issue of the radial approach, the generalized directional distance function is introduced. The common limitation is that the Shephard distance function may overestimate the effectiveness when some slacks exist [42]. Using the directional weight vector (g), conventional DDF can overcome this overestimation. It can be expressed in mathematical form as follows:

$$\vec{D}\left(x^1, x^2, x^3, B, C; g\right) = \sup\{\beta : \left(\left(x^1, x^2, x^3, B, C\right) + g.\beta\right) \in T\} \tag{3}$$

On the other hand, the non-radial efficiency measure is more generalized and personalized for E&E studies due to its advantage to overcome the many limitations of the radial efficiency functions, with a more generalized form that takes into account undesirable outputs [37,43,44]. Thus, our study also adopts the non-radial DDF (NDDF), which is expressed mathematically as follows:

$$\vec{D}\left(x^1, x^2, x^3, B, C; g\right) = \sup\{W^T\beta : \left(\left(x^1, x^2, x^3, B, C\right) + g.\text{diag}(\beta)\right) \in T\} \tag{4}$$

The symbol *'diag'* in the Equation (4) denotes diagonal matrices, which are related to the numbers of all variables and can be expressed as follows: $W^T = (W_{x^1}, W_{x^2}, W_{x^3}, W_B, W_C)^T$ denotes a normalized weight vector; g denotes an explicit directional vector which can be expressed as $g = \left(-g_{x^1}, -g_{x^2}, -g_{x^3}, g_B, -g_C\right)$; $\beta = (\beta_{x^1}, \beta_{x^2}, \beta_{x^3}, \beta_B, \beta_C)T \geq 0$ represents individual inefficiency measures by denoting the scaling vector for all variables. In order to avoid the diluting effect of non-energy variables, we set the directional vector of capital and labor to be zero, because capital and labor do not directly affect CO_2 emissions [45]. Therefore, the directional vector of this study is g = $\left(0, 0, -g_{x^3}, g_B, -g_C\right)$ and the weight vector is (0, 0, 1/3, 1/3, 1/3). We can, therefore, define NDDF value in the mathematical form [46]:

$$\vec{D}\left(x^1, x^2, x^3, B, C; g\right) = \max W_{x^3}\beta_{x^3} + W_B\beta_B + W_C\beta_C$$

$$\sum_{n=1}^{N} Z_n x^1_n \leq x^1_{n\prime}, \ \sum_{n=1}^{N} Z_n x^2_n \leq x^2_{n\prime}, \ \sum_{n=1}^{N} Z_n x^3_n \leq x^3_{n\prime} - \beta_{x^3}\, g_{x^3}\, \sum_{n=1}^{N} Z_n B_n$$

$$\leq B_{n\prime} + \beta_B g_B \sum_{n=1}^{N} Z_n C_n = C_{n\prime} - \beta_C g_C \tag{5}$$

$$Z_n \geq 0, \ n = 1, 2, \dots, N$$

$$\beta_{x^1}, \beta_{x^2}, \beta_{x^3}, \beta_B, \beta_C \geq 0.$$

In Equation (5), we can modify the directional vector g, according to our different political objectives of reducing CO_2 emissions. If $\vec{D}$ $(x^1, x^2, x^3, B, C; g) = 0$, this would indicate that a specific FDI or domestic firm under the evaluation is located in the best-practice production frontier in the g direction [37]. We also formulated the static total-factor CO_2 emissions performance index (TCPI) as the ratio of the actual carbon intensity to the potential target carbon intensity [43]. If β_C^* and β_B^* are the optimal solutions that correspond to undesirable output CO_2 emissions and desirable output sales turnover in Equation (5), then, the TCPI can be described as follows,

$$\text{TCPI} = \frac{\left(C - \beta_C^* C / \left(B + \beta_B^* B\right)\right)}{C/B} = \frac{1 - \beta_C^*}{1 + \beta_B^*} \tag{6}$$

In this study, Equation (6) can be used to measure the performance of CO_2 emissions from FDI and domestic firms and measure the maximum possible reductions in CO_2 intensity over the 2012–2018 period. If the TCPI is higher, then the CO_2 emission performance will be better. Besides, the TCPI lies between zero and unity; therefore, the best CO_2 emission performance of FDIs and domestic firms will be located along the frontier when the TCPI is equal to unity.

3.3. Non-Radial Metafrontier Malmquist CO_2 Performance Index (NMMCPI)

Based on the metafrontier Malmquist index, some authors have proposed a *Non-radial Metafrontier Malmquist CO_2 Performance Index (NMMCPI)* to measure the dynamic total-factor CO_2 emission performance [47,48]. Here, the MMCPI is defined as the difference between the weighted average rates of change of (negative) inputs, (positive) outputs, and (negative) outputs. The *NMMCPI* measures the change in TCPI during period t and t + 1. It is useful to evaluate environmental performance change over time. Thus, we adopted the non-radial metafrontier Malmquist CO_2 Performance Index (NMMCPI) to examine the environmental efficiency of FDI and domestic firms in India.

To define and decompose the NMMCPI in this study, we need to define the three sets of environmental technologies, i.e., contemporaneous environmental technology, intertemporal environmental technology, and global environmental technology. We determine these sets as follows [49]. The first sets of environmental technologies called the contemporaneous environmental technology of R_h can be expressed as $T_{R_h}^C = \left\{\left(x^{1t}, x^{2t}, x^{3t}, B^t, C^t\right) : \left(x^{1t}, x^{2t}, x^{3t}\right) \text{ can produce } \left(B^t, C^t\right)\right\}$ where $t = 1 \ldots T$ For a particular period t and group R_h, it constructs production technology set in Equation (2). The second set of environmental technologies is an intertemporal environmental technology and its R_h an be expressed as $T_{R_h}^I = T_{R_h}^1 \cup T_{R_h}^2 \cup \ldots \cup T_{R_h}^T$. This is constructed from observations for group R_h which consists of a single technology over the whole period. This implies that the observations for an intertemporal environmental technology are assumed to be unable to access different intertemporal environmental technologies easily. For the distinct intertemporal technologies, we assumed that there are h groups subsequently. Finally, global environmental technology is considered in the third set of environmental technologies that can be expressed in the form: $T^G = T_{R_1}^I \cup T_{R_2}^I \cup \ldots \cup T_{R_H}^I$, which is constructed from all observations for FDI and domestic firms groups over the 2012–2018 period. This implies that global environmental technology encompasses all intertemporal environmental technologies. For the sake of analysis, we assume that every observation (theoretically and potentially) can access global technology through its innovations.

Based on these three decomposed technologies, our environmental technologies under the non-radial directional distance functions in Equation (4) can be expanded as follows. First, based on contemporaneous environmental technology $\left(T_{R_h}^C\right)$ we define contemporaneous NDDF of a specific group R_h as $\vec{D} C (.) = \sup \left\{w^T \beta^C : \left((x^1, x^2, x^3, B, C) + g * \text{diag}(\beta_C)\right) \epsilon T_{R_h}^C\right.$ Second, we define intertemporal NDDF as $\vec{D} I (.) = \sup \left\{w^T \beta^I : \left((x^1, x^2, x^3, B, C) + g * \text{diag}(\beta^I)\right) \epsilon T_{R_h}^I\right.$ of a specific group R_h based on the intertemporal environmental technology $T_{R_h}^I$ Following global environmental technology, we define global NDDF as $\vec{D} G (.) = \sup \left\{w^T \beta^G : \left((x^1, x^2, x^3, B, C) + g *\right.\right.$ $\text{diag}(\beta^G)) \epsilon T_{R_h}^G$ Therefore, the following six different NDDFs could be solved in order to decompose and compute the NMMPI as follows: $\vec{D} C = (x^{1S}, x^{2S}, x^{3S}, B^S, C^S)$, $\vec{D} I = (x^{1S}, x^{2S}, x^{3S}, B^S, C^S)$, and $\underset{D}{\to} G = (x^{1^S}, x^{2^S}, x^{3^S}, B^S, C^S)$, where $S = t, t + 1$. With these three technologies, NDDF can be solved using Equation (5) as follows:

$$\underset{D}{\to} d = (x^{1^S}, x^{2^S}, x^{3^S}, B^S, C^S; g) = \max W_{x3}\beta^d_{x3} + W_B\beta^d_B + W_c\beta^d_c$$

$$\sum_{con} Z^S_n x^{1^S}_n \leq x^1{}_{n\prime} \quad \sum_{con} Z^S_n x^{2^S}_n \leq x^2{}_{n\prime}$$

$$\sum_{con} Z^S_n x^{3^S}_n \leq x^3{}_{n\prime} - \beta^d_{x3}g_{x3} \quad \sum_{con} Z^S_n B^S_n \geq B{}_{n\prime} + \beta^d_B g_B \quad \sum_{con} Z^S_n C^S_n \geq C{}_{n\prime} - \beta^d_C g_C \tag{7}$$

$$Z^S_n \geq 0, \ \beta^d \geq 0$$

where the superscript d on $\underset{D}{\to} d$ (.) can be a contemporaneous, intertemporal, and global environmental technology, which represents the type of NDDF. Besides, the symbol "con" under $\sum$ designates the construction conditions of three environmental technologies. If we want to construct the contemporaneous NDDF, we required the conditions d≡C and con ≡ {n ϵR_h to build the intertemporal NDDF, we need the conditions set d≡I and con ≡ {n ϵR_h s ϵ1, 2, T]}; and the requirements should be d≡G con ≡ {n ϵ[$R_1 \cup$ [$R_2 \cup \ldots \cup R_H$], s ϵ1,2, ... , T]}, for the global NDDF. Using Equation (7), we can solve the six different NDDF models in Equation (6).

$$\text{TCPI}^d\left(x^{1^S}, x^{2^S}, x^{3^S}, B^S, C^S\right) = \left[\frac{\left(C - \beta^{d^*}_C C / \left(B + \beta^{d^*}_B B\right)\right)}{C/B} \right]^S = \left(\frac{1 - \beta^{d^*}_C}{1 + \beta^{d^*}_B} \right)^S \tag{8}$$

where $S = t, t + 1$ and $d = (C, I, G)$. We define the NMMCPI, based on the formulation of the metafrontier Malmquist index within the framework of the global environmental technology set [46]. It can be expressed as the ratio.

$$\text{NMMCPI}\left(x^{1^S}, x^{2^S}, x^{2^S}, B^S, C^S\right) = \frac{\text{TCPI}^G\left(x^{1^{t+1}}, x^{2^{t+1}}, x^{3^{t+1}}, B^{t+1}, C^{t+1}\right)}{\text{TCPI}^G\left(x^{1^t}, x^{2^t}, x^{3^t}, B^t, C^t\right)} \tag{9}$$

NMMCPI measures the variations of the TCPI from Equation (9), on the T^G for the period between t and t + 1. Now, NMMCPI can be decomposed into a technical efficiency change (EC) index of CO_2 emissions, a best-practice gap change (BPC) index of CO_2 emission reduction technologies, and a technology gap change (TGC) as follows [47,48]:

$$\text{NMMCPI}\left(x^{1^S}, x^{2^S}, x^{3^S}, C^S\right) = \frac{\text{TCPI}^G\left(.^{t+1}\right)}{\text{TCPI}^G\left(.^t\right)}$$

$$= \left\{ \frac{\text{TCPI}^C\left(.^{t+1}\right)}{\text{TCPI}^C\left(.^t\right)} \right\} * \left\{ \frac{\text{TCPI}^I\left(.^{t+1}\right)/\text{TCPI}^C\left(.^{t+1}\right)}{\text{TCPI}^I\left(.^t\right)/\text{TCPI}^C\left(.^t\right)} \right\} * \left\{ \frac{\text{TCPI}^G\left(.^{t+1}\right)/\text{TCPI}^I\left(.^{t+1}\right)}{\text{TCPI}^G\left(.^t\right)/\text{TCPI}^I\left(.^t\right)} \right\}. \tag{10}$$

$$= \left[\frac{\text{TE}^{t+1}}{\text{TE}^t} \right] * \left[\frac{\text{BPR}^{t+1}}{\text{BPR}^t} \right] * \left[\frac{\text{TGR}^{t+1}}{\text{TGR}^t} \right] = \text{EC} * \text{BPC} * \text{TGC}$$

In the Equation (10) the term efficiency change or EC index measures the "catching-up" effect, which measures the technical efficiency changes of CO_2 emissions for a specific FDI or a domestic firm for two time periods (t, t + 1) within a specific group. The EC Index captures how close an FDI or a domestic firm is toward contemporaneous environmental technology. If EC > 1, there is an efficiency gain, and if the EC < 1, there is a loss of efficiency [39]. The best practice gap change or BPC index is the change in the best practice gap in two adjacent periods between contemporaneous environmental technology and intertemporal environmental technology. Here, if BPC > 1, the contemporaneous environmental technological frontier in period t + 1 is closer to intertemporal environmental technology than in period t, and if BPC < 1, the contemporaneous technological frontier is further from the intertemporal environmental technological frontier [45]. BPC is also seen as the innovation effect since innovation allows a shift of the frontier. The Technology Gap Change or TGC Index is the change in technological leadership during the two periods between the intertemporal environmental technology frontier and the global environmental technology frontier for reducing CO_2 emissions. For example, TGC> (or <) 1 indicates that a technical gap between a specific group of intertemporal technology and global technology is reduced (increased). Therefore, TGC measures the effect of technological leadership for a given group.

Figure 3 shows NMMCPI and its decomposed components. Here, contemporaneous environmental technology is presented as $T^C_{R_1}$ and $T^{C+1}_{R_1}$ in the group R_1 or the periods t and t + 1. Intertemporal environmental technology is presented as $T^I_{R_1}$ for group R_1 and global environmental technology for two groups is presented as $T^G_{R_1}$ [50]. Here, the intertemporal technological set is the envelope of all contemporaneous technological set in the current period of a particular group, and the global technological set is the envelope of all intertemporal technologies set. Here, the observed FDI and domestic firms for the two periods t and t + 1 are a_1 and a_2 because it is a case of two groups (R_1, R_2) and two periods (t, t + 1), respectively.

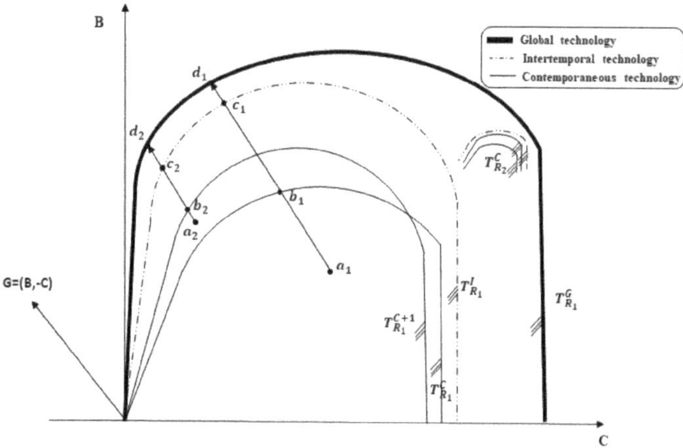

Figure 3. Graphic illustration of non-radial metafrontier Malmquist CO_2 performance index (NMMCPI) and its decomposition components. Source: Zhang and Choi, 2014 [37], Oh et al., 2010 [49].

4. Analysis and Discussion

4.1. Data Collection and Preprocessing

Since this study aims to compare the environmental efficiency of FDI firms with domestic firms, we collected data on 50 firms belonging to both groups from 2012 to 2018. Concerning the industrial sector, the following sectors were chosen: chemicals, constructions, pharmaceuticals, power, automobile, infrastructure, and others. We selected these sectors because they have the largest inflow of FDI to India, subject to the percentage of total inflows in 2018 (see Figure 2). Additionally, we selected 25 FDI companies and 25 domestic companies in India, based on the availability of firm data and FDI equity inflows by sector in India as of 2018. These firms could be considered the representative companies of these sectors, which have high CO_2 emissions [51]. The service sector and telecommunications sectors were excluded because of their low emission volume and the scarcity of data. Based on the argument on Table 1, we collected three input variables, capital (x^1) labor (x^2) and energy (x^3) and two output variables, sales turnover (B) and CO_2 emissions (C). All the data concerning capital, labor, energy, sales turnover, and CO_2 were collected in the annual report of each FDI and domestic firm between 2012 and 2018. We selected seven-year data from 2012 to 2018 because the classification in this study was based on two types of the policy paradigm shift in India to reduce CO_2 emissions. These two policies are energy and carbon taxes under the United Nations Framework Convention on Climate

Change, 2010, and Nationally Determined Contributions (NDCs) under the Paris Agreement, 2016. According to THE HINDU in 2018, environmental regulations in India have very few impacts on FDI firms [13]. The World Resources Institute India declared in June 2019 that after biomass combustion, companies are the second largest contributor to PM2.5 in India [52]. Concerning this, we wanted to analyze the environmental efficiency of FDI and domestic firms in India based on the policy paradigm shift in India.

For the capital input, we extracted the data on fixed assets from the published annual reports of each FDI and domestic firms from 2012 to 2018. We collected the employees per head of each company, as shown in the annual report of each DMU. For energy input data, since there is no right amount of energy consumption provided by each firm in their report, we had to convert the power and fuel consumption into total energy consumption equivalent value from the annual report published by the firms. For sales turnover output, we collected the revenues of the firms generated by the operations. For the CO_2 emissions data, we extracted the CO_2 values using the macro level of the FDI and domestic firm's data of power and fuel consumption rate [37,53]. These descriptive statistics (all the variables) are shown in Table 2.

Table 2. Descriptive statistics of input and output variables from 2012–2018.

Group	Variable	Input/Output	Unit	Mean	Std. Deviation	Maximum	Minimum
FDI	Capital	Input	Million rupees	16,134.19	29,101.62	145,220.00	103.85
	Employee	Input	Per person	3387.12	4968.39	24,491.00	48.00
	Energy	Input	Gj	362,115.59	654,115.20	3,195,000.00	3116.16
	Sales Turnover	Desirable output	Million rupees	49,033.27	90,114.70	493,699.73	813.75
	CO_2 Emissions	Undesirable output	Tons	17,782.89	36,434.82	218,541.00	111.40
Domestic	Capital	Input	Million rupees	173,821.11	4,315,63.83	3,004,470.00	549.20
	Employee	Input	Per person	10,790.33	14,906.18	71,826.00	133.00
	Energy	Input	Gj	1,981,662.00	3,012,483.17	13,180,794.43	17,142.97
	Sales Turnover	Desirable output	Million rupees	305,544.82	738,513.58	3,990,530.00	1082.51
	CO_2 Emissions	Undesirable output	Tons	100,795.48	159,748.60	636,676.79	612.86

Table 3 illustrate the correlation matrix of input and output variables of FDI and domestic firms from 2012–2018. The result shows that the correlation between all types of input and output variables is positive. Capital and labor are positively linked to sales turnover because they are representative variables to explain production; in particular, capital and turnover show a highly significant relationship. On the other hand, energy shows a positive relationship for both sales turnover and CO_2 emissions; in particular, energy and CO_2 emissions show a highly significant relationship [54,55]. In this study, the result verifies the suitability of analyzing the data from an environmental point of view.

Table 3. Correlation matrix of input and output variables.

Variables	Capital	Employee	Energy	Sales Turnover	CO_2
Capital	1.00				
Employee	0.27	1.00			
Energy	0.50	0.50	1.00		
Sales turnover	0.82	0.22	0.29	1.00	
CO_2 emission	0.48	0.41	0.96	0.26	1.00

4.2. Result and Its Implications

Table 4 shows the average value of the NMMCPI index and its decomposition for FDI and the domestic firm in India. Due to the firm's confidentiality, we used the firm's *id* in our results in Table A1.

Table 4. Comparison of average NMMCPI and its decomposition.

Year	FDI				Domestic			
	NMMCPI	EC	BPC	TGC	NMMCPI	EC	BPC	TGC
2012–2013	1.0046	0.9972	1.0457	0.9634	0.9925	0.9554	1.0075	1.0311
2013–2014	0.9989	0.9932	1.0316	0.9749	1.0029	0.9766	1.0075	1.0193
2014–2015	0.9969	1.0189	0.9649	1.0140	1.0101	1.0481	0.9800	0.9834
2015–2016	1.0019	0.9970	1.0044	1.0005	1.0129	1.0094	1.0054	0.9981
2016–2017	1.0007	1.0190	0.9466	1.0374	1.0120	1.0618	0.9854	0.9672
2017–2018	1.0062	1.0065	1.0007	0.9990	1.0585	1.0174	1.0281	1.0120
Average	1.0015	1.0053	0.9990	0.9982	1.0148	1.0115	1.0023	1.0019

First, we can see that the overall NMMCPI of the FDI firms remains stable with a slightly lower growth rate of 0.15%, while domestic firms show 1.48%. This result implies that domestic firms improve their performance in implementing the environmental regulatory regime during the study period, compared to FDI firms in India. In terms of the average EC index of FDI firms, the growth rate is 0.53%, indicating a lower growth rate in efficiency during the study period. At the same time, domestic firms show an annual increase of 1.15%, implying that domestic firms do better for the best catch-up effect, suggesting the more proactive movement towards the contemporary environmental technological frontier. In terms of the average BPC index, FDI firms show the value lower than unity (0.9990), while domestic firms show a 0.23% increase. Since the BPC index captures the "innovation effect," this result implies that domestic firms expanded their production level under the environmental regulatory regime. The average annual TGC index of FDI and domestic firms is 0.9982 and 1.0019, respectively. This implies a lack of technological leadership among FDI firms in India during the study period. The gap between the intertemporal frontier and the global frontier has narrowed for FDI firms, as the TGC index is a measure of changes in technological leadership for a given group.

Figure 4 shows the trends in the NMMCPI for FDI and domestic firms during 2012–2018. Regarding the dynamic perspective over time, we can see that the performance of FDI firms in annual total factor CO_2 emissions shows a lower growth rate during the study period, which is close to unity. This could be due to the lack of policy effect on FDI firms in India because policy in India has a soft signal for the foreign market [13]. Therefore, there is a missing link in the role of the Indian government for FDI firms. We suggest that the Indian government should be aware of this phenomenon and take certain measures to reduce the differences between local firms and FDI firms. Like FDI firms, domestic firms also show a low growth rate in the performance of total factor CO_2 emissions from 2012 to 2017, which is 1.0120. However, after 2016–2017, the graph shows an upward trend, which is above the unity of 5.85% in 2018. This J-curve effect means that there is an effect of government regulation on domestic firms in India. This may be due to the implementation of the policy that the Indian government has adopted. Indian government set a very clear and strong CO_2 reduction target in 2016, called Nationally Determined Contributions (NDC), under the Paris Climate Agreement. Consequently, all of the firms that emit the most CO_2 are under pressure to reduce their CO_2 emissions. Due to the higher regulatory costs, this type of strict regulation could affect a firm's performance at the initial level. That is why the performance of domestic firms was observed highly enhanced after 2016–2017. Even if there are no strong regulations yet for the FDI firms in India, they should show their efforts to reduce CO_2 emissions under the mixed economy such as India. It also supports Porter's hypothesis that if a country has tighter environmental regulations, it will increase the efficiency of that particular country and encourage innovation for a more environmentally friendly production process [53].

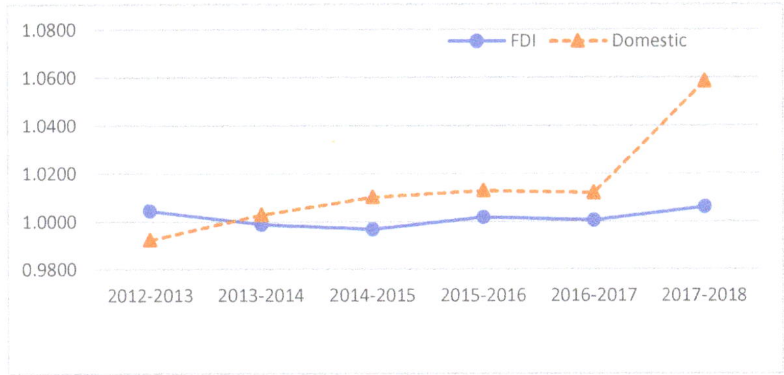

Figure 4. Change in the NMMCPI for FDI and domestic firms during 2012–2018.

Figure 5 shows the trends in the EC index of FDI and domestic firms from 2012–2018. In this figure, the FDI shows an M-shaped trend. The EC index of FDI firms for 2012–2014 shows a value less than or equal to unity, indicating the loss of catch-up performance. However, the EC index alternately goes above or below, for the 2014–2018 period, but remains close to unity. This means that the catch-up performance of FDI firms remains unchanged during the study period. This needs to be investigated by Indian policymakers to determine why there is a lower growth rate efficiency of FDI firms in India. Policymakers should, therefore, come up with appropriate policy solutions that will be easily adopted by FDI firms in India, as some foreign firms are not very sensitive to local regulations. The EC index of domestic firms for the period 2012–2014 shows a value of less than unity. This indicates that during these years, the catch-up performance decreased. However, after 2013–2014, the EC index was higher than unity, which suggests that their catch-up performance in reducing CO_2 emissions has improved. Given that, in a specific group, the EC index measures to what extent an FDI and domestic firm increase its efficiency. There is a gain in efficiency if EC > 1, and there is a loss in efficiency if EC < 1 [39]. Therefore, in terms of the catch-up effect, domestic firms perform better than FDI firms. This means that domestic firms in India have improved efficiency by the catch-up effect of the best practice frontier during the period 2012–2018. This may be because most of the FDI companies come from market-oriented economies and are therefore not very serious about emission reduction policies. Therefore, Indian policymakers should adopt more robust, transparent, and predictable policies for FDI companies in India.

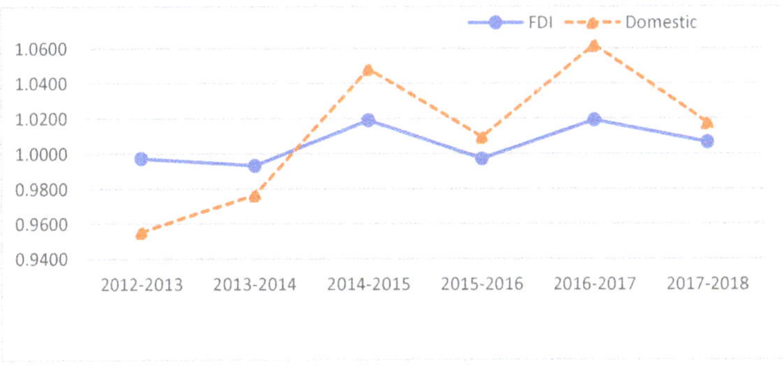

Figure 5. Change in the efficiency change (EC) index.

Figure 6 shows the trends in the BPC index of FDI and domestic firms during the period 2012–2018. From Figure 6, we can see in the BPC index that FDI and domestic firms show a similar trend. For both the FDI and domestic firms, the BPC index was higher than unity during the 2012–2014 and 2017–2018 periods, which suggests a benefit from better innovation performance as well as fast upgrading of equipment and technology. However, during the 2014–2015 and 2016–2017 periods, the BPC index of FDI and domestic firms declined, while 2016–2017 was the lowest growth rate in FDI firms. The reasons could be due to the slowdown in Indian economic growth in 2016–2017, reflecting lower growth of industry and other sectors due to several factors [56]. Overall, domestic firms perform better than FDI firms in terms of the innovation effect in the Indian firm scenario. We suggest that policymakers should support innovative companies and offer them a reward for sustainable development. Additionally, other companies should consider particular innovative companies as their benchmark for creating sustainability.

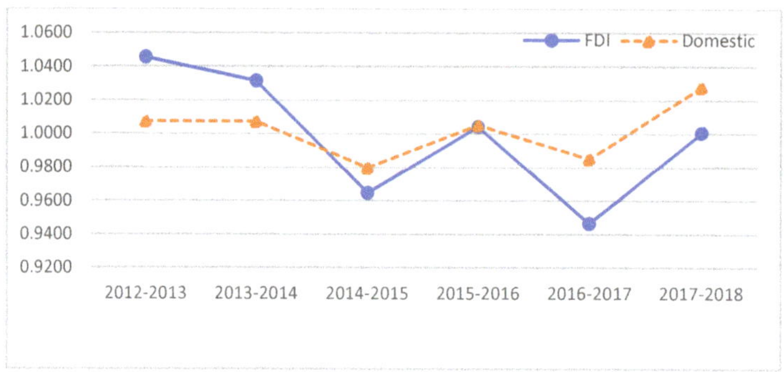

Figure 6. Change in the best-practice gap change (BPC) index.

As shown in Figure 7, the TGC index of FDI and domestic firms shows the contradictory trend between them. For example, in 2012–2014 and 2017–2018, the performance of FDI was less than unity, indicating a decrease in the performance of technology leadership. At the same time, the performance of domestic firms in terms of technology leadership is greater than unity in the same year. However, in 2014–2017, the results show that the performance of FDI was greater than unity, indicating an improvement in the performance of technology leadership and vice-versa for the domestic firms. This suggests that there has been conflicting technological leadership performance by FDI and domestic firms during the study period. Compared to unity, if TGC> (or <) 1, there is a decrease (increase) in the technology gap between the intertemporal technology and the global technology for a specific group. Although the average annual value of domestic firms in the TGC index is a little higher than that of FDI firms, dynamic trends show that there is an increase in the performance of FDI firms during the study period from 2012–2018. Policymakers should understand technological revolutions and encourage companies to initiate and lead the commercialization of technological advancements and coordinate their use of technology to achieve sustainable goals.

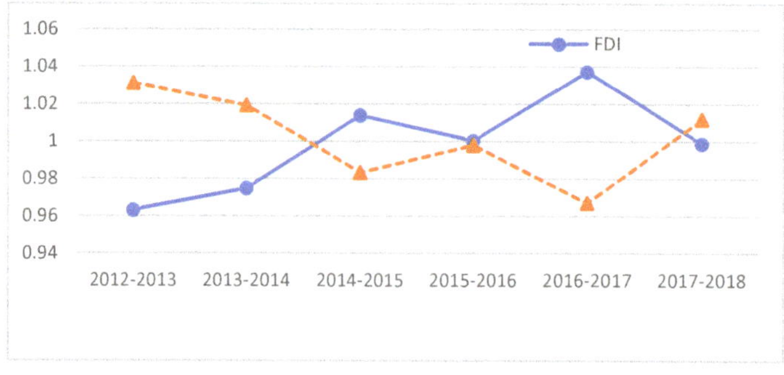

Figure 7. Change in technology gap change (TGC) index.

5. Conclusions

In this study, we analyzed the environmental efficiency of both FDI firms and domestic firms in India for the seven years (2012–2018), based on NMMCPI. This methodology could incorporate group heterogeneity; it could provide each firm with more optimized implications and solutions.

Our main findings and suggestions are summarized as follows: first, according to the NMMCPI index, the result implies that domestic firms have improved their performance in the implementation of the environmental regulatory regime, compared to FDI firms in India. This could be due to the lack of policy effects on FDI firms in India. Therefore, there is a missing link in the role of the Indian government for FDI firms. We suggest that the Indian government should be aware of this phenomenon and take certain measures to reduce the differences between local firms and FDI firms. Second, based on our study, we found that domestic firms showed better catch-up performance (EC) and better innovation performance (BPC) than FDI firms in India. This may be since most of the FDI firms come from market-oriented economies, and thus they are not very serious for the emission abatement policies. Therefore, Indian policy-makers should adopt stronger transparent and predictable policies for the FDI firms in India. Otherwise, it is just a tiger drawn in the paper for the FDI firms with negotiable powers with the local government. Third, although the average annual value of domestic firms in the TGC index is higher than that of FDI firms, the overall dynamics of the TGC index of FDI firms show an increasing trend. In contrast, domestic firms decline from the period 2012–2018. Many global companies are not sustainable in their host country due to their weak accountability of global standards in the host economy and lack of surveillance of local government. Finally, many local governments expect too much of FDI firms with the same level of regulations as domestic firms. Still, this is not strong enough for global companies. Therefore, they must understand that their bad behavior can have a significant negative effect not only on the host country but also on all other local markets in the world, especially developing countries. It is time for FDI firms to shift their paradigm of local management in the host country into sustainable and inclusive economic perspectives.

Due to the scarcity of data and the low volume of emission, we excluded the service sector and telecommunications sectors, which represented the highest FDI inflows to India in 2018. Our future research will include both the service sector and the telecommunications sectors to analyze the efficiency of FDI and domestic firms in India. In addition to our model, we could use regression analysis for local and global companies to thoroughly test Porter's hypothesis through statistical analysis on the determinants of CO_2 emission performance. On the other hand, this paper could be further extended with the MNMCPI bootstrap for total factor CO_2 emission performance and its decompositions to perform statistical inference.

Author Contributions: The authors contributed to each part of the paper by: conceptualization, Y.C.; methodology and software, H.L.; validation, Y.C.; formal analysis, J.D.; investigation, J.D.; resources and data collection, J.D.; writing—original draft preparation, J.D.; writing—review and editing, Y.C.; supervision, Y.C.; project administration, Y.C.; funding acquisition Y.C. All authors have read and agreed to the published version of the manuscript.

Funding: This research received no external funding.

Appendix A

Table A1. The average value of the NMMCPI for FDI and domestic firms.

DMU	Group	NMMCPI	EC	BPC	TGC
FDI 1	FDI	1.0001	1.0008	1.0217	0.9781
FDI 2	FDI	0.9997	1.0057	0.9971	0.9969
FDI 3	FDI	1.0015	1.0056	1.0001	0.9958
FDI 4	FDI	1.0004	1.0024	0.9705	1.0283
FDI 5	FDI	0.9989	0.9999	0.9973	1.0017
FDI 6	FDI	0.9964	0.9997	0.9897	1.0071
FDI 7	FDI	1.0029	1.0143	0.9932	0.9955
FDI 8	FDI	1.0007	1.0022	1.0345	0.9652
FDI 9	FDI	1.0013	1.0071	1.0474	0.9492
FDI 10	FDI	1.0152	1.0383	1.0150	0.9633
FDI 11	FDI	1.0014	1.0041	0.9984	0.9989
FDI 12	FDI	1.0004	1.0013	1.0019	0.9972
FDI 13	FDI	1.0285	1.0389	1.0503	0.9426
FDI 14	FDI	1.0001	1.0016	0.9893	1.0093
FDI 15	FDI	1.0000	1.0004	0.9911	1.0086
FDI 16	FDI	1.0002	1.0011	0.9982	1.0009
FDI 17	FDI	1.0000	1.0004	0.9899	1.0098
FDI 18	FDI	0.9876	0.9805	0.9991	1.0081
FDI 19	FDI	1.0003	1.0011	0.9650	1.0354
FDI 20	FDI	1.0023	1.0115	0.9945	0.9964
FDI 21	FDI	0.9999	1.0051	0.9942	1.0006
FDI 22	FDI	1.0003	1.0016	0.9790	1.0201
FDI 23	FDI	1.0009	1.0031	0.9977	1.0001
FDI 24	FDI	0.9997	1.0002	0.9598	1.0414
FDI 25	FDI	0.9996	1.0062	0.9999	0.9935
	FDI	**1.0015**	**1.0053**	**0.9990**	**0.9978**
Dom1	Domestic	1.0535	1.0000	0.9993	1.0542
Dom2	Domestic	1.0011	1.0041	1.0023	0.9947
Dom3	Domestic	0.9465	0.9523	1.0126	0.9815
Dom4	Domestic	1.0003	1.0020	0.9981	1.0002
Dom5	Domestic	1.0012	1.0040	0.9990	0.9982
Dom6	Domestic	0.9915	0.9900	1.0137	0.9880
Dom7	Domestic	1.0012	1.0081	1.0063	0.9869
Dom8	Domestic	0.9949	0.9821	0.9986	1.0145
Dom9	Domestic	1.0524	0.9892	1.0111	1.0522
Dom10	Domestic	1.0182	1.0000	0.9813	1.0376
Dom11	Domestic	1.0908	1.0662	0.9973	1.0258
Dom12	Domestic	0.9978	0.9962	0.9990	1.0026
Dom13	Domestic	1.0148	0.9709	1.0055	1.0395
Dom14	Domestic	1.0024	1.0136	0.9988	0.9901
Dom15	Domestic	1.0033	1.0130	0.9996	0.9908
Dom16	Domestic	1.0002	1.0001	0.9992	1.0009
Dom17	Domestic	0.9995	1.0017	1.0163	0.9818
Dom18	Domestic	0.9985	0.9995	1.0102	0.9889
Dom19	Domestic	1.0148	1.0575	1.0158	0.9447
Dom20	Domestic	1.0003	1.0041	0.9916	1.0047
Dom21	Domestic	1.1657	1.1630	0.9951	1.0073
Dom22	Domestic	0.9563	0.9837	0.9987	0.9734
Dom23	Domestic	1.0001	1.0008	0.9978	1.0015
Dom24	Domestic	1.0650	1.0846	1.0162	0.9663
Dom25	Domestic	0.9997	0.9998	0.9941	1.0058
	Domestic	**1.0148**	**1.0115**	**1.0023**	**1.0013**

References

1. Dillon, J. Another Climate Milestone Falls at Mauna Loa Observatory. Available online: https://scripps.ucsd.edu/programs/keelingcurve/2018/06/07/another-climate-milestone-falls-at-mauna-loa-observatory/ (accessed on 7 August 2020).
2. Myllyvirta, L. Analysis: India's CO_2 Emissions Growth Poised to Slow Sharply in 2019. Available online: https://www.carbonbrief.org/analysis-indias-co2-emissions-growth-poised-to-slow-sharply-in-2019 (accessed on 7 August 2020).
3. The World Bank. Available online: https://data.worldbank.org/indicator/EN.ATM.CO2E.KT?locations=IN (accessed on 21 September 2020).
4. FDI Statistics. Available online: https://dipp.gov.in/sites/default/files/FDI_Factsheet_27May2019.pdf (accessed on 12 August 2020).
5. Debroy, B.; Nayyar, D. Make in India. Available online: https://www.makeinindia.com/home (accessed on 7 August 2020).
6. Kılıçarslan, Z.; Dumrul, Y. Foreign Direct Investments and CO_2 Emissions Relationship: The Case of Turkey. *Bus. Econ. Res. J.* **2017**, *8*, 647–660. [CrossRef]
7. Ran, N.; Majmudar, U. Sustainable Finance: Trends for 2020. Available online: https://economictimes.indiatimes.com/blogs/ResponsibleFuture/sustainable-finance-trends-for-2020/ (accessed on 7 August 2020).
8. Aravossis, K.; Kapsalis, V.C.; Kyriakopoulos, G.; Xouleis, T.G. Development of a Holistic Assessment Framework for Industrial Organizations. *Sustainability* **2019**, *11*, 3946. [CrossRef]
9. Kim, S. CO_2 emissions, foreign direct investments, energy consumption, and GDP in developing countries: A more comprehensive study using panel vector error correction model. *Korean Econ. Rev.* **2019**, *35*, 5–24.
10. Blalock, G.; Gertler, P.J. Welfare gains from Foreign Direct Investment through technology transfer to local suppliers. *J. Int. Econ.* **2008**, *74*, 402–421. [CrossRef]
11. Guo, Z.F.; Lin, J.H.; Luo, J. An Empirical Analysis of the Regional Influence of FDI on Energy Efficiency in China. *Adv. Mater. Res.* **2013**, *684*, 626–629. [CrossRef]
12. To, A.H.; Ha, D.T.-T.; Nguyen, H.M.; Vo, D.H. The Impact of Foreign Direct Investment on Environment Degradation: Evidence from Emerging Markets in Asia. *Int. J. Environ. Res. Public Health* **2019**, *16*, 1636. [CrossRef]
13. Rao, M. The Impact of Environment on FDI. Available online: https://www.thehindu.com/opinion/op-ed/the-impact-of-environment-on-fdi/article24017203.ece (accessed on 7 August 2020).
14. Copeland, B.R.; Taylor, M.S. North-South Trade and the Environment. *Q. J. Econ.* **1994**, *109*, 755–787. [CrossRef]
15. Forslid, R.; Okubo, T.; Ulltveit-Moe, K.H. Why are firms that export cleaner? International trade, abatement and environmental emissions. *J. Environ. Econ. Manag.* **2018**, *91*, 166–183. [CrossRef]
16. Mielnik, O.; Goldemberg, J. Foreign direct investment and decoupling between energy and gross domestic product in developing countries. *Energy Policy* **2002**, *30*, 87–89. [CrossRef]
17. Yao, X.; Guo, C.; Shao, S.; Jiang, Z. Total-factor CO_2 emission performance of China's provincial industrial sector: A meta-frontier non-radial Malmquist index approach. *Appl. Energy* **2016**, *184*, 1142–1153. [CrossRef]
18. Fan, M.; Shao, S.; Yang, L. Combining global Malmquist–Luenberger index and generalized method of moments to investigate industrial total factor CO_2 emission performance: A case of Shanghai (China). *Energy Policy* **2015**, *79*, 189–201. [CrossRef]
19. Solarin, S.A.; Al-Mulali, U. Influence of foreign direct investment on indicators of environmental degradation. *Environ. Sci. Pollut. Res.* **2018**, *25*, 24845–24859. [CrossRef] [PubMed]
20. Lin, B.; Chen, X. Evaluating the CO_2 performance of China's non-ferrous metals Industry: A total factor meta-frontier Malmquist index perspective. *J. Clean. Prod.* **2019**, *209*, 1061–1077. [CrossRef]
21. Lin, B.; Fei, R. Regional differences of CO_2 emissions performance in China's agricultural sector: A Malmquist index approach. *Eur. J. Agron.* **2015**, *70*, 33–40. [CrossRef]
22. Liu, Q.; Wang, S.; Zhang, W.; Zhan, D.; Li, J. Does foreign direct investment affect environmental pollution in China's cities? A spatial econometric perspective. *Sci. Total Environ.* **2018**, *613*, 521–529. [CrossRef]
23. Salim, R.; Yao, Y.; Chen, G.; Zhang, L. Can foreign direct investment harness energy consumption in China? A time series investigation. *Energy Econ.* **2017**, *66*, 43–53. [CrossRef]

24. Sarkodie, S.A.; Strezov, V. Effect of foreign direct investments, economic development and energy consumption on greenhouse gas emissions in developing countries. *Sci. Total Environ.* **2019**, *646*, 862–871. [CrossRef]

25. Pan, X.; Guo, S.; Han, C.; Wang, M.; Song, J.; Liao, X. Influence of FDI quality on energy efficiency in China based on seemingly unrelated regression method. *Energy* **2020**, *192*, 116463. [CrossRef]

26. Wang, S. Impact of FDI on energy efficiency: An analysis of the regional discrepancies in China. *Nat. Hazards* **2016**, *85*, 1209–1222. [CrossRef]

27. Mastromarco, C.; Simar, L. Effect of FDI and Time on Catching Up: New Insights from a Conditional Nonparametric Frontier Analysis. *J. Appl. Econ.* **2014**, *30*, 826–847. [CrossRef]

28. Yue, S.; Yang, Y.; Hu, Y. Does Foreign Direct Investment Affect Green Growth? Evidence from China's Experience. *Sustainability* **2016**, *8*, 158. [CrossRef]

29. Song, M.; Tao, J.; Wang, S. FDI, technology spillovers and green innovation in China: Analysis based on Data Envelopment Analysis. *Ann. Oper. Res.* **2013**, *228*, 47–64. [CrossRef]

30. Yang, X.; Li, C. Industrial environmental efficiency, foreign direct investment and export ——Evidence from 30 provinces in China. *J. Clean. Prod.* **2019**, *212*, 1490–1498. [CrossRef]

31. Lei, M.; Zhao, X.; Deng, H.; Tan, K.C. LeeDEA analysis of FDI attractiveness for sustainable development: Evidence from Chinese provinces. *Decisi. Support. Syst.* **2013**, *56*, 406–418. [CrossRef]

32. Zang, C.Q.; Liu, Y. Analysis on total-factor energy efficiency and its influencing factors of Shandong thinking about environmental pollution. *China Popul. Resour. Environ.* **2012**, *22*, 8.

33. Monaheng, M.P.; Donghui, Z.; Zaman, Q.U. The Relationship between FDI and Economic Performance (BRICS). *Eur. Online J. Nat. Soc. Sci.* **2019**, *8*, 148–157.

34. Luo, J.; Cheng, K. The Influence of FDI on Energy Efficiency of China: An Empirical Analysis Based on DEA Method. *Appl. Mech. Mater.* **2013**, *291*, 1217–1220. [CrossRef]

35. Lee, H.; Choi, Y. Greenhouse gas performance of Korean local governments based on non-radial DDF. *Technol. Forecast. Soc. Chang.* **2018**, *135*, 13–21. [CrossRef]

36. Färe, R.; Grosskopf, S. *New Directions: Efficiency and Productivity*; Springer Science & Business Media: Berlin, Germany, 2006; Volume 3.

37. Zhang, N.; Zhou, P.; Choi, Y. Energy efficiency, CO_2 emission performance and technology gaps in fossil fuel electricity generation in Korea: A meta-frontier non-radial directional distance functionanalysis. *Energy Policy* **2013**, *56*, 653–662. [CrossRef]

38. Faere, R.; Grosskopf, S.; Lovell, C.A.K.; Pasurka, C. Multilateral Productivity Comparisons When Some Outputs are Undesirable: A Nonparametric Approach. *Rev. Econ. Stat.* **1989**, *71*, 90. [CrossRef]

39. Zhang, N.; Choi, Y. Total-factor carbon emission performance of fossil fuel power plants in China: A metafrontier non-radial Malmquist index analysis. *Energy Econ.* **2013**, *40*, 549–559. [CrossRef]

40. Pyatt, G.; Shephard, R.W. Theory of Cost and Production Functions. *Econ. J.* **1972**, *82*, 1059. [CrossRef]

41. Chambers, R.G.; Chung, Y.; Färe, R. Benefit and Distance Functions. *J. Econ. Theory* **1996**, *70*, 407–419. [CrossRef]

42. Zhang, N.; Choi, Y. A note on the evolution of directional distance function and its development in energy and environmental studies 1997–2013. *Renew. Sustain. Energy Rev.* **2014**, *33*, 50–59. [CrossRef]

43. Zhou, P.; Ang, B.; Wang, H. Energy and CO_2 emission performance in electricity generation: A non-radial directional distance function approach. *Eur. J. Oper. Res.* **2012**, *221*, 625–635. [CrossRef]

44. Fukuyama, H.; Weber, W.L. A directional slacks-based measure of technical inefficiency. *Socio-Econ. Plan. Sci.* **2009**, *43*, 274–287. [CrossRef]

45. Choi, Y.; Lee, H.S. Heterogeneity and its policy implications in GHG emission performance of manufacturing industries. *Carbon Manag.* **2018**, *9*, 347–360. [CrossRef]

46. Choi, Y.; Lee, H.S. Are Emissions Trading Policies Sustainable? A Study of the Petrochemical Industry in Korea. *Sustainability* **2016**, *8*, 1110. [CrossRef]

47. Zhang, N.; Choi, Y. A comparative study of dynamic changes in CO_2 emission performance of fossil fuel power plants in China and Korea. *Energy Policy* **2013**, *62*, 324–332. [CrossRef]

48. Wang, Q.; Zhang, H.; Zhang, W. A Malmquist CO_2 emission performance index based on a metafrontier approach. *Math. Comput. Model.* **2013**, *58*, 1068–1073. [CrossRef]

49. Oh, D.-H. A metafrontier approach for measuring an environmentally sensitive productivity growth index. *Energy Econ.* **2010**, *32*, 146–157. [CrossRef]

50. Oh, D.-H.; Lee, J.-D. A metafrontier approach for measuring Malmquist productivity index. *Empir. Econ.* **2009**, *38*, 47–64. [CrossRef]

51. INFOGRAPHIC: India's Energy-Related CO_2 Emissions. Available online: https://energy.economictimes.indiatimes.com/news/coal/infographic-indias-energy-related-co2-emissions/72277641 (accessed on 21 September 2020).

52. Singh, A. Why Do Indian Businesses Need to Take Air Pollution Seriously? Available online: https://wri-india.org/blog/why-do-indian-businesses-need-take-air-pollution-seriously (accessed on 7 August 2020).

53. Lee, M.; Jin, Y. The substitutability of nuclear capital for thermal capital and the shadow price in the Korean electric power industry. *Energy Policy* **2012**, *51*, 834–841. [CrossRef]

54. Choi, Y.; Yu, Y.; Lee, H.S. A Study on the Sustainable Performance of the Steel Industry in Korea Based on SBM-DEA. *Sustainability* **2018**, *10*, 173. [CrossRef]

55. Choi, Y.; Lee, H.; Mastur, A. Are Sustainable Development Policies Really Feasible? Focused on the Petrochemical Industry in Korea. *Sustainability* **2019**, *11*, 3980. [CrossRef]

56. Indian Economic Growth Slowed Down in 2016–2017: Government. Available online: https://economictimes.indiatimes.com/news/economy/indicators/indian-economic-growth-slowed-down-in-2016-17-government/articleshow/62294363.cms?from=mdr (accessed on 7 August 2020).

 *sustainability*

Communication

Climate Change, Agriculture, and Energy Transition: What Do the Thirty Most-Cited Articles Tell Us?

Dmitry A. Ruban [1,2,*] , Natalia N. Yashalova [3] , Olga A. Cherednichenko [4] and Natalya A. Dovgot'ko [4]

[1] K.G. Razumovsky Moscow State University of Technologies and Management (the First Cossack University), Zemlyanoy Val Street 73, 109004 Moscow, Russia
[2] Department of Management, Higher School of Business, Southern Federal University, 23-ja Linija Street 43, 344019 Rostov-on-Don, Russia
[3] Department of Economics and Management, Business School, Cherepovets State University, Sovetskiy Avenue 10, Cherepovets, 162600 Vologda Region, Russia; natalij2005@mail.ru
[4] Department of Economic Theory and Agrarian Economics, Stavropol State Agrarian University, Zootekhnicheskiy Lane 12, 355017 Stavropol, Russia; chered72@mail.ru (O.A.C.); ndovgotko@yandex.ru (N.A.D.)
*   Correspondence: ruban-d@mail.ru

Received: 1 September 2020; Accepted: 25 September 2020; Published: 28 September 2020

Abstract: The thirty journal articles dealing with the relationship between climate change and agriculture (the latter is treated in general, i.e., as an industry) and which have gained >1000 citations are thought to be sources of the most precious information on the noted relationship. They were published between 1994 and 2011. Many are authored by West European and North American experts. The most-cited articles are attributed to three major themes and eight particular topics, and the best-explored topic is the influence of climate change on agriculture. Moreover, they provide some essential information about the strong relation of both agriculture and climate change to energy transition. The general frame characterizing complex interactions of climate change and agriculture development is proposed on the basis of the most-cited works, but it needs further detail, improvement, and update. The considered articles are basic sources with historical importance.

Keywords: bibliographical survey; climatic factor; energy industry; sustainable development; world agriculture

1. Introduction

Global climate change has intensified in the last third of the 20th century, and it remains one of the most important environmental problems [1,2]. This phenomenon, as well as its multiple consequences for biodiversity, several planetary systems, and ecological processes have been addressed by the Intergovernmental Panel on Climate Change (IPCC) [2] and discussed, particularly, by Adger et al. [3], Bellard et al. [4], Crowley [5], Giorgi [6], Parmesan and Yohe [7], Patz et al. [8], Ramanathan and Carmichael [9], Solomon et al. [10], Trenberth [11], and Walther et al. [12] and more recently by Edmonds et al. [13], Huang et al. [14], Konapala et al. [15], Molina et al. [16], and Sanderson and O'Neill [17]. These works (among many others) provide clear evidence that human activity challenges the global environment in the form of the climate change, and this challenge affects society and its sustainable development. Adaptation, mitigation, and, more generally, creative thinking are required to address this challenge.

Undoubtedly, all major industries of the world economy are and will be disturbed by climate change, but agriculture seems to be the most vulnerable from them because its productivity depends on climate conditions; moreover, agricultural development has a heavy environmental impact (see the literature reviewed below). The relevant knowledge, which has accumulated during several decades of intense international research, is outstandingly rich. According to the bibliographical database Scopus, boasting significant completeness and embracing thousands of scientific journals, there are about 40,000 works discussing various aspects of the relationship between climate change and agriculture (see the parameters of such a literature search below). If so, every researcher in this field (especially the newcomer) faces a serious challenge: the number of the sources is too big to treat them efficiently with qualitative literature reviews or even bibliometric approaches. Moreover, frequent additions to this knowledge undermine the utility of such reviews soon after their publication. To concentrate on more or less 'narrow' lines of this research is another possible approach [18,19], but the relationship between climate change and agriculture is too complex to avoid tracing far-going and very general consequences and feedbacks. For instance, energy transition is evidently related to both agriculture and climate change, and their joint action seems to be highly complex. So, a general (even very general) picture would be strongly demanded by both the newcomers and the senior researchers searching for themes for high-impact writing.

Among about 40 thousands of literature sources, there are a few works that are thought to be the most important, trustful, and influential. Their outstanding value is demonstrated by their numerous citations in the other papers. Evidently, these works seem to be the most important, and, thus, it appears to be reasonable to check what these works are and which issues they address. The objective of the present paper is to re-visit the thirty most-cited articles dealing with the agriculture–climate change relationship, with three particular aims. First, it is intended to collect the relevant bibliographical information and to provide the general characteristics of the most-cited works. Second, it appears necessary to link logically different pieces of the outstanding knowledge available 'here and there' in the literature. Third, it is intended to discuss whether the most-cited works provide a more or less comprehensive vision of the relationship between climate change and agriculture. This paper is not a subject review and does not aim at comprehensive characteristics of the noted relationship. Actually, it is based on a bibliographical survey and qualitative content analysis, with details explained below. This paper starts with the general characteristics of the selected literature. Special attention is paid to the relevance of the agriculture–climate change relationship to the problem of energy transition. This problem is emphasized because one of the aims of the energy transition is climate change mitigation and agriculture, on the one hand, depends on the energy transition, and, on the other hand, it is able to contribute to this (e.g., via biofuel production). Then, the main research themes linked to the relationship between climate change and agriculture are identified. Finally, the general contribution of the most-cited articles to the understanding of this relationship is discussed.

2. Bibliographical Survey

2.1. Methodological Remarks

The literature describing the relationship between climate change and agriculture constitutes the material of the present study. The online bibliographical database Scopus is used to select the most-cited articles dealing with this relationship. This database is preferred because of its significant coverage (especially after 2005) and easy functionality. The search is generally realized with the basic keywords, and the procedure is explained below.

The analysis itself consists of two procedures. First, the basic characteristics of the selected articles are examined. These include the number of citations (total and average annual), the year of publication, the number of authors, the authors' country affiliations, and the bibliometric parameters of the journals where the articles appeared. These characteristics shed light on the research that resulted in the most-cited articles. Indeed, there may be some papers with the average annual number of citations higher than those of the most-cited articles selected for the purposes of this study. This issue is not addressed in the present paper because its analysis would require separate study; moreover, it is evident that the ordinary users of Scopus pay attention to the total number of citations only. Second, the content of the selected literature is addressed qualitatively. The general/global versus regional/country focus is established. Then, the topics of the articles are outlined (this is an intuitive and somewhat subjective procedure, but it is essential for literature categorization). Articles with a similar focus are attributed to the same topics. All topics are analyzed critically to find their logical connections, which allow 'delineation' of the major themes consisting from one to several topics. This analysis is tentative and does not pretend to be fully universal. Nonetheless, the literature categorization provides useful material for quick orientation within the general problem of the agriculture–climate change relationship (as it is reflected by the most-cited works). The relevance of the considered articles to this problem is examined: the most-cited works can bear important notions on this relationship linked to any other issue (low relevance), mix the extensive treatment of this relationship with the other knowledge (medium relevance), and be fully dedicated to the relationship (high relevance). This relevance indicates the degree of development of each theme in the most-cited papers.

2.2. Selection of Articles

The on-line bibliographical database 'Scopus' is searched to find all sources bearing the words 'climate change(s)' and 'agriculture(-al)' in their titles, abstracts, and key words. The entire time span represented in this database is taken into account. The number of these sources is ~38,000 (as of 7 July 2020). Naturally, these differ by the number of their citations in the other works. Among these sources, there are <50 journal articles that were cited more than 1000 times. These seem to be the cornerstone publications that are in the focus of the present study. The latter deals with only those papers that consider agriculture in general, i.e., not those papers that pay attention to the very particular issues (e.g., livestock practices) and omit notions to agriculture in their titles, abstracts, or key words. Alternatively, the number of possible papers to filter would be larger, but it appears unrealistic to treat all of them, as well as to make the sample fully homogeneous. Nonetheless, a tentative experimental search in the database has shown that the words "agriculture" or "agricultural" exist often in the papers addressing very 'narrow' aspects.

Initially, all journal articles (45 items) dealing with the relationship between climate change and agriculture and with a number of citations higher than 1000 are chosen. However, it is evident that these include several sources where the above-mentioned words appear 'occasionally' and in any other context. The content of each initially chosen paper is checked 'manually' in order to select only those articles that really deal with the noted relationship. A total of 30 articles are detected this way (Table 1). These articles still differ by the number of citations and the relevance to climate change and agriculture, and these issues are addressed below. Nonetheless, these seem to be really outstanding publications.

Table 1. The articles considered in the present study.

Article	Year	Number of Citations in Scopus (as of 7 July 2020)		Number of Authors	Number of Countries from the Authors' Affiliations
		Total	Average Annual		
[20]	2010	1026	114	3	1 (UK)
[21]	2005	2079	149	33	9 (Belgium, Denmark, Finland, France, Germany, Italy, Spain, Switzerland, USA)
[22]	2009	1126	113	28	1 (UK)

Table 1. *Cont.*

Article	Year	Number of Citations in Scopus (as of 7 July 2020)		Number of Authors	Number of Countries from the Authors' Affiliations
		Total	Average Annual		
[23]	1997	1474	67	3	2 (UK, USA)
[24]	2008	2585	235	5	1 (USA)
[25]	2005	6071	434	19	2 (UK, USA)
[26]	2011	3017	377	21	4 (Canada, Germany, Sweden, USA)
[27]	2007	1128	94	6	1 (France)
[28]	2010	4613	513	10	1 (UK)
[29]	2003	1082	68	7	1 (Switzerland)
[30]	2003	1252	78	2	1 (USA)
[31]	2004	3360	224	1	1 (USA)
[32]	2004	1669	111	1	1 (USA)
[33]	2003	1332	83	3	1 (Belgium)
[34]	2008	1497	136	6	1 (USA)
[35]	2011	1606	201	3	1 (USA)
[36]	2010	3113	346	3	1 (Portugal)
[37]	2004	1012	67	5	4 (Austria, Spain, UK, USA)
[38]	2004	1255	84	9	2 (China, Philippines)
[39]	2010	1564	174	16	3 (China, France, UK)
[40]	2010	2324	258	6	4 (Germany, Switzerland, UK, USA)
[41]	1999	1196	60	2	1 (USA)
[42]	2009	1850	185	29	8 (Australia, Belgium, Denmark, Germany, the Netherlands, Sweden, UK, USA)

Table 1. *Cont.*

Article	Year	Number of Citations in Scopus (as of 7 July 2020)		Number of Authors	Number of Countries from the Authors' Affiliations
		Total	Average Annual		
[43]	2009	4608	461	29	8 (Australia, Belgium, Denmark, Germany, the Netherlands, Sweden, UK, USA)
[44]	1994	1033	41	2	2 (UK, USA)
[45]	2009	1079	108	2	1 (USA)
[46]	2005	1060	76	35	11 (Belgium, Finland, France, Germany, the Netherlands, South Africa, Spain, Sweden, Switzerland, UK, USA)
[47]	2010	1767	196	8	1 (Switzerland)
[48]	2001	2203	122	10	1 (USA)
[49]	2011	2435	304	4	1 (USA)

2.3. Basic Characteristics of the Articles

The selected articles dealing with the relationship between climate change and agriculture have received from 1012 (minimum value) to 6071 (maximum value) citations (Table 1), with the mean value of 2047 and the median value of 1585. Expectedly, the number of the articles with the higher number of citations decreases exponentially (Figure 1). Although one can suppose that the 'older' papers have gained naturally more citations than the 'younger' papers, this is not so in fact; for instance, the papers with the noted maximum and minimum values were published very closely (Table 1). Moreover, there were important papers before 1994 (when the 'oldest' of the considered articles that reached the threshold of 1000 citations was published), but these 'too early' works, surprisingly, have not gained 1000 citations (probably, they were published before the 'critical mass' of the literature allowed extensive citation records or the scarcity of online bibliographical resources and open-access mode of publishing limited their promotion among scientists). The average annual number of citations of the considered articles differs substantially, from <50 to >500 per year (Table 1). Notably, the work with the highest number of total citations [25] and the work with the highest average number of annual citations [28] are not the same. Nonetheless, the correlation between the total and average annual numbers of citations is strong (the correlation coefficient value is 0.91). The information from Table 1 also stresses that the number of citations does not depend on the year of their appearance. Some sources of the same year differ by the number of citations strikingly.

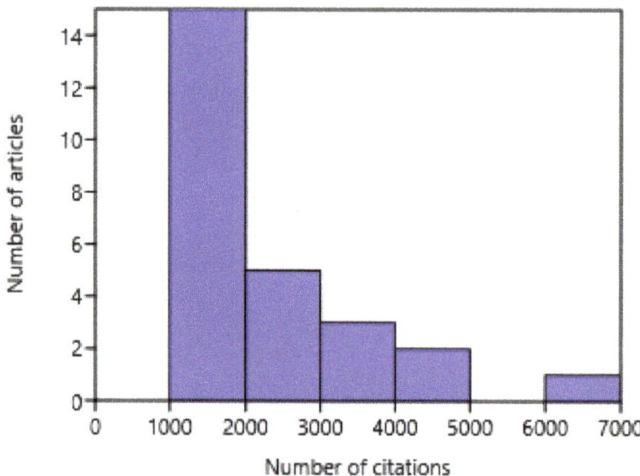

Figure 1. Frequency distribution of citations of the considered articles.

The present study deals with the almost unlimited time span (it is limited by the only entire time span of the resources of Scopus, which range from pre-19th century to 2021). However, the selected articles were published within the time period of 1994–2011 (Table 1). Notably, the two most 'productive' intervals were 2003–2005 and 2009–2011, when the majority of the most-cited papers appeared (Figure 2). Many most-cited articles appeared when the entire research field had started to grow (Figure 2). As a result, these papers became the primary, contemporaneous sources for citations in the other works. Together with the subsequent growth of the research, the specialists saw the already highly cited sources that, thus, were cited even more. It is also notable that no paper published after 2011 attracted more than 1000 citations, although the number of the potentially citing works increased substantially. Two possible explanations are the following. First, together with the increase in the number of the relevant publications, the authors of the following works faced a necessity to choose which one to cite. As a result, each new article, even if reporting results of outstanding importance, had less chance to be cited. Second, it is also possible that the scientific progress required a shift from too general, world-scale, and thus influential papers to those dealing with more particular issues in particular places of the world, and such papers were cited less intensively. Anyway, the correspondence of the publishing output and the citation dynamics is highly complex and non-linear, and, thus, it cannot be excluded that the concentration of the most-cited articles on the two noted time intervals is occasional.

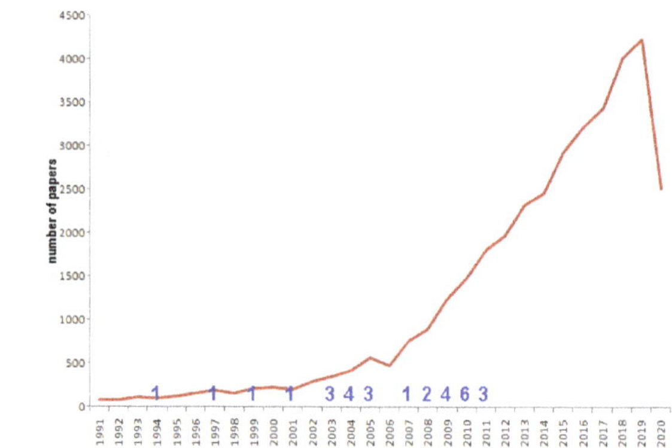

Figure 2. Dynamics of publishing papers dealing with the relationship between climate change and agriculture (brown line—all papers, blue numbers—the number of the most-cited papers by years). The information for 2020 is strongly incomplete, which explains the sharp drop.

The most-cited articles differ significantly by their author teams (Table 1). Although the majority have two and more authors (the only two papers have single author), the number of the authors is either relatively small or big (Figure 3). The mean number of authors is 10, and the median number of authors is 6. This means that the considered articles are the products of collective rather than individual work. In other words, groups (even 'pools') of experts and, speaking more broadly, relatively extensive research networks were required to generate such valuable knowledge. This inference also stresses the role of scientists' consensus on global climate change as the environmental challenge of the world agriculture. It is interesting to state that the papers with a single author boast a number of citations above the median (Table 1), although the correlation coefficient between the number of authors and the number of citation is low (0.23 in the case of total citations and 0.24 in the case of average annual citations).

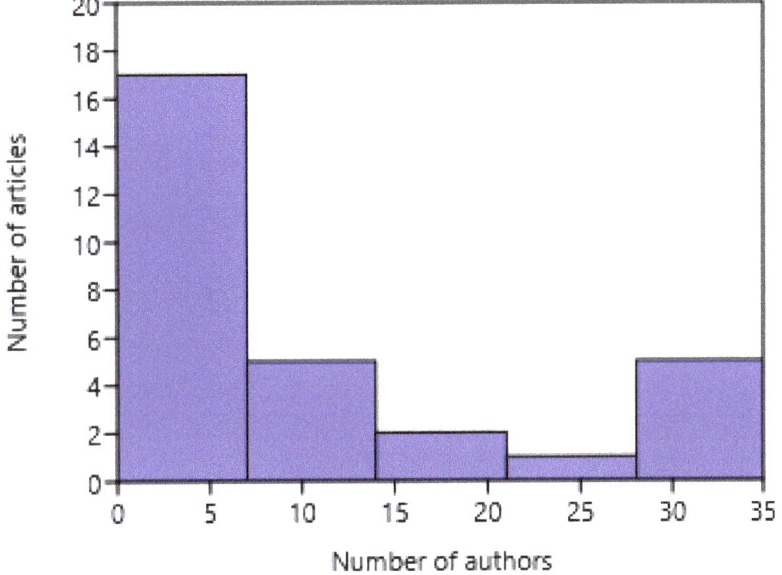

Figure 3. Frequency distribution of the author teams size in the considered articles.

Of big interest is how international (in regard to the authors' affiliations) was the research that resulted in the most-cited articles. The bibliographical information allows ambivalent judgments (Table 1). On the one hand, many articles were written by the representatives of one–two countries (often, these are the UK and the USA) (Figure 4). On the other hand, there are articles authored by truly international teams (Table 1), and experts from a total of 19 countries authored the considered articles taken entirely (Figure 5). One should also note that the authors represent chiefly Western Europe and North America (Figure 5). Generally, it appears that the most-cited articles dealing with the relationship between climate change and agriculture resulted from the limited collaboration of experts representing different countries (when existing, this is chiefly North American–European collaboration). Notably, international collaboration does not affect the number of citations, because the correlation coefficient between the number of countries indicated in the author affiliations and the number of citations is very low (0.08 in the case of total citations and 0.09 in the case of average annual citations).

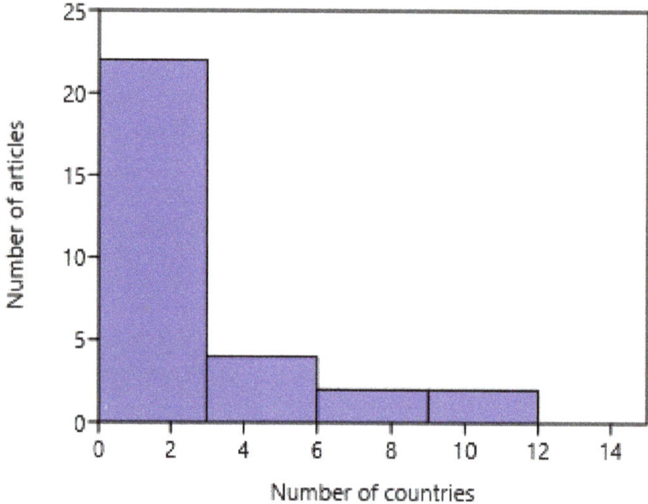

Figure 4. Frequency distribution of the number of countries indicated in author affiliations in the considered articles.

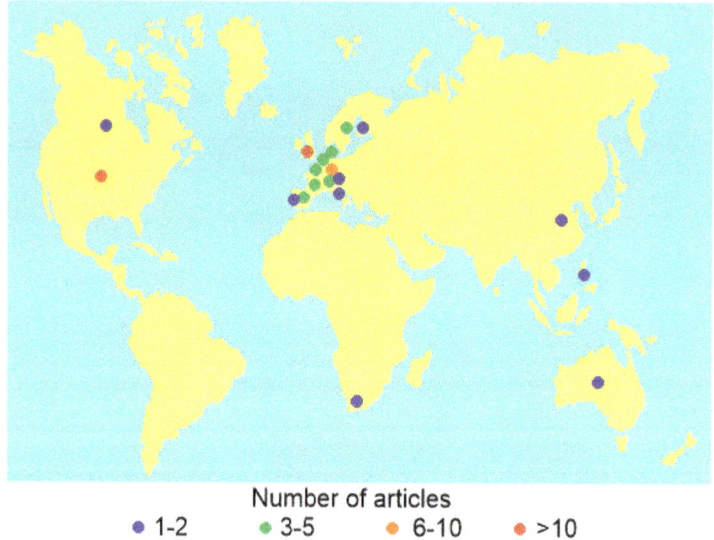

Figure 5. Geography of the authorship (country affiliations of the authors) of the considered articles.

The thirty articles considered in this study appeared in fifteen journals (Table 2). Two of them, namely *Science* and *Nature* hosted half of all articles. Although all fifteen journals boast outstanding metrics (high impact factor, high CiteScore value, and top quartile), they differ substantially. For instance, the impact factor ranges between 3.3 and 60.4 (Table 2), which complicates the relationship between journal prestige and the number of citations of the papers published there. Apparently, publishing in a high-prestige journal (every journal listed in Table 2 is high-prestige) is enough to make the paper highly citable, irrespective of whether this journal has higher or lower metrics relative to the other prestige journals.

Table 2. Bibliometric characteristics of the journals where the considered articles were published.

Journal	Number of Articles	Bibliometrics (for 2019)			
		Web of Science		Scopus	
		Impact Factor	Quartile	CiteScore	Quartile
Annual Review of Environment and Resources	1	8.065	1	15.8	1
Annual Review of Plant Biology	1	19.54	1	32.8	1
Ecology and Society	1	3.89	1	7.5	1
Earth-Science Reviews	1	9.724	1	15.0	1
Geoderma	1	4.848	1	7.6	1
Global Biogeochemical Cycles	1	4.608	1	9.4	1
Global Environmental Change	1	10.466	1	17.0	1
International Journal of Life Cycle Assessment	1	4.307	1, 2	8.5	1
Nature	7	42.778	1	51.0	1
Plant and Soil	1	3.299	1, 2	5.9	1
Proceedings of the National Academy of Sciences of the United States of America	3	9.412	1	15.7	1
Renewable and Sustainable Energy Reviews	1	12.11	1	25.5	1
Science	8	41.845	1	45.3	1
The Lancet	1	60.392	1	73.4	1
Trends in Ecology and Evolution	1	14.764	1	22.3	1

3. Relevance to Energy Transition

The agriculture–climate change relationship has an important dimension that requires special attention, also in regard to the reviewed literature sources. Energy transition is a fundamental and multidimensional problem of modern civilization [50–54]. Particularly, its solution is aimed at sustainable development of energy systems and industry in the whole, broad implementation of clean, innovative energy-related technologies, more attention to renewable energy sources, etc. Undoubtedly, the energy transition is not only important with regard to the global climate change and sustainable world agriculture (and food security), but it is strongly related to these two planetary-scale issues. The direct relevance of energy transition to the global climate change and the policies of its mitigation on municipal, national, regional, and global levels is explained by Hoppe and van Bueren [55], Fazey et al. [56], Newell and Bulkeley [57], Scheffran et al. [58], and Urban [59]. The strong relation of agriculture to the energy transition is also discussed. Probably, the most important contribution was made by Sutherland et al. [60] who explain how management of land and ecosystems in agriculture relates it to the renewable energy transition, although this requires significant policy support; moreover, these specialists note that agriculture competes with the electricity industry over access to natural resources. Importantly, this study also demonstrates the full complexity of the energy–agriculture–climate change nexus.

The literature reviewed in this work 'touches' the relevance of the agriculture–climate change relationship, and not superficially. This issue is considered in several works. First, the rise of energy demand together with the world agriculture development is noted by Foley et al. [25]. This means that agriculture can be judged a factor challenging the energy transition. Second, several works focus on energy crops and biofuel production. Fargione et al. [24] explain that whether low-carbon energy sources and biofuels contribute to climate change mitigation or not depends on agricultural practices, and extensive clearance of forests and grasslands for energy crops results in too big carbon dioxide emissions. Lal [31] focuses on the positive side of energy crops growing and lists this among the other agricultural opportunities to mitigate the climate change. Mata et al. [36] stress the importance of microalgae for biofuel production. This means that the new agricultural practice is able to facilitate the energy transition with evident benefits for climate change mitigation. Generally, as biofuel production is thought to be an instrument towards energy transition [61–63], this means that the success of implementation of this instrument strongly depends on the mode of agricultural development. In other words, agriculture plays a critical role for achievement of the energy transition in regard to biofuels and the relevant policies.

Contrary to what will be said below about some thematic deficiencies of the most-cited articles addressing the agriculture–climate change relationship, it appears that the selected piece of literature, i.e., the thirty most-cited articles shed light on the relevance of this relationship to energy transition. This issue is considered adequately and in a very balanced way. This means that the most-cited works would be more or less enough for newcomers interested in the energy–agriculture–climate change nexus. This finding can be also interpreted so that this nexus is of crucial importance for the understanding of the relationship between climate change and agriculture. Nonetheless, the nexus has a lot of particular aspects that are addressed in some important, but less-cited articles. For instance, Bardi et al. [64] raise the theoretical question of agriculture adaptation to the energy transition and, particularly, to the use of sustainable energy sources; Chai et al. [65] indicate the possible importance of agriculture-related methane emissions as a new energy source; Safriel [66] hypothesizes that climate change and the relevant desertification may create not only challenges, but also opportunities for solar energy production and agricultural development. Importantly, the general ideas of the energy–agriculture–climate change nexus and the above-mentioned particular aspects contribute to the understanding of how strongly this nexus determines sustainable development in the modern world. It is worth noting the social application (also in regard to sustainability) of the relevant issues [67].

4. General Research Themes and Key Findings

The most-cited articles dealing with the relationship between climate change and agriculture are very different with regard to their content. Below, attention is paid to their geographical focus, themes and topics, main findings, and relevance. These are detected on the basis of the critical reading of each paper and the qualitative analysis of the context in which climate change and agriculture are considered.

The most-cited articles are either of a general nature and address planetary-scale issues or region- or country-focused. The former prevail, which is expected (Figure 6). Moreover, it is surprising that there are some papers focusing on particular countries and, nonetheless, attracting such a big number of citations. This means that there are issues of outstanding importance that can be analyzed deeply on the basis of case examples. The papers with a regional or country focus deal with the USA, Brazil, all of Europe, and China.

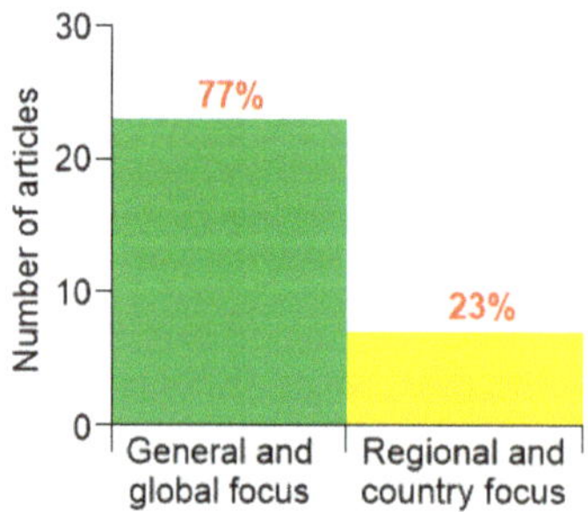

Figure 6. The focus of the considered articles.

The content of the articles is rather diverse. Three major themes can be outlined, namely, technologies, links, and adaptations (Table 3). These themes correspond to the main contexts in which climate change and agriculture are considered in the most-cited articles. The themes are represented by several particular topics (Table 3). Apparently, the most attention is paid to climate change influences on agriculture. The development of agriculture as a factor of climate change is considered in a slightly smaller part of the works. The thematic affinity of each considered paper is summarized in Table 4.

Table 3. Categorization of the content of the considered articles.

Abbreviation	Theme	Topic	Number of Articles
T1	Technologies (approaches, working schemes, procedures, etc.)	T1.1. Technologies addressing climate change and agriculture	5
		T1.2. Technologies addressing agriculture	2
T2	Links (what influences on what)	T.2.1. Climate change interacts with agriculture	3
		T.2.2. Agriculture influences on climate change	6
		T.2.3. Climate change influences on agriculture	9
		T.2.4. Linking through general frame	4
		T.2.5. Joint action of climate change and agriculture	1
T3	Adaptations (pragmatic vision of the problem)	T3.1. Incentives	1

Table 4. Thematic affinity of the considered articles.

Article	Theme	Topic	Relevance to the Agriculture–Climate Change Relationship
[20]	T1	T1.1	High
[21]	T2	T2.1	High
[22]	T1	T1.2	Low
[23]	T2	T2.1	Medium
[24]	T1	T1.1	High
[25]	T2	T2.2	Medium
[26]	T1	T1.2	High
[27]	T2	T2.2	Low
[28]	T2	T2.3	Low
[29]	T2	T2.4	Low
[30]	T2	T2.2	Low
[31]	T1	T1.1	High
[32]	T1	T1.1	High
[33]	T2	T2.5	Medium
[34]	T3	T3.1	High
[35]	T2	T2.3	High
[36]	T1	T1.1	Low
[37]	T2	T2.3	High
[38]	T2	T2.3	High
[39]	T2	T2.3	High
[40]	T2	T2.3	Medium
[41]	T2	T2.2	Medium
[42]	T2	T2.4	Low
[43]	T2	T2.4	Low
[44]	T2	T2.3	High
[45]	T2	T2.3	High
[46]	T2	T2.4	Low
[47]	T2	T2.3	Medium
[48]	T2	T2.2	Medium
[49]	T2	T2.2	High

Theme 1 (Technologies) is represented by two topics, one of which deals with approaches in both climate change and agriculture (T1.1), and the other topic focuses on agricultural approaches, although with reference to climate change (T1.2). Five papers contribute to the topic T1.1. Atkinson et al. [20] consider biochar application to soils. On the basis of the example from the Amazon Basin, they argue that highly stable organic black carbon waste increases the agronomic fertility of soils and contributes simultaneously to carbon sequestration. The authors extrapolate this knowledge to temperate regions. Fargione et al. [24] address a serious dilemma of biofuel production. On the one hand, biofuel allows reduction of greenhouse gas emissions. On the other hand, the relevant agricultural activities require land clearing, which triggers increase in the amounts of carbon dioxide to the atmosphere. The authors propose two solutions, namely, biofuel production from waste biomass and biofuel production on degraded/abandoned agricultural lands. In his two articles, Lal [31,32] addresses soil carbon sequestration as a tool to mitigate climate change. He indicates that the efficacy of this tool depends, particularly, on agricultural practices, and the relevant technologies are 'win–win', allowing to increase soil productivity and to reduce carbon dioxide emissions. Mata et al. [36] focus on microalgae for biodiesel production. Their cultivation can be linked to agricultural activities and facilitates renewable energy production, carbon sequestration, and improvement in agricultural food. Two papers contribute to the topic T1.2. Costello et al. [22] call for improvement in agricultural practices to increase carbon biosequestration. Foley et al. [26] review various solutions that would help agriculture to minimize its environmental impact and, particularly, its contribution to climate change. Essentially, the topic T1.2 is very similar to the topic T1.1. However, the authors of the T1.2 papers pay more attention to a 'green' shift in agriculture. Taking the entire theme T1 critically, it is possible to judge that it highlights many interesting opportunities, but it does not make clear how extensive these agriculture- and climate-friendly solutions may be and how long would be the road from their initial implementation to reaching the planetary-scale effects. The willingness and the readiness of the farmers and the other stakeholders to implement such advanced technologies appear to be questionable. Nonetheless, the local contribution to agricultural and climate change benefits with the mentioned approaches is undisputable. It should be stressed that the majority of the articles of the T1 theme demonstrate high relevance to the relationship between climate change and agriculture (Table 4), which implies significant development of this theme.

The theme T2 (Links) includes five topics, each of which corresponds to a definite causal relationship between climate change and agricultural development, namely their interaction (T2.1), agriculture as a factor of climate change (T2.2), climate change as a factor of agriculture (T2.3), phenomena linking through any general frame (T2.4), and joint action (T2.5). This theme is the most diverse, and it embraces 73% of the considered articles. The topic T2.1 is represented by two papers. Ciais et al. [21] explain that heatwaves in temperate regions (like the heatwave that struck Europe in 2003) cause reduction in ecosystem productivity (also agricultural), increase in carbon dioxide release to atmosphere, and reversal of carbon sequestration. Drake et al. [23] document the plant response to the elevated concentrations of carbon dioxide in the atmosphere and argue for an increase in nutrient use efficiency. These specialists extrapolated the noted effects to agricultural systems and stressed their fundamental changes. Six papers can be attributed to the topic T2.2. Foley et al. [25] note that the agricultural land expansion has limited the ability of natural ecosystems to regulate climate. Fontaine et al. [27] propose that agricultural practices are able to destabilize carbon pool in soils and, thus, to contribute to greenhouse gas emissions. Kalnay and Cai [30] model land-use changes linked to urbanization and agricultural activities, and these changes are able to contribute sufficiently to climate warming. Ramankutty and Foley [41] investigate changes in the global croplands during three centuries and claim that such changes could influence climate. Tilman et al. [48,49] pay attention to extraordinary environmental and climate change effects of the agricultural development and the relevant land clearing. These specialists call for improvement in agricultural practices in the both 'rich' and 'poor' countries. The topic T2.3 includes 30% of all considered articles. Godfray et al. [28] state that climate change is among the threats for the agricultural production and food security. Lobell et al. [35] consider climate change as a significant negative factor of agriculture. They demonstrate that the decline in the global maize and wheat production in the 1980s–2000s was driven by climate change, and the latter can minimize positive effects of agricultural innovations. Parry et al. [37] consider several scenarios of the agricultural response to the climate change. Particularly, they find that the latter will strengthen regional differences of crop yields, with the negative socio-economic consequences (e.g., crop price increase). Peng et al. [38] explain that global warming triggers higher night temperatures decreasing rice yields. Piao et al. [39] conclude that it is impossible to state definite effects of climate change on agriculture in China in the past decades. Potts et al. [40] find that climate change and agricultural development are related, particularly, through pollinators, and the decline of the latter due to several factors, including climate change, challenges crop production. Rosenzweig and Parry [44] suppose that climate change will not alter significantly global food production, but such an influence will be really major in developing countries. Schlenker and Roberts [45] forecast a significant drop in crop yields (corn, soybeans, and cotton) in the USA in the 21st century because of the modeled climate change. Finally, Seneviratne et al. [47] propose that the understanding of the soil moisture–climate interactions helps to understand the perspectives of agriculture under climate change.

The topic T2.4 is represented by four papers. The works by Jolliet et al. [29], Rockstrom et al. [42,43], and Schroter et al. [46] indicate the relevance of climate change and the agricultural development to the major environmental challenges of modern society and put these into the general frame of sustainability-related mechanisms, factors, and consequences. In other words, the noted works link climate change and agriculture through a general frame. Undoubtedly, these papers are thought to be highly influential, although their focus is much broader than the relationship between the two phenomena. The topic T2.5 boasts the only paper by Lambin et al. [33] who consider cropland changes, agricultural intensification and expansion, land clearing, and some other processes that accompany climate change. Importantly, these specialists distinguish the climate change–agriculture interactions from the influences of these phenomena taken separately. Generally, the entire theme T2 sheds light on the causes of the links between climate change and agriculture. However, three peculiarities of the relevant literature deserve criticism. First, the links are treated too generally, and these are 'secondary-order' inferences from more general studies. Second, the topics are illustrated with representative, but very particular examples, and, thus, these topics do not boast comprehensive coverage. Third, there are conclusions based on modeling—on the one hand, the models need permanent update and refinement; on the other hand, no model escapes simplifications and, thus, these need to be tested by the factual information. Apparently, all principal mechanisms of the relationship between climate change and agriculture are considered within the theme T2, but this knowledge seems to be enough for the only very general conceptual frame. Notably, only 35% of the articles attributed to the theme T2 boast high relevance (Table 4), which means certain underdevelopment or even 'marginalization' of this theme in the most-cited articles considered in the present study. Importantly, the articles attributed to the topic 2.3 are not only the most numerous, but also the most comprehensive and boast high relevance (Table 4). This means that the climate change factor of the agriculture development is largely well understood.

The theme T3 (Adaptations) includes the only topic T3.1 dealing with incentives, and this topic is represented by the only paper of high relevance (Table 4). Lobell et al. [34] show that South Asia and Southern Africa need adaptation to climate change to avoid agriculture failures and disturbance to their food security. They call for adequate risk attitudes and investment activities. The importance of this work is undisputable, but the entire theme T3 seems to be restricted in the most-cited articles.

5. Discussion

5.1. General Vision

The thirty most-cited articles considered in this study contribute substantially to a general frame of the relationship between climate change and agriculture. First, these works prove that the global environmental challenge has two sides, namely, the climate change influences and the agricultural influences (Figure 7). These influences can be both positive and negative, and these are realized via different mechanisms. Second, the relationship is highly complex because of feedbacks (Figure 7). Third, the articles highlight not only challenges to sustainability, but also opportunities to achieve this sustainability (Figure 7). One of the most important opportunities is linked to carbon sequestration in soil that can be facilitated by agricultural practices. Many considered articles also imply that the agriculture–climate change relationship cannot be understood comprehensively if consideted alone. In fact, many other mechanisms and phenomena (biological, socio-economical, etc.) should be taken into account.

The critical analysis of the themes of the considered papers (see above) indicates some incompleteness (these are found in the only most-cited literature, not in all ~38,000 publications). Several potentially important topics are not found in the analyzed literature. For instance, these include (and not limited to) livestock farming in the changing conditions, climate change influences on distribution channels of agricultural products (and food trade networks), new challenges for local communities and labor force under the climate change–agriculture interactions, cost of climate-friendly agricultural innovations, new agricultural opportunities after climate amelioration, tax policy modification, etc. (these topics are selected provisionally as nothing more than examples and with regard to the authors' research experience; many other topics can be proposed). Indeed, the literature on these topics is available (Table 5), but these works are not among the most-cited. Some sources were available decades ago, but these are also not among the most-cited. When the amount of literature is really vast, it is very probable that the researchers (especially the newcomers) tend to base their own ideas on the most respected works. When the latter do not provide some information, the chances for the rise of the relevant topics are lower. A possible solution would come from the authoritative books (e.g., [68–70]). However, the book distribution channels are significantly more limited than those of journal articles, and not all potentially interested specialists may easily acquire the necessary book. Moreover, searching for highly cited papers in the on line bibliographical databases appears to be a technically easier and logical option for a newcomer than struggling with book collecting. Additionally, as shown above (Figure 2), the most-cited articles are older than 2012. This means they do not bear certain up-to-date knowledge.

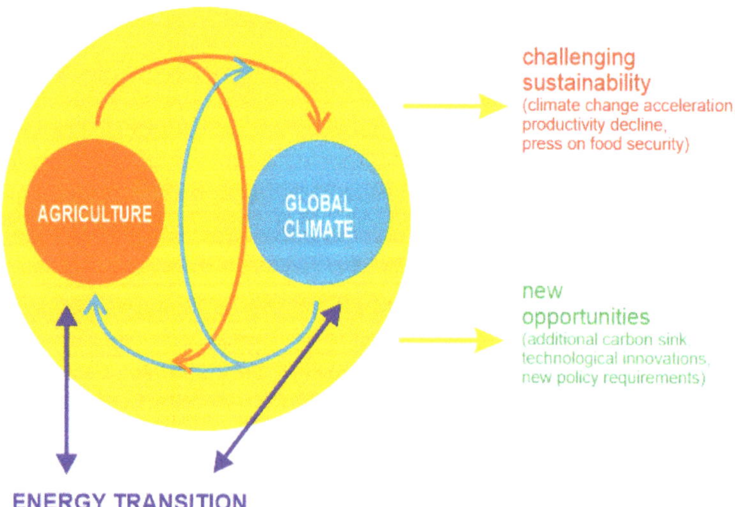

Figure 7. The general frame for judgment of the relationship between climate change, agriculture, and energy transition, as follows from the most-cited articles. Arrows within the circle represent influences, arrows outside the circles represent outcomes.

Table 5. Selected topics extending the vision of the agriculture–climate change relationship.

Topic	Some Findings	Literature Sources	Year of Publishing; Total Number of Citations
Livestock farming and climate change	Negative effects on animal husbandry with negative socio-economic effects (e.g., increase in food price)	[71]	2020; 1
Climate change and distribution of agricultural products	European meat and dairy supply chains are linked to greenhouse gas emissions	[72]	2020; 1
Climate change–agriculture interactions as challenge to local communities and labor force	High percentages of labor forces employed in agriculture of an Australian region as a driver of local vulnerability to climate change	[73]	2018; 0
Economical and political aspects of climate-friendly agricultural innovations	The Climate-Smart Agriculture Prioritization Framework (CSA-PF) in Mali as example of successful adaptation scheme	[74]	2017; 25
Agricultural opportunities due to climate change	Extending potential for wheat cultivation and higher wheat yields in some areas of Russia	[75]	2018; 6
Tax policy modification in regard to agriculture and climate change	Nitrogen tax implementation in Germany; more generally, new taxes in the face of new challenges, similarly to raising new taxes together with Internet service growth	[76] (see also [77] for general reference)	2020; 0

5.2. Utlity of the Most-Cited Articles

The results of the present study allow for discussion of the utility of the most-cited articles for the understanding of the relationship between climate change and agriculture. Three lines of evidence should be taken into account. First, the considered articles bear enough knowledge to construct the conceptual frame of this relationship (Figure 7) and to realize the diversity of the latter (Table 3). Second, not all articles demonstrate high relevance to climate change and agriculture (Table 4), and the knowledge they provide may be incomplete (see above). The best-understood is the influence of climate change on agriculture. Third, the most-cited articles reflect the opinion of experts from a limited number of countries, not the entire international research community (Figure 5), even if the scope of many articles is global. With regard to this evidence, it is possible to state that the most-cited articles are really basic, 'classical' for the understanding of the relationship between climate change and agriculture on the planetary scale, but their utility in modern research is somewhat limited. Nonetheless, these articles are of evident historical importance, and they indicate the principal research directions.

Previous research [78–80] has examined the role of the highly-cited papers in the modern science. Some of their conclusions match the findings of the present study. Marx et al. [78] analyzed the bibliographical records for climate change and noted some really outstanding papers, many of which are of historical importance. The present study also indicates that the most-cited papers appeared at the time of research acceleration (Figure 2), i.e., their historical role is outstanding. Liu et al. [79] found that the highly cited papers in food science are often authored by US and Chinese specialists. Ma et al. [80] investigated the highly cited papers from environmental science. They also found that the number of these papers accelerates and that American and Chinese experts are especially productive. These findings coincide with the outcome of the present study only partly: the most-cited papers are not new and, thus, no acceleration in their number is visible (Figure 2), and the number of the Chinese works among them is limited (Figure 5). Generally, the three above-mentioned papers, as well as the present study prove the utility of the highly cited papers in modern research. However, with regard to the agriculture–climate change relationship, this utility appears to be lesser that in the other cases.

Currently, a discussion of the importance of the open-access mode of publishing articles for attraction of the other researchers' attention and citations goes on [81–83]. Undoubtedly, such a mode can increase the influence of some articles. As for the literature considered for the purposes of the present study, these works appeared before the mid-2010s when, apparently, the distribution and the importance of open access was limited. However, when some sources gained importance, their copies appeared for free in Internet. Such 'open access' contributes to the current promotion of these papers and may stimulate the growth of their citations.

6. Conclusions

The present study combines the knowledge of the relationship between climate change and agriculture from the thirty most-cited papers and examines the utility of the latter. Three very general conclusions are possible on the basis of the findings of this study.

First, the thirty most-cited articles revealing the relationship between climate change and agriculture were published within a limited time interval and chiefly by specialists from Western Europe and North America. This implies certain biases. However, these articles appeared in high-reputation journals, which is the sign of their outstanding importance (at least, historical).

Second, energy transition is reflected in several articles (the different functions of agriculture are shown), which indicates a particular, but highly important issue. More generally, the content analysis of the works implies that the complex interactions between global climate change and agricultural development have as many negative effects as new opportunities for sustainable development.

Third, the most-cited articles can be judged as fundamental for the understanding of the relationship between climate change and agriculture, but the knowledge from only them, when taken entirely, does not avoid incompleteness. Nonetheless, the most-cited articles provide some essential knowledge from a borderless research field (especially for the newcomers). Moreover, the most-cited articles form a valuable basis for the understanding of the energy–agriculture–climate nexus.

The outcomes of the present paper imply that the conceptualization of the relationship between climate change and agriculture is yet to become comprehensive in the most-cited articles. This conclusion means new synthetic works by the world-leading experts are necessary to review this relationship on a modern basis. Undoubtedly, a lot of knowledge is already available, but it is scattered over a big number of works. In other words, specialists in climate change and agriculture need to make efforts for systematization and proper communication of the available information, which is a task for theoretical research. The researchers also need to train their vision to not miss some potentially valuable, even if less-cited works, either old or new.

Author Contributions: Conceptualization, N.N.Y. and O.A.C.; formal analysis, D.A.R. and N.A.D.; writing—original draft preparation, D.A.R. and N.N.Y. All authors have read and agreed to the published version of the manuscript.

Funding: The reported study was funded by RFBR, project number 20-010-00375 (project "Formation methodology and development of organizational and economic mechanism for achieving sustainable development goals in the national agri-food system").

Acknowledgments: The authors gratefully thank the special issue editors E. Karasmanaki (Greece) and G. Tsantopoulos (Greece) for their kind invitation and a lot of support, as well as the reviewers for their helpful recommendations.

Conflicts of Interest: The authors declare no conflict of interest.

References

1. Houghton, J. *Global Warming. The Complete Briefing*; Cambridge University Press: Cambridge, UK, 2009.
2. IPCC. *Climate Change 2014: Synthesis Report*; IPCC: Geneva, Switzerland, 2014.

3.   Adger, W.N.; Arnell, N.W.; Tompkins, E.L. Successful adaptation to climate change across scales. *Global Environ. Chang.* **2005**, *15*, 77–86. [CrossRef]

4.   Bellard, C.; Bertelsmeier, C.; Leadley, P.; Thuiller, W.; Courchamp, F. Impacts of climate change on the future of biodiversity. *Ecol. Lett.* **2012**, *15*, 365–377. [CrossRef] [PubMed]

5.   Crowley, T.J. Causes of climate change over the past 1000 years. *Science* **2000**, *289*, 270–277. [CrossRef] [PubMed]

6.   Giorgi, F. Climate change hot-spots. *Geophys. Res. Lett.* **2006**, *33*, L08707. [CrossRef]

7.   Parmesan, C.; Yohe, G. A globally coherent fingerprint of climate change impacts across natural systems. *Nature* **2003**, *421*, 37–42. [CrossRef]

8.   Patz, J.A.; Campbell-Lendrum, D.; Holloway, T.; Foley, J.A. Impact of regional climate change on human health. *Nature* **2005**, *438*, 310–317. [CrossRef] [PubMed]

9.   Ramanathan, V.; Carmichael, G. Global and regional climate changes due to black carbon. *Nat. Geosci.* **2008**, *1*, 221–227. [CrossRef]

10. Solomon, S.; Plattner, G.-K.; Knutti, R.; Friedlingstein, P. Irreversible climate change due to carbon dioxide emissions. *Proc. Natl. Acad. Sci. USA* **2009**, *106*, 1704–1709. [CrossRef]

11. Trenberth, K.E. Changes in precipitation with climate change. *Clim. Res.* **2011**, *47*, 123–138. [CrossRef]

12. Walther, G.-R.; Post, E.; Convey, P.; Menzel, A.; Parmesan, C.; Beebee, T.J.C.; Fromentin, J.-M.; Hoegh-Guldberg, O.; Bairlein, F. Ecological responses to recent climate change. *Nature* **2002**, *416*, 389–395. [CrossRef]

13. Edmonds, H.K.; Lovell, J.E.; Lovell, C.A.K. A new composite climate change vulnerability index. *Ecol. Indic.* **2020**, *117*, 106529. [CrossRef]

14. Huang, J.; Yu, H.; Han, D.; Zhang, G.; Wei, Y.; Huang, J.; An, L.; Liu, X.; Ren, Y. Declines in global ecological security under climate change. *Ecol. Indic.* **2020**, *117*, 106651. [CrossRef]

15. Konapala, G.; Mishra, A.K.; Wada, Y.; Mann, M.E. Climate change will affect global water availability through compounding changes in seasonal precipitation and evaporation. *Nat. Commun.* **2020**, *11*, 3044. [CrossRef]

16. Molina, C.; Akçay, E.; Dieckmann, U.; Levin, S.A.; Rovenskaya, E.A. Combating climate change with matching-commitment agreements. *Sci. Rep.* **2020**, *10*, 10251. [CrossRef] [PubMed]

17. Sanderson, B.M.; O'Neill, B.C. Assessing the costs of historical inaction on climate change. *Sci. Rep.* **2020**, *10*, 9173. [CrossRef]

18. Jussi, L.; Sanna, L.; Markku, O. Climate change mitigation and agriculture: Measures, costs and policies—A literature review. *Agric. Food Sci.* **2020**, *29*, 110–129.

19. Praveen, B.; Sharma, P. A review of literature on climate change and its impacts on agriculture productivity. *J. Public Aff.* **2019**, *19*, e1960. [CrossRef]

20. Atkinson, C.J.; Fitzgerald, J.D.; Hipps, N.A. Potential mechanisms for achieving agricultural benefits from biochar application to temperate soils: A review. *Plant Soil* **2010**, *337*, 1–18. [CrossRef]

21. Ciais, P.; Reichstein, M.; Viovy, N.; Granier, A.; Ogue, J.; Allard, V.; Aubinet, M.; Buchmann, N.; Bernhofer, C.; Carrara, A.; et al. Europe-Wide reduction in primary productivity caused by the heat and drought in 2003. *Nature* **2005**, *437*, 529–533. [CrossRef]

22. Costello, A.; Abbas, M.; Allen, A.; Ball, S.; Bell, S.; Bellamy, R.; Friel, S.; Groce, N.; Johnson, A.; Kett, M.; et al. Managing the health effects of climate change. Lancet and University College London Institute for Global Health Commission. *Lancet* **2009**, *373*, 1693–1733. [CrossRef]

23. Drake, B.G.; Gonzalez-Meler, M.A.; Long, S.P. More efficient plants: A Consequence of Rising Atmospheric CO_2? *Ann. Rev. Plant Biol.* **1997**, *48*, 609–639. [CrossRef] [PubMed]

24. Fargione, J.; Hill, J.; Tilman, D.; Polasky, S.; Hawthorne, P. Land clearing and the biofuel carbon debt. *Science* **2008**, *319*, 1235–1238. [CrossRef] [PubMed]

25. Foley, J.A.; DeFries, R.; Asner, G.P.; Barford, C.; Bonan, G.; Carpenter, S.R.; Chapin, F.S.; Coe, M.T.; Daily, G.C.; Gibbs, H.K.; et al. Global consequences of land use. *Science* **2005**, *309*, 570–574. [CrossRef] [PubMed]

26. Foley, J.A.; Ramankutty, N.; Brauman, K.A.; Cassidy, E.S.; Gerber, J.S.; Johnston, M.; Mueller, N.D.; O'Connell, C.; Ray, D.K.; West, P.C.; et al. Solutions for a cultivated planet. *Nature* **2011**, *478*, 337–342. [CrossRef]

27. Fontaine, S.; Barot, S.; Barr, P.; Bdioui, N.; Mary, B.; Rumpel, C. Stability of organic carbon in deep soil layers controlled by fresh carbon supply. *Nature* **2007**, *450*, 277–280. [CrossRef]

28. Godfray, H.C.J.; Beddington, J.R.; Crute, I.R.; Haddad, L.; Lawrence, D.; Muir, J.F.; Pretty, J.; Robinson, S.; Thomas, S.M.; Toulmin, C. Food security: The challenge of feeding 9 billion people. *Science* **2010**, *327*, 812–818. [CrossRef]

29. Jolliet, O.; Margni, M.; Charles, R.; Humbert, S.; Payet, J.; Rebitzer, G.; Rosenbaum, R. IMPACT 2002+: A New Life Cycle Impact Assessment Methodology. *Int. J. Life Cycle Assess.* **2003**, *8*, 324–330. [CrossRef]

30. Kalnay, E.; Cai, M. Impact of urbanization and land-use change on climate. *Nature* **2003**, *423*, 528–531. [CrossRef]

31. Lal, R. Soil carbon sequestration impacts on global climate change and food security. *Science* **2004**, *304*, 1623–1627. [CrossRef]

32. Lal, R. Soil carbon sequestration to mitigate climate change. *Geoderma* **2004**, *123*, 1–22. [CrossRef]

33. Lambin, E.F.; Geist, H.J.; Lepers, E. Dynamics of land-use and land-cover change in tropical regions. *Ann. Rev. Environ. Resour.* **2003**, *28*, 205–241. [CrossRef]

34. Lobell, D.B.; Burke, M.B.; Tebaldi, C.; Mastrandrea, M.D.; Falcon, W.P.; Naylor, R.L. Prioritizing climate change adaptation needs for food security in 2030. *Science* **2008**, *319*, 607–610. [CrossRef] [PubMed]

35. Lobell, D.B.; Schlenker, W.; Costa-Roberts, J. Climate trends and global crop production since 1980. *Science* **2011**, *333*, 616–620. [CrossRef]

36. Mata, T.M.; Martins, A.A.; Caetano, N.S. Microalgae for biodiesel production and other applications: A review. *Renew. Sustain. Energy Rev.* **2010**, *14*, 217–232. [CrossRef]

37. Parry, M.L.; Rosenzweig, C.; Iglesias, A.; Livermore, M.; Fischer, G. Effects of climate change on global food production under SRES emissions and socio-economic scenarios. *Global Environ. Chang.* **2004**, *14*, 53–67. [CrossRef]

38. Peng, S.; Huang, J.; Sheehy, J.E.; Laza, R.C.; Visperas, R.M.A.; Zhong, X.; Centeno, G.S.; Khush, G.S.; Cassman, K.G. Rice yields decline with higher night temperature from global warming. *Proc. Natl. Acad. Sci. USA* **2004**, *101*, 9971–9975. [CrossRef] [PubMed]

39. Piao, S.; Ciais, P.; Huang, Y.; Shen, Z.; Peng, S.; Li, J.; Zhou, L.; Liu, H.; Ma, Y.; Ding, Y.; et al. The impacts of climate change on water resources and agriculture in China. *Nature* **2010**, *467*, 43–51. [CrossRef] [PubMed]

40. Potts, S.G.; Biesmeijer, J.C.; Kremen, C.; Neumann, P.; Schweiger, O.; Kunin, W.E. Global pollinator declines: Trends, impacts and drivers. *Trends Ecology Evol.* **2010**, *25*, 345–353. [CrossRef] [PubMed]

41. Ramankutty, N.; Foley, J.A. Estimating historical changes in global land cover: Croplands from 1700 to 1992. *Global Biogeochem. Cycles* **1999**, *13*, 997–1027. [CrossRef]

42. Rockstrom, J.; Steffen, W.; Noone, K.; Persson, A.; Chapin, F.S., III; Lambin, E.; Lenton, T.M.; Scheffer, M.; Folke, C.; Schellnhuber, H.J.; et al. Planetary boundaries: Exploring the safe operating space for humanity. *Ecol. Soc.* **2009**, *14*, 32. [CrossRef]

43. Rockstrom, J.; Steffen, W.; Noone, K.; Persson, E.; Chapin, F.S.; Lambin, E.F.; Lenton, T.M.; Scheffer, M.; Folke, C.; Schellnhuber, H.J.; et al. A safe operating space for humanity. *Nature* **2009**, *461*, 472–475. [CrossRef] [PubMed]

44. Rosenzweig, C.; Parry, M.L. Potential impact of climate change on world food supply. *Nature* **1994**, *367*, 133–138. [CrossRef]

45. Schlenker, W.; Roberts, M.J. Nonlinear temperature effects indicate severe damages to U.S. crop yields under climate change. *Proc. Natl. Acad. Sci. USA* **2009**, *106*, 15594–15598. [CrossRef] [PubMed]

46. Schroter, D.; Cramer, W.; Leemans, R.; Prentice, I.C.; Araujo, M.B.; Arnell, N.W.; Bondeau, A.; Bugmann, H.; Carter, T.R.; Gracia, C.A.; et al. Ecology: Ecosystem service supply and vulnerability to global change in Europe. *Science* **2005**, *310*, 1333–1337. [CrossRef] [PubMed]

47. Seneviratne, S.I.; Corti, T.; Davin, E.L.; Hirschi, M.; Jaeger, E.B.; Lehner, I.; Orlowsky, B.; Teuling, A.J. Investigating soil moisture-climate interactions in a changing climate: A review. *Earth-Sci. Rev.* **2010**, *99*, 125–161. [CrossRef]

48. Tilman, D.; Fargione, J.; Wolff, B.; D'Antonio, C.; Dobson, A.; Howarth, R.; Schindler, D.; Schlesinger, W.H.; Simberloff, D.; Swackhamer, D. Forecasting agriculturally driven global environmental change. *Science* **2001**, *292*, 281–284. [CrossRef]

49. Tilman, D.; Balzer, C.; Hill, J.; Befort, B.L. Global food demand and the sustainable intensification of agriculture. *Proc. Natl. Acad. Sci. USA* **2011**, *108*, 20260–20264. [CrossRef]

50. Bridge, G.; Bouzarovski, S.; Bradshaw, M.; Eyre, N. Geographies of energy transition: Space, place and the low-carbon economy. *Energy Policy* **2013**, *53*, 331–340. [CrossRef]

51. Chlebna, C.; Mattes, J. The fragility of regional energy transitions. *Environ. Innov. Soc. Transit.* **2020**, *37*, 66–78. [CrossRef]
52. Meadowcroft, J. What about the politics? Sustainable development, transition management, and long term energy transitions. *Policy Sci.* **2009**, *42*, 323–340. [CrossRef]
53. Sovacool, B.K. How long will it take? Conceptualizing the temporal dynamics of energy transitions. *Energy Res. Soc. Sci.* **2016**, *13*, 202–215. [CrossRef]
54. Verbong, G.; Geels, F. The ongoing energy transition: Lessons from a socio-technical, multi-level analysis of the Dutch electricity system (1960–2004). *Energy Policy* **2007**, *35*, 1025–1037. [CrossRef]
55. Hoppe, T.; van Bueren, E. Guest editorial: Governing the challenges of climate change and energy transition in cities. *Energy Sustain. Soc.* **2015**, *5*, 19. [CrossRef]
56. Fazey, I.; Schäpke, N.; Caniglia, G.; Patterson, J.; Hultman, J.; van Mierlo, B.; Säwe, F.; Wiek, A.; Wittmayer, J.; Aldunce, P.; et al. Ten essentials for action-oriented and second order energy transitions, transformations and climate change research. *Energy Res. Soc. Sci.* **2018**, *40*, 54–70. [CrossRef]
57. Newell, P.; Bulkeley, H. Landscape for change? International climate policy and energy transitions: Evidence from sub-Saharan Africa. *Clim. Policy* **2017**, *17*, 650–663. [CrossRef]
58. Scheffran, J. Adaptive management of energy transitions in long-term climate change. *Comput. Manag. Sci.* **2008**, *5*, 259–286. [CrossRef]
59. Urban, F. Climate-Change mitigation revisited: Low-Carbon energy transitions for China and India. *Dev. Policy Rev.* **2009**, *27*, 693–715. [CrossRef]
60. Sutherland, L.-A.; Peter, S.; Zagata, L. Conceptualising multi-regime interactions: The role of the agriculture sector in renewable energy transitions. *Res. Policy* **2015**, *44*, 1543–1554. [CrossRef]
61. Dalmat, Y.-M. Biofuels, an asset for energy transition and independence. *Option/Bio* **2020**, *31*, 10.
62. Osunmuyiwa, O. Politics of Energy Transitions: A decade after Nigeria's biofuels crusade, a tale of non-commercialization and lost opportunities. *Environ. Policy Gov.* **2017**, *27*, 632–646. [CrossRef]
63. Stokes, L.C.; Breetz, H.L. Politics in the U.S. energy transition: Case studies of solar, wind, biofuels and electric vehicles policy. *Energy Policy* **2018**, *113*, 76–86. [CrossRef]
64. Bardi, U.; El Asmar, T.; Lavacchi, A. Turning electricity into food: The role of renewable energy in the future of agriculture. *J. Clean. Prod.* **2013**, *53*, 224–231. [CrossRef]
65. Chai, X.; Tonjes, D.J.; Mahajan, D. Methane emissions as energy reservoir: Context, scope, causes and mitigation strategies. *Progress Energy Combust. Sci.* **2016**, *56*, 33–70. [CrossRef]
66. Safriel, U. Deserts and desertification: Challenges but also opportunities. *Land Degrad. Dev.* **2009**, *20*, 353–366. [CrossRef]
67. Schwartzman, D.; Schwartzman, P. A rapid solar transition is not only possible, it is imperative! *Afr. J. Sci. Technol. Innov. Dev.* **2013**, *5*, 297–302. [CrossRef]
68. Choudhary, K.K.; Kumar, A.; Singh, A.K. (Eds.) *Climate Change and Agricultural Ecosystems*; Woodhead Publishing: Cambridge, UK, 2019.
69. Iizumi, T.; Hirata, R.; Matsuda, R. (Eds.) *Adaptation to Climate Change in Agriculture*; Springer: Singapore, 2019.
70. Kang, M.S.; Banga, S.S. *Combating Climate Change: An Agricultural Perspective*; CRC Press: Boca Raton, FL, USA, 2013.
71. Gomez-Zavaglia, A.; Mejuto, J.C.; Simal-Gandara, J. Mitigation of emerging implications of climate change on food production systems. *Food Res. Int.* **2020**, *134*, 109256. [CrossRef]
72. aan den Toorn, S.I.; Worrell, E.; van den Broek, M.A. Meat, dairy, and more: Analysis of material, energy, and greenhouse gas flows of the meat and dairy supply chains in the EU28 for 2016. *J. Ind. Ecol.* **2020**, *24*, 601–614. [CrossRef]
73. Smith, E.F.; Lieske, S.N.; Keys, N.; Smith, T.F. The socio-economic vulnerability of the Australian east coast grazing sector to the impacts of climate change. *Reg. Environ. Chang.* **2018**, *18*, 1185–1199. [CrossRef]
74. Andrieu, N.; Sogoba, B.; Zougmore, R.; Howland, F.; Samake, O.; Bonilla-Findji, O.; Lizarazo, M.; Nowak, A.; Dembele, C.; Corner-Dolloff, C. Prioritizing investments for climate-smart agriculture: Lessons learned from Mali. *Agric. Syst.* **2017**, *154*, 13–24. [CrossRef]
75. Di Paola, A.; Caporaso, L.; Di Paola, F.; Bombelli, A.; Vasenev, I.; Nesterova, O.V.; Castaldi, S.; Valentini, R. The expansion of wheat thermal suitability of Russia in response to climate change. *Land Use Policy* **2018**, *78*, 70–77. [CrossRef]

76. Henseler, M.; Delzeit, R.; Adenäuer, M.; Baum, S.; Kreins, P. Nitrogen Tax and Set-Aside as Greenhouse Gas Abatement Policies under Global Change Scenarios: A Case Study for Germany. *Environ. Resour. Econ.* **2020**, *76*, 299–329. [CrossRef]
77. Artemenko, D.; Aguzarova, F.; Artemenko, G.; Novoselov, K.; Vertakova, Y. Media resources in education: The taxation aspect. In Proceedings of the 33rd International Business Information Management Association Conference, IBIMA 2019: Education Excellence and Innovation Management through Vision 2020, Granada, Spain, 10–11 April 2019; pp. 4372–4376.
78. Marx, W.; Haunschild, R.; Thor, A.; Bornmann, L. Which early works are cited most frequently in climate change research literature? A bibliometric approach based on Reference Publication Year Spectroscopy. *Scientometrics* **2017**, *110*, 335–353. [CrossRef]
79. Liu, B.; Chen, L. A Bibliometric Study on Highly Cited Papers of Food Science. *J. Chin. Inst. Food Sci. Technol.* **2020**, *20*, 308–318.
80. Ma, Q.; Li, Y.; Zhang, Y. Informetric analysis of highly cited papers in environmental sciences based on essential science indicators. *Int. J. Environ. Res. Public Health* **2020**, *17*, 3781. [CrossRef]
81. Gaulé, P.; Maystre, N. Getting cited: Does open access help? *Res. Policy* **2011**, *40*, 1332–1338. [CrossRef]
82. Holmberg, K.; Hedman, J.; Bowman, T.D.; Didegah, F.; Laakso, M. Do articles in open access journals have more frequent altmetric activity than articles in subscription-based journals? An investigation of the research output of Finnish universities. *Scientometrics* **2020**, *122*, 645–659. [CrossRef]
83. Li, Y.; Wu, C.; Yan, E.; Li, K. Will open access increase journal CiteScores? An empirical investigation over multiple disciplines. *PLoS ONE* **2018**, *13*, e0201885. [CrossRef]

 *sustainability*

Article

A Multicriteria Approach for Assessing the Impact of ICT on EU Sustainable Regional Policy

Christiana Koliouska * and Zacharoula Andreopoulou

Laboratory of Forest Informatics, Faculty of Forestry and Natural Environment,
Aristotle University of Thessaloniki, P.O. Box 247, 54124 Thessaloniki, Greece; randreop@for.auth.gr
* Correspondence: ckolious@for.auth.gr; Tel.: +30-694-691-7724

Received: 26 March 2020; Accepted: 3 June 2020; Published: 15 June 2020

Abstract: As a global actor, the European Union (EU) plays a leading role in international efforts to promote sustainable development globally. All sustainable objectives and targets need Information and Communication Technologies (ICTs) as key catalysts, since ICTs constitute tools of unprecedented power which help people to face the growing challenges of rising population, poverty, epidemics and climate change. Policy makers in the EU are increasingly putting ICTs into relations with sustainable regional development. This paper aims to study and assess the impact of ICT on the EU regional policy in terms of sustainable development by applying the multicriteria approach, PROMETHEE II, using the software Visual PROMETHEE. The criteria that were used in this research are the criteria that both the European Commission and member states define to assess the ICT implications of new EU legislation since 2010. The results revealed that the impact of ICT on EU sustainable regional policy has gotten stronger in the last two decades.

Keywords: ICT; EU regional policy; sustainable development; multicriteria approach; policy assessment

---

1. Introduction

Information and Communication Technology (ICT) advancements not only bring new opportunities, but also bring to light new risks for the achievement of sustainable development (SD) goals [1]. ICT proved to accelerate the worldwide socio-technological progress through knowledge transfer, marketing goods or services, network externalities and the development of cooperative relationships [2]. The contribution of ICT in addressing the major challenges of sustainable energy, climate change and sustainable development is highlighted by researchers, entrepreneurs, decision-makers and policy-makers [3]. The integration of ICT into social development and economic growth provides better opportunities for enhancing competitiveness and satisfying human needs [4,5].

The initiatives required to help ICT enable transformation to sustainable development and global competitiveness are as follows [6]:

- Improve the four inter-related dimensions of ICT (the 4C Framework): Computers, Connectivity, Content, (human) Capacity.
- The successful application ICT for sustainable development depends on the scalability and the sustainability.

  - ICT constitutes just a means of achieving sustainable development.
  - Active efforts are required to foster global inclusion.
  - Sustainable ICT should be economically feasible and create end-user value.
  - ICT for sustainable development research and practice should be participatory, collaborative and empowering for the solutions to be globally consistent and relevant.

- Sustainable ICT should become a globally recognized and funded industry.

  - Inspire and effectively engage all relevant stakeholders to have a shared vision to implement sustainable ICT.
  - Develop metrics to quantify the level of success and effectiveness and consider new academic rigor.
  - Focus on the challenges of modernization.
  - Plan innovative models for research and development.

ICT is supposed to trigger the co-evolutionary process that will meet sustainable development goals in the EU through the efficient use of natural resources [7]. National and regional agencies involved in activities related to the agenda about digital inclusion and web accessibility in rural and underserved areas should adopt a new technology, engage community participation and incorporate sustainable development as essential factors to link ICT with community development and prosperity [8]. ICT solutions constitute vehicles for sustainable development in a more effective and cost-efficient way, as they have great potential to assist poor people to improve the quality of their life [9].

Information and knowledge are essential to achieve regional development and economic growth, since they are performed as primal components of socio-economic activities for strategic management in developing countries [10]. According to OECD Report Greener and Smarter in 2010, ICT is a crucial component for green growth and green economy [11] as it promotes smart growth, "development that is economically sound, environmentally friendly and supportive of community livability–growth that enhances our quality of life" [12,13]. Realtime data streaming will support the process of effectively monitoring the impact of regulations regarding the environment and society [14]. The multimedia distributed systems such as ICT, communication networks and smart media are widely used for electronic data interchange (EDI) [15]. The international community has integrated many goals in public policy to ensure environmental sustainability since it constitutes part of global socio-economic well-being [16,17]. The sustainable management of natural resources plays a vital role in the achievement of global sustainability goals [18]. The aim of sustainable development is to "promote the human well-being, meet the basic needs of the poor and protect the welfare of future generations (intra- and inter- generational justice), preserve environmental resources and global life-support systems (respecting limits), integrate economics and natural environment in decision-making, and encourage public involvement in development processes" [19,20]. However, the current framework for sustainable development is completely different from the framework developed over past years, as new factors (e.g., multinational corporate companies and civil society organizations) and modern ICT (computerized communication channels, such as Internet) influence the environmental and socio-economic aspects of the development [21].

According to the EU Green Paper on Innovation in 1995, the key factors to boosting innovation potential are the following: environmental policy, regional industrial policy, technology policy, education and training policy, research policy (RP), competition policy, Small and Medium Enterprises (SMEs) policy and taxation policy [22]. In today's digital era, the EU encourages the international cooperation in research and innovation as a key success factor in sustainability [23,24]. Sustainable regional and local development constitutes an integrated approach to the planning and development of our regions [25] but cooperation among different authorities is essential [26,27]. Regional Policy aims to create employment opportunities, support business agility and business competitiveness, while promoting sustainable economic growth to all European regions [28]. EU regional policy funding focuses on four categories [29]:

- Information and Communication Technology (ICT).
- Research and Innovation (R&I).
- Enhancing the competitiveness of Small and Medium-sized Enterprises (SMEs).
- Transitioning to a low-carbon economy.

The main aim of the current policies and strategies for regional development is the flow of money from the rich nations to the poor nations, as well as the support to confront regional challenges through funding programs [30]. Aiming to create empowering partnerships among EU regions and EU member states in the context of these policies' implementation, the EU adopts some practices, such as monitoring and policy evaluation, in order to enhance the effectiveness of the policies [30]. The European Commission's involvement in regional development can be traced back to 1957 when the Treaty of Rome required the community to ensure "harmonious" development by reducing regional differences and the backwardness of less-favored regions [31].

According to the latest Eurobarometer regarding "Citizens' awareness and perceptions of EU regional policy" in 2017, almost 80% of the EU citizens were convinced that EU regional policy investments had a positive effect on their region or city, while almost half of the EU citizens supported the idea that EU regional investments should continue [32]. In order to be effective, a policy, such as regional policy, necessarily involves multiple partners operating at different spatial scales and different governance levels [33]. There are four structural funds: the European Regional Development Fund (ERDF), which is intended to finance large infrastructure projects and has the largest weight in the budget; the European Social Fund (ESF), which is the main financial instrument assisting the EU in realizing the strategy and the main objectives of its implementing policy; the European Agricultural Guidance and Guarantee Fund (EAGGF—Guidance Section), which accelerates the reformation of the agricultural structure; the Financial Instrument for Fisheries Guidance (FIFG), which is the specific fund for the reform of the structure of the fisheries sector [34]. The structural actions represent approximately one-third of whole the budget of the EU [34]. Infrastructure expenditures by are addressed to low-growth, low-employment and low-productivity regions [35].

It is widely accepted that EU regional policy has been an important contributing factor to the promotion of EU political regionalism and decentralization [36]. This is primarily as a result of the Funds of Partnership principle, taking into consideration that there are competent regional authorities and they get actively involved in the developing and implementation of regional aid [36]. Whereas a succession of income transfers may lead to a series of possible reductions in regional disparities, it should not be confused with the process of convergence on the regional level, which would be the successful implication of regional policies [37]. An emphasis on endogenous growth models has developed, and in connection with this, the intention to enhance the competitiveness among EU regions in contrast with simply getting involved in redistributive activities characterized by former regional policy interventions [38].

The European Regional Policy Research Consortium (EoRPA), originally launched in 1978, and was funded by government departments in Austria, Finland, France, Germany, Italy, the Netherlands, Norway, Poland, Sweden, Switzerland and the United Kingdom. It involves the monitoring and analysis of national regional policies in 30 European countries, and the study of the inter-relationships between EU Cohesion policy and EU Competition policy control of state aid [39].

The objective of this study is to investigate the impact of ICT on EU regional policy in the terms of sustainable development by applying the multicriteria approach, PROMETHEE II, using the software, Visual PROMETHEE. This approach was used for the assessment of the impact of ICT and for the ranking of the EU regional policies.

2. Materials and Methods

EU regional policies are retrieved from the official European Union website (www.europa.eu). The first step was to record the regulations, the directives, the decisions, the communications and other acts regarding the regional issues. EU sustainable regional policies form the alternatives. After the collection of the policies, a two-dimensional table was developed in order to find out the existence or lack of criteria that both the European Commission and member states have defined and proposed in order to assess the ICT implications of new EU legislation since 2010 [40]. These criteria constitute the variables X1, X2, …, X12 (Table 1). Variable X1 refers to the requirement of the design of information rich processes by the legislation, while variable X2 refers to the requirement of the design of new business processes. Variable X3 represents the requirement of large amounts of data gathering in these processes and variable X4 represents the requirement of collaboration between ICT systems of multiple DGs or institutions/organizations. Variable X5 is about the fact that the legislation concerns ICT systems or that ICT is a supporting function of the legislation. The first five criteria/variables describe the level of dependence on ICT solutions of the EU regional policies and the weight of each is 1. The rest of the criteria/variables describe the levels of complexity of the ICT solutions and the weight of each is 0.83. The weights of the criteria were defined according to the method used by the European Commission to assess the ICT implications of EU legislation [40], giving the same importance to the level of dependence on the ICT implications and the level of complexity of the ICT implications. Variable X6 refers to whether the legislation requires new ICT solutions or existing applications can fulfill the requirements, while variable X7 refers to the existence of legacy systems which might hamper the implementation. Variable X8 concerns the imposition of authentication requirements by the legislation and variable X9 concerns the requirement of large amounts of data exchange between member states and/or the Commission. Variable X10 is about the required lead-time of the implementation (urgency), variable X11 is about the requirement of new interoperability specifications and variable X12 is about the imposition of high security requirements on the ICT solution by the initiative. The total amount of criteria achieved by each EU regional policy was also studied.

Table 1. Criteria [41].

Category	Variable	Criteria
Dependence on the ICT solutions	X_1	Does the legislation require the design of information rich processes?
	X_2	Does the legislation require the design of new business processes?
	X_3	Are large amounts of data gathering required in these processes?
	X_4	Is collaboration between ICT systems of multiple DGs or institutions/organizations required?
	X_5	Is the legislation concerning ICT systems or is ICT a supporting function of the legislation?
Complexity of the ICT solutions	X_6	Does the legislation require new ICT solutions or can existing applications fulfill the requirements?
	X_7	Are there any legacy systems which might hamper the implementation?
	X_8	Does the legislation impose authentication requirements?
	X_9	Is a large amount of data exchange between member states and/or the Commission required?
	X_{10}	What is the required lead-time of the implementation (urgency)?
	X_{11}	Are new interoperability specifications required?
	X_{12}	Does the initiative impose high security requirements on the ICT solution?

Furthermore, the impact of ICT was assessed, and the EU regional policies were ranked using the multicriteria analysis, PROMETHEE II. PROMETHEE constitutes a prescriptive methodology which enables the decision-maker to rank the actions regarding his preferences [42]. Two rankings are calculated: the partial ranking is calculated mainly by undisputable preferences (PROMETHEE I), while the complete ranking, which is probably weaker, is obtained according to the decision-maker's requirements, too (PROMETHEE II) [42]. The PROMETHEE IV method solves a choice problematic for an infinite set of actions. It uses the same outranking relation, but the flows are defined on a compact subset of R" [43].

The PROMETHEE method includes four processes [44]:

(1) Define the preferences: this function deals with the decision-maker's preference for an alternative x_k with respect to another alternative x_l regarding a criterion.
(2) Calculate the preference index: this index is used to compare the alternatives in pairs, quantitatively taking into consideration all the defined criteria.
(3) Construct valued outranking graph: outgoing and incoming flows are determined by means of relevant preference indices.
(4) Rank alternatives according to the valued outranking graph: determination of the weights is an important step in most multi-criteria methods [45].

The method normalizes the weights of the criteria in order for their sum to be equal to 1.0 (100%) [44]. The PROMETHEE II method is described thoroughly in Brans and Mareschal (2005) [46] and in Andreopoulou et al. (2017) [47].

PROMETHEE provides the researcher with rankings of the alternatives and GAIA with a graphical representation of the decision problem [48]. The GAIA analysis is based on the uni-criterion net flows [42]. GAIA uses the principal components analysis (PCA), a well-known dimension-reduction technique for statistical data analysis [49].

PROMETHEE II methodology was selected in order to evaluate the impact of ICT on EU sustainable regional policies and to rank the policies because [50]:

- there is now so much sensitivity of the estimated relation in small changes.
- the results can be easily interpreted and discussed.
- the use of the superiority relation is applied when the alternatives (sustainable regional policies) have to be ranked from the alternative with the highest score to the alternative with the lowest score.
- the assessment and ranking process of complicated cases of sustainable regional policies is suitable for the application of PROMETHEE II methodology in the way that it seems to be closer to reality.

3. Results

The research through the official European Union website (www.europa.eu) resulted in the retrieval of 50 regional policies, which are presented in the Appendix A. In total, 19 out of the 50 EU regional policies have been established since 2010 (when the European Commission began to assess the ICT implications of EU legislation). Figure 1 presents the partial rankings of the 19 EU regional policies based on the computation of the two preference flows (Phi+ and Phi-).

Figure 2 presents the complete rankings of the 19 EU regional policies, which are based on the total net preference flow (Phi). According to the results of PROMETHEE I and PROMETHEE II, COM(2012)19 "Waste Electrical and Electronic Equipment (WEEE)" is preferred to all other regional policies. COM(2012)19 has the highest score on Phi (0.5841), followed by Reg.2015/207 "The models for the progress report, submission of the information on a major project, the joint action plan, the implementation reports for the Investment for growth and jobs goal, the management declaration, the audit strategy, the audit opinion and the annual control report and the methodology for carrying out the cost-benefit analysis" and COM(2014)473 "Sixth report on economic, social and territorial cohesion: investment for jobs and growth", while COM(2015)118 "The Agreement on the European Economic Area" has the lowest score. The ranking of PROMETHEE I is confirmed by the ranking of PROMETHEE II.

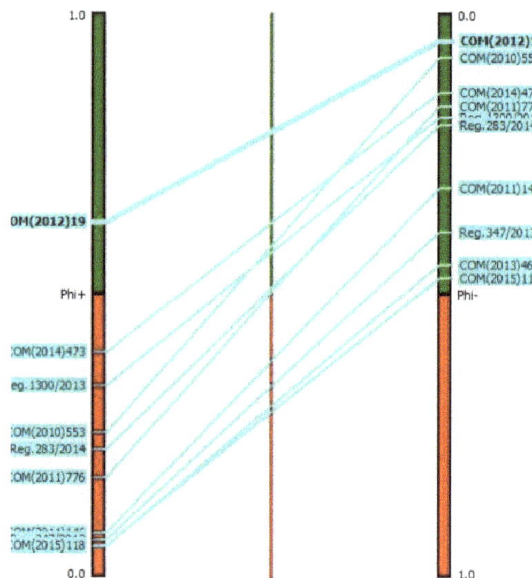

Figure 1. Partial ranking of EU regional policies.

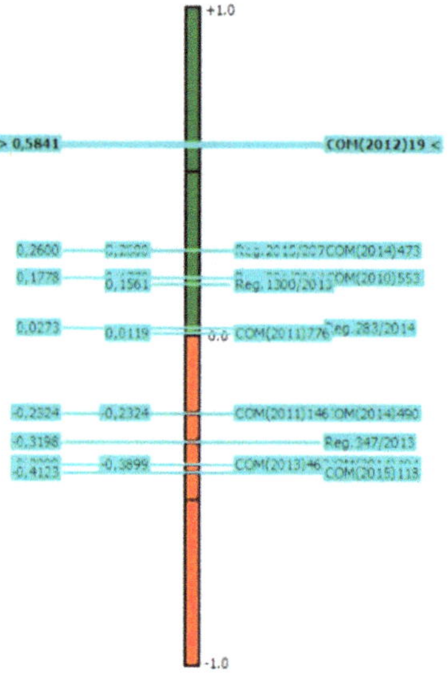

Figure 2. Complete ranking of EU regional policies.

Figure 3 shows the PROMETHEE diamond, which depicts in a better way the two rankings (Phi+ and Phi-), while the vertical dimension represents the total net flow (Phi) by the complete ranking process. It can be observed that all action cones are located on the left axis (Phi+), which means that the total net flow of the regional policies is less than 1.

Table 2 shows the Phi scores of all the EU regional policies. The values calculated for the total net flows (Phi) present a large spectrum of values between +0.5841 and −0.4123, and that shows a great difference concerning "superiority" between the first and the last case in the ranking of EU regional policies according to the impact of ICT.

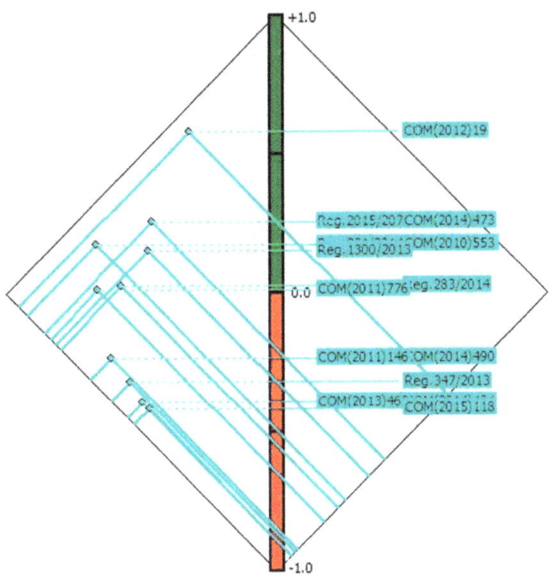

Figure 3. PROMETHEE Diamond.

Table 2. Preference flows.

	Regional Policy	**Phi+**	**Phi-**	**Phi**
1	COM(2012)19	0.631	0.0469	0.5841
2	Reg.2015/207	0.3992	0.1391	0.26
3	COM(2014)473	0.3992	0.1391	0.26
4	Reg.1301/2013	0.2556	0.0778	0.1778
5	Reg.1304/2013	0.2556	0.0778	0.1778
6	Reg.1299/2013	0.2556	0.0778	0.1778
7	Reg.1303/2013	0.2556	0.0778	0.1778
8	Reg.231/2014	0.2556	0.0778	0.1778
9	COM(2010)553	0.2556	0.0778	0.1778
10	Reg.1300/2013	0.3401	0.184	0.1561
11	Reg.283/2014	0.2256	0.1983	0.0273
12	COM(2011)776	0.1752	0.1633	0.0119
13	COM(2014)490	0.077	0.3094	−0.2324
14	COM(2011)146	0.077	0.3094	−0.2324
15	Reg.347/2013	0.0667	0.3865	−0.3198
16	Reg.240/2014	0.0547	0.4445	−0.3899
17	COM(2014)494	0.0547	0.4445	−0.3899
18	COM(2013)463	0.0547	0.4445	−0.3899
19	COM(2015)118	0.0564	0.4687	−0.4123

In Figure 4, the GAIA plane is displayed. The red axis is the decision axis, which indicates the direction for the best solution according to the weight vectors on the GAIA plane. Because the direction of the decision axis is in the same direction as the variables X2 "The requirement of the design of new business processes" and X4 "The requirement of collaboration between ICT systems of multiple DGs or institutions/organizations", it can be expected that the PROMETHEE II ranked actions to be stronger on this variable and potentially weaker on variables X8 "The imposition of authentication requirements by the legislation" and X11 "The requirement of new interoperability specifications". COM(2010)553 "Regional Policy contributing to smart growth in Europe 2020" and Reg.1300/2013 "Cohesion Fund and repealing Council Regulation (EC) No 1084/2006" are very close to each other and they have similar actions, whereas the other policies are in the opposite direction. It can be concluded that they are different from other actions.

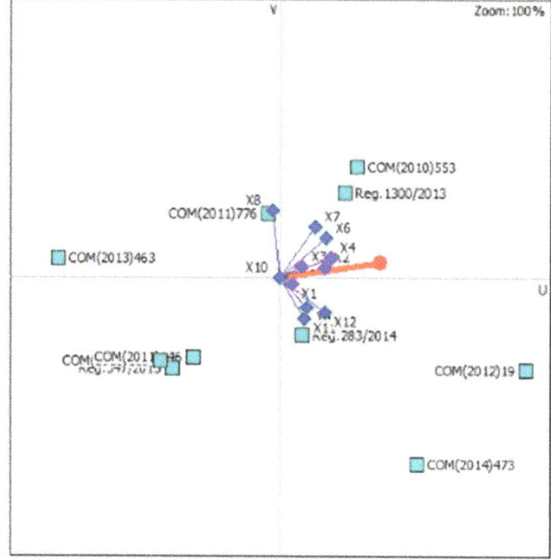

Figure 4. PROMETHEE GAIA plane.

4. Discussion

Research into the official EU website retrieved 50 regional policies, while 19 out of them have been established since 2010 (when the European Commission began to assess the ICT implications of EU legislation). The fulfillment of these 12 criteria, used by the European Commission, was studied in order to depict the impact of ICT on EU sustainable regional policy. Multicriteria decision analysis and GAIA analysis, are presented to identify the impact of ICT on EU sustainable regional policies.

The results and conclusions are summarized as follows:

- PROMETHEE II analysis shows that ICT has the strongest impact on COM(2012)19 "Waste Electrical and Electronic Equipment (WEEE)", while Reg.2015/207 "The models for the progress report, submission of the information on a major project, the joint action plan, the implementation reports for the Investment for growth and jobs goal, the management declaration, the audit strategy, the audit opinion and the annual control report and the methodology for carrying out the cost- benefit analysis" is ranked in the second position.

- The same analysis shows that ICT has the weakest impact on COM(2015)118 "The Agreement on the European Economic Area".

- Most EU sustainable regional policies adopt ICT solutions, as most of them contribute positively to their total net flows (Phi).

- The large spectrum of the values of the total net flows (Phi) shows a great difference concerning "superiority" between the first and the last EU regional policies, according to the impact of ICT.

- GAIA analysis depicts that COM(2012)19 has the strongest impact of ICT and COM(2015)118 has the weakest impact of ICT.

- MCDA and GAIA analyses provide similar results in terms of scenario ranking.

- "The requirement of the design of new business processes" and "The requirement of collaboration between ICT systems of multiple DGs or institutions/organizations" are found to be the most robust criteria in the PROMETHEE II ranking.

- "The imposition of authentication requirements by the legislation" and "The requirement of new interoperability specifications" are found to be the weakest criteria.

As most EU sustainable regional policies present positive total net flows, this research confirms that the impact of ICT on EU sustainable regional policy has been getting stronger since 2010. The applied methodology constitutes an efficient planning tool at EU level for the policy makers for the assessment of sustainable regional policies, based on the impact of ICT. Furthermore, the findings of this research can be a supportive tool for the policy makers, as the superior EU regional policies can be used as benchmarks for future policies in terms of sustainable development. ICT adoption shows an enormous potential for accelerating the progress towards SDGs (Sustainable Development Goals), while at the same time it can improve the quality of life of people in fundamental ways [51]. However, it would be very interesting to apply different methods for multiple criteria decision-making because of some disadvantages of PROMETHEE method, such as the paradigm of the underlying method, the determination of the weights and the rank reversal problem [52].

Author Contributions: C.K. and Z.A. conceived of the presented idea. C.K. contributed to the design and the implementation of the research, to the analysis of the results and to the xriting of the manuscript. All authors have read and agreed to the published version of the manuscript.

Funding: This research received no external funding.

Conflicts of Interest: The authors declare no conflict of interest.

Appendix A

Table A1. Sustainable regional policies.

	Code	Title
1	Reg.1083/2006	General provisions on the European Regional Development Fund, the European Social Fund and the Cohesion Fund.
2	Reg.1081/2006	European Social Fund.
3	Reg.1082/2006	A European grouping of territorial cooperation (EGTC).
4	Reg.1290/2005	Financing of the common agricultural policy.
5	Reg.2012/2002	Establishing the European Union Solidarity Fund.
6	Reg.1301/2013	The European Regional Development Fund and on specific provisions concerning the Investment for growth and jobs goal.
7	Reg.1304/2013	The European Social Fund and repealing Council Regulation (EC) No 1081/2006.
8	Reg.1299/2013	Specific provisions for the support from the European Regional Development Fund to the European territorial cooperation goal.
9	Reg.1303/2013	Common provisions on the European Regional Development Fund, the European Social Fund, the Cohesion Fund, the European Agricultural Fund for Rural Development and the European Maritime and Fisheries Fund and laying down general provisions on the European Regional Development Fund, the European Social Fund, the Cohesion Fund and the European Maritime and Fisheries Fund.
10	Reg.1300/2013	Cohesion Fund and repealing Council Regulation (EC) No 1084/2006.
11	Reg.347/2013	Guidelines for trans-European energy infrastructure.
12	Reg.283/2014	Guidelines for trans-European networks in the area of telecommunications infrastructure.
13	Reg.1828/2006	Rules for the implementation of Council Regulation (EC) No 1083/2006 laying down general provisions on the European Regional Development Fund, the European Social Fund and the Cohesion Fund and of Regulation (EC) No 1080/2006 of the European Parliament and of the Council on the European Regional Development Fund.

Table A1. *Cont.*

	Code	Title
14	Reg.1445/2007	Common rules for the provision of basic information on Purchasing Power Parities and for their calculation and dissemination.
15	Reg.1059/2003	Establishment of a common classification of territorial units for statistics (NUTS).
16	Reg.231/2014	Establishing an Instrument for Pre-accession Assistance (IPA II).
17	Reg.1085/2006	Establishing an Instrument for Pre-Accession Assistance (IPA).
18	Reg.240/2014	The European code of conduct on partnership in the framework of the European Structural and Investment Funds.
19	Reg.2015/207	The models for the progress report, submission of the information on a major project, the joint action plan, the implementation reports for the Investment for growth and jobs goal, the management declaration, the audit strategy, the audit opinion and the annual control report and the methodology for carrying out the cost-benefit analysis.
20	COM(2014)473	Sixth report on economic, social and territorial cohesion: investment for jobs and growth.
21	COM(2009)295	Classification of certain goods in the Combined Nomenclature.
22	COM(2008)371	Fifth progress report on economic and social cohesion Growing regions, growing Europe.
23	COM(2007)273	Fourth progress report on economic and social cohesion Growing regions, growing Europe.
24	COM (2004)107	Arsenic, cadmium, mercury, nickel and polycyclic aromatic hydrocarbons in ambient air.
25	COM(2006)281	The Growth and Jobs Strategy and the Reform of European cohesion policy Fourth progress report on cohesion.
26	COM(2005)192	Third progress report on cohesion: Towards a new partnership for growth, jobs and cohesion.
27	COM(2007)798	Member States and Regions delivering the Lisbon strategy for growth and jobs through EU cohesion policy, 2007–2013.
28	COM(2008)876	Cohesion Policy: investing in the real economy.
29	COM(2008)616	Green Paper on Territorial Cohesion: Turning Territorial Diversity into Strength.
30	COM(2006)30	Time to move up a gear: The new partnership for growth and jobs.
31	COM(2003)690	A European initiative for growth, investing in networks and knowledge for growth and jobs. Final Report to the European Council.
32	COM(2014)490	The urban dimension of EU policies–Key features of an EU urban agenda.
33	COM(2010)553	Regional Policy contributing to smart growth in Europe 2020.
34	COM(2006)675	A contribution by the Swiss Confederation to the European Union military operation in support of the United Nations Organization Mission in the Democratic Republic of the Congo (MONUC) during the election process (Operation EUFOR RD Congo).
35	COM(2006)385	Cohesion Policy and cities: the urban contribution to growth and jobs in the regions.
36	COM (2009)103	Insurance against civil liability in respect of the use of motor vehicles, and the enforcement of the obligation to insure against such liability.
37	COM(2015)118	The Agreement on the European Economic Area.
38	COM(1999)54	Amending Council Directive 66/402/EEC on the marketing of cereal seed.
39	COM(97)172	Amending the boundaries of the less-favored areas in the Federal Republic of Germany within the meaning of Council Directive 75/268/EEC.
40	COM(2002)709	A framework for target-based tripartite contracts and agreements between the Community, the States and regional and local authorities.
41	COM(2003)585	Structural indicator.
42	COM(2003)811	Dialogue with associations of regional and local authorities.

Table A1. *Cont.*

	Code	Title
43	COM(2011)146	Reform of the EU State Aid Rules on Services of General Economic Interest.
44	COM(2012)19	Waste Electrical and Electronic Equipment (WEEE).
45	COM(2014)494	Guidelines on the application of the measures linking effectiveness of the European Structural and Investment Funds to sound economic governance according to Article 23 of Regulation (EU) 1303/2013.
46	COM(2011)776	Seventh progress report on economic, social and territorial cohesion.
47	COM(2013)463	Eighth progress report on economic, social and territorial cohesion. The regional and urban dimension of the crisis.
48	Dec.2006/702	Community strategic guidelines on cohesion.
49	Dec.1336/97	A series of guidelines for trans-European telecommunications networks.
50	Direct.2008/57	The interoperability of the rail system within the Community (Recast).

References

1. Hilty, L.M.; Seifert, E.K.; Treibert, R. *Information Systems for Sustainable Development*; IGI Global: Hershey, PA, USA, 2005.
2. Mohamed, M.; Murray, A.; Mohamed, M. The role of information and communication technology (ICT) in mobilization of sustainable development knowledge: A quantitative evaluation. *J. Knowl. Manag.* **2010**, *14*, 744–758. [CrossRef]
3. Hilty, L.M. *Information Technology and Sustainability: Essays on the Relationship between Information Technology and Sustainable Development*; BoD–Books on Demand: Norderstedt, Germany, 2011.
4. Andreopoulou, Z.; Samathrakis, V.; Louca, S.; Vlachopoulou, M. *E-Innovation for Sustainable Development of Rural Resources during Global Economic Crisis*; IGI Global: Hershey, PA, USA, 2013.
5. Koliouska, C.; Andreopoulou, Z.; Zopounidis, C.; Lemonakis, C. E-commerce in the Context of Protected Areas Development: A Managerial Perspective Under a Multi-Criteria Approach. In *Multiple Criteria Decision Making*; Springer International Publishing: Cham, Switzerland, 2017; pp. 99–111.
6. Tongia, R.; Subrahmanian, E.; Arunachalam, V. *Information and Communications Technology for Sustainable Development: Defining a Global Research Agenda*; No. Information and Communications Technology for Sustainable Development: Defining a Global Research Agenda; Allied Publishers Pvt. Ltd.: Bangalore, India, 2005.
7. Yao, X.; Watanabe, C.; Li, Y. Institutional structure of sustainable development in BRICs: Focusing on ICT utilization. *Technol. Soc.* **2009**, *31*, 9–28. [CrossRef]
8. Armenta, Á.; Serrano, A.; Cabrera, M.; Conte, R. The new digital divide: The confluence of broadband penetration, sustainable development, technology adoption and community participation. *Inf. Technol. Dev.* **2012**, *18*, 345–353. [CrossRef]
9. Mehta, S.; Kalra, M. Information and Communication Technologies: A bridge for social equity and sustainable development in India. *Int. Inf. Libr. Rev.* **2006**, *38*, 147–160. [CrossRef]
10. Pade, C.I.; Mallinson, B.; Sewry, D. An exploration of the categories associated with ICT project sustainability in rural areas of developing countries: A case study of the Dwesa project. In *Proceedings of the 2006 Annual Research Conference of the South African Institute of Computer Scientists and Information Technologists on IT Research in Developing Countries*; South African Institute for Computer Scientists and Information Technologists: Pretoria, South Africa, 2006; pp. 100–106.
11. Koliouska, C.; Andreopoulou, Z. Classification of ICT in EU Environmental Strategies. *J. Environ. Prot. Ecol.* **2016**, *17*, 1385–1392.
12. Benedict, M.A.; McMahon, E.T. Green infrastructure: Smart conservation for the 21st century. *Renew. Resour. J.* **2002**, *20*, 12–17.
13. Koliouska, C.; Andreopoulou, Z. Exploring the use of smart services in Forestry. *J. Environ. Prot. Ecol.* **2019**, *20*, 1434–1439.

14. Ahmedi, F.; Ahmedi, L.; O'Flynn, B.; Kurti, A.; Tahirsylaj, S.; Bytyçi, E.; Salihu, A. InWaterSense: An Intelligent Wireless Sensor Network for Monitoring Surface Water Quality to a River in Kosovo. *Int. J. Agric. Environ. Inform.* **2018**, *9*, 39–61. [CrossRef]

15. Khosravi, M.R.; Rostami, H.; Samadi, S. Enhancing the Binary Watermark-Based Data Hiding Scheme Using an Interpolation-Based Approach for Optical Remote Sensing Images. *Int. J. Agric. Environ. Inf. Syst.* **2018**, *9*, 53–71. [CrossRef]

16. Andreopoulou, Z.S.; Manos, B.; Polman, N.; Viaggi, D. *Agricultural and Environmental Informatics, Governance, and Management: Emerging Research Applications*; IGI Global: Hershey, PA, USA, 2011.

17. Andreopoulou, Z.; Koliouska, C.; Galariotis, E.; Zopounidis, C. Renewable energy sources: Using PROMETHEE II for ranking websites to support market opportunities. *Technol. Forecast. Soc. Chang.* **2018**, *131*, 31–37. [CrossRef]

18. Calzada-Infante, L.; Lopez-Narbona, A.M.; Nunez-Elvira, A.; Orozco-Messan, J. Assessing the Efficiency of Sustainable Cities Using an Empirical Approach. *Sustainability* **2020**, *12*, 2618. [CrossRef]

19. Meadowcroft, J. Sustainable development: A new (ish) idea for a new century? *Political Stud.* **2000**, *48*, 370–387. [CrossRef]

20. Grin, J.; Rotmans, J.; Schot, J. *Transitions to Sustainable Development: New Directions in the Study of Long Term Transformative Change*; Routledge: Abingdon-on-Thames, UK, 2010.

21. Elliott, J.A. *An Introduction to Sustainable Development*; Routledge: Abingdon-on-Thames, UK, 2012.

22. Kaufmann, A.; Wagner, P. EU regional policy and the stimulation of innovation: The role of the European Regional Development Fund in the objective 1 region Burgenland. *Eur. Plan. Stud.* **2005**, *13*, 581–599. [CrossRef]

23. Dohse, D.; Fornahl, D.; Vehrke, J. Fostering place-based innovation and internationalization–the new turn in German technology policy. *Eur. Plan. Stud.* **2018**, *26*, 1137–1159. [CrossRef]

24. Koliouska, C.; Andreopoulou, Z. The Typology for ICT adoption of EU environmental policies. *Wseas Trans. Comput.* **2019**, *18*, 46–55.

25. Koliouska, C.; Andreopoulou, Z.; Misso, R.; Borelli, I.P. Regional sustainability: National Forest Parks in Greece. *Int. J. Agric. Environ. Inf. Syst.* **2017**, *8*, 29–40. [CrossRef]

26. Barnett, M.L.; Henriques, I.; Husted, B.W. Governing the void between stakeholder management and sustainability. *Adv. Strateg. Manag.* **2018**, *38*, 121–143.

27. Andreopoulou, Z.; Koliouska, C. Benchmarking Internet Promotion of Renewable Energy Enterprises: Is Sustainability Present? *Sustainability* **2018**, *10*, 4187. [CrossRef]

28. European Commission. The EU's Main Investment Policy. 2015. Available online: http://ec.europa.eu/regional_policy/en/policy (accessed on 7 February 2020).

29. European Union. Regional Policy. 2018. Available online: https://europa.eu/european-union/topics/regional-policy_en (accessed on 7 February 2020).

30. MLIT. An Overview of Spatial Policy in Asian and European Countries. 2016. Available online: http://www.mlit.go.jp/ (accessed on 7 February 2020).

31. Michie, R.; Fitzgerald, R. The evolution of the structural funds. In *The Coherence of EU Regional Policy: Contrasting Perspectives on the Structural Funds*; Bachtler, J., Turok, I., Eds.; Routledge: Abingdon-on-Thames, UK, 2013.

32. Interreg Europe. Perception and Awareness of EU Regional Policy on the Rise. 2017. Available online: https://www.interregeurope.eu/ (accessed on 7 February 2020).

33. McCann, P.; Ortega-Argilés, R. Smart specialisation, entrepreneurship and SMEs: Issues and challenges for a results- oriented EU regional policy. *Small Bus. Econ.* **2016**, *46*, 537–552. [CrossRef]

34. Alegre, J.G. An evaluation of EU regional policy. Do structural actions crowd out public spending? *Public Choice* **2012**, *151*, 1–21. [CrossRef]

35. Becker, J.; Fuest, C. EU regional policy and tax competition. *Eur. Econ. Rev.* **2010**, *54*, 150–161. [CrossRef]

36. Baun, M. EU regional policy and the candidate states: Poland and the Czech Republic. *J. Eur. Integr.* **2002**, *24*, 261–280. [CrossRef]

37. Funck, B.; Pizzati, L. *European Integration, Regional Policy and Growth*; The World Bank: Washington, DC, USA, 2003.

38. Hart, M. Evaluating EU regional policy: How might we understand the causal connections between interventions and outcomes more effectively? *Policy Stud.* **2007**, *28*, 295–308. [CrossRef]

39. Bachtler, J.; Mendez, C. *EU Cohesion Policy and European Integration: The Dynamics of EU Budget and Regional Policy Reform*; Routledge: Abingdon-on-Thames, UK, 2016.
40. European Commission. *Method for Assessing ICT Implications of EU Legislation*. 2010. Available online: http://ec.europa.eu/idabc/servlets/Doc792e.pdf?id=32704 (accessed on 7 February 2020).
41. Koliouska, C.; Andreopoulou, Z.; Golumbeanu, M. The Contribution of ICT in EU Development Policy: A Multicriteria Approach. In *Advances in Operational Research in the Balkans*; Springer: Cham, Switzerland, 2020; pp. 111–123.
42. Mareschal, B.; De Smet, Y. Visual PROMETHEE: Developments of the PROMETHEE & GAIA multicriteria decision aid methods. In *Industrial Engineering and Engineering Management*; IEEM: Beijing, China, 2009; pp. 1646–1649.
43. Brans, J.P.; Vincke, P.; Mareschal, B. How to select and how to rank projects: The PROMETHEE method. *Eur. J. Oper. Res.* **1986**, *24*, 228–238. [CrossRef]
44. Yu, X.; Xu, Z.; Ma, Y. Prioritized multi-criteria decision making based on the idea of PROMETHEE. *Procedia Comput. Sci.* **2013**, *17*, 449–456. [CrossRef]
45. Tsolaki-Fiaka, S.; Bathrellos, G.D.; Skilodimou, H.D. Multi-Criteria Decision Analysis for an Abandoned Quarry in the Evros Region (NE Greece). *Land* **2018**, *7*, 43. [CrossRef]
46. Brans, J.P.; Mareschal, B. PROMETHEE methods. In *Multiple Criteria Decision Analysis: State of the Art Surveys*; Springer: New York, NY, USA, 2005; pp. 163–186.
47. Andreopoulou, Z.; Koliouska, C.; Zopounidis, C. *Multicriteria and Clustering: Classification Techniques in Agrifood and Environment*; Springer: Cham, Switzerland, 2017.
48. Macharis, C.; Brans, J.P.; Mareschal, B. The GDSS promethee procedure. *J. Decis. Syst.* **1998**, *7*, 283–307.
49. Mareschal, B. *Visual PROMETHEE 1.4 Manual*; Visual PROMETHEE: Brussels, Belgium, 2013; p. 1.
50. Zopounidis, C. *Analysis of Financing Decisions with Multiple Criteria*; Anikoula Publications: Thessaloniki, Greece, 2001.
51. International Telecommunications Union. ICTs for a Sustainable World. 2018. Available online: https://www.itu.int (accessed on 7 February 2020).
52. Macharis, C.; Springael, J.; De Brucker, K.; Verbeke, A. PROMETHEE and AHP: The design of operational synergies in multicriteria analysis: Strengthening PROMETHEE with ideas of AHP. *Eur. J. Oper. Res.* **2004**, *153*, 307–317. [CrossRef]

 *sustainability*

Article

Efficiency Enhancement of Gas Turbine Systems with Air Injection Driven by Natural Gas Turboexpanders

Ali Rafiei Sefiddashti [1], Reza Shirmohammadi [2,3,*] and Fontina Petrakopoulou [3]

[1] Faculty of Mechanical Engineering, Babol Noshirvani University of Technology, Babol 47148-71167, Iran; alirafieisd@gmail.com

[2] Department of Renewable Energies and Environment, Faculty of New Sciences & Technologies, University of Tehran, Tehran 14174-66191, Iran

[3] Department of Thermal and Fluid Engineering, University Carlos III of Madrid, 28911 Madrid, Spain; fpetrako@ing.uc3m.es

* Correspondence: r.shirmohammadi1987@gmail.com or r.shirmohammadi@ut.ac.ir or rshirmoh@uc3m.es

Abstract: The fuel source of many simple and combined-cycle power plants usually comes from a nearby natural gas transmission pipeline at a pressure from 50 to over 70 bar. The use of a turboexpander instead of throttling equipment offers a promising alternative to regulate the pressure of natural gas introduced to the power plant. Specifically, it helps recover part of the available energy of the compressed gas in the transmission pipeline, increase the power output and efficiency of the gas turbine system, and decrease the fuel use and harmful emissions. In this paper, the addition of such a turboexpander in a gas pressure-reduction station is studied. The recovered power is then used to drive the compression of extra air added to the combustion chamber of a heavy-duty gas turbine. The performance of this configuration is analyzed for a wide range of ambient temperatures using energy and exergy analyses. Fuel energy recovered in this way increases the output power and the efficiency of the gas turbine system by a minimum of 2.5 MW and 0.25%, respectively. The exergy efficiency of the gas turbine system increases by approximately 0.36% and the annual CO_2 emissions decrease by 1.3% per MW.

Keywords: gas turbine; air injection; turboexpander; performance enhancement; emission reduction

Citation: Sefiddashti, A.R.; Shirmohammadi, R.; Petrakopoulou, F. Efficiency Enhancement of Gas Turbine Systems with Air Injection Driven by Natural Gas Turboexpanders. *Sustainability* **2021**, *13*, 10994. https://doi.org/10.3390/su131910994

Academic Editors: Georgios Tsantopoulos and Evangelia Karasmanaki

Received: 27 August 2021
Accepted: 27 September 2021
Published: 3 October 2021

Publisher's Note: MDPI stays neutral with regard to jurisdictional claims in published maps and institutional affiliations.

1. Introduction

A prominent technology today for the energy conversion of fossil fuels, such as natural gas (NG) and oil, are the gas turbine systems. These machines reach high energy conversion efficiencies due to technological progress and advanced materials in their design and construction. Nevertheless, the associated environmental impact of these machines plays a key role in climate change, highlighting the necessity of energy efficiency improvement policy in power plants and energy policies overall [1]. Currently, such policies motivate governments to improve the efficiency of gas turbines by further recuperating thermal energy from the exhaust gasses to produce steam and drive a steam turbine [2]. However, this kind of relatively high-investment cost of solutions force private companies to seek cheaper solutions [3,4].

In a simple gas turbine system, the temperature and pressure of the ambient air increases by passing through the compressor. After mixing with the fuel and the ignition, the high-pressure combustion products reach the highest operating temperature. The hot combustion product (gases) is expanded in the turbine, moving the rotating blades, and consequently rotating the turbine shaft to provide power for rotating the compressor and the generator [5]. The amount of required power for the compressor depends on the inlet volumetric flow of the air; more power is required to compress the same mass flow of air of lower density to a given outlet pressure.

A means to decrease the inlet air temperature and boost the turbine output recommended by most gas turbine manufacturers is the use of cooling equipment. Cooling equipment includes evaporative coolers, fogging, and chillers that significantly increase the capital cost of the plant. Although cooling systems improve operation, their efficacy is highly dependent on ambient temperature [6] and humidity. Steam injection into the combustion chamber for power enhancement is another method, but it requires large quantities of demineralized water, and is linked to combustion and other operational challenges.

Another measure to increase the generated power of gas turbine systems is the compressed air injection (CAI), i.e., the injection of additional pressurized air into the combustion chamber or at the compressor outlet. This additional air flow requires then more fuel to maintain the inlet temperature of the expander. Nevertheless, in such applications, the fuel increase pales in comparison to the gas turbine power increase. The significant power increase is due to the higher mass flow in the turbine, and consequently, the increased work generated in comparison to the compressor's required work. This leads to an overall enhancement of the thermal efficiency of the gas turbine. Nakhamkin et al. [7] proposed injecting compressed air in a highly efficient electrically driven compressor upstream of the combustion chamber. The air can be injected through the ports of steam injection that are already available in some commercial gas turbines. CAI also helps to increase the lifetime of the gas turbine by reducing the inlet temperature of the turbine without a reduction in the power generation. Akita et al. showed that the reduction of firing temperature with air injection by approximately 110 °C increases the maintenance intervals and reduces the maintenance costs by a factor of two in both cases [8]. Typically, up to 10% of a gas turbine's airflow at ISO conditions (temperature = 15 °C, relative humidity = 60%, and pressure = 101.3 kPa) can be used for injection purposes. However, avoiding compressor surge and the torque limit of the shaft restrict the maximum retrieved air at any given ambient temperature. An electrical motor or an efficient reciprocating engine may drive an intercooled compressor that compresses ambient air and adds it to the compressor outlet [9]. Internal combustion engines are less sensitive to temperature and humidity, maintaining their nominal power output and efficiency over a broader range of ambient conditions. Hence, some companies designed a series of standardized building block modules which can be connected together to operate at high injection air flows [10]. Combined diesel-engine gas turbine systems enable distributed power generation plants to attain high thermal efficiencies while enjoying the operational advantages of both diesel engines and gas turbines [11]. Abudu et al. evaluated the implication of the steady-state injection of compressed air into two multi-spool gas turbines for power enhancement. The steady-state analysis demonstrated that with an 8% flow injection, a power increase of at least 16% is obtained [12]. Gas turbines also play a key role in synchronous power generation and back-up systems for intermittent renewable systems. Igie et al. [13] considered the extraction of compressed air from a single-shaft gas turbine to store energy when surplus power is available and then the reinjection of the pressurized air at peak demand. CAI can constitute thus an alternative solution for energy storage, required by most renewable power sources.

Although a wide range of fuels can be used in gas turbines, compressed natural gas is the most common fuel used. Natural gas is transported through pipelines over long distances. The pressure of the natural gas must be significantly decreased before it is supplied to the combustion chamber of the gas turbine system. The pressure reduction of the natural gas that usually occurs in throttling valves is accompanied by substantial energy and exergy losses [14].

Today, many researchers study energy recovery devices for the decompression of high-pressure natural gas. The amount of energy that can be recovered depends on various parameters including both operating conditions (pressure difference, temperature, and mass flow) and design parameters (efficiency, capacity, performance map, etc.) [15–19]. Furthermore, the quality of NG (in terms of hydrate formation) is also crucial [14]. Many authors, such as Morgese et al. [20], propose an optimization design procedure of a turboexpander by considering fluid dynamic and technical requirements. Recovery of waste energy of the gas stations can also be used for both producing power and freshwater with a potentially substantial effect on the reduction of greenhouse gases and air emissions [21]. Golchoobian et al. [22] investigated the feasibility of using a turboexpander coupled with a refrigeration cycle to decrease the inlet temperature of air and increase the generated power. Although many studies evaluate waste energy recovery from pressure-reducing stations and air injection into the combustion chamber separately, the combined use of waste energy to inject air into gas turbine combustion chambers is still missing. This paper aims to address this research gap with energy and exergy analyses of a hybrid system of a gas turbine including a natural gas turboexpander and air injection for performance enhancement. Lastly, since the capacity and operating conditions of pressure-reducing stations in power plants vary moderately, important parameters and their effects are studied in this work as well.

2. Process Description

The pressure-reducing station is the endpoint of the natural gas transmission system. There, the pressure of the delivered gas is decreased to the final domestic or industrial consumer [23]. These stations have usually two or three parallel pressure regulator lines to provide redundancy in case of changing filters and for safety purposes. There are several pieces of equipment on each uniform line but their arrangement or configuration in each station may change based on ambient and operating conditions. The common elements of all stations, and probably the most important, are the control or reduction valves that maintain the pressure downstream of the station constant. In some stations due to the ambient conditions or the high-pressure reduction ratio, a heating element, such as a bath heater, is provided to reduce the risk of hydrate formation from the Joule–Thomson effect.

Currently, a commercially available alternative technology to throttling is axial or radial turbines (also called turboexpanders) coupled with a generator to convert mechanical energy into electrical energy. Figure 1a shows the schematic placement of a turboexpander unit in the bypass line which can be isolated by two shut-off valves. In some arrangements, it is necessary to preheat the high-pressure gas because of the throttling process before the entrance of the turbine.

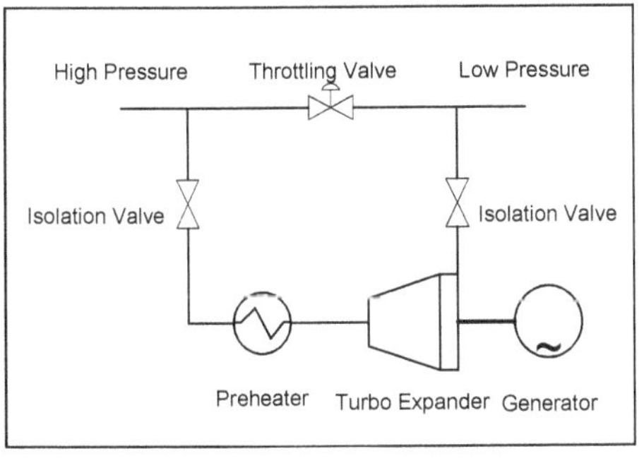

(a)

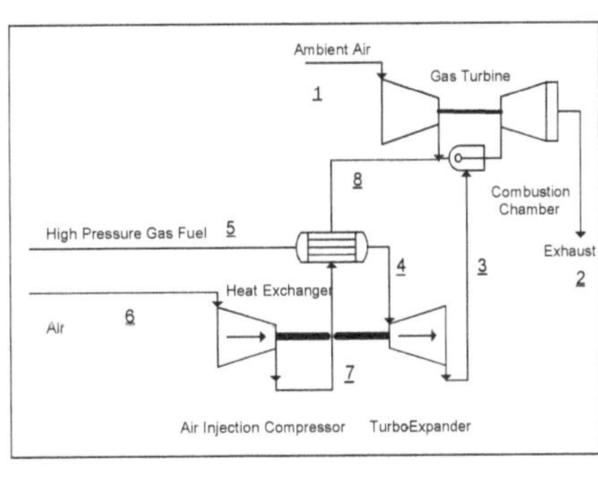

(b)

Figure 1. (**a**) Typical schematic view of a turboexpander arrangement in pressure-reducing stations; (**b**) schematic view of the proposed hybrid system.

In this work, a novel cycle for the arrangement of a turboexpander in a gas station of simple and combined-cycle power plants is proposed as illustrated in Figure 1b. The turboexpander is connected to a compressor that compresses ambient air. The air that is led to the combustion chamber of the gas turbine system after it is passed through a heat exchanger to preheat the high-pressure gas. This reduces the risk of hydrate formation in the turboexpander. Injecting extra air to the combustion chamber increases the mass flow of the GT and produces more power. Moreover, using this arrangement eliminates the need for a natural gas bath heater and generator, increases the plant efficiency, and decreases the plant's capital investment in comparison to individual turboexpander energy-recovery systems.

3. Methodology

To determine the effects of the air-injection system on the performance of the chosen gas turbine, a computer code was developed in the engineering equation solver (EES). The calculation procedure of the EES code is summarized in Figure 2. This code calculates the thermodynamic properties and off-design performance of the gas turbine with and without the high-pressure injection system. Another model was simulated using the Thermoflex software to validate the in-house code results. Operational compatibility between the turbine and the compressor of the gas turbine (matching calculations) depends on mass flow compatibility, pressure ratio (work), and rotational speed [24]. The characteristic curves of mass flow, pressure ratio, and efficiency with rotational speed of the compressor, turbine, and combustion chamber were obtained for the gas turbine model V94.2. It should be noted that the demonstrated flow chart has been developed on the assumption that the turbine inlet temperature (TIT) remains constant. This assumption depends on the control system mode of the gas turbine, and it can be adjusted for other GT control modes. Considering constant TIT and compatibility of speed and flow for a single-shaft machine, the pressure ratio and other performance characteristics of the gas turbine were determined.

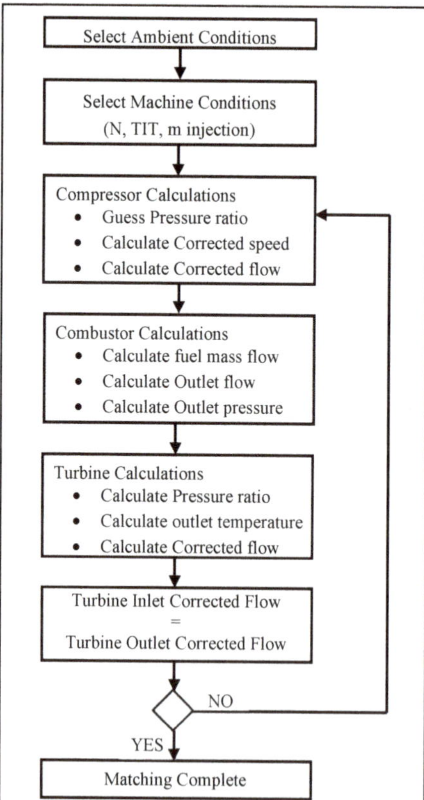

Figure 2. Procedure of component matching of the gas turbine system.

In this study, the effect of the proposed air-injection system was studied on the heavy-duty gas turbine of Siemens V94.2, a model widely used in power plants. V94.2 is a single-shaft gas turbine with a rated power of 162 MW. This turbine incorporates a 16-stage compressor, two large silo-type combustion chambers, and a four-stage turbine. Performance data (including the compressor and turbine data) for the simulation were found in various references and official original equipment manufacturer (OEM) websites of Siemens and Alstom [25–28]. The design performance characteristics of the gas turbine are presented in Table 1. Calculated performance parameters with the EES code, including power and efficiency at various ambient temperatures, agree with published OEM data with an accuracy of more than 98%.

Table 1. Design performance characteristics of the simulated GT.

Parameter	Value	Unit
Frequency	50	Hz
Gross power output	161.7	MW
Gross efficiency	34.8	%
Heat rate	10,350	kJ/kWh
Exhaust temperature	542	°C
Exhaust mass flow	518	kg/s
Pressure ratio	11.8	-
Fuel mass flow	9.47	kg/s

The pressure ratio of the gas turbine is a function of the compressor pressure ratio and the pressure drop within the combustor.

$$P_{03}/P_{02} = 1 - \Delta P_{CC}/P_{02} \tag{1}$$

where P_{02} and P_{03} are the pressures at the inlet and outlet of the combustor, respectively. The mass flow that passes through the turbine is equal to the outlet mass flow of the compressor plus the fuel flow and the additional compressed air:

$$\dot{m}_3 = \dot{m}_1 + \dot{m}_F + \dot{m}_{inj}. \tag{2}$$

The turbine and compressor shafts were coupled together to assure compatible rotational speed.

$$\frac{N}{\sqrt{T_{03}}} = \frac{N}{\sqrt{T_{01}}} \times \frac{\sqrt{T_{01}}}{\sqrt{T_{03}}}. \tag{3}$$

In most gas turbines, the TIT is constant during the operation due to metallurgical limitations. Although there are various definitions and positions to measure the TIT (T_{03}), in this study it was considered constant so that for given ambient conditions, the square root of the temperature ratio was constant as well. Moreover, the non-dimensional flow term expresses the compatibility of the flow between the compressor and the turbine as follows:

$$\frac{\dot{m}_3\sqrt{T_{03}}}{P_{03}} = \frac{\dot{m}_1\sqrt{T_{01}}}{P_{01}} \times \frac{P_{01}}{P_{02}} \times \frac{P_{02}}{P_{03}} \times \frac{\sqrt{T_{03}}}{\sqrt{T_{01}}} \times \frac{\dot{m}_3}{\dot{m}_1} \tag{4}$$

where $\dot{m}_1$ is the inlet mass flow of the compressor, $\dot{m}_3$ is the inlet mass flow of the turbine, T_{01} is the ambient temperature at the inlet of the compressor, P_{01} is the pressures at the inlet of the compressor. The adiabatic work of the compressor can be calculated with Equation (5):

$$W_{compressor} = \dot{m}_1 \times Cp_A \times (T_{02} - T_{01}) \tag{5}$$

where Cp_A is the specific heat capacity of the air at constant pressure and T_{02} is the outlet temperature of the compressor.

The actual compressor outlet temperature (T_{02}), considering its isentropic efficiency (η_c), can be estimated with the following equation:

$$T_{02} = T_{01} + \frac{T_{01}}{\eta_c}\left(\left(\frac{P_{02}}{P_{01}}\right)^{\frac{\gamma-1}{\gamma}} - 1\right)$$

(6)

where η_c is the compressor efficiency and γ the air-specific heat ratio.

The outlet pressure of the combustion chamber (P_{03}) is also calculated from the compressor's delivery pressure (P_{02}) and the pressure drop of the air in the combustor (ΔP_{CC}). For most available combustors it is in the range of 0.03–0.05 of the inlet pressure [29].

$$P_{03} = P_{02} - \Delta P_{CC}$$

(7)

With constant blade dimensions and negligible changes in efficiency, higher inlet mass flow will lead to an off-design operation of the turbine. A similar equation to the compression process is used for the calculation of the turbine's expansion work (W_T) by considering the total mass flow of the gas calculated with Equation (2):

$$W_T = \dot{m}_3 \times Cp_G \times (T_{04} - T_{03})$$

(8)

where Cp_G. is the specific heat of the exhaust gas and T_{04} is the temperature at the outlet of the turbine.

The actual turbine outlet temperature (T_{04}). can be estimated with Equation (9), considering the isentropic efficiency of the turbine (η_T):

$$T_{04} = T_{03} - \eta_T \times T_{03}\left(1 - \left(\frac{P_{04}}{P_{03}}\right)^{\frac{\gamma-1}{\gamma}}\right)$$

(9)

The net or useful work of the gas turbine can be obtained by subtracting the consumed work of the compressor from the produced work of the turbine.

$$W_{Net} = \eta_m \times W_T - W_c.$$

(10)

where η_m is the mechanical efficiency of the gas turbine.

Similar equations can be used to determine the mass flow of the additional compressed air in the turboexpander at different conditions.

$$\eta_m \times W_{TE} = W_{AIC}$$

(11)

where η_m. is the combined mechanical efficiency, W_{TE}. the produced work of the turboexpander, and W_{AIC} the shaft power of the air-injection compressor.

In this study, the effect of various parameters on the mass flow of injected air were investigated. The mass flow of the fuel that expands in the turboexpander plays a key role on the mass flow of the injected air. Based on OEM data of several industrial gas turbine models, up to 5% of the main gas turbine inlet flow can be injected into the combustion chamber safely [30]. Table 2 presents selected parameters used to model the proposed system.

Table 2. Considered assumptions for model calculation.

Parameter	Value	Unit
Heat exchanger effectiveness	90	%
Fuel gas pressure	40–70	bar
Fuel gas temperature	0–45	°C
Extra air compressor efficiency	90	%
Turboexpander efficiency	80	%
Ambient air relative humidity	60	%
Ambient air	15	°C
Ambient pressure	1.013	bar

To evaluate the environmental performance of the proposed system and compare it to that of a conventional system, the amount of generated CO_2, CO, and NOx have been calculated. The emitted CO_2 was calculated using the combustion and equilibrium reactions. Empirical relations proposed in [31] are used to determine the emission of CO and NOx, using adiabatic flame temperature in the primary zone of the combustion chamber as follows [32]:

$$\begin{cases} T_{ad} = A\sigma^{\alpha} \exp\left(\beta(\sigma+\lambda)^2\right)\pi^x\theta^y\psi^z \\ x = a_1 + b_1\sigma + c_1\sigma^2 \\ y = a_2 + b_2\sigma + c_2\sigma^2 \\ y = a_3 + b_3\sigma + c_3\sigma^2 \end{cases} \tag{12}$$

where θ is a dimensionless temperature, π is a dimensionless pressure, σ is the fuel to air equivalent ratio, and ψ is the H/C atomic ratio. Parameters A, α, β, λ, a_i, b_i, and c_i are constants, depending on σ. and θ, available in [33]. Accordingly, by using adiabatic flame temperature, the produced CO and NOx can be estimated based on the following empirical equations in grams per kilogram of fuel flow:

$$\dot{m}_{NOx} = \frac{1.5 \times 10^{15}\tau^{0.5}\exp(-7110/T_{ad})}{P_2^{0.05}(\Delta P_{cc}/P_2)^{0.5}} \tag{13}$$

$$\dot{m}_{CO} = \frac{0.179 \times 10^9 \exp(7800/T_{ad})}{P_2^2\tau(\Delta P_{cc}/P_2)^{0.5}}. \tag{14}$$

where P_2 is the pressure at the inlet of the combustor, ΔP_{cc} is the dimensionless pressure loss in the combustion chamber, and τ is the residence time in the combustion zone (considered constant at 0.02 s).

As mentioned, the validation of the energy model of the gas turbine with injection was carried out with the Thermoflow software—commercially available thermal engineering software for analyzing the performance of thermodynamic cycles. In this validation process, the total pressure loss of intake and exhaust were assumed to be 10 and 5 mbar, respectively. The air compressor was fed with power from the natural gas turboexpander. A schematic of the Thermoflow model is shown in Figure 3.

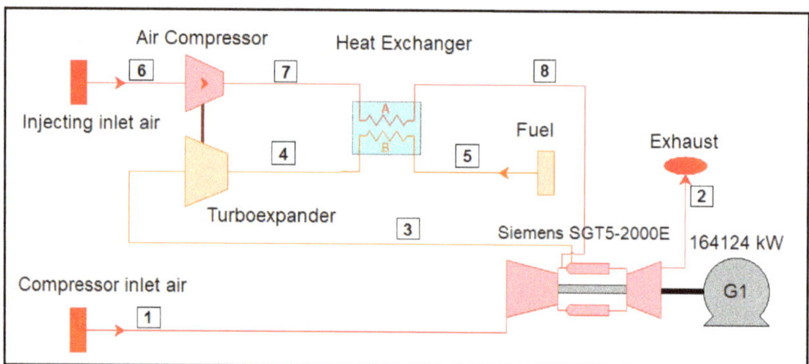

Figure 3. Thermoflow model for validating the EES code.

The validation results are reported in Table 3, where the gas turbine power and efficiency were calculated with and without CAI. It is seen that there is generally good agreement between the EES code and the Thermoflow results, with acceptable errors for both power and efficiency.

Table 3. Validation of the EES energy code.

Parameter	EES Code	Thermoflow	Error
Power w/o CAI	161.4 MW	161.8 MW	0.25%
Power with CAI	163.6 MW	164.1 MW	0.30%
Efficiency w/o CAI	34.45	34.51%	0.17%
Efficiency with CAI	34.62	34.7%	0.23%

By applying the laws of thermodynamics within component k, exergy destruction is obtained, which is a relation between the fuel and product exergy as follows [34]:

$$\dot{Ex}_{D,k} = \dot{Ex}_{F,k} - \dot{Ex}_{P,k} \tag{15}$$

where $\dot{Ex}_{F,k}$ and $\dot{Ex}_{P,k}$. are the fuel and product exergy of each component, respectively, and $\dot{Ex}_{D,k}$ is the exergy destruction within component k. Exergy loss is not defined at the component level, as it is only relevant for the overall process [35–37]. All exergy calculations of streams are based on the sum of chemical and physical exergies as follows [38]:

$$\dot{Ex} = \dot{Ex}^{PH} + \dot{Ex}^{CH} \tag{16}$$

The exergetic efficiency of each thermodynamic component is calculated as:

$$\varepsilon_{GT} = \frac{\dot{Ex}_{P,k}}{\dot{Ex}_{F,k}} = 1 - \frac{\dot{Ex}_{D,k}}{\dot{Ex}_{F,k}} \tag{17}$$

All components are analyzed based on their exergy destruction and exergy efficiency, determined by the definition of exergy of the fuel and exergy of the product of each component, as shown in Table 4.

Table 4. Definitions of the fuel and product exergy for each component.

Components	Fuel Exergy, $\dot{Ex}_{F,k}$	Product Exergy, $\dot{Ex}_{P,k}$
Gas Turbine	$\dot{Ex}_{F,GT} = \dot{Ex}_1 + \dot{Ex}_3 + \dot{Ex}_8 - \dot{Ex}_2$	$\dot{Ex}_{P,GT} = \dot{W}_{GT}$
Turbo Expander	$\dot{Ex}_{F,TE} = \dot{Ex}_4 - \dot{Ex}_3$	$\dot{Ex}_{P,TE} = \dot{W}_{TE}$
Air Injection Compressor	$\dot{Ex}_{F,AIC} = \dot{W}_{AIC}$	$\dot{Ex}_{P,AIC} = \dot{Ex}_6 - \dot{Ex}_7$
Heat Exchanger	$\dot{Ex}_{F,HX} = \dot{Ex}_8 - \dot{Ex}_7$	$\dot{Ex}_{P,HX} = \dot{Ex}_4 - \dot{Ex}_3$

The calculated values of the thermodynamic parameters and the total rate of exergy at various points of the system are shown in Figure 3 and listed in Table 5. To facilitate comparison, these values are shown for both systems with and without the turboexpander air-injection system.

Table 5. Thermodynamic data of the streams.

Stream	M (kg/s)	P (bar)	T (C)	h (kJ/kg)	s (kJ/kgC)	$\dot{Ex}$ (kW)
			Without turboexpander air injection			
1	512.7	1.013	15	−10.13	0.1435	1584.7
2	522	1.013	544.4	568.4	1.338	145,426
3	9.39	60	25	50047	−2.266	492,640
			With turboexpander air injection			
1	512.7	1.013	15	−10.13	0.1435	1584.7
2	525.8	1.013	543.6	567.6	1.337	146,236
3	9.45	17	16.62	50,029	−1.56	493,698
4	9.45	58.8	88.5	50193	−1.747	495,667
5	9.45	60	25	50,047	−2.266	495,560
6	3.65	1.013	15	−10.13	0.1435	11.34
7	3.65	17.1	417.3	407.7	0.232	1447
8	3.65	17	55	27.5	−0.5415	898

4. Results

In a single-shaft machine, the air injection does not affect rotational speed of the engine or the compressor airflow. With the inlet temperature and the fuel input of the turbine fixed, the cycle pressure ratio of the system must increase. Figure 4a depicts a simplistic interpretation of the effect of air injection on the T-s diagram of the Brayton cycle. This figure, based on a semi-perfect gas model, shows that one of the main effects of air injection is the increase of the pressure ratio. With fixed turbine blade design, higher flow rates through the combustor result in a higher turbine pressure ratio and for a fixed compressor inlet pressure, the compressor pressure ratio increases. Subsequently, with fixed turbine inlet temperature, the turbine work output and exhaust temperature increase. It should be mentioned that the power output and the efficiency of the gas turbine decreases with higher ambient temperature, due to the lower density and, subsequently, the lower compressor mass flow. Power and, to some extent, efficiency can be restored through the injection of compressed air because the work required by the turbocompressor is covered with the turboexpander. At higher ambient temperatures, the power output of the turbine decreases at a rate of 1 MW per degree of centigrade (Figure 4b). Injecting approximately 5% of the turbine's exhaust mass flow at ISO conditions (or merely 25 kg/s) results in rapidly increasing the power output by around 11% (16 MW).

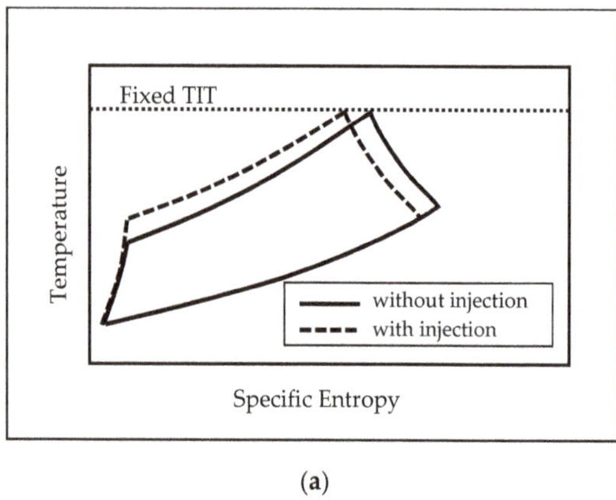

(a)

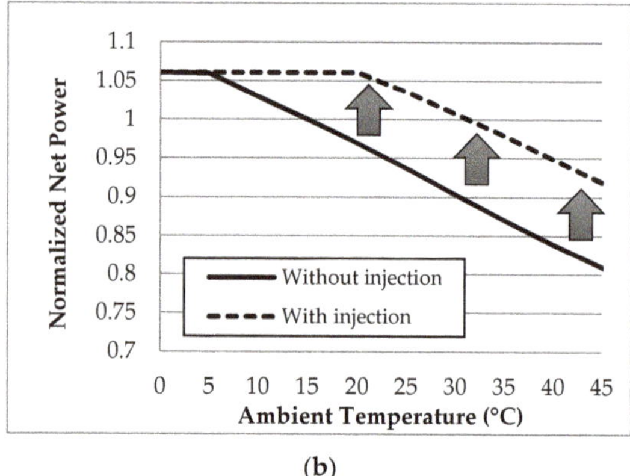

(b)

Figure 4. (**a**) Effect of air injection on a single-shaft gas turbine cycle; (**b**) air injection impact on the output power of the V94.2 turbine.

Performance enhancement of the gas turbine leads to overall fuel savings. Although the used fuel increases for a range of ambient temperatures due to the increased air mass flow, the performance of the gas turbine improves considerably. The latter has a strong impact on the overall consumption of fuel and, consequently, on the generated emission, as also shown in Figure 5a. As seen, the proposed system results in approximately 200 kg/h of fuel savings, while the CO_2 emissions reduce by 500–600 kg/h (or 4000–4800 tons/year) for a wide range of ambient temperatures.

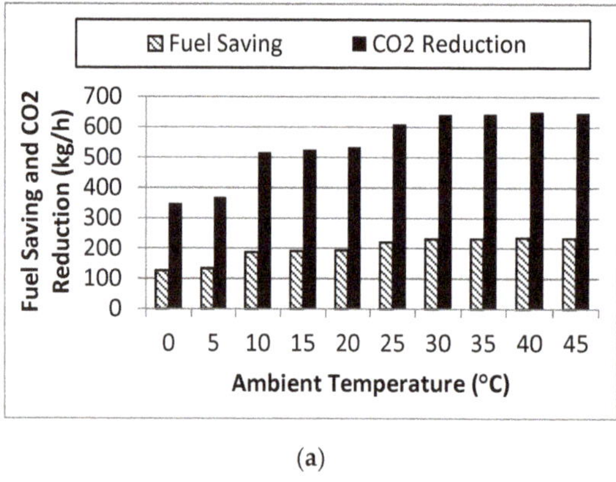

(a)

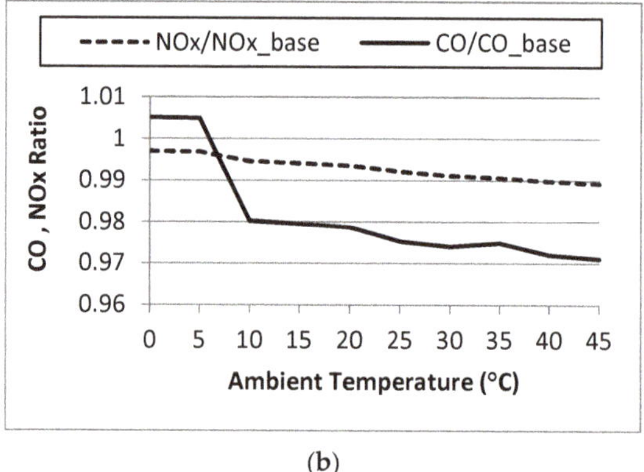

(b)

Figure 5. (**a**) Fuel savings and CO_2 emission reduction; (**b**) normalized emission values of NOx and CO at various ambient temperatures.

The generation of the air pollutants CO and NOx per megawatt have been calculated using Equations (13) and (14). Two factors play an important role in the generated emissions: the combustor inlet pressure and the relative fuel savings (per megawatt of produced power). Figure 5b demonstrates the variation of CO and NOx emissions relative to the conventional gas turbine system. As it is seen, the ratio of pollutants per megawatt are lower than those of the conventional system for most ambient temperatures studied. At lower ambient temperatures, the proposed system results in a marginal increase of the CO emissions due to the decrease of the inlet temperature of the turbine (constant maximum power of GT and lower combustor inlet temperature). However, the NOx and CO emissions reduce at higher temperatures by about 1% and 2%. Considering 8000 h of GT operating per year and average ambient temperature of 25 °C, the overall fossil fuel savings and CO_2 emission reductions are estimated at about 1600 and 4800 tons per year, respectively.

As mentioned before, the required fuel of the gas turbine determines the recovered energy and the mass flow of injected air. Figure 6a illustrates the variation of fuel flow in the V94.2 gas turbine versus the ambient temperature. The pressure and temperature of the gas transmission pipelines are assumed to be 60 bar and 25 °C, respectively. The required fuel mass flow decreases as the ambient temperature increases due to the control system of the GT that maintains the inlet temperature of the GT constant. Injecting high-pressure air into the combustion chamber increases the mass flow of the exhaust and, subsequently, the required fuel. As shown in Figure 6a, the proposed system does not improve the performance at lower ambient temperatures, due to mechanical limitations of the GT. However, at lower ambient temperatures, the constant power of the GT and the increasing exhaust mass flow result in a decrease in the fuel mass. In other words, the GT control system decreases the TIT to maintain the power constant at lower temperatures that results in higher efficiencies. The high pressure of roughly around 9 kg/s of fuel can be recovered and used to inject about 3–4 kg/s of air into the combustion chamber. This amount of air is less than 0.8% of the air flow of the GT, and hence, has no drawback on the stability of the gas turbine.

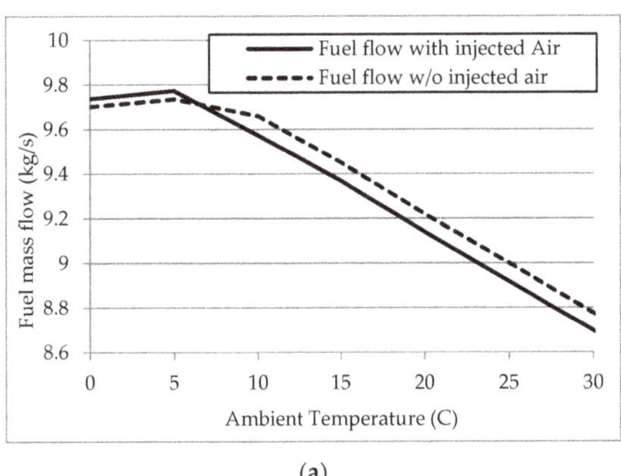

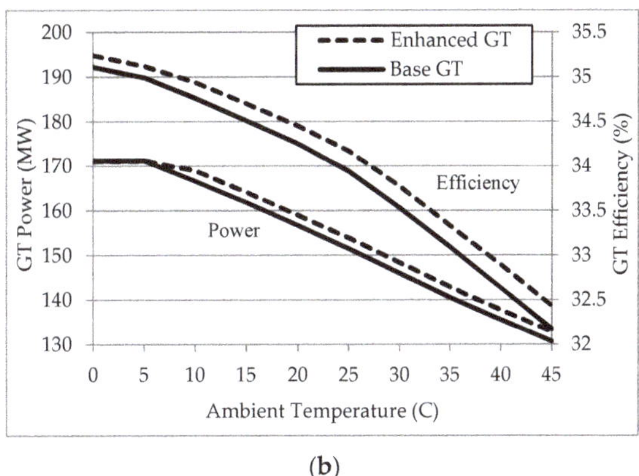

Figure 6. Variation of the (**a**) consumed fuel and the (**b**) power output of the gas turbine with and without air injection at various ambient temperatures.

Injecting high-pressure air into the combustion chamber can enhance the performance of the gas turbine system. As shown in Figure 6b, recovering the available fuel energy in the studied V94.2 gas turbine increases the output power by approximately 2.5 MW for a wide range of ambient temperatures, and similarly, the efficiency can increase by about 0.25%. At lower ambient temperatures (about 5 °C), air injection has no major impact on the gas turbine power due to GT mechanical and maximum power limitations, but it still improves the efficiency by somewhat decreasing the required fuel flow. Although here, one gas reducing station is included in the analysis, more than one station usually exists in real power plants. Therefore, in most real cases, more high-pressure air can be generated for injection into the combustion chamber. The potential energy recovery from gas can thus provide the required energy to compress 3% to 5% more air into the combustion chamber.

It is estimated that in conventional pressure-reducing stations, roughly up to 40% of the energy of the consumed fuel can be recovered to supply high-pressure air. Since air injection can result in a decline of the surge margin, OEM recommends air injection with a mass flow lower than 3% of the compressor's inlet flow [39]. The impact on power and efficiency of the amount of injected air into the V94.2 gas turbine is shown in Figure 7. It is seen that adding 1% more air into the combustion chamber can increase the power and the efficiency by about 2% and 0.75%, respectively. The addition of compressed air into the GT leads to a slightly higher compressor pressure ratio (Figure 8a). A 3% air-injection ratio increases the compressor pressure ratio by about 3%. Although the temperature of the inlet fuel of the turboexpander affects the outlet pressure, it plays a minor role and can be considered negligible.

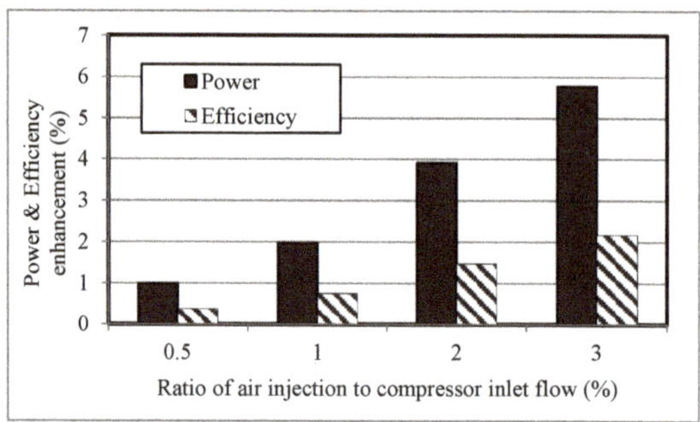

Figure 7. Performance variation of the V94.2 gas turbine with air-injection ratio.

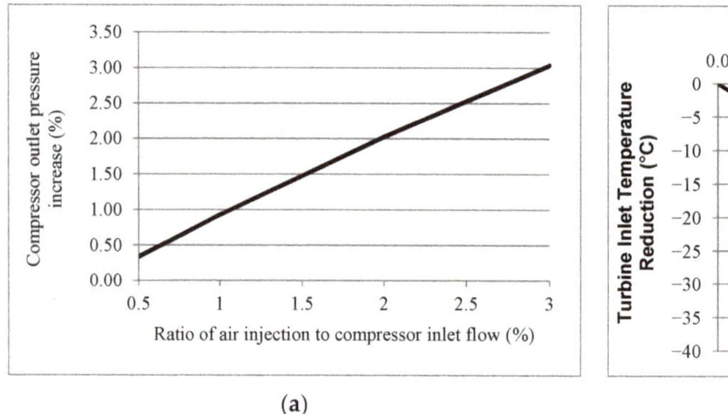

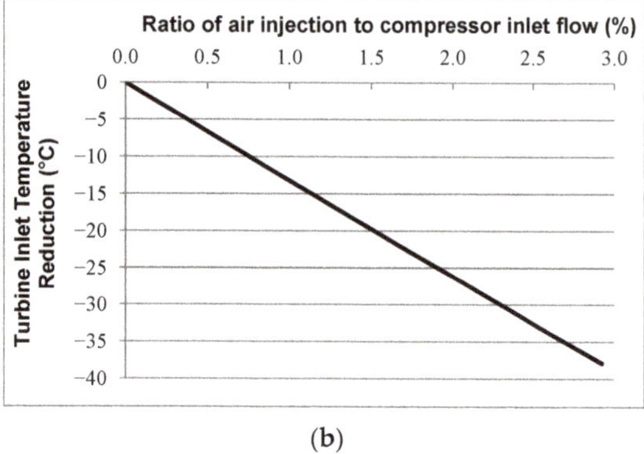

(a)   (b)

Figure 8. Variation of the (**a**) GT pressure ratio and the (**b**) inlet temperature of the turbine versus the air-injection ratio with constant power output.

As seen in Figure 8b, air injection may be used to reduce the inlet temperature of the turbine as well. Turbine inlet temperature reduction has a great impact on extending the lifetime of gas turbines and increases the maintenance intervals and the overall GT life cycle costs. Approximately, adding 1% extra air into the combustion chamber may result in a 12 °C reduction of the TIT keeping the power output constant.

As mentioned, the amount of energy that can be recovered by the turboexpander depends on various parameters including the expander pressure ratio and the temperature of the fuel at the inlet of the expander. Figure 9a shows the power produced with the turboexpander based on the expander's operating parameters.

The amount of compressed air that can be supplied to the gas turbine can be estimated by considering the power output of the turboexpander in conjunction to the air compressor. As seen in Figure 9b, the mass flow of compressed air is directly related to both the working pressure ratio and the inlet temperature of the turboexpander. To compare the two systems with and without the air-injection unit, a component-level exergy analysis was performed, and the results are presented in Table 6. As seen, with the proposed modification of the gas turbine, exergy efficiency increases by approximately 0.36%. In addition, the exergy destruction of the gas turbine with the turboexpander system is approximately 2 MW lower than that of the gas turbine without the turboexpander.

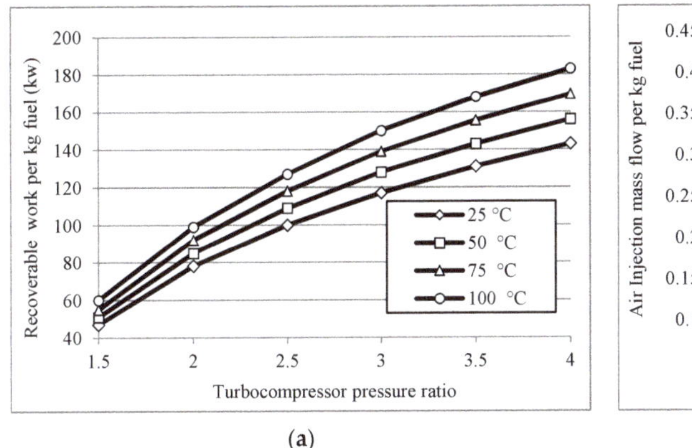

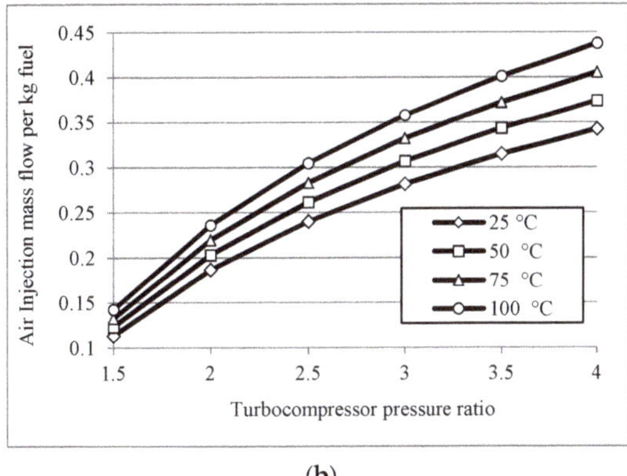

Figure 9. Effect of the turbocompressor pressure ratio and the fuel temperature on (**a**) the recoverable work and (**b**) the amount of compressed air per kg of fuel.

Table 6. Exergy efficiency and destruction of each component.

Components	Fuel Exergy (kW)	Product Exergy (kW)	Exergy Destruction (kW)	Exergy Efficiency (%)
Gas Turbine without TE	348,680	161,800	186,880	46.40
Gas Turbine	350,494	164,100	184,846	46.82
Turboexpander	1969	1536	433	77.98
Air Injection Compressor	1546	1447	99	93.60
Heat Exchanger	549	107	442	19.50
Total Gas Turbine with TE	350,919	164,100	185,832	46.76

The Sankey diagram of exergy flows can provide important information of the operation of an energy system. The Sankey diagram showing the distribution of exergy flows of the proposed system is presented in Figure 10. In this diagram, the exergy destruction flows are shown in red. As seen, the exergy destruction of the GT systems accounts for about one third of the total exergy input to the turbine, mainly associated with irreversibilities within the combustion process. Moreover, the total exergy destruction of other components (including AIC, HX, and TE) is less than 1 MW.

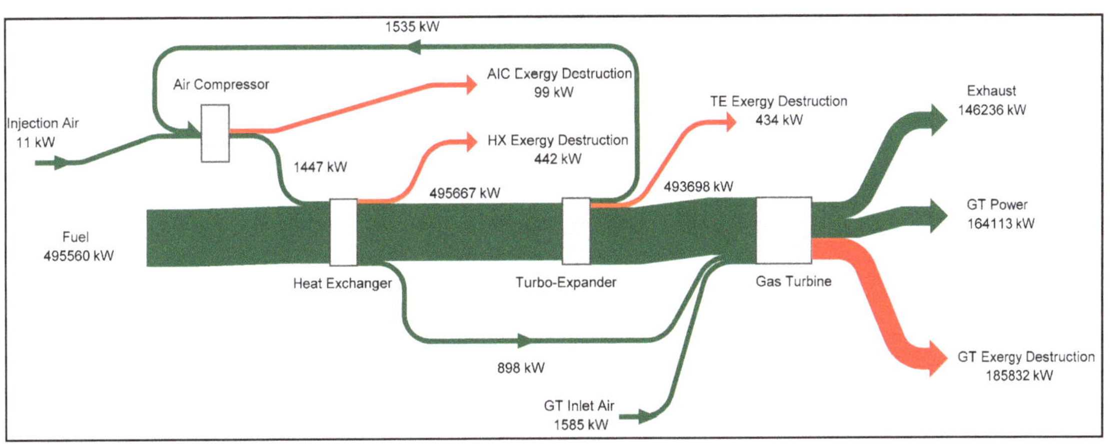

Figure 10. Exergy flow Sankey diagram of the turboexpander air-injection system.

The total exergy destruction of the air compressor, heat exchanger, and expander vary largely with the fuel transmission pressure. Specifically, it is found that the sum of exergy destruction of these three components increases from 678 to 1266 kW for fuel pressures from 40 to 90 bar. The bar diagram in Figure 11a presents the ratio of exergy destruction of these components of the proposed system with the fuel feed pressure. Figure 11a shows that the exergy destruction of the turboexpander increases with increasing fuel pressure, while the exergy destruction of the heat exchanger presents the opposite trend.

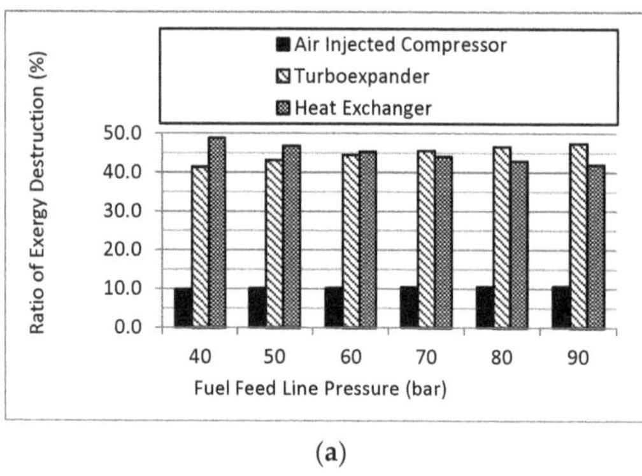

(a)

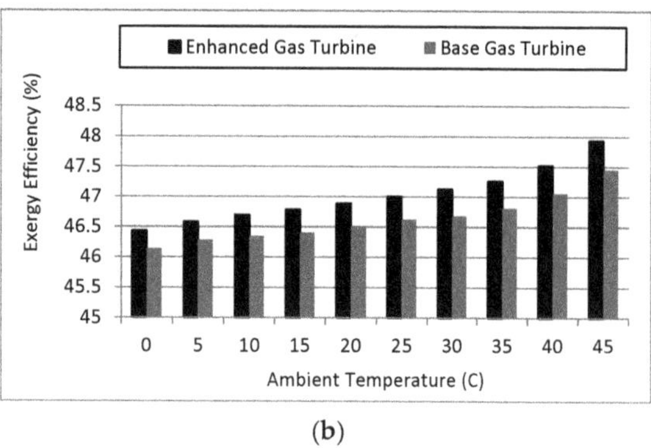

(b)

Figure 11. (**a**) Variation of the exergy destruction ratio of components with the inlet pressure of the fuel; (**b**) variation of the exergy efficiency of the GT with ambient temperature, with and without air injection.

The exergy efficiency of the system with and without the turboexpander increases directly with the ambient temperature (Figure 11b). However, the efficiency enhancement of the proposed system is higher at elevated temperatures and varies from about 0.3% to 0.5% with increasing ambient temperature from 0 to 45 °C. Hence, at elevated ambient temperatures, this system shows a higher efficiency than at lower ambient temperatures.

5. Conclusions

In this article, a turboexpander was introduced in a conventional high-pressure natural gas pressure-reduction station. The power recovered from the expansion of the natural gas was used to compress and introduce extra air into the combustion chamber of a heavy-duty gas turbine V94.2 of Siemens for performance enhancement.

The exergy analysis revealed that the exergy destruction of the gas turbine with the new turboexpander system is approximately 2 MW lower than that of the conventional system without a turboexpander. In other words, the proposed system results in an increase in the overall exergy efficiency of the gas turbine of approximately 0.36%. The recovery of the potential energy of the fuel led to an increase of the power output and efficiency of the gas turbine by 2.5 MW and 0.25%, respectively. In addition, the proposed system led to considerable fuel savings and reduced generated pollutants. Considering 8000 h of operating per year, annual fuel savings of at least 2 million cubic meters and an annual CO_2 reduction of 4000–4800 tons (depending on site conditions) are estimated. Finally, the NOx and CO emissions of the system decrease by about 1% and 2%, respectively.

Overall, it was shown that a single-shaft gas turbine can benefit from this hybridization not only as a strategy to increase the output power and efficiency of the gas turbine but also as an innovative way to recover energy and reduce the required fuel and emissions. It is noteworthy that this hybrid system results in better performance at higher ambient temperatures, when compared to the conventional gas turbine. The amount of recoverable work depends on the fuel feeding line and pressure ratio of the turboexpander. It is estimated that in conventional pressure-reducing stations, roughly up to 40% of the energy of the consumed fuel can be recovered. This power can be used to supply high-pressure air. However, consulting with the gas turbine manufacturer is recommended for injecting air with a flow rate higher than 3% the compressor's inlet flow. Another important point is that the proposed system can be used in gas turbines to lower the inlet temperature of the turbine by at least 10 degrees to extend the lifetime of gas turbine parts when more power is not required.

Author Contributions: Investigation, A.R.S.; resources, A.R.S.; data curation, A.R.S.; writing—original draft preparation, A.R.S.; software, A.R.S.; visualization, A.R.S.; validation, A.R.S.; formal analysis, A.R.S.; conceptualization, A.R.S., R.S.; methodology, A.R.S., R.S., F.P.; writing—review and editing, A.R.S., R.S., F.P.; supervision, R.S., F.P. All authors have read and agreed to the published version of the manuscript.

Funding: This research received no external funding.

Institutional Review Board Statement: Not applicable.

Informed Consent Statement: Not applicable.

Data Availability Statement: Not applicable.

Acknowledgments: Reza Shirmohammadi would like to acknowledge the Erasmus + International Credit Mobility (KA107-2020 project), Alianza 4 Universidades, and International Affairs at University of Tehran and Carlos III University of Madrid. Fontina Petrakopoulou would like to thank the Spanish Ministry of Science, Innovation and Universities and the Universidad Carlos III de Madrid (Ramón y Cajal Programme, RYC-2016-20971).

Conflicts of Interest: The authors declare no conflict of interest.

Abbreviations

Nomenclature	
T	Temperature ($^\circ$C)
p	Pressure (kPa)
h	Specific enthalpy (kJ/kg)
s	Entropy (kJ/kgK)
$\dot{m}$	Mass flow rate (kg/s)
$\dot{Ex}$	Exergy rate (kW)
$\dot{Q}$	Heat transfer rate (kW)
$\dot{W}$	Work rate (kW)
N	Shaft speed (1/s)
Cp	Specific heat at constant pressure (kJ/kg)
ε	Exergy efficiency
η	Isentropic efficiency
γ	specific heat ratio
T_{ad}	Adiabatic flame temperature
σ	Fuel to air equivalent ratio
π	Dimensionless pressure
θ	Dimensionless temperature
ψ	H/C atomic ratio
Subscripts and Superscripts	
TE	Turboexpander
AIC	Air injection compressor
CC	Combustion chamber
Q	Heat

W	Work
A	Air
F	Fuel
T	Turbine
C	Compressor
G	Gas
M	Mechanical
INJ	Injected Air
D	Destruction
L	Loss
PH	Physical
CH	Chemical
Acronyms	
TE	Turboexpander
CC	Combustion chamber
GT	Gas turbine
TIT	Turbine inlet temperature
CAI	Compressed air injection
OEM	Original equipment manufacturer
NG	Natural gas

References

1. Shirmohammadi, R.; Aslani, A.; Ghasempour, R. Challenges of carbon capture technologies deployment in developing countries. *Sustain. Energy Technol. Assess.* **2020**, *42*, 100837. [CrossRef]
2. Petrakopoulou, F.; Morozyuk, T.; Carassai, A. Conventional and advanced exergetic analyses applied to a combined cycle power plant. *Fuel Energy Abstr.* **2012**, *41*, 146–152. [CrossRef]
3. Petrakopoulou, F.; Robinson, A.; Olmeda-Delgado, M. Impact of climate change on fossil fuel power-plant efficiency and water use. *J. Clean. Prod.* **2020**, *273*, 122816. [CrossRef]
4. Petrakopoulou, F. On the economics of stand-alone renewable hybrid power plants in remote regions. *Energy Convers. Manag.* **2016**, *118*, 63–74. [CrossRef]
5. Khoshgoftar Manesh, M.H.; Kabiri, S.; Yazdi, M.; Petrakopoulou, F. Thermodynamic evaluation of a combined-cycle power plant with MSF and MED desalination. *J. Water Reuse Desalin.* **2020**, *10*, 146–157. [CrossRef]
6. Sharifi, S.; Nozad Heravi, F.; Shirmohammadi, R.; Ghasempour, R.; Petrakopoulou, F.; Romeo, L.M. Comprehensive thermodynamic and operational optimization of a solar-assisted LiBr/water absorption refrigeration system. *Energy Rep.* **2020**, *6*, 2309–2323. [CrossRef]
7. Nakhamkin, M.; Pelini, R.; Patel, M.; Wolk, R. Power Augmentation of Heavy Duty and Two-Shaft Small and Medium Capacity Combustion Turbines with Application of Humid Air Injection and Dry Air Injection Technologies. In Proceedings of the ASME: Power Conference, Baltimore, MD, USA, 30 March–1 April 2004; pp. 301–306.
8. Akita, E.; Gomi, S.; Cloyd, S.; Nakhamkin, M.; Chiruvolu, M. The Air Injection Power Augmentation Technology Provides Additional Significant Operational Benefits. In Proceedings of the ASME Turbo Expo: Power for Land, Sea, and Air, Montreal, QC, Canada, 14–17 May 2007; pp. 1079–1083.
9. Gay, R.R.; van der Linden, S. Power Augmentation using Air Injection, an Alternative solution to Peak Power Demands—using the large installed base of existing GT & CC power plants. *Conference Electr. Power* **2007**. Available online: http://www.espcinc.com/library/Electric_Power_2007_Conference_Paper_on_Air_Injection.pdf (accessed on 23 September 2021).
10. Arias Quintero, S.; Auerbach, S.; Kraft, R. Performance Improvement of Gas Turbine With Compressed Air Injection for Low Density Operational Conditions. In Proceedings of the ASME Turbo Expo: Turbine Technical Conference and Exposition, Düsseldorf, Germany, 16–20 June 2014.
11. El-Awad, M. Energy and Exergy Analysis of a combined Diesel-Engine Gas-Turbine System for Distributed Power Generation. *Int. J. Therm. Environ. Eng.* **2013**, *5*, 31–39. [CrossRef]
12. Abudu, K.; Igie, U.; Roumeliotis, I.; Hamilton, R. Impact of gas turbine flexibility improvements on combined cycle gas turbine performance. *Appl. Therm. Eng.* **2021**, *189*, 116703. [CrossRef]
13. Igie, U.; Abbondanza, M.; Szymański, A.; Nikolaidis, T. Impact of compressed air energy storage demands on gas turbine performance. *Proc. Inst. Mech. Eng. Part A J. Power Energy* **2020**, *235*, 850–865. [CrossRef]
14. Kuczynski, S.; Łaciak, M.; Olijnyk, A.; Szurlej, A.; Włodek, T. Techno-economic assessment of turboexpander application at natural gas regulation stations. *Energies* **2019**, *12*, 755. [CrossRef]
15. Lim, J.; Kim, E.; Seo, Y. Dual inhibition effects of diamines on the formation of methane gas hydrate and their significance for natural gas production and transportation. *Energy Convers. Manag.* **2016**, *124*, 578–586. [CrossRef]
16. Jelodar, M.T.; Rastegar, H.; Ahyaneh, H.A. Modeling turbo expander systems. *Simulation* **2013**, *89*, 234–248. [CrossRef]

17. Farzaneh-Gord, M.; Rahbari, H.R. Response of natural gas distribution pipeline networks to ambient temperature variation (unsteady simulation). *J. Nat. Gas Sci. Eng.* **2018**, *52*, 94–105. [CrossRef]

18. Barone, G.; Buonomano, A.; Calise, F.; Palombo, A. Natural gas turbo-expander systems: A dynamic simulation model for energy and economic analyses. *Therm. Sci.* **2018**, *22*, 2215–2233. [CrossRef]

19. Ashouri, E.; Veysi, F.; Shojaeizadeh, E.; Asadi, M. The minimum gas temperature at the inlet of regulators in natural gas pressure reduction stations (CGS) for energy saving in water bath heaters. *J. Nat. Gas Sci. Eng.* **2014**, *21*, 230–240. [CrossRef]

20. Morgese, G.; Fornarelli, F.; Oresta, P.; Capurso, T.; Stefanizzi, M.; Camporeale, S.M.; Torresi, M. Fast design procedure for turboexpanders in pressure energy recovery applications. *Energies* **2020**, *13*, 3669. [CrossRef]

21. Deymi-Dashtebayaz, M.; Dadpour, D.; Khadem, J. Using the potential of energy losses in gas pressure reduction stations for producing power and fresh water. *Desalination* **2021**, *497*, 114763. [CrossRef]

22. Golchoobian, H.; Taheri, M.H.; Saedodin, S. Thermodynamic analysis of turboexpander and gas turbine hybrid system for gas pressure reduction station of a power plant. *Case Stud. Therm. Eng.* **2019**, *14*, 100488. [CrossRef]

23. GhasemiKafrudi, E.; Amini, M.; Habibi, M.R. Application of Turbo-Expander to Greenhouse Gas and Air Pollutant Emissions Reduction Using Exergy and Economical Analysis. *Iran. J. Chem. Eng.* **2017**, *14*, 32–47.

24. Galyas, A.B.; Tihanyi, L.; Szunyog, I.; Kis, L. Investigation of pressure regulator replacement by turbo expander in hungarian gas transfer stations. *Acta Tecnol.* **2018**, *4*, 5–13. [CrossRef]

25. Tahani, M.; Masdari, M.; Salehi, M.; Ahmadi, N. Optimization of wet compression effect on the performance of V94.2 gas turbine. *Appl. Therm. Eng.* **2018**, *143*, 955–963. [CrossRef]

26. Ameri, M.; Tahvildar, B. V94. 2 Gas Turbine Thermodynamic Modeling for Estimation of Power Gained by Fog System in Iran Power Plants. *Int. J. Energy Eng.* **2011**, *1*, 33–43.

27. Kowalczyk, B.; Kowalczyk, C.; Rolf, R.M.; Badyda, K. Model of an ANSALDO V94.2 gas turbine from Lublin Wrotków Combined Heat and Power Plant using GateCycle T M software. *J. Power Technol.* **2014**, *94*, 190–195.

28. Siemens Energy sgt5-2000e. Available online: https://www.siemens-energy.com/global/en/offerings/power-generation/gas-turbines/sgt5-2000e.html. (accessed on 23 September 2021).

29. Saravanamuttoo, H.I.H.; Rogers, G.F.C.; Cohen, H. *Gas Turbine Theory*, 5th ed.; Prentice Hall: Hoboken, NJ, USA, 2001.

30. Nakhamkin, M.; van der Linden, S. Integration of a Gas Turbine (GT) With a Compressed Air Storage (CAES) Plant Provides the Best Alternative for Mid-Range and Daily Cyclic Generation Needs. In Proceedings of the ASME Turbo Expo: Power for Land, Sea, and Air, Munich, Germany, 8–11 May 2000.

31. Gulder, O.L. Flame Temperature Estimation of Conventional and Future Jet Fuels. *J. Eng. Gas Turbines Power* **1986**, *108*, 376–380. [CrossRef]

32. Lazzaretto, A.; Toffolo, A. Energy, economy and environment as objectives in multi-criterion optimization of thermal systems design. *Energy* **2004**, *29*, 1139–1157. [CrossRef]

33. Oyedepo, S.O.; Fagbenle, R.O.; Adefila, S.S.; Alam, M.M. Thermoeconomic and thermoenvironomic modeling and analysis of selected gas turbine power plants in Nigeria. *Energy Sci. Eng.* **2015**, *3*, 423–442. [CrossRef]

34. Bagheri, B.S.; Shirmohammadi, R.; Mahmoudi, S.M.S.; Rosen, M.A. Optimization and comprehensive exergy-based analyses of a parallel flow double-effect water-lithium bromide absorption refrigeration system. *Appl. Therm. Eng.* **2019**, *152*, 643–653. [CrossRef]

35. Petrakopoulou, F.; Boyano, A.; Cabrera, M.; Tsatsaronis, G. Exergoeconomic and exergoenvironmental analyses of a combined cycle power plant with chemical looping technology. *Int. J. Greenh. Gas Control* **2011**, *5*, 475–482. [CrossRef]

36. Bejan, A.; Tsatsaronis, G.; Moran, M.J. *Thermal Design and Optimization*, 1st ed.; Wiley-Interscience: New York, NY, USA, 1995; 560p.

37. Petrakopoulou, F.; Tsatsaronis, G.; Morosuk, T. CO_2 capture in a chemical looping combustion power plant evaluated with an advanced exergetic analysis. *Environ. Prog. Sustain. Energy* **2014**, *33*, 1017–1025. [CrossRef]

38. Shirmohammadi, R.; Aslani, A.; Ghasempour, R.; Romeo, L.; Petrakopoulou, F. Process design and thermoeconomic evaluation of a CO2 liquefaction process driven by waste exhaust heat recovery for an industrial CO2 capture and utilization plant. *J. Therm. Anal. Calorim.* **2021**, *145*, 1585–1597. [CrossRef]

39. Wang, T.; Stiegel, G.J. *Integrated Gasification Combined Cycle (IGCC) Technologies*, 1st ed.; Elsevier Science: Amsterdam, The Netherlands, 2016; 928p, ISBN 9780081001851.

Article

Assessment of the Greek National Plan of Energy and Climate Change—Critical Remarks

Efthimios Zervas [1,*], Leonidas Vatikiotis [1], Zoe Gareiou [1], Stella Manika [1] and Ruth Herrero-Martin [2]

[1] School of Applied Arts and Sustainable Design, Hellenic Open University, 263 35 Patra, Greece; leonidas.vatikiotis@gmail.com (L.V.); gareiou@latpee.eap.gr (Z.G.); manika.stella@ac.eap.gr (S.M.)

[2] Department of Thermal and Fluid Engineering, School of Industrial Engineering, Universidad Politecnica de Cartagena, 30202 Cartagena, Spain; Ruth.Herrero@upct.es

* Correspondence: zervas@eap.gr; Tel.: +30-2610-367-566

Abstract: The Greek National Energy and Climate Plan was validated by the Greek Governmental Committee of Economic Policy on 23 December 2019. The decisions included in this plan will have a significant impact on the Greek energy mix as the production of electricity from lignite combustion ceases in 2028, when lignite will be replaced by natural gas (NG) and renewable energy sources (RES). This work presents an assessment of the Greek National Energy and Climate Plan by analyzing its pros and cons. The main critiques made are focused on the absence of risk analysis and alternative scenarios, the proposed energy mix, the absence of other alternatives on the energy mix and energy storage, the low attention given to energy savings (transport, buildings), the future energy prices, and the economic and social impacts. This analysis shows that delaying this transition for some years, to better prepare it by taking into consideration the most sustainable paths for that transition, such as using more alternatives, is the best available option today.

Keywords: climate change; energy and climate plan; energy price; energy transition; Greece; lignite; natural gas; RES

Citation: Zervas, E.; Vatikiotis, L.; Gareiou, Z.; Manika, S.; Herrero-Martin, R. Assessment of the Greek National Plan of Energy and Climate Change—Critical Remarks. *Sustainability* **2021**, *13*, 13143. https://doi.org/10.3390/su132313143

Academic Editors: Georgios Tsantopoulos and Evangelia Karasmanaki

Received: 30 September 2021
Accepted: 22 November 2021
Published: 27 November 2021

Publisher's Note: MDPI stays neutral with regard to jurisdictional claims in published maps and institutional affiliations.

1. Introduction

Climate change is one of the main current environmental problems [1] and its mitigation requires a great effort from scientists to find adapted solutions, from policy makers to find adapted policy measures, and from different stakeholders to apply them. One of these measures is the transition from the production of electricity via coal combustion to more efficient or renewable energy sources.

Coal was and is still today one of the major sources for electricity production in Europe, as it accounts for 22.9% of the total final energy production in EE27 in 2017 [2]. However, the target set by the European Green Deal is to decrease greenhouse gas emissions by 40% in EE27 in 2030, compared to the 1990 emissions [3], and to reach climate neutrality (80–95%) in the EE countries in 2050 [4]. In that sense, coal's participation in the EE energy mix has to decrease to 12% by 2050, with the complete elimination of oil as a power-generating source [5]. To achieve these goals, several countries set up measures to decrease coal's participation in their energy mix. Greece is one of them, as electrical energy production from lignite was 29.3% in 2019 [6].

All EU countries recently released National Energy and Climate Plans [7]. The Greek plan [8] sets the goal of greenhouse gas emissions in Greece for 2030 and the main actions proposed to achieve this goal. It is an important milestone for the current national policy on energy and climate, as it sets out climate goals at the heart of development policy in Greece and actions to protect the environment. This plan sets several very ambitious goals. However, the Greek NECP is one of the most critical for several reasons: Greece was heavily impacted by the recent economic crisis, and for this reason both economic growth and available funds are limited; also, Greece is very dependent on local lignite production and imported oil; thus, the radical change of the energy mix in just a few years is very challenging.

The aim of this work is to address the points of this plan that could pose some issues for the future, and to serve as a guideline for other European NECPs in the cases of similar issues. This work follows the general structure of several similar policy works on energy regulations, policies, or research agendas that can be found in the literature [9–13]. It should be noted that this work assesses some major points in a high-level critique of this plan and that this work is clearly of an application nature. For each one of these points, a specific analysis will be conducted to precisely quantify the economic, social, and environmental pros and cons. These detailed analyses will be presented in future dedicated works.

2. Presentation of the Mains Points of the Greek National Energy and Climate Change Plan

The purpose of the NECP is described in detail in the introduction of the plan (published on the website for the Greek Ministry of Environment and Energy [14]). More specifically, it is stated that the National Plan for Energy and Climate is, for the Greek government, a strategic plan for the issues of climate and energy. A detailed roadmap for the achievement of specific energy and climate objectives by 2030 is given. The NECP presents and analyzes policy priorities and measures in a wide range of development and economic activities for the benefit of Greek society, making this text a reference for the next decade.

It is noted that the NECP is a tool for national policy in the field of energy and for the mitigation of climate change, which highlights the priorities of, and development opportunities in, Greece. The aim of the NECP is to be the main tool for the establishment of national economic, energy, and climate policies over the next decade, taking into account the recommendations of the European Commission and the UN's Sustainable Development Goals. The strategic goals of the Greek government in the field of energy and climate are set until 2030, and they aim to contribute decisively to the necessary energy transition in the most competitive way for the national economy, achieving a drastic reduction in greenhouse gas emissions. In this way, Greece can emerge as one of the member states with the most ambitious climate and energy goals through a comprehensive and coherent program of measures and policies for both 2030 and 2050.

The NECP sets specific targets for greenhouse gas emissions from Greece in 2030 and determines the Greek energy mix until that year by increasing the participation of natural gas and renewable energy sources [8], as well as by increasing energy efficiency. At the same time, the NECP identifies the policies and measures that are necessary for achieving those goals, analyzes the evolution of the Greek energy system until 2030, and reports the investment needs and the different impacts on society, the economy, and the environment.

In summary, the objectives of this plan [8] are the following:

- Greece will cease the production of electricity from lignite combustion in 2023, except for the Ptolemaida V thermal plant, which will close in 2028;
- Lignite will be substituted with imported natural gas;
- The RES participation in the Greek energy mix will be 35% in 2030 (60% in the case of electricity production);
- For the year 2030, GHG emissions will be reduced by 43%, compared to the 1990 levels, and 56%, compared to the 2005 levels;
- The improvement of energy efficiency of the final energy consumption will be 38% by 2030.

3. Methodology

The Greek National Energy and Climate Plan is analyzed from two points of view: first, the general concept of this plan is analyzed; then, a specific analysis is conducted for each one of its parts. For every one of these points, the general methodological approaches are analyzed and the specific problematic points are revealed. Concrete proposals are formulated for these approaches and points. Then, an analysis of the specific parts of the plan is performed; as previously, the general methodological approaches are analyzed and the specific problematic points are revealed, followed by concrete proposals. In both cases, alternatives are suggested for several points, following a high level of argumentation.

More specifically, the critiques analyzed in this article concern the application of a single scenario and the absence of updating and alternatives, the de-lignification of energy production and the proposed future energy mix (natural gas and renewable energy sources), the absence of other alternatives for the energy mix/production (biomass, gasification of lignite, carbon storage, and energy storage), vehicle electrification and public transport, the energy savings in buildings, the local production of products and dietary habits, the cost of energy and its economic and social consequences, the impact on labor issues, employees education and training, the employees and citizen information, and the regional and sectoral plans.

Several proposals, based on the general needs of the Greek economy, the protection of the environment, and the mitigation of climate change are presented for the above issues. Their general feasibility is also shown. These proposals aim to address the economic development of Greece, the decrease of energy poverty, with the parallel respect of the environment and the mitigation of climate change. As mentioned above, only a high-level critique is addressed in this article; a specific and detailed study is necessary for each one of the proposals presented here, and these studies will be presented in the near future.

4. Results—Assessment of the Plan

4.1. Implementation of a Single Scenario and the Absence of Alternative Scenarios

The submitted National Energy and Climate Plan is, with a few exceptions, a linear implementation of a single scenario. However, it should be noted that the energy and climate sectors are highly unpredictable. Taking this high variability into consideration, the National Plan should include a risk assessment and multiple alternative scenarios. However, the possible cases of deviation from, or even the failure of, this linear implementation are not analyzed, and no alternative scenarios are given.

Obviously, scientific progress leads to better materials, more efficient technologies, and finally to a decrease in the cost of energy production. In the field of energy, many technological achievements have been developed in recent years, such as the extraction of marine shale gas [15] or new photovoltaic materials, such as the recent progress in chlorinated organic photovoltaic materials, the discovery of two-dimensional photovoltaic materials accelerated by machine learning, or their new applications, such as in highways for signal systems, for agricultural and livestock purposes due to the need for water during periods of intense sunshine, for charging car and boat batteries, etc. [16] However, while technology has a positive effect on energy production, either in terms of efficiency or price, geopolitical events can affect the price and availability of energy in the opposite direction. The price of natural gas, which will be one of the major components of the future Greek energy mix, can be very significantly affected by these events in the future. The relationship between the geopolitical situation and energy is interdependent, as geopolitical changes can affect energy markets and, conversely, energy trends can disrupt geopolitical dynamics [17].

The 1973 oil crisis is the easiest example. Geopolitically, the Middle East, an important oil-producing region, is a politically unstable region with two ongoing conflicts, one in Iraq and one in Syria. Additionally, the political and economic developments in China have a global impact. Libya is another hotbed of instability in the close-to-Greece region; moreover, the relationship between Greece and Turkey is often very tense. Other global tensions, however, especially those concerning oil-producing countries (such as Iran or Venezuela), have a significant impact on energy prices.

Figure 1 shows the change in the price of crude oil for the period of 1946–2021. This figure shows a course of sharp changes and alternating increases and decreases in crude oil price; this price shows a range of variations, from less than USD 20 to near USD 180 during this time period. This sharply changing picture is also reflected in Figure 2, which shows the changes in the price of crude oil during only the last two years (January 2020–October 2021).

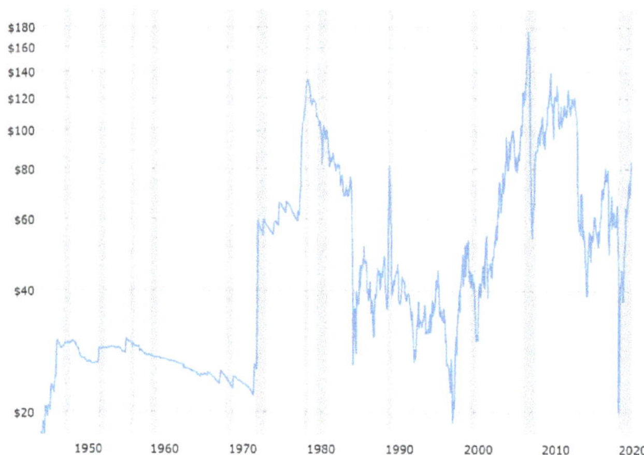

Figure 1. Change in the price of crude oil from 2010 to today (source: https://www.macrotrends.net/1369/crude-oil-price-history-chart, (accessed on 5 September 2021)).

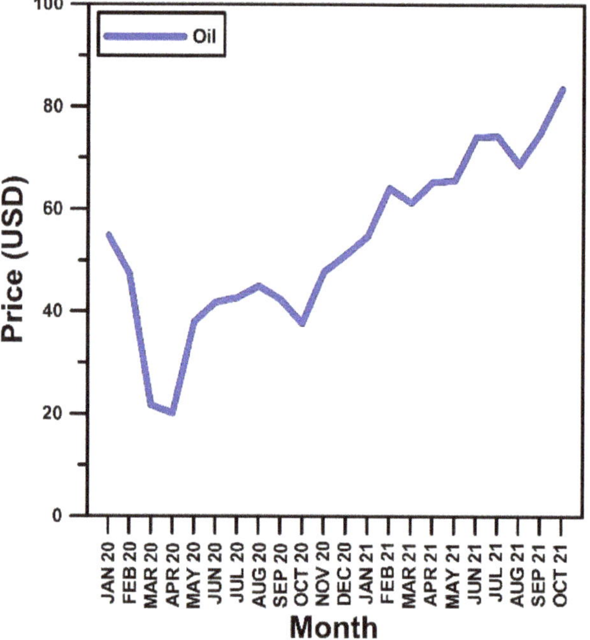

Figure 2. Change in the price of crude oil in the period January2020–October 2021 (Source of the data: https://www.macrotrends.net/1369/crude-oil-price-history-chart, (accessed on 5 September 2021).

While unexpected health crises, such as the current crisis of COVID-19, along with new technological advances and geopolitical developments play an important role in the global supply chain and the energy markets [18], it is a fact that the situation in the international market becomes even more unpredictable given the uncharted course of the coronavirus, the announcements about mutations, the preventive measures taken, as well as the course of vaccinations. These conditions directly determine consumption, which, in turn, affects the supply–demand balance and the course of prices.

Figure 3 shows a well-known figure of the evolution of crude oil prices the last 20 years, indicating some major global political events [19]. This figure shows the great impact of these events on the price of crude oil.

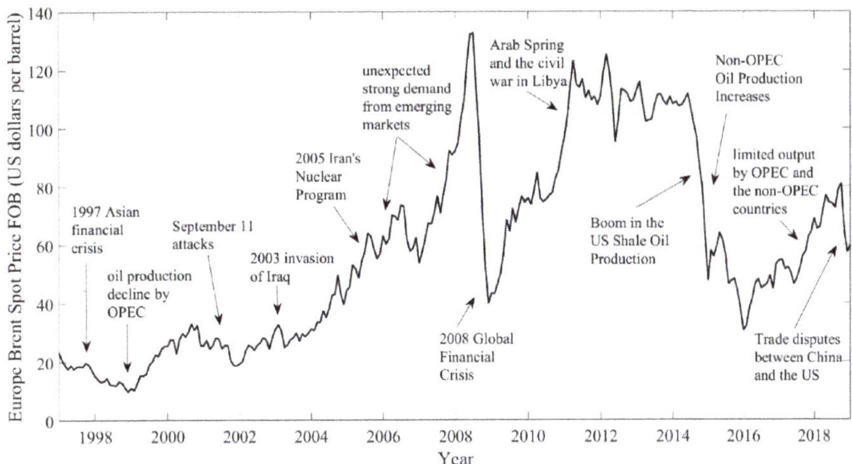

Figure 3. Evolution of crude oil price during the last 20 years. The major political events are indicated (source: [https://doi.org/10.1016/j.ribaf.2020.101357], (accessed on 24 November 2021)).

Recent research has shown the correlation between daily new cases or total number of deaths due to communicable diseases and adverse stock returns in the Chinese equity market [20] and presented the negative, direct, and indirect effects of COVID-19 on global markets [21]. Meanwhile, even the way that COVID-19 is covered by the media, or that panic is caused by them, seems to affect stock market volatility [22].

In addition to the previous figures, Figure 4 shows the index of prices for the three fossil fuels.

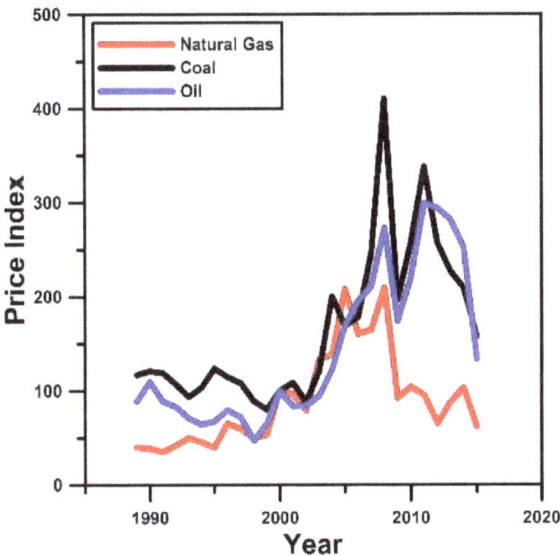

Figure 4. Price index for the three fossil fuels (source of data: https://ourworldindata.org/fossil-fuels, (accessed on 5 September 2021)).

The above figures show the instability and uncertainty of the price of fuels; however, the linear application of only one scenario cannot take into consideration these changes.

At the same time, the Intergovernmental Panel on Climate Change (IPCC) has recognized that many parameters affect the emissions of greenhouse gas and, therefore, there is considerable uncertainty for the prediction of future emissions. For this reason, six groups of greenhouse gas emission scenarios have been proposed (A1T, A1B, A1F1, A2, B1, and B2) with a total of 40 scenarios, the implementation of which give different future temperatures in the Earth's atmosphere (Figure 5) [23].

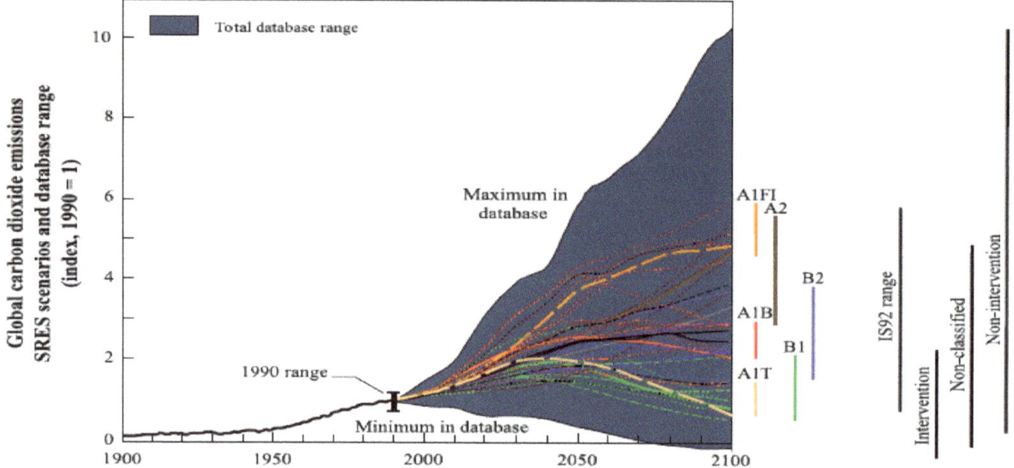

Figure 5. Index of carbon dioxide emissions produced by energy and industry for the various IPCC scenarios (1990 = 1) (source: [23]).

Based on the previous analysis, two additional works must be conducted to complement the NECP. The first one concerns the inclusion of alternative scenarios to cover the probability that unpredictable factors inhibit the implementation of the initial planning. The alternative scenarios should take into account both the positive and negative eventualities, such as:

- The technological developments that will improve the efficiency and effectiveness of technology and reduce the price of energy produced;
- Economic issues, such as the evolution of the country's GDP, the price of gas, of RES, of electricity, etc.;
- Possible schedule delays due to unexpected factors and events (such as a new pandemic or a major natural disaster in Greece, e.g., an earthquake), changes in institutional, economic, or social parameters, etc.;
- Geopolitical issues that may affect the availability of an uninterrupted gas supply in Greece, or a great change in energy prices, etc.

The second necessary work is to have a specific provision for the regular reporting of the progress of the plan's actions and, based on this reporting, the establishment of an annual update of the objectives of the plan, of the policies to be set, and of the actions to be implemented. It should be noted that this regular reporting and the regular update of the objectives of the plan are not mentioned in the current plan.

4.2. De-Lignification and Energy Mix

The plan proposes the complete de-lignification of the country in 2028 and the increase of the participation of natural gas and RES in Greece's energy mix. This option has some positive and some negative points. The main positive point is the reduction of greenhouse gas production. However, this advantage is not so obvious. Natural gas produces 201.96 kg of CO_2/MWh, compared to 363.6 kg of lignite [24]; or, put differently, natural gas produces 55% of the GHG produced by lignite. However, methane has a 100-year global warming potential, 25 times that of CO_2, or 72 times for a 20-year period [25], and leaks of methane from the production, transport, and consumption of NG are significant [26]. For the above reasons, the gain in GHG emissions will be much smaller than initially estimated, as already reported [27,28]. The additional risk of accidents in the natural gas circuit also decreases this difference. In conclusion, leaks of 5% of methane can completely cancel this difference.

Aside from the previous statement, there are also some other issues about the proposed energy mix. The first one is the deterioration of the country's trade balance, both from the import of raw materials for the installation of RES and natural gas installations and also from imports of natural gas to replace a domestic product, lignite. The other negative point is the energy dependence of Greece on an imported fuel instead of a locally produced one, leading to higher risks of national energy autonomy.

It is clear that the deterioration of the country's trade balance will affect the entire Greek economy, and heavy actions should be proposed to mitigate this. However, the analysis of these actions and/or of alternative scenarios is not performed in this plan.

It should be also noted that countries that use natural gas to a large extent in their energy mix, such as the Netherlands, have decided to become independent of it in the coming years [29]. Taking this into consideration, natural gas can be considered only as a transition solution, and the cost of replacing lignite with natural gas, which will be also replaced in two or three decades, is not examined in the NECP.

The proposed energy mix also has several uncertain elements, such as the final price of the energy, which can be much higher than the current one, or the uncertainty of finding domestic hydrocarbon reservoirs, as the NECP estimates that the domestic fuel production will increase from 281ktoe in 2020 to 536ktoe in 2030. However, this last estimation is not consolidated, and no alternatives are examined in case of failure.

Even if a close economic analysis of the substitution of lignite by natural gas is out of the scope of this work, some elements are given here. The current cost of electricity production from lignite is EUR 105/MWh (EUR 35 is the direct cost and EUR 70 is the emission price) [30], while the corresponding cost for natural gas is EUR 75/MWh, and for RES is EUR 135.6/MWh [31].

However, these values, and the great instability of prices shown in Figures 1–4, indicate that the complete abandonment of domestic energy sources (lignite) and the use of only imported fossil fuels (oil and natural gas) can have a very high cost for the energy in Greece and, as a consequence, for the entire Greek economy.

Therefore, a more detailed examination of the possibility of continuing to exploit lignite, an available domestic fuel, for a longer time, instead of completely substituting it for imported natural gas (at least until the cost of energy produced from RES decreases significantly) is proposed. This can be done by using modern, more environmentally efficient technologies, by combining lignite combustion with biomass combustion or gasification technology, and using synthetic gas and, in addition, carbon storage technology, as will be exposed here.

4.3. Biomass Combustion

The development of more renewable energy sources started in a more systematic manner after the oil crisis of 1973. During this period, scientists adopted a systematic approach to energy and coined the term biomass [32]. Biomass is a renewable energy source because the CO_2 released from its combustion is bound to the plants for their development. Therefore, its use as a fuel can have a positive impact on the overall GHG balance. Due to this positive impact, the use of biomass as fuel has increased during the last years. In addition, biomass is abundant, which is why biomass energy has become the world's fourth largest energy source today, following coal, oil, and natural gas, indicating its significant economic, societal, and environmental potential [33].

In several countries affected by the economic crisis after 2008, or even in the case of citizens with economic hardship in economically developed countries, the shift in the use of biomass as a heating fuel is mainly due to its lower price, or to the ability to burn materials that had not previously found a suitable route of exploitation through domestic combustion. Several citizens of low economic status have used pruning or even organic waste as heating fuel during the last years [34,35].

However, the combustion of biomass has a significant negative impact: the high emission of pollutants, mainly of particulate matter (PM) [36]. The increased use of biomass for domestic heating in recent years has led to a very poor air quality in Greek and many European cities, especially in winter. This poor air quality has serious consequences for both quality of life and human health, and these consequences will strongly appear in the next years. The European Respiratory Society has already highlighted the serious effects of biomass-burning on human health in cities in developed countries, and recommends limiting its use [37]. However, the poor air quality comes from the domestic combustion of biomass, where no pollution control system exists. Central power plants using biomass and equipped with pollution-control systems are widely available. Biomass co-firing has already received wide acceptance in many European countries, mostly in the northern and central parts of Europe, such as the United Kingdom, Germany, and the Nordic countries [38]. From this point of view, the use of biomass combustion to produce electrical energy for central power plants can be a very serious alternative to lignite combustion. This alternative is not proposed in this plan. Very roughly, the following data can show the feasibility of this alternative.

The consumption of lignite in Greece is 4.5 million tons of oil equivalent [39] or 46 million tons of lignite [40]. The typical thermal power of wood is quite similar to that of lignite, of course depending on the wood type [41]. The total timber production in Greece was almost 1.1 million m^3 in 2013 [42]; considering an average density of wood of $600\,kg/m^3$ [41], almost 0.7 million tons of wood was produced in Greece in 2013. However, in other neighboring or European countries with an equivalent or smaller surface area than Greece, the production is several times higher: 6.1 million m^3 in Bulgaria, 5.5 in Croatia, 15.3 in the Czech Republic, 7.6 in Estonia, 7.0 in Lithuania, 6.0 in Hungary, 17.4 in Austria, 8.0 in Slovakia, etc. [42]. Moreover, the forest cover of Greece is about 3.9 million hectares, with 3.5 million available for wood production [43]. In addition to the previous data, 52 thousand acres of forests were burned in Greece in 2018. Considering a wood density of about 10 m^3/acre (although, depending on the tree species, it can reach up to 40 m^3/acre), the total volume of burned forests corresponds to 0.5–4 times the annual timber production in Greece [44,45]. The data for the forest fires of 2021 are even worse, as 1.55 million acres of the total forest area was burned in Greece [46], which corresponds to 4.2% of the total forest area (of 36.8 million acres) [47].

It is important to mention that funds allocated in 2021 for fire protection was only EUR 1,700,000, which corresponds to only 10% of the costs requested by the relevant institutions [48]. This indicates that, with a very small increase of this fund, a significantly decreased amount of forest fires will occur in the future, allowing for the better exploitation of forests for timber to be used as fuel, rather than being devastated by fires.

The above data and calculations are approximate and, of course, a more detailed analysis is necessary. However, the above data show that a ten-times increase in the total timber production in Greece in the coming years could be an achievable goal. This production could be specifically focused on the mountainous areas of Western Macedonia, where the majority of the lignite mines are found today, but also on the many mountainous/semi-mountainous areas of Greece that are currently bare of forests and could accommodate special fast-growing tree plantations. This amount could replace about 15% of the current lignite consumption and will have several advantages, such as:

- Zero contribution to the emission of CO_2, because the CO_2 produced from biomass combustion is absorbed by the plants for their development;

- For equal energy production, natural gas produces about 55% of the emissions of CO_2 from lignite; however, the replacement of 15% of the quantity of lignite with wood corresponds to the net production of carbon dioxide equal to 85% of the original production. This decrease can be achieved with less effort than the complete transition to natural gas (the disadvantages of this transition were shown previously);
- Biomass will be a domestic product and, thus, the dependence on energy imports will be lower;
- The stimulation of jobs in Greece, and especially in the province, instead of the current decrease in the number of jobs due to the closure of the existing thermal plants;
- The development of a cutting-edge technology and the creation of a Greek know-how that can be exported; this will have multiple benefits for Greek society and economy;
- The possible co-combustion of waste will be another option;
- The protection and upgrade of Greece's rural and mountainous environment by stimulating ecosystems, reducing erosion, changing microclimate areas to less-warm, etc.

The decentralized electricity production, i.e., the creation of plants in many areas, e.g., 1–2 per county, will be more efficient due to the shorter transportation distance of biomass. This will also have a positive impact on the control of particles emissions, as the pollution will not be emitted in the same area.

The above (approximate) analysis shows that this route, albeit complementary to the import of natural gas, should be better exploited. In this case, the future use of lignite, combined with carbon capture technologies, may be more advantageous for the Greek economy than the transition to natural gas. A comprehensive study is necessary to take into account all the pros and cons of this alternative. Of course, a comprehensive technical and economic study should be carried out, including the external costs of this alternative, ensuring that there are no major environmental nuisances or degradations [49].

The above analysis shows that the further exploitation of biomass, a domestic product, can replace lignite to a certain extent with zero-equivalent CO_2 emissions, and it is therefore proposed that this is taken into consideration.

4.4. Gasification of Lignite

The gasification of lignite and the production of fuel gas is another alternative; however, this alternative is not considered by this plan. This technology, used in several parts of the world in the past, is found in recession after the mid-20th century due to the high competitiveness of oil and gas prices, but is again on the rise because of the necessity to reduce greenhouse gas emissions [50].

It is therefore proposed that the use of this alternative technology is explored in more technological, economic, and environmental detail. The gaseous fuel could be used for electricity generation or in large central plants (industry, hot water production, etc.), or even be considered as an addition to the domestic natural gas network.

The above process could continue the use of lignite, in combination with the use of biomass and carbon capture technologies, and continue to produce energy with domestic raw materials, lower costs for the Greek economy, and lower CO_2 emissions. A comprehensive technical, economic, and environmental study is again necessary to take into consideration all the pros and cons of this alternative.

4.5. CO_2 Storage

The use of CO_2 storage technology is not mentioned in this plan and is proposed only at one point, concerning future research actions.

It is true that this technology has, so far, been used worldwide in a limited number of facilities [51]. However, many countries, such as Canada, or the Netherlands in the port of Amsterdam [52], invest significantly in this technology. The plan proposes a research action for this technology, but only after the end of lignite production. In this case, the implementation of carbon storage technology will be of very limited value.

An immediate examination of the technological and economic uses of this technology, in combination with the continued use of lignite and in biomass combustion, is proposed here. This alternative can possibly have lower CO_2 emissions than the use of natural gas combustion [53]. In addition, a combination of an existing domestic source (lignite) with a new one (biomass), and a combination of a mature technology and infrastructure (lignite combustion) with a new one (biomass combustion and carbon storage) will be used.

4.6. Energy Storage

It is well known that the production of electricity from RES does not necessarily go hand-in-hand with consumption. The highest production of energy from photovoltaics occurs during the sunshine hours of the day, falling to zero during evening hours, when a peak in consumption occurs. The energy produced by wind turbines is quite unstable, as it depends on the windy hours. Moreover, the distribution of winds in the year and in space is also of high variability. In contrast to that, energy production from thermal plants, using fossil fuels or biomass combustion, and from hydropower plants, can be adapted to energy consumption.

Therefore, in order to efficiently use the energy produced from certain RES, such as wind turbines and photovoltaics, it is necessary to store the energy produced during the low consumption hours in order to use it during the high consumption hours. The main available energy storage techniques are pumped storage hydropower (using the pumping of water from a reservoir of low elevation to one of high elevation during low consumption hours and then allowing the flow of water from the high to low reservoir for the production of electric energy during peak hours), batteries, and hydrogen [54].

This plan mentions storage in batteries or in gas production (e.g., hydrogen), without giving specific data, but pumped storage hydropower is not included. However, pump storage could be an efficient way of storing energy. In addition, the creation of new water reserves could be very beneficial for agricultural purposes. The storage of energy in batteries on the level of an entire country can be quite problematic, as the cost of these batteries may be too high. Moreover, the environmental consequences of this very high amount of necessary batteries are not negligible [55,56].

Therefore, it is proposed that the alternative of pumped storage hydropower, instead of the battery storage that is proposed in this plan, is developed.

4.7. Vehicle Electrification

One of the actions of the NECP to reduce the use of fossil fuels is to increase the electrification of vehicles. The plan presents an estimation for the development of electric mobility in Greece until 2030. However, the estimated numbers are rather high.

The total market for new passenger cars is projected to increase from 103,431 units in 2018 (reference year) to 275,133 units in 2030, which corresponds to an annual increase of 8–11%, which is rather high. It should be noted that sales of 280–320,000 units/year took place in Greece during the period of 2000–2006.

However, the economic development of that past period cannot be compared to the current economic situation of Greek households. In addition, the plan estimates that Greece's GDP will be approximately the same as in 2008, when the economic crisis started, only by 2030. It should be noted that the rest of the European countries will have recovered much earlier from this economic crisis and will be at much higher corresponding levels of GDP in 2030. Having gone through a severe economic crisis, with declining incomes, high unemployment, and a large exodus abroad, especially of young scientists, it is probably very difficult to have such a large increase in the market for new passenger cars. Additional components that support this argument are:

- The market of passenger cars is much more saturated than in the period of 2000–2006;
- The implementation of sustainable mobility should lead to an increase in the use of public transport, so the needs for passenger cars will be lower in the future;

- This plan estimates an additional reduction in the country's population, as a result of either a decrease in births or migration abroad. It should be noted that the number of births per woman in Greece was only 1.4 in 2017, a value that is one of the lowest not only in Europe, but also worldwide. Low birth rates and migration of the Greek population abroad are both signs of economic recession and not of high economic growth;
- A high increase in the prices of residences (both for acquisition and rent) has occurred since 2017. Is should be noted that the index of the prices of dwellings constantly increased from 1997 to 2008 and then constantly decreased until 2017, due to the economic crisis. It is estimated that this increase will continue in the coming years [57]. This increase absorbs a higher and higher percentage of the income of households, and is rather competitive to the automobile markets, as it reduces disposable income for the purchase of a passenger car.

Moreover, the estimated rate of electric vehicle penetration (24–30% by 2030) may be overestimated. Electric vehicles, from almost non-existent today, are projected to have a quite-high penetration in 2030. Given the current available technology for electric cars (such as the number of kilometers that an electric vehicle can travel, sufficient for urban travel but not always for long-distance, or battery life, etc.), the necessary infrastructure to be created to recharge a car's battery, especially in public places, and the higher prices of electric cars compared to conventional vehicles, the above objective for the penetration of these cars may not be met so early. Several researchers already expressed their reservations about the announced rapid introduction of electric cars to the market [58]. It should also be noted that, as an additional difficulty, the battery-charging infrastructure of electric cars in public places is quite problematic in Greek cities, due to the general insufficient width of sidewalks and, moreover, to the high lack of parking availability in all Greek cities.

Due to the higher price of electric vehicles compared to conventional ones, high financial incentives for their purchase will probably be required, and this will be another additional charge for the Greek national budget and the Greek National Balance, as all these vehicles are imported.

For the above reasons, the existence of several scenarios with alternatives is more than necessary.

4.8. Public Transportation

Although there is a specific chapter in this plan on the electrification of passenger cars, the increase of public transportation in Greece is not taken into consideration. It is well known that the use of public transportation emits lower CO_2 emissions than the use of passenger cars [59], and this difference is even higher in a complete product life-cycle analysis, with all the external costs taken into consideration, since the total impact of policies or measures in the long term are unclear [60].

Greece has one of the lowest percentages of train-passenger kilometers, and the second lowest percentage of railways for the transportation of goods in the EU [61,62]. Moreover, Greece showed a very high decrease in the share of public transport in total passenger traffic, from 28% in 2000 to 18% in 2018 [62].

The shift to public transport is therefore of paramount importance, and this action should be immediately taken into consideration in this plan.

4.9. Energy Savings

The remarks concerning the energy savings are analyzed as a function of the type of the building: public administration buildings or residential buildings.

4.9.1. Energy Savings in Public Administration Buildings

The plan provides an annual energy upgrade of 3% of the total surface of the buildings of the central public administration. Some facts should be mentioned here. The first is that the ages of Greek public administration buildings are quite high [63]. In addition, many of them are listed; therefore, they require a specific process for their restoration and the targeted energy results cannot be achieved easily. Moreover, the procedures for such upgrades in the public sector are very time-consuming. These facts show that the target of the energy upgrade of 3% of the total surface each year is very probably unattainable, at least during the first years. On the other hand, there is an urgent need to upgrade much more than one-third of the total buildings' area by 2030.

The above shows that it is initially necessary to radically review and accelerate the current procedures for the energy upgrade of public buildings, of course with the necessary protection of listed buildings, and set a more ambitious target. The energy-saving measures must be first implemented, as saving energy is one of the most efficient measures to decrease CO_2 emissions.

4.9.2. Energy Savings in Residential Buildings

In Greece, there were about 3 million households and more than 6 million residences in 2019 [63]. More than 55% of these residences were built before 1980, i.e., they have very poor energy performance [63]. The plan proposes the energy upgrade of 60,000 residences per year, a number that corresponds to less than 10% of all residences by 2030. This percentage is obviously very small.

If the estimated economic growth is taken into account (the country's GDP will be by 2030 equal to that of 2008), it seems that the disposable income of citizens for energy upgrades will not be very high. This statement indicates that either there will be financial difficulty in upgrading many buildings, or that large public funds will be required for subsidizing this upgrade. For an estimated cost of EUR 10,000–15,000 for a residence of 80–100m^2 (depending on the climatic zone, the age of the residence, the exposure, etc.), the total cost will be more than EUR 60 billion. For comparison, the public revenues, spending, and the Program of Public Investment of the Greek state was, in 2019, EUR 53.02, 57.79, and 6.75 billion, respectively, and the GDP of Greece was EUR 192.75 billion in 2020 [64].

The plan also proposes an increase of the use of domestic natural gas. Taking into consideration the current coverage of Greece in the use of natural gas (only 5.4% of the residences used natural gas from 2011–2012 [65]) and the large and time-consuming projects required to increase the natural gas network, there are high reservations for the rapid penetration of the use of natural gas. Therefore, it is necessary to change the priorities and practices followed so far in order to achieve this goal. At this point, we can again express the previous comment concerning the choice to use domestic natural gas while other countries choose to abandon it.

4.10. Local Production of Products, Dietary Habits

There are several additional measures to decrease the emission of greenhouse gases. One of them is local/global production/consumption. The large penetration of globalization led to the high increase of the transportation of products on a global scale. However, a very effective action to decrease CO_2 emissions is to enhance the local production/consumption of products, as their transportation is significantly decreased. This policy has proven to be one of the most effective policies/actions to reduce greenhouse gas emissions from product transport (e.g., "food-kilometers") in the case of food transport [66].

However, this action is completely missing from this plan. In addition, this policy can strengthen the Greek economy and Greek businesses, and help with the creation of new jobs, especially in small cities or suburban areas. The dynamic integration of this policy is proposed in this plan. Specific policies and actions must be implemented immediately, as this policy will bring only positive results.

The food sector is responsible for a large proportion of greenhouse gas emissions, stemming from the production of primary food products, their process, transport, etc. [67]. Greece has an average consumption of 3353 kcal/cap/day calories in 2017, against the 2000 calories a day for women and 2500 for men that is recommended by the WHO [68]. Meat consumption in Greece is 76.7 Kg/cap/year [69], compared to 63.12 Kg/cap/year in

Europe in 2013 [70]. The decrease in extra calories and meat consumption can be two very efficient methods for the decrease in the greenhouse gas emissions of a country [67], with a very low cost and, moreover, with several other significant health advantages. However, there is no mention in this plan about these, or similar, alternative and low-cost measures for the decrease of greenhouse gases emissions.

4.11. Energy Price, Social Impact, Energy Poverty

The causes of energy poverty are low incomes, high fuel prices, and the poor thermal conditions of houses [71,72]. A total of 35% of Greeks have debts to energy bills (first place among the member countries of the European Union) and the percentage of Greeks who cannot keep their home warm is very high, ranking Greece in third place among the member states of the European Union [73].

It is, therefore, absolutely necessary that the actions of the plan should focus on mitigating energy poverty. However, there are some reservations about the effectiveness of the proposed actions. Reducing energy poverty requires either a high increase in income or a high decrease in the price of energy. However, the plan does not foresee either of those two options. Based on projected GDP growth in Greece, the Greek GDP will reach that of 2008, i.e., before the economic crisis, only in 2030.

The plan estimates a decrease in the price of RES in the coming years without providing more information about the final prices or about how this decrease will take place. It should be noted that it is very difficult to obtain the real cost of energy production of RES from official sources and, thus, this point cannot be verified.

It should be noted that the complete de-lignification and change of the energy mix of Greece in such a short period of time carries the risk of a significant increase in the final price of electricity. In recent years, from 2006 to 2017, there has been an increase of 28% in the price of electricity for medium-sized industries and 177% for households [74]. The current worldwide increases in fuel and energy prices are another example. Therefore, several reservations can be expressed about the announced decrease of energy prices in the coming years. In addition, the plan does not present alternative scenarios if the projected final energy price does not occur, nor possible actions or legal shields in case of speculative trends from the liberalized energy market.

Therefore, the decrease of energy poverty mentioned in the plan seems to be very difficult to achieve. It is proposed here that more generous, but also more specific measures to deal with this phenomenon are adopted. The control of household energy prices is the first of those measures.

Is should be noted that the environmental and social impacts from the energy transition to the main lignite area of Greece, that of Western Macedonia, are analyzed in another work [75].

4.12. Education, Public Awareness

The change of the energy mix of Greece with the shift to RES and natural gas, the gradual change of the fleet of vehicles to electric cars, and the actions of energy-upgrading buildings, etc., will bring significant changes in many technical professions directly related to the above issues.

It is obvious that not only these professions must be protected from any downgrading, but also that a substantial improvement of their role as well as their working conditions should take place. To upgrade these professions, the role of education and training should be enhanced, in addition to institutional and legislative upgrading. Education and training, both conventional and lifelong, and both in presence and distance, need to be upgraded to issues related to energy, the environment, and climate change. These issues need to be more strongly integrated at all levels of education, starting from the lowest, even that of primary school. However, in Greece, during the last years, there is a severe lack of technical workers to face the necessary technical works of the energy transition, and this is not taken into consideration in this plan.

Additionally, it is already known that well-informed and sensitized citizens can implement environmental protection actions much better than those who are not properly or fully informed [76,77]. Several points of the plan refer to citizens/consumers' actions on energy and climate change; however, these points are quite unclear. A specification of the increase of public information and awareness on issues of the safe and proper use of energy, energy savings, environmental protection, and climate change is proposed.

4.13. Regional and Sectoral Plans

The plan proposes individual regional plans to better implement the objectives of the project in the regions of Greece. According to the National Statistical Authority of Greece, the regions of Greece have an imbalanced contribution to the country's GDP [78]. Special mention should be given to regions with a small contribution, so as not to exclude these regions or to have them find themselves at second speed from the actions to be developed.

However, in addition to regional targeting, a sectoral dimension of this plan is missing. The establishment of a sectoral plan is probably more important than those of regional plans and should be established very soon.

5. Discussion

The main pillars of a National Energy and Climate Plan should first take into consideration the future energy mix of the country to mitigate climate change. However, energy is one of the main pillars of society and economy. For this reason, available energy, for example, without power interruptions or blackouts, and energy at an affordable price must be ensured so that the economy can function efficiently, but also to protect the disposable income of citizens and small enterprises. The high price of energy is the main reason for energy poverty, a very significant social problem in Greece, but also in many other countries in Europe or worldwide.

As a first step, a policy maker should guarantee that his proposal is efficient. If, for some reason, this proposal cannot be implemented, efficient alternatives should be applied. It is very strange that the Greek NECP lacks a risk analysis and proposes no alternatives to cover the probable cases of failure. This is a major shortcoming of this plan and should be restored as soon as possible, especially taking the very high increase of energy prices during 2020 into consideration. Greece, especially, is characterized as the most expensive wholesale electricity market in Europe, with the wholesale price at EUR 157 per MWh, a 70% increase since the beginning of the year [79].

Energy mixes and alternative technologies are the other critical points of this plan. The very rapid de-lignification of the country and the transition to natural gas can lead to a very problematic situation. The first point is that the decrease in greenhouse gas emissions will not be as high as is expected, mainly due to leaks of natural gas and its high global warming potential [26–28]. Additionally, the deterioration of the country's trade balance from the transition to natural gas is not taken into consideration in this plan. However, this is a major point, as the trade balance of Greece is, generally, highly negative. For example, the trade balance of Greece reached its highest point in 2015 (EUR -1.8 billion), during the debt crisis due to the collapse of aggregate demand, while it reached its lowest point on 2020 (EUR -12.52 billion) because of the collapse of tourism. Moreover, the trade balance of Greece becomes more negative when the GDP increases [80].

The energy dependence of Greece is also not taken into consideration, as a local product, lignite, is substitute with an imported one. A very specific analysis, using several scenarios of energy prices, should have been conducted to prove the advantages of this transition. However, this analysis is not shown in this plan. It is also clear that natural gas will be used only for some decades, and this statement adds a supplementary cost for the replacement of the infrastructure created only for a limited period of time.

More sustainable alternatives, such as the combustion of biomass, the gasification of lignite, or carbon storage, are not considered in this plan. However, the precise economic, environmental, and social evaluation of these alternatives, using different scenarios, should have been completed first. The same is valid in the case of energy storage, which is necessary due to the high future penetration of RES, where no comparative analysis, with different scenarios, is provided. Pumped storage hydropower, one of the established techniques for energy storage, is not compared with the other techniques to prove that storage in batteries has a lower cost, is more efficient, and more environmentally friendly.

The plan proposes a gradual shift to electric passenger cars. However, the targets set are overestimated and cannot be reached very easily. Moreover, the increase in the use of public transportation, one of the most efficient methods for decreasing CO_2 emissions from the transport sector, is not taken into consideration in this plan.

Energy saving is one of the most efficient ways to decrease greenhouse gas emissions. However, the proposed energy savings in public buildings and private residencies are too low and, moreover, will not be achieved very easily. Energy-saving measures must be first implemented before examining a radical change to an energy mix, which will probably not be adequate for the new consumption of energy.

Other alternatives for the mitigation of greenhouse gas emissions, such as the enhancement of local production/consumption, or the very low action of changing dietary habits, which also have several other advantages for human health, are not taken into consideration in this plan. In addition, the establishment of a circular economy is not strong enough in this plan.

The social impacts of this energy transition are weakly taken into consideration in this plan. Some examples are the provision for technical workers, and their education/training, or the social awareness and information of citizens.

The absence of sectoral plans is another shortcoming of this plan; these plans should be established as soon as possible.

Finishing the list of main shortcomings, the future low energy price is not guaranteed in this plan. It is mentioned that future energy prices will be lower than the current ones without providing, however, a precise analysis. Even if there are some actions to tackle energy poverty, achieving this with a low economic growth, high energy prices, and a low percentage of dwellings renovated each year is impossible.

From the previous analysis, a more detailed examination of the possibility of continuing to exploit lignite, an available domestic product, for a longer time, instead of its complete substitution with imported natural gas, (at least until the cost of energy produced from RES decreases significantly) is therefore proposed. This can be done by using modern, more environmentally efficient technologies, by combining lignite combustion with biomass combustion or gasification technology, and by using synthetic gas and, in addition, carbon storage technology, as is exposed here.

There is also a very strong necessity to ensure the availability of all forms of energy at an affordable price. It is proposed that special care is taken to verify the projected reduction in the prices of energy and that actions, institutional initiatives, and other arrangements are taken so that the final price of energy does not increase and, additionally, so that any speculative trends are avoided.

6. Conclusions

The Greek National Energy and Climate Plan set the priorities of the Greece in the terms of Energy and Climate. A complete de-lignification by 2028 is proposed. The future energy mix will be mainly composed of natural gas and RES.

This plan has several shortcomings: there is no risk analysis and no alternative scenarios are proposed; the participation of other energy sources, such as biomass or the gasification of lignite, is not considered; energy storage is mainly focused on batteries and hydrogen, while pump energy storage is not considered; the targets set for the electrification of the passenger car fleet are too difficult to achieve, and at the same time, there is no provision for the enhancement of sustainable mobility by increasing the participation of public transportation; energy savings in buildings are not so ambitious; the tackling of energy poverty is almost impossible; and the same goes for the control of energy prices.

It is suggested that this transition be delayed for some years, taking into consideration the most sustainable paths for transitioning, such as using more alternatives.

Author Contributions: Conceptualization, E.Z.; methodology, E.Z. and L.V.; investigation, L.V., Z.G. and S.M.; resources, L.V., Z.G., S.M. and R.H.-M.; data curation, Z.G. and S.M.; writing—original draft preparation, Z.G. and S.M.; writing—review and editing, E.Z., L.V. and R.H.-M.; supervision, E.Z. All authors have read and agreed to the published version of the manuscript.

Funding: This research received no external funding.

Institutional Review Board Statement: Not applicable.

Informed Consent Statement: Not applicable.

Data Availability Statement: Not applicable.

Conflicts of Interest: The authors declare no conflict of interest.

References

1. IPCC. Available online: https://www.ipcc.ch/ (accessed on 8 September 2021).
2. DCD. Available online: https://www.datacenterdynamics.com/en/opinions/what-fuel-is-used-by-power-stations-in-europe-and-the-us-today/ (accessed on 8 September 2021).
3. IEA. Available online: https://www.iea.org/reports/european-union-2020 (accessed on 5 September 2021).
4. European Commission. *Energy Roadmap 2050*; Publications Office of the European Union: Luxembourg, 2012; pp. 1–24.
5. IISD. Available online: https://sdg.iisd.org/news/wind-and-solar-will-provide-50-of-electricity-in-2050-bnef-report-finds/ (accessed on 10 September 2021).
6. Energypedia. Available online: https://energypedia.info/wiki/Greece_Energy_Situation#Electricity (accessed on 7 September 2021).
7. European Commission. Available online: https://ec.europa.eu/info/energy-climate-change-environment/implementation-eu-countries/energy-and-climate-governance-and-reporting/national-energy-and-climate-plans_en (accessed on 24 October 2021).
8. Ministry of Environment and Energy. *National Energy and Climate Plan (NECP)*; Ministry of Environment and Energy: Athens, Greece, 2019. Available online: https://ypen.gov.gr/energeia/esek/ (accessed on 5 September 2021).
9. Bampatsou, C.; Zervas, E. Critique of the Regulatory Limitations of Exhaust CO_2 Emissions from Passenger Cars in European Union. *Energy Policy* **2011**, *39*, 7794–7802. [CrossRef]
10. Martin, N.; Bishop, J.; Choudhary, R.; Boies, A. Can UK passenger vehicles be designed to meet 2020 emissions targets? A novel methodology to forecast fuel consumption with uncertainty analysis. *Appl. Energy* **2015**, *157*, 929–939. [CrossRef]
11. Liu, Y.; Liu, Y.; Chen, J. The impact of the Chinese automotive industry: Scenarios based on the national environmental goals. *J. Clean. Prod.* **2015**, *96*, 102–109. [CrossRef]
12. Palm, J. The Transposition of Energy Communities into Swedish Regulations: Overview and Critique of Emerging Regulations. *Energies* **2021**, *14*, 4982. [CrossRef]
13. Jain, M.; Sharma, G.D.; Mahendru, M. Can I Sustain My Happiness? A Review, Critique and Research Agenda for Economics of Happiness. *Sustainability* **2019**, *11*, 6375. [CrossRef]
14. Official Journal of Greek Government. Available online: https://ypen.gov.gr/wp-content/uploads/2020/11/%CE%A6%CE%95%CE%9A-%CE%92-4893.2019.pdf (accessed on 8 September 2021).
15. Wei, Y.; Jia, A.; Wang, J.; Qi, Y.; Jia, C. Current Technologies and Prospects of Shale Gas Development in China. In *Shale Gas. New Aspects and Technologies*; Al-Juboury, A., Ed.; Intechopen: London, UK, 2018. [CrossRef]
16. Sacco, A.; Bella, F.; Bianco, S.; Bongiovanni, R. Photocured Polymer Electrolyte Membranes for Dye-Sensitized Solar Cells. In *Photovoltaics: Synthesis, Applications and Emerging Technologies*; Gill, M., Ed.; Nova Science Publishers: Hauppauge, NY, USA, 2014; pp. 53–72.
17. Ladislaw, S.; Leed, M.; Watlon, M. *New Energy, New Geopolitics: Balancing Stability and Leverage*; CSIS: Washington, DC, USA, 2014; ISBN 978-1-4422-2835-1.
18. Shaikh, I. Impact of COVID-19 pandemic on the energy markets. *Econ Chang. Restruct* **2021**, 1–52. [CrossRef]
19. Lin, B.; Bai, B. Oil prices and economic policy uncertainty: Evidence from global, oil importers, and exporters' perspective. *Res. Int. Bus. Financ.* **2021**, *56*, 101357. [CrossRef]
20. Al-Awadhi, A.M.; Alsaifi, K.; Al-Awadhi, A.; Alhammadi, S. Death and contagious infectious diseases: Impact of the COVID-19 virus on stock market returns. *J. Behav. Exp. Financ.* **2020**, *27*, 100326. [CrossRef] [PubMed]
21. Papadamou, S.A.F.; Kenourgios, D.; Dimitriou, D. Direct and Indirect Effects of COVID-19 Pandemic on Implied Stock Market Volatility: Evidence from Panel Data Analysis. *MPRA* **2020**. Available online: https://mpra.ub.uni-muenchen.de/id/eprint/100020 (accessed on 9 May 2021).
22. Haroon, O.; Rizvi, S.A.R. COVID-19: Media coverage and financial markets behavior—A sectoral inquiry. *J. Behav. Exp. Financ.* **2020**, *27*, 100343. [CrossRef] [PubMed]

23. Nakicenovic, N.; Swart, R. *Emission Scenarios, IPCC*; Cambridge University Press: Cambridge, UK, 2000; p. 6.
24. Our Word in Data. Available online: https://ourworldindata.org/grapher/carbon-dioxide-emissions-factor (accessed on 23 September 2021).
25. Forster, P.; Ramaswamy, V.; Artaxo, P.; Berntsen, T.; Betts, R.; Fahey, D.W.; Haywood, J.; Lean, J.; Lowe, D.C.; Myhre, G.; et al. Changes in Atmospheric Constituents and in Radiative Forcing. In *Climate Change 2007: The Physical Science Basis. Contribution of Working Group I to the Fourth Assessment Report of the Intergovernmental Panel on Climate Change*; Solomon, S., Qin, D., Manning, M., Chen, Z., Marquis, M., Averyt, K.B., Tignor, M., Miller, H.L., Eds.; Cambridge University Press: Cambridge, UK; New York, NY, USA, 2007.
26. Oliveira, B.T.; Schulz, R.; McGlade, C.; Gould, T. *Methane Emissions from Oil and Gas*; IEA: Paris, France, 2019.
27. Our Word in Data. Available online: https://ourworldindata.org/fossil-fuels (accessed on 16 September 2021).
28. UNFCCC. Available online: https://unfccc.int/process/transparency-and-reporting/greenhouse-gas-data/greenhouse-gas-data-unfccc/global-warming-potentials (accessed on 18 September 2021).
29. Potter, P. The Netherlands to go completely gas-free in the future. *The Holland Times*, 28 June 2018.
30. Energypress. Available online: https://energypress.gr/news/sta-1356-eyro-meso-kostos-kathe-paragomenis-mwh-ape-2020-poso-syneisferei-kathe-pigi-eisrois (accessed on 10 September 2021).
31. Kathimerini. Available online: https://www.kathimerini.gr/economy/561503764/oi-lignitikes-monades-einai-aparaitites/ (accessed on 10 September 2021).
32. Cablevey. Available online: https://cablevey.com/the-history-of-biomass-as-a-renewable-energy-source/ (accessed on 12 September 2021).
33. Demirbas, A. Progress and recent trends in biofuels. *Progr. Energy Combust. Sci.* **2007**, *33*, 1–18. [CrossRef]
34. Zervas, E.; Drimili, E.; Gareiou, Z. Annoyance from winter atmospheric pollution created from wood burning for heating. In Proceedings of the ERS International Congress 2019, Vienna, Austria, 7–9 September 2019.
35. Drimili, E.; Zervas, E. Opinions of the residents of Athens on the heating of their homes in the winters since the beginning of the crisis. In Proceedings of the 1st Online Conference on Energy Poverty, Athens, Greece, 11 November 2020.
36. Chen, J.; Li, C.; Ristovski, Z.; Milic, A.; Gu, Y.; Islam, M.S.; Wang, S.; Hao, J.; Zhang, H.; He, C.; et al. A review of biomass burning: Emissions and impacts on air quality, health and climate in China. *Sci. Total Environ.* **2018**, *579*, 1000–1034. [CrossRef] [PubMed]
37. Sigsgaard, T.; Forsberg, B.; Annesi-Maesano, I.; Blomberg, A.; Bølling, A.; Boman, C.; Bønløkke, J.; Brauer, M.; Bruce, N.; Héroux, M.E.; et al. Health impacts of anthropogenic biomass burning in the developed world. *Eur. Resp. J.* **2015**, *46*, 1577–1588. [CrossRef] [PubMed]
38. Tzelepi, V.; Zeneli, M.; Kourkoumpas, D.-S.; Karampinis, E.; Gypakis, A.; Nikolopoulos, N.; Grammelis, P. Biomass Availability in Europe as an Alternative Fuel for Full Conversion of Lignite Power Plants: A Critical Review. *Energies* **2020**, *13*, 3390. [CrossRef]
39. Eurostat. Available online: https://ec.europa.eu/eurostat/statistics-explained/index.php/Glossary:Tonnes_of_oil_equivalent_%28toe%29 (accessed on 15 September 2021).
40. EURACOAL. Available online: https://euracoal.eu/info/country-profiles/greece/ (accessed on 17 September 2021).
41. Francescato, V.; Bergomi, L.Z.; Metschina, C.; Schneld, C.; Krajnc, N.; Koscik, K.; Nocentini, G.; Stranieri, S. *Wood Fuel Manual*; Biomass Trade Cent; AIEL—Italian Agroforestry Energy Association: Legnaro, Italy, 2008.
42. Eurostat. Available online: https://ec.europa.eu/eurostat/statistics-explained/index.php?title=File:Wood_production,_2000%E2%80%932015_(thousand_m%C2%B3)_YB17.png (accessed on 17 September 2021).
43. Eurostat. Available online: https://ec.europa.eu/commission/presscorner/detail/en/STAT_08_146 (accessed on 18 September 2021).
44. FAO. Available online: http://www.fao.org/3/y1997e/y1997e07.htm (accessed on 16 September 2021).
45. NRCAN. Available online: https://www.nrcan.gc.ca/our-natural-resources/forests-forestry/state-canadas-forests-report/howmuch-forest-does-canada-have/indicator-wood-volume/16399 (accessed on 14 September 2021).
46. Dasarxeio. Available online: https://dasarxeio.com/2021/08/27/101205/comment-page-1/ (accessed on 12 September 2021).
47. To Vima. Available online: https://www.tovima.gr/2021/03/05/society/dasiko-to-60-tis-ellinikis-epikrateias/ (accessed on 8 September 2021).
48. Dasarxeio. Available online: https://dasarxeio.com/2021/04/19/96134/ (accessed on 12 September 2021).
49. Greek Fireservice. Available online: https://www.fireservice.gr/el_GR/synola-dedomenon (accessed on 12 September 2021).
50. Globalsyngas. Available online: https://www.globalsyngas.org/resources/the-gasification-industry/ (accessed on 11 September 2021).
51. NRCAN. Available online: https://www.nrcan.gc.ca/energy/publications/16226 (accessed on 14 September 2021).
52. Port of Rotterdam. Available online: https://www.portofrotterdam.com/en/news-and-press-releases/port-authority-gasunie-and-ebn-studying-feasibility-of-ccs-in-rotterdam (accessed on 3 September 2021).
53. Rhodes, J.S.; Keith, D.W. Biomass with capture: Negative emissions within social and environmental constraints: An editorial comment. *Clim. Chang.* **2008**, *87*, 321–328. [CrossRef]
54. Energy.gov. Available online: https://www.energy.gov/eere/water/pumped-storage-hydropower (accessed on 21 September 2021).
55. Dehghani-Sanij, A.R.; Tharumalingam, E.; Dusseault, M.B.; Fraser, R. Study of energy storage systems and environmental challenges of batteries. *Renew. Sustain. Energy Rev.* **2019**, *104*, 192–208. [CrossRef]

56. McManus, M. Environmental consequences of the use of batteries in sustainable systems: Battery production. In Proceedings of the 2nd International Conference on Microgeneration and Related Technologies, Glasgow, UK, 4–6 April 2011; pp. 4–6.

57. Bank of Greece, Index of Prices of Dwellings (Historical Series). Available online: https://www.bankofgreece.gr/en/statistics/real-estate-market/residential-and-commercial-property-price-indices-and-other-short-term-indices (accessed on 19 September 2021).

58. Temple, J. Why the Electric-Car Revolution May Take a Lot Longer Than Expected. MIT Technology Review. Available online: https://www.technologyreview.com/2019/11/19/65048/why-the-electric-car-revolution-may-take-a-lot-longer-than-expected (accessed on 19 September 2019).

59. Shapiro, R.J.; Hassett, K.A.; Arnold, F.S. *Conserving Energy and Preserving the Environment: The Role of Public Transportation*; American Public Transportation Association: Washington, DC, USA, 2002.

60. Kii, M. Reductions in CO_2 emissions from passenger cars under demography and technology scenarios in Japan by 2050. *Sustainability* **2020**, *12*, 6919. [CrossRef]

61. Sipotra. Available online: https://www.sipotra.it/wp-content/uploads/2019/06/Passenger-transport-statistics.pdf (accessed on 22 September 2021).

62. Odyssee-mure. Available online: https://www.odyssee-mure.eu/publications/efficiency-by-sector/transport/transport-eu.pdf (accessed on 22 September 2021).

63. ELSTAT, 2019, Greece with Numbers, ELSTAT. Available online: https://www.statistics.gr/ (accessed on 27 September 2021).

64. Ministry of Finances of Greece. Available online: https://www.minfin.gr/documents/20182/7655501/%CE%95%CE%99%CE%A3%CE%97%CE%93%CE%97%CE%A4%CE%99%CE%9A%CE%97+%CE%95%CE%9A%CE%98%CE%95%CE%A3%CE%97.pdf/d100d4de-02c3-4651-be19-58020e917e84 (accessed on 20 September 2021).

65. European Commission. Available online: https://ec.europa.eu/energy/sites/ener/files/documents/20142207.78-93.pdf (accessed on 14 September 2021).

66. Weber, C.; Matthews, S. Food-Miles and the Relative Climate Impacts of Food Choices in the United States. *Environ. Sci. Technol.* **2008**, *42*, 3508–3513. [CrossRef] [PubMed]

67. Salamaliki, C.; Matsouki, N.; Drimili, E.; Vatikiotis, L.; Zervas, E. CO_2 benefits from the change of the Greek dietary habits. In Proceedings of the 1st International Conference on Environmental Design, Athens, Greece (Virtual), 24–25 October 2020.

68. NHS. Available online: https://www.nhs.uk/common-health-questions/food-and-diet/what-should-my-daily-intake-of-calories-be/ (accessed on 18 September 2021).

69. Euronews. Available online: https://www.euronews.com/2019/02/10/which-european-countries-eat-the-most-meat (accessed on 17 September 2021).

70. Our Word in Data. Available online: https://ourworldindata.org/meat-production/ (accessed on 17 September 2021).

71. Csiba, K. Overview. In *Energy Poverty Handbook*; Csiba, K., Ed.; European Union: Brussels, Belgium, 2016.

72. Energy Poverty. Available online: https://www.energypoverty.eu/publication/measuring-energy-poverty-greece (accessed on 19 September 2021).

73. Eurostat, 2019, Ageing Europe, Eurostat. Available online: https://ec.europa.eu/eurostat/web/products-statistical-books/-/KS-02-19-681 (accessed on 27 September 2021).

74. Vatikiotis, L. *Energy Poverty in Small and Medium Enterprises*; Research Texts; IME GSEVEE: Athens, Greece, 2019; p. 64.

75. Zervas, E.; Vatikiotis, L.; Gareiou, Z. Proposals for an environmental and social just transition for the post-lignite era in Western Macedonia, Greece. In Proceedings of the 2nd International Conference on Environmental Design, Athens, Greece (Virtual), 23–24 October 2021.

76. Gilg, A.; Barr, S. Behavioural Attitudes Towards Water Saving? Evidence from a Study of Environmental Actions. *Ecol. Econ.* **2006**, *57*, 400–414. [CrossRef]

77. Drimili, E.; Gareiou, Z.; Vranna, A.; Poulopoulos, S.; Zervas, E. An integrated approach to public's perception of urban water use and ownership of water companies during a period of economic crisis. Case study in Athens, Greece. *Urban Water J.* **2019**, *16*, 334–342. [CrossRef]

78. ELSTAT, 2020, Regional Accounts: Gross Value Added for the Year 2017, Press Release. Available online: https://www.statistics.gr/documents/20181/a3067d2d-82fd-703b-a600-8c2ad811885b (accessed on 29 September 2021).

79. Kathimerini. Available online: https://www.ekathimerini.com/economy/1167134/steep-hikes-in-electricity-bills/ (accessed on 18 September 2021).

80. ELSTAT. Available online: https://www.statistics.gr/en/statistics/-/publication/SEL30/- (accessed on 18 September 2021).

sustainability

MDPI

Article

South Korean Public Acceptance of the Fuel Transition from Coal to Natural Gas in Power Generation

Hyung-Seok Jeong, Ju-Hee Kim and Seung-Hoon Yoo *

Department of Energy Policy, Graduate School of Convergence Science,
Seoul National University of Science & Technology, Seoul 01811, Korea;
azar76@seoultech.ac.kr (H.-S.J.); jhkim0508@seoultech.ac.kr (J.-H.K.)
* Correspondence: shyoo@seoultech.ac.kr; Tel.: +82-2-970-6802

Abstract: South Korea has set up a plan to convert 24 coal-fired power plants into natural gas-fired ones by 2034 in order to reduce carbon dioxide (CO_2) emissions. This fuel transition can succeed only if it receives the public support. This article seeks to investigate the public acceptance of the fuel transition. For this purpose, data on South Koreans' acceptance of the fuel transition were gathered on a nine-point scale from a survey of 1000 people using face-to-face individual interviews with skilled interviewers visiting households. The factors affecting acceptance were identified and examined using an ordered probit model. Of all the interviewees, 73.6 percent agreed with and 12.2 percent opposed the fuel transition, respectively, agreement being about six times greater than opposition. The model secured statistical significance and various findings emerged. For example, people living in the Seoul Metropolitan area, people who use electricity for heating, people with a low education level, young people, and high-income people were more receptive of the fuel transition than others. Moreover, several implications arose from the survey in terms of enhancing acceptance.

Keywords: coal; natural gas; CO_2 emissions; public acceptance; ordered probit model

check for
updates

Citation: Jeong, H.-S.; Kim, J.-H.;
Yoo, S.-H. South Korean Public
Acceptance of the Fuel Transition
from Coal to Natural Gas in Power
Generation. *Sustainability* **2021**, *13*,
10787. https://doi.org/10.3390/
su131910787

Academic Editors:
Georgios Tsantopoulos and
Evangelia Karasmanaki

Received: 17 August 2021
Accepted: 23 September 2021
Published: 28 September 2021

Publisher's Note: MDPI stays neutral
with regard to jurisdictional claims in
published maps and institutional affiliations.

1. Introduction

Since coal has a higher carbon content than other fossil fuels such as oil and natural gas (NG), it emits more carbon dioxide (CO_2), a greenhouse gas, during the combustion process. As a result, countries around the world are making various efforts to reduce coal power generation to prevent climate change [1,2]. In other words, energy transition is being pushed around the world to change from coal, which is high-carbon energy, to low-carbon or zero-carbon energy [3–5].

For example, the United States will shut down one-fifth of all coal-fired power plants by 2025 [6,7]. Germany plans to abolish all coal-fired power plants by 2038 by enacting the so-called 'de-coal law' and establishing a three-stage de-coal schedule [8,9]. As of December 2019, the plan is to reduce the capacity of 43.9 GW coal-fired power plants in Germany to 30 GW by 2022, 17.8 GW by 2030, and zero by 2038. In China, NG-fired power plants, replacing coal-fired power plants, are expected to grow at an average annual rate of 10% from 2025 to 2030, with a capacity of 235.7 GW by 2030 [10]. Japan also has a plan to abolish 100 of its 140 coal-fired power plants by 2030 [11].

South Korea is no exception to this trend [12]. The Ninth Basic Plan for Electricity Demand and Supply (2020–2034), which was finalized at the end of December 2020, proposed a goal to reduce the proportion of coal-fired power generation from 40.4 percent in 2019 to 29.9 percent in 2034 [13]. This is quite challenging as it will abolish 30 coal-fired power plants, half of the total of 60 coal-fired power plants in operation, by 2034. Of the 30 coal-fired power plants that will be abolished, six will be shut down completely, while the remaining 24 will be converted to NG-fired power plants. In producing the same amount of electricity, an NG-fired power plant emits less than half of the greenhouse gases emitted by a coal-fired power plant. In addition, if this trend continues, the remaining

30 coal-fired power plants that are supposed to operate without being abolished by 2034 are expected to be converted into NG-fired power plants or abolished as early as possible.

South Korea's total CO_2 emissions in 2018 stood at 728 million tons, ranking eighth in the world; its per capita CO_2 emissions ranked higher, at sixth in the world. More specifically, the electricity and heat sector, industry sector, transportation sector, and commercial and other sector accounted for 287 million tons (39.4%), 243 million tons (33.4%), 98 million tons (13.5%), and 100 million tons (13.7%), respectively. Total CO_2 emissions in 2018 reached 149 percent of the amount in 1990, but CO_2 emissions in the electricity and heat sector reached 480 percent of the amount in 1990. That is to say, since the increase in CO_2 emissions in the electricity and heat sector is due to the increase in coal-fired power generation, reducing coal-fired power generation has emerged as the most basic challenge to reducing CO_2 emissions [14,15].

Under the Paris Agreement signed in 2015, South Korea submitted a nationally de-termined contribution (NDC) with a goal of reducing its 2018 greenhouse gas emissions by 26.3% by 2030 to the United Nations Framework Convention on Climate at the end of December last year. Currently, raising the reduction target on the NDC is discussed in preparation for the 26th Conference of the Parties to be held in Glasgow, United Kingdom in November 2021. For the successful implementation of the NDC, the power generation sector should abate its 2018 greenhouse gas emissions (269.6 million tons of CO_2e) to 192.7 million tons of CO_2e by 2030. To this end, the South Korean government intends to change the mix of the power generation sources in two directions [13].

First, South Korea will increase the share of renewable energy (RE) generation from 6.5% in 2019 to 20.8% by 2030. In particular, current global trends suggest that RE has become more reliable and cost-effective [16]. In South Korea, RE-related costs have been gradually decreasing. However, not a small amount of subsidies are still required for RE generation projects. This is because the price of RE facilities continues to fall, but the costs of compensation for local residents are increasing significantly as the acceptance of the residents in areas where RE facilities are being installed is getting worse. Therefore, the implementation of the NDC is almost impossible with the expansion of RE alone.

Second, replacing some of the coal-fired power plants with NG-fired ones will be promoted. Some argue that a significant portion of the coal-fired power plants being abolished should be substituted with RE. However, there are many limitations in terms of intermittency and variability to significantly expanding RE immediately. Moreover, unfortunately, South Korea's power grid is not linked to that of neighboring countries, and its technological capabilities and market systems are not fully equipped to cope with the variability of RE. Thus, most of the coal-fired power plants that are abolished will be replaced by NG-fired ones for the time being. Of course, since RE will expand significantly in the long term, NG-fired power plants are expected to serve as bridges in the era of energy transition [17–19].

Three major problems have been raised in this regard, causing some controversy. First, South Korea is one of the highest-cost countries in the world to consume NG, as it must liquefy, transport, vaporize, and use NG produced overseas. In other words, converting power generation fuel from coal to NG could increase costs and eventually lead to higher electricity bills [20], which could increase the burden on the people and weaken industrial competitiveness. Second, almost all of the NG consumed in the country depends on imports, and due to its high dependence on the Middle East, such as Qatar and Oman, NG supply and price stability are weak depending on international political circumstances [21]. Therefore, it is pointed out that it may be not desirable to expand the use of NG in terms of improving energy security [22]. Third, comparing the number of people employed at coal-fired power plants with those employed at NG-fired power plants of equal capacity, the fuel transition is expected to reduce the number of jobs as the latter is about half of the former.

Because the above three issues have become disputable problems and no one is willing to take responsibility, public consensus on fuel transition is insufficient due to a lack of public debate and discussion [23]. In some cases, the fuel transition may not be carried out and may only have a declarative meaning, or it could run aground in the face of public opposition. Thus, it is necessary to draw clear implications after determining the public's acceptance of the fuel transition from coal to NG at this point in time [24,25]. This is because it is the people who bear the social costs of fuel transition, and energy policymakers desperately need information about their acceptance.

With countries around the world struggling to reduce greenhouse gas emissions, the policy of converting power generation fuels from coal to NG is a global trend [26]. This fuel transition can be seen from two perspectives. First, both coal and NG are fossil fuels, but the fuel transition is inevitable because in generating the same amount of electricity combustion of NG emits much less both CO_2 and particulate matters than that of coal. For example, de Gouw et al. [27] reported that emissions of greenhouse gases and air pollutants decreased by 56% and 40–44%, respectively, when converting fuel from coal to NG. Eventually, in order to reduce emissions of air pollutants as well as greenhouse gases, the transition of power generation fuel from coal to NG will accelerate [14,28]. As mentioned earlier, the representative countries that are pursuing this fuel transition are the United States, Germany, China, and Japan [26].

Second, because NG also emits CO_2 during combustion, NG plays a role as a bridge energy that maintains an intermediate stage to a complete RE society [18,19,29–31]. Of course, since NG is also a fossil fuel, the expansion of its use may be limited at some point in the future. However, it is expected that its role as a bridge energy will expand [32,33]. Regardless of which of the two perspectives is more important, the fuel transition will continue for the time being.

However, so far as the authors know, there is no literature that analyzes the public acceptance of the fuel transition and the factors affecting it. Since the fuel transition leads to an increase in power generation costs, the public acceptance of the transition needs to be clearly examined. This study aims to meet this need. In other words, this study intends to add this analysis using the specific case of South Korea to the literature at a time when coal as a power-generation fuel is being replaced by NG, but investigation of the public acceptance of the replacement and the factors influencing it remains scarce.

The public acceptance covered in this paper is not about technology or the fuel itself. The fuel transition can have several effects, but the most important effect is an increase in electricity bills. Therefore, the main target of the public acceptance dealt with in this study can be said to be the increase in electricity rates caused by the fuel transition. This is because abolishing coal-fired power plants and converting them into NG-fired power plants will reduce greenhouse gas emissions while increasing electricity rates. In the South Korean situation, NG is more expensive than coal.

After simply examining the public acceptance of the transition of power generation fuel from coal to NG, this study attempts to analyze the factors that affect its acceptance. To this end, a survey of 1000 people nationwide was conducted and the results reported. To the best of the authors' knowledge, since this attempt is the first in South Korea and there are few cases internationally, this research is believed to be able to contribute significantly to the related literature. The remainder of this paper consists of four sections. The next section reports a brief literature review, and the history and the present situation of coal-fired power generation in South Korea. Section 3 presents a description of the materials and methods adopted in this work. Section 4 explains and discusses the results. Section 5 deals with conclusions.

2. Brief Literature Review and History of Coal-Fired Power Generation in South Korea

2.1. Literature Review

Public acceptance, as well as expert opinion, is an important consideration in the establishment and implementation of energy policies [34]. Therefore, a number of studies have been conducted to analyze the public acceptance of various RE sources and energy policies. For example, the public acceptance of hydrogen technology [35], hydrogen charging stations [36,37], solar energy [38], wind energy [38–41], geothermal energy [42], hydroelectric energy [43], energy infrastructure [44,45], various energy sources [46], RE cooperatives [47], RE policies [48], and energy transition policy [4] have been analyzed in the literature.

Some of these studies identified factors influencing public acceptance and then examined them as independent variables. For instance, Huijts et al. [37] used variables on individual perceptions such as individual behavior, perception of the effectiveness of hydrogen charging stations, subjective norms, personal norms, trust in industries, trust in local governments, awareness of environmental issues, and awareness of fairness in hydrogen charging station deployment. Mistur [46] employed health level, political orientation, ideology, gender, education, age, race, religion, and residential area as variables. Tabi and Westenhagen [43] adopted gender, age, income, education, political views, residential areas, and membership of environmental organizations as variables.

Kim et al. [40] utilized gender, education, age, income, residence in the metropolitan area, home solar power retention, average monthly electricity consumption per household, perception of the proportion of NG-fired power generation in the nation's total energy generation, and political tendencies as variables. Fischer et al. [47] included risk-taking perception, patience, political orientation, environmental perception, age, gender, education, income, rural area residence, and West Germany as variables. Kim et al. [4] deployed gender, age, education, income, residence in the metropolitan area, electric heating, cooktop use, environmental awareness, and prior knowledge of the renewable energy 100% campaign as variables. Furthermore, Venkatesh et al. [49] formulated the unified theory of acceptance and use of technology, and analyzed user acceptance of information technology. They found that the socio-demographic factors and individual experience affect individual acceptance of information technology.

Looking closely at these factors, they are largely divided into four categories. The first category is the socioeconomic variables of the respondent. Most of the studies that analyzed the public acceptance considered variables such as gender, age, education, and income of respondents. The second category is related to respondents' perception and includes respondents' perception of energy policy, RE-related technology, or prior knowledge of the object to be investigated. The third category is the variables concerning the respondent's living environment, the location of the respondent's residence, and the characteristics of the respondent's house. The fourth category relates to the characteristics of respondents. For example, whether respondents engage in environmental group activities or use eco-friendly products. The variables used in previous studies are summarized in Table 1.

Table 1. Factors affecting public acceptance used in previous studies.

Factors	Sources
Gender	Kim et al. [40], Tabi and Wuestenhagen [43], Mistur [46], Fischer et al. [47], Kim et al. [4], Venkatesh et al. [49], Seo et al. [19]
Age	Kim et al. [40], Tabi and Wuestenhagen [43], Mistur [46], Fischer et al. [47], Kim et al. [4], Venkatesh et al. [49], Seo et al. [19]
Education	Kim et al. [40], Tabi and Wuestenhagen [43], Mistur [46], Fischer et al. [47], Kim et al. [4], Seo et al. [19], Kim et al. [28]
Income	Kim et al. [40], Tabi and Wuestenhagen [43], Fischer et al. [47], Kim et al. [4], Seo et al. [19], Kim et al. [28]
Residential area	Kim et al. [40], Tabi and Wuestenhagen [43], Mistur [46], Fischer et al. [47], Kim et al. [4], Seo et al. [19]
Personal life characteristics	Huijts et al. [37], Kim et al. [40], Tabi and Wuestenhagen [43], Kim et al. [4], Seo et al. [19]
Health and faith	Mistur [46]
Personal Perception	Huijts et al. [37], Kim et al. [40], Tabi and Wuestenhagen [43], Mistur [46], Fischer et al. [47], Kim et al. [4], Venkatesh et al. [49], Seo et al. [19]

2.2. History of Coal-Fired Power Generation in South Korea

Coal fired power plants, along with nuclear power plants, have contributed signif icantly to enhancing the industrial competitiveness of the export-driven South Korean economy by serving as a source of low electricity bills over the past 30 years. With the country's successful localization of coal-fired power plants with a capacity of 500 MW, coal-fired power plants have not only played a role in offering a cheap and stable power supply, but also helped create jobs. South Korea suffered a nationwide rolling power outage on 15 September 2011 due to a lack of power supply facilities. A stable power supply emerged as an important task, and the Sixth Basic Plan for Electricity Demand and Supply (2013–2027), announced in February 2013, reflected the new construction of 10.5 GW capacity of coal-fired power plants [50].

In the course of establishing the Seventh Basic Plan for Electricity Demand and Supply (2015–2029), announced in July 2015, reducing the proportion of coal-fired power plants was discussed for the first time to reduce emissions of CO_2 and particulate matter [51]. However, the trend of expanding coal-fired power plants remained, as a stable electricity supply was considered more important than the environment. Until April 2017, a policy of continuously expanding coal-fired power generation had been implemented since the cost of coal-fired power generation was lower than that of NG-fired power generation or that of RE generation. In particular, the Ministry of Trade, Industry and Energy—a South Korean government department in charge of electric power policy—judged that maintaining cheap electricity bills through coal-fired power plants was more important than supplying eco-friendly power.

However, with the launch of a new government advocating energy transition in May 2017, policies began to be implemented to control the pace of the increase in coal-fired power plants. The coal-fired power plants under construction would be completed to stabilize the supply and demand of electricity, but all of the planned new coal-fired power plants would be scrapped, and older coal-fired power plants that reach 30 years of operation would be abolished [52]. The coal-fired power plants which were abolished were old and had a small capacity of 500 MW, but the coal-fired power plants under construction were the latest model, with a capacity of more than 1000 MW. Therefore, the speed of increase in the number of coal-fired power plants was reduced, but it was not planned to reduce the overall capacity of the coal-fired plants significantly.

Since then, severe particulate matter problems occurred in March and April 2019. People's anxiety about particulate matter overwhelmed other social problems. In April 2019, the National Climate and Environment Council was launched as a state agency to take responsibility for and deal with particulate matter under the President's direct control. Ban Ki-Moon, who served as the Secretary-General of the United Nations, was appointed chairman of the Council, to take charge of international cooperation to reduce particulate matter. The Korea Ministry of Environment announced that about 15 percent of particulate matter generated in the country was emitted from coal-fired power plants, and the issue of how to deal with coal-fired power plants was seriously discussed.

As a first step, the Council proposed the introduction of a particulate matter seasonal management system that would stop or minimize the operation of coal-fired power plants for four months from December 2019 to March 2020. The Korea Ministry of Trade, Industry and Energy opposed the introduction of the system, citing a surge in electricity demand for heating during winter, but the system was introduced and implemented in the name of reducing particulate matter. Under the system, coal-fired power plants should not operate as much as possible. If coal-fired power plants are inevitably operated to meet the increased demand for electricity, their power generation should be lowered to 80% of normal operation. This is because it is impossible for South Korea's coal-fired power plants to reduce their power generation to less than 80% due to their design. Therefore, unlike other countries that aimed to reduce CO_2 emissions, South Korea has begun to push for reducing coal-fired power generation to abate serious particulate matter emissions.

In November 2020, the year after the Council was launched, it proposed to remove all coal-fired power plants by 2045. Considering Germany's push to eliminate coal-fired power generation in 2038, a vote was taken among selected people in the nation with three proposed dates to cease coal-fired power generation: 2040, 2045 and 2050; the majority chose

2045. These people, called national representatives, were 500 people selected from all over the country by the Council in terms of age, income, gender, and region. In the meantime, China declared its intention to reach carbon neutrality by 2060 in September 2020, and Japan declared its intention to reach carbon neutrality by 2050 in October 2020. South Korea, geopolitically located between the two countries, also declared carbon neutrality by 2050 in October 2020. The specific carbon neutrality route and implementation means will be determined through further discussions, which are currently active.

The Ninth Basic Plan for Electricity Demand and Supply (2020–2034), which began in January 2019, was finalized and announced in December 2020, about two years later [13]. The reason it took such a long time was to reflect the greatly strengthened goal to reduce CO_2 emissions compared to the Eighth Basic Plan for Electricity Demand and Supply (2017–2031). The Ninth Basic Plan for Electricity Demand and Supply (2020–2034) calls for an additional reduction in CO_2 emissions in the power generation sector of 34.1 million tons by 2030 compared to the Eighth Basic Plan for Electricity Demand and Supply (2017–2031).

Due to time constraints, the Ninth Plan did not reflect carbon neutrality by 2050, but the Tenth Plan, which is scheduled to be finalized at the end of 2022, decided to reflect it. The Ninth Plan decided to abolish coal-fired power plants when they reached the age of 30 and to introduce a price-bidding mechanism on coal-fired power plants from April 2022 by setting an upper limit on coal-fired power generation [13]. The amount of coal-fired power generation decided annually will decrease year by year, and South Korea's coal-fired power plants will be shut down in the near future.

3. Materials and Methods

3.1. How to Investigate Public Acceptance

Public acceptance of a particular policy pursued by the government means the quantity of people who agree with the implementation of the policy. The implementation could gain momentum if many people vote for it. However, it is difficult to secure momentum for the policy if many people disagree with its implementation. Thus, as mentioned in Section 2, there are quite a number of studies analyzing public acceptance of a newly introduced policy or technology in the field of energy. Of course, people's approval is not the only consideration in pursuing a particular policy, but it must be one of the most important considerations [53].

The first thing to do here is to determine the methodology for analyzing public acceptance. This study aims to collect data by asking about public acceptance through a survey of a large number of people selected at random. Surveys are an effective means of collecting people's opinions directly [4,34,40]. The study will investigate whether, in order to reduce CO_2 emissions, they are in favor of or against the policy of abolishing coal-fired power generation early and replacing the corresponding capacity with NG-fired generation. Not only are the pros and cons examined, but they are also tallied. In fact, comprehensively aggregating pros and cons is a very simple task. A more complex and meaningful task is to derive the implications by analyzing the determinants of those pros and cons, which is also performed in this work.

First, various issues related to survey data collection, such as the survey target, sampling method, sample size, and survey method, are reviewed below, given that a survey is used for this study. Next, how to construct specific questions to identify public acceptance is explained. Finally, an econometric model is presented that can derive implications by analyzing the determinants while considering the nature of the data collected on public acceptance.

3.2. How to Gather the Data through a Survey

Regarding the collection of data, the survey method, survey target, sampling method, sample size, and survey area are determined. First, the survey method used in this study is a costly person-to-person individual interview with households. Of course, other low-cost survey methods were also available, such as postal interviews, telephone interviews, and internet interviews; however, person-to-person individual interviews were essential to fully explain the background to the transition of power generation fuel from coal to NG. In the case of postal interviews, there is no guarantee that people will properly look at the enclosed data, and the collection rate of questionnaires is extremely low in South Korea. In the case of telephone interviews, it is difficult to fully explain background information to respondents. While internet interviews have the advantage of being the lowest cost of the survey methods, they can cause problems that make random sampling difficult, leading to an increased probability of sample selection bias.

Second, the survey target in this study was selected as adults aged between 20 and 65 years. Of course, the opinions of people under 20 or over 65 may be important, but to obtain a responsible answer, the survey target was limited to those who could engage in economic activities. In South Korea, one graduates from high school at the age of 19 and becomes a true adult from the age of 20, becoming a university student or getting a job, and engaging in economic activities. In addition, people aged 65 usually retire from economic activity. Meanwhile, the proportions of men and women in the survey were the same.

Third, the sample size in this study was 1000. If a relatively large sample size is used, it is desirable in that it reflects many people's opinions, but it also creates disadvantages that increase the cost of the survey. In the end, it was necessary to determine the sample size that could collect people's opinions with some representation and avoid the problem of a sharp increase in survey costs. In this regard, Arrow et al. [54] pointed out that a sample size of 1000 may be appropriate to collect people's opinions. The Korea Development Institute [55] also proposed that the sample size be 1000 when conducting a survey for public sector decision-making. The sample size of 1000 used in this study is consistent with the suggestions made in these studies.

Fourth, the interviews were conducted by experienced interviewers belonging to a professional survey company. The authors first fully discussed the content of the questionnaire with the company's supervisors. Next, the supervisors trained the interviewers. In the course of the training, interviewers practiced questioning each other with the questionnaire. Interviewers who completed the training visited households and had them complete the questionnaire. The interviewers checked that there was no response to the main questions in the questionnaire and asked respondents to correct and supplement the questionnaire if necessary. Respondents who completed the questionnaire received simple household items such as a shopping bag, toothpaste, or a portable sewing kit as gifts.

Fifth, out of a total of 17 provinces in South Korea, 16 provinces were targeted, excluding Jeju Island. This was because Jeju-do, an island far from the mainland, had the lowest population among the 17 provinces, while the unit price of the survey was the highest. For these reasons, Jeju-do is usually excluded when conducting person-to-person individual interview surveys in South Korea. The Korea Development Institute [55] also suggests excluding Jeju-do when conducting a national opinion survey.

3.3. How to Prepare the Questionnaire

The final version of the questionnaire used in this study consisted of three main parts. After explaining the purpose of the survey, the first part asked about basic perceptions of several things. The second part asked about acceptance of the policy of converting 24 currently operating coal-fired plants into equal-capacity NG-fired power plants by 2034. The third part contained questions about the general characteristics of respondents. Answers to these questions are considered as candidates for factors affecting acceptance. The questions were about residential area, heating system, household income, personal income, education level, age, gender, etc.

For the second part, which is a key part of the questionnaire, how to ask questions about public acceptance had to be decided. For example, Ono and Tsunemi [56] used four views: "absolutely disagree," "slightly disagree," "slightly agree," and "absolutely agree." However, as a result of requesting a preliminary survey of 30 people from a professional survey company, two points were raised. First, those who participated in the preliminary survey asked to add a "neutral" view. In practice, the most widely used Likert scale adopts five levels. For example, it would be appropriate to use the five options "strongly disagree," "disagree," "neutral," "agree," and "strongly agree."

Second, participants required more granularity in the five views. For example, instead of two views, "strongly disagree" and "disagree," it was proposed to use more granular views. Thus, these two were divided into four: "absolutely disagree," strongly disagree," "disagree," and "slightly disagree." The same was true of "strongly agree" and "agree." In fact, the analytic hierarchy process, developed by Saaty [57] and widely used in decision analysis, suggests the use of a nine-point scale rather than a five-point scale in value judgment. Consequently, this research finally used a total of nine views, from opposition to affirmation, as measures of acceptance. In other words, 1 to 9 correspond to "absolutely disagree," "strongly disagree," "disagree," "slightly disagree," "neutral," "slightly agree," "agree," "strongly agree," and "absolutely agree."

3.4. How to Identify the Factors Affecting Public Acceptance

Public acceptance will be affected by a variety of characteristics of respondents. This research considers a total of 11 variables in relation to these characteristics. The names and definitions of these are shown in Table 2. In addition, basic statistics such as average and standard deviation are included in the table. In determining these 11 variables, the previous studies that investigated some factors affecting the public acceptance presented in Table 1 were referred to as important. As shown in Table 1, the main variables used in previous studies were income, residential area, gender, age, education, personal life characteristics, health and faith, and personal perception. These variables are generally composed of three categories: the characteristics of respondent households, the individual characteristics of respondents, and the perception and judgment of respondents. Therefore, the 11 variables presented in Table 2 are similarly divided into three categories.

Table 2. Information on variables in the model.

Variables	Definitions	Mean	Standard Deviation
Metro	Dummy for interviewees living in the Seoul Metropolitan area (0 = no; 1 = yes)	0.534	0.499
Heating	Dummy for interviewee households using electricity for heating (0 = no; 1 = yes)	0.013	0.113
Income	Dummy for interviewee households' monthly income being larger than KRW 4.88 million (USD 5.75 thousand) (0 = no; 1 = yes)	0.478	0.500
Education	Interviewees have more than twelve years' education (0 = no; 1 = yes)	0.633	0.482
Age	Interviewees' age	48.009	9.417
Know1	Dummy for interviewees knowing about energy transition policy well before the survey (0 = no; 1 = yes)	0.408	0.492
Know2	Dummy for interviewees knowing about hydrogen vehicles well before the survey (0 = no; 1 = yes)	0.323	0.468
Environment	Interviewees' subjective judgment about which is more important: jobs or the environment (0 = jobs; 1 = environment)	0.456	0.498
Forest	Dummy for interviewees being in favor of the utilization of unused forest biomass (0 = no; 1 = yes)	0.482	0.500
Fsolar	Dummy for interviewees being in favor of the expansion of floating solar power facilities (0 = no; 1 = yes)	0.510	0.500
H2-car	Dummy for interviewees being in favor of the expansion of hydrogen vehicles (0 = no; 1 = yes)	0.359	0.480

The first category is the characteristics of respondent households. This category contains three variables: Metro, Heating, and Income. The Metro, Heating, and Income variables are a dummy for the interviewee's living in the Seoul Metropolitan area (0 = no; 1 = yes), a dummy for the interviewee households' using electricity for heating

(0 = no; 1 = yes), and a dummy for the interviewee household's monthly income being larger than KRW 4.88 million (USD 5.75 thousand) (0 = no; 1 = yes), respectively. The Income variable is defined as a dummy that identifies whether the interviewee household income is greater or less than the average value of the sample. That is, considering that the average monthly household income in the sample is KRW 4.88 million, the Income variable has a value of one if the interviewee household's income is greater than KRW 4.88 million and zero otherwise.

In the second category, two variables, Education and Age, were used as individual characteristics. The Education and Age variables are dummies for interviewees having more than twelve years' education (0 = no; 1 = yes) and interviewees' age in years, respectively. Other personal characteristic variables, such as the gender of the respondents and whether they are household owners, were also considered candidates for reflection, but were eventually excluded from Table 2 as the analysis indicated that they had little effect on acceptance.

Six variables related to respondents' recognition and judgment were utilized as the third category. Know1, Know2, Environment, Forest, Fsolar, and H2-car are dummies for interviewees knowing about the energy transition policy well before the survey (0 = no; 1 = yes), a dummy for interviewees knowing about hydrogen vehicles well before the survey (0 = no; 1 = yes), interviewees' subjective judgment on which is more important between jobs and the environment (0 = jobs; 1 = environment), a dummy for interviewees being in favor of the utilization of unused forest biomass (0 = no; 1 = yes), a dummy for interviewees being in favor of expanding floating solar power facilities (0 = no; 1 = yes), and a dummy for interviewees being in favor of expanding hydrogen vehicles (0 = no; 1 = yes), respectively.

3.5. How to Model the Data

Two important points should be taken into account in establishing a model in which acceptance is a dependent variable and the factors affecting this acceptance are independent variables. First, the observed dependent variable has a range of only 1 to 9—that is, the minimum is 1 and the maximum is 9. Therefore, the range is not the whole real number, but a natural number between 1 and 9. Second, the acceptance values are ordinal, not cardinal. Looking at the score, the measure of acceptance, it is clear that the larger the number, the greater the acceptance. However, this value is not cardinal. Performing classical regression without reflecting these two points can produce misleading analysis results.

To simplify the analysis, Ono and Tsunemi's [56] work transformed the collected data on a four-point scale into a two-point scale for pros and cons, and then applied a discrete logit model to the transformed data. However, adopting this approach results in the loss of important information from data collected on a nine-point scale. Therefore, it is necessary to apply a model that fully utilizes the collected data. The ordered probit model is useful in dealing with data on acceptance evaluated on the Likert scale as in this study [58,59]. The model defines a latent variable distributed over the whole real number instead of an observed variable on a nine-point scale and sets it as a dependent variable. In addition, the likelihood function reflects the relationship between the observed and latent variables.

The ordered probit model used in this research can be formulated as follows. For respondent i ($i = 1, \cdots, I$), the latent variable variable, A_i^*, and the observed variable, A_i, are:

$$\begin{cases} A_i^* = y_i'\beta + \omega_i \\ A_i = J \text{ if } \sigma_{J-2} < A_i^* \leq \sigma_{J-1} \text{ for } J = 1, \cdots, 9 \end{cases} \tag{1}$$

where y_i is a vector of a constant term and the variables given in Table 2, y_i' means the transpose matrix of y_i, β is a vector of the parameters matching y_i, ω_i is the disturbance term, and σ is a threshold value that is not known and should be estimated.

However, following the usual practice in the literature, $\sigma_{-1} = -\infty$, $\sigma_0 = 0$, and $\sigma_8 = \infty$ are assumed. Furthermore, the disturbance term is assumed to be distributed as normal with a standard deviation of one. Therefore, the probability that the observed variable has one value of 1 to 9 can be induced as:

$$\text{Prob}(y_i = J) = F\left(\frac{\sigma_{J-1} - y_i'\beta}{1}\right) - F\left(\frac{\sigma_{J-2} - y_i'\beta}{1}\right) \tag{2}$$

where $F(\cdot)$ indicates the standard normal cumulative distribution function. The finally derived likelihood function is:

$$L = \prod_{i=1}^{I} \prod_{J=1}^{9} D_t^s \text{Prob}(y_i = J) \tag{3}$$

where $D_t^s = 1(y_i = J)$ for $J = 1, 2, \cdots, 9$ where $1(\cdot)$ is a function that returns one if the argument is true and zero otherwise. The maximum likelihood estimates can be obtained by finding the parameter values that maximize Equation (3).

4. Results and Discussion

4.1. Data

The authors sought to focus on three points in conducting field surveys. First, scientific sampling should reflect the characteristics of the population. Second, experienced professional interviewers should obtain reliable responses from respondents. Third, the person-to-person interviews should be carried out maintaining the distancing rules in the pandemic situation caused by COVID-19. To meet these three points, a professional survey company took charge of the entire process of the survey. The survey was conducted for one month from mid-March to mid-April 2021. Judging from the comments of the interviewers belonging to the company, respondents responded to the survey without any difficulties. In particular, people were actively involved in the survey because of the controversy over coal-fired power generation due to issues such as particulate matter and CO_2 emissions.

The results from interviewees selecting one of the nine views are presented in Table 3. Four views—"absolutely disagree," "strongly disagree," "disagree," and "slightly disagree"—can be aggregated as "disagree with implementing the fuel transition;" while four views—"slightly agree," "agree," "strongly agree," and "absolutely agree"—can be aggregated as "agree with implementing the fuel transition". Of the 1000 respondents, 122 opposed the fuel transition and 736 supported the fuel transition, the latter (73.6%) being about six times the former (12.2%). Overall, therefore, the approval rate was higher than the disapproval rate. A total of 142 respondents said "neutral." It was interesting that 14.2 percent of people were neutral or indifferent to fuel transition.

Table 3. Summary of responses regarding acceptance of replacing coal-fired power plants with natural gas-fired ones.

Responses	Frequency	Percentage (%)
Absolutely disagree	2	0.2
Strongly disagree	16	1.6
Disagree	63	6.3
Slightly disagree	41	4.1
Neutral	142	14.2
Slightly agree	140	14.0
Agree	383	38.3
Strongly agree	153	15.3
Absolutely agree	60	6.0
Totals	1000	100.0

4.2. Estimation Results of the Model

The estimation results of the ordered probit model are given in Table 4. R^2, which indicates goodness of fit, was 0.165. In the analysis using cross-sectional data, the value of R^2 is usually low [60,61]. In particular, Gans [62] pointed out that it is a kind of norm for R^2 to be between 0.1 and 0.2 when analyzing data obtained from the survey. Thus, it can be seen that the value of 0.165 for R^2 obtained in this study is not particularly low. A likelihood ratio test can be applied for the specification test of the model. In this case, the null hypothesis is that all estimated coefficients except the constant term are zero—that is, the model is mis-specified. The computed likelihood ratio test statistic was 175.62, which corresponds to a p-value of 0.000. Thus the statistical significance of this model, determined at a significance level of 1 percent, is ascertained.

Table 4. Estimation results of the ordered probit model.

Variables [a]	Coefficient Estimates	t-Values
Constant	2.6779	8.30 *
Metro	0.4915	6.92 *
Heating	0.5862	1.99 *
Income	0.1262	1.86 [#]
Education	−0.1381	−1.71 [#]
Age	−0.0085	−2.08 *
Know1	0.1689	2.47 *
Know2	0.1306	1.81 [#]
Environment	0.2248	3.33 *
Forest	0.2658	3.53 *
Fsolar	0.2096	2.63 *
H2-car	0.1712	2.23 *
σ_1	0.7698	3.74 *
σ_2	1.4839	6.83 *
σ_3	1.7282	7.92 *
σ_4	2.3110	10.51 *
σ_5	2.7474	12.45 *
σ_6	3.9078	17.50 *
σ_7	4.7369	20.68 *
Sample size		1000
Log-likelihood		−1671.95
Peusdo-R^2		0.165
Log-likelihood ratio test statistic (p-value)		175.62 (0.000)

[a] The variables are described in Table 2. * and [#] indicate statistical significance at the 5% and 10% levels, respectively. σ's are parameters to be estimated in the model.

All of the σ values used to define the observed variable, A_i, were estimated to be positive, having statistical significance at the 1 percent level. Moreover, the estimation results of the coefficients corresponding to all the variables defined in Table 4 had statistical significance at the 10 percent level. Thus, the estimation results of the ordered probit model are significant. Interestingly, the model provides reasonably good performance. The application of the ordered probit model in this study was an appropriate strategy.

The estimated coefficients for the Metro, Heating, Income, Know1, Know2, Environment, Forest, Fsolar, and H2-car variables had a positive sign. The positive sign of each estimated coefficient implies that the greater the value of each variable, the greater the acceptance of the fuel transition. For example, those who lived in the Seoul Metropolitan area, those who used electricity for heating, those whose household income was high, those who knew about the energy transition policy before the survey, those who knew about hydrogen vehicles before the survey, those who considered the environment more important than jobs, those who were in favor of utilizing unused forest biomass, those who were in favor of expanding floating solar power facilities, and those who were in favor of expanding hydrogen vehicles were more prone to the fuel transition than others.

On the other hand, the sign of the estimated coefficients for the Education and Age variables was negative. A negative sign indicates that the greater the value of each variable, the lower the acceptance of the fuel transition. Respondents with more than twelve years' education and those who were older than 48 years were less receptive to the fuel transition than others.

4.3. Discussion of the Results

After data on acceptance of the fuel transition were collected on a nine-point scale, the determinants of acceptance were identified and analyzed. The results derived from this work can contribute to the literature, having various implications in terms of both research and policy. First of all, the usefulness of applying an ordered probit model to ordinal data collected on a nine-point scale was ascertained. The specification test confirmed the statistical significance of the model, and both the threshold values appearing in the model and the estimated coefficients for the eleven variables of interest were statistically significant. Thus, it is possible to derive various implications from the results.

The fact that the approval rate for the fuel transition (73.6%) exceeded the disapproval rate (12.2%) by about six times was a positive finding in promoting the fuel transition in South Korea. In order to achieve the goal of reducing CO_2 emissions declared to the international community, the country should carry out the fuel transition continuously and robustly while maintaining a stable power supply [13,14,28]. This finding can be recognized as an encouraging sign for the country. Without public support, the fuel transition cannot succeed. It is also worth noting that 14.2 percent of the respondents considered themselves neutral or indifferent.

Interestingly, three interviewee household characteristic variables had a significant impact on acceptance. First, those living in the Seoul Metropolitan area were more receptive to the fuel transition than those who were not. The Seoul Metropolitan area is home to about half of the population, an important area that determines public opinion in South Korea. Therefore, it is quite difficult to implement policies that residents in the area oppose. The finding that the acceptance of residents in the area of the fuel transition is secure is quite encouraging in promoting the fuel transition.

Second, those who use electricity for heating were more receptive than those who do not. The main fuel for residential heating in South Korea is city NG since electricity for heating is more expensive than city NG. Nevertheless, some people use electricity instead of city NG for heating because they prefer electricity to other fuels. This is because electricity is partially made from RE, and even if it is produced using fossil fuels, fossil fuel-fired power plants greatly reduce particulate matter through air pollutant reduction facilities, while city NG boilers are not equipped with these facilities. In other words, since people who use electricity for heating have a high interest in the environment, they seem to make more positive judgments on the fuel transition.

Third, the household income of the respondent was positively correlated with acceptance. The transition of power generation fuel from coal to NG will inevitably lead to higher electricity bills. From an individual household's point of view, income is limited, and an increase in electricity costs means spending on other goods should be reduced. In fact, the high-income group is more likely to accept an increase in electricity bills than others. In particular, low-income people may be opposed to an increase in electricity bills caused by the fuel transition rather than to the fuel transition itself. Therefore, even if the fuel transition is made, measures will need to be in place to alleviate the burden by continuously applying the electricity rate discount system for low-income people.

Moreover, two individual characteristics of interviewees had a significant impact on acceptance. The education level of the respondent had a negative impact on acceptance of the fuel transition, while older interviewees were less receptive to the fuel transition than younger interviewees. In fact, in South Korea, the higher the level of education and the older the person is, the more they tend to settle for the present situation. This is because fuel transition can not only increase electricity bills, but also reduce jobs and cause problems in the stability of electricity generation fuel supply. To help increase the acceptance of fuel transition among people with a high education level or older age, the fuel transition should be promoted in more persuasive and appealing ways.

Three issues concerning the fuel transition were addressed in the introduction: rising electricity bills due to increased power generation costs, reduced fuel supply stability due to increased NG use, and job losses [20–23]. Interviewees responded after hearing full explanations of the issues during the survey, and it was found that the approval rate for the fuel transition was six times higher than the opposition rate. However, if these three issues become a reality, the public acceptance of the fuel transition may drop significantly, which could place South Korea in a difficult situation. It may happen that coal-fired power generation facilities are reduced but NG-fired power facilities are not increased in time, which could give rise to a crisis in the supply and demand of electricity. Therefore, the three issues need to be discussed here in conjunction with qualitative implications obtained from respondents in the process of the survey.

First, since the unit cost of NG-fired power generation is higher than that of coal-fired power generation, the fuel transition will bring about an increase in electricity bills, which could negatively affect acceptance of the fuel transition. As of March 2021, South Korea's unit cost of NG-fired power generation was 99.72 KRW per kWh, about 10 percent higher than that of coal-fired power generation, which is 90.19 KRW per kWh. However, because NG prices are linked to oil prices, rising international oil prices could have a significant impact on NG prices, which could widen the gap. Usually, NG prices soar during periods of high oil prices and plunge or stabilize during periods of low oil prices. Recently, as oil prices have been rising, voices opposing the fuel transition have already begun to appear.

In addition, since NG will serve as a bridge in an era of energy transition, NG prices are likely to rise further in the future as demand for NG increases around the world. As coal-fired power generation decreases and NG-fired power generation continues to increase, the cost burden for electricity distribution companies is expected to increase. This could eventually lead to higher electricity bills, which could place a burden on consumers. Therefore, it is necessary to persuade the public by fully informing them that fuel transition is inevitable and that they must endure an increase in electricity bills to implement the fuel transition. Fuel transition will gain momentum only when there is a social consensus that electricity produced from NG-fired power plants is inevitably more expensive than that from coal-fired ones, just as organic products are more expensive than regular products.

Second, given the geopolitical situation of South Korea, NG is more vulnerable to supply instability than coal. Almost all of the NG consumed in the country is imported from abroad. In particular, the country relies heavily on the Middle East, including Qatar and Oman, as NG suppliers. Therefore, if political instability occurs in the Middle East, South Korea's NG-fired power plants may have to be shut down. In addition, liquefied NG produced in the Middle East is transported to South Korea, and the country's NG procurement costs are the highest in the world due to excessive liquefaction, transportation, and vaporization costs.

On the other hand, as coal is imported from all over the world without the need for liquefaction, it has higher supply stability than NG and its procurement cost is also low. Thus, it is quite important to secure a stable supply of NG in order for fuel transition to succeed. In Europe, NG is supplied stably mainly in the form of pipeline NG, but South Korea is introducing NG through liquefied NG carriers only. Currently, efforts are being made to import NG produced from other regions besides the Middle East. For example, South Korea is trying to increase NG imports from Australia, the United States, Indonesia, and Mozambique. These efforts must yield significant results. If South Korea fails to secure stability in NG supply, it will be difficult for the fuel transition to succeed and receive public support.

Third, if jobs are lost due to the fuel transition, this can lead to serious social conflict. Currently, a 500 MW coal-fired power plant in South Korea has a total of 168 to 200 employees in the operating and maintenance sectors, while the number of workers in an NG-fired power plant of the same capacity is between 90 and 110. In other words, the latter is about half the former. Consequently, converting all 24 (12.6 GW) coal-fired power plants to NG-fired power plants could result in the loss of between 2000 and 2400 jobs.

Since all 24 coal-fired power plant operators are public companies, not private ones, responding to the fuel transition by reducing working hours and increasing the retraining of staff may not actually reduce jobs; however, this approach will be neither long-term nor stable.

The decline in the number of jobs will not only increase unemployment, making workers' lives difficult, but also cause considerable damage to the local economies where power plants are located. The 24 coal-fired power plants are located on the shores of rural areas, not urban areas, because they are a form of "not in my backyard" facility and need to use seawater as cooling water. Since most of them are located in underdeveloped areas, they contribute greatly to the local economy through local taxes, donations, public utility expenditures, and procurement of resources within the region. However, because it is advantageous for NG-fired power plants to be located in urban areas or industrial complexes adjacent to the demand for electricity, they are likely to change their location during the transition. Consequently, the areas where coal-fired power plants used to be located could face economic difficulties due to the fuel transition.

Due to these problems, during a recent visit to Chungnam Province, where the largest number of coal-fired power plants are located, President Moon Jae-In stressed that "The energy transition, which replaces existing coal-fired power plants with NG-fired ones, should be done in a just way so that no-one loses a job and the local economy is not damaged." Therefore, the just and fair transition of fuel is an important issue that South Korea will face in the future. There should be in-depth consideration of this and immediate preparation of concrete measures to implement this in practice. The measures currently proposed include the following.

- Send workers on master's and doctorate courses in graduate schools for long-term re-training;
- Deploy minimum management personnel for coal-fired power plants that are not normally operated but are used as reserve power resources in case of an emergency in which demand increases;
- Relocate existing employees working at coal-fired power plants to new business sectors, such as RE plants and fuel cell generation plants, after sufficient re-education;
- Subsidize living expenses and education expenses for many years so that a worker can transfer to a completely different field.

5. Conclusions

In Introduction section, the authors looked at previous studies dealing with the transition of power generation fuel from coal to NG. An important implication was that this transition is inevitable to reduce CO_2 and air pollutant emissions, and is a strategy adopted by a number of countries. South Korea is no exception. The previous policy to expand coal-fired power generation has been changed to a reduction policy since May 2017. It was clearly pointed out that the use of NG will be expanded as an intermediate stage because it is difficult to immediately engage in a significant expansion of RE. It was also mentioned that the speed of fuel transition will be faster for the implementation of carbon neutrality by 2050. In the end, concerning coal-fired power generation, the current status and prospects of South Korea are consistent with those of other countries presented in several previous studies.

In Section 2, previous studies that analyzed energy-related public acceptance were examined. It was found that, to the best of the authors' knowledge, this study was the first to analyze the public acceptance of converting power generation fuel from coal to NG. In particular, it is an important discovery, differentiated from previous research, that the public acceptance of the fuel transition policy is secured to some extent and that many people support the transition even though there is an increase in cost. In addition, this study not only identified various factors affecting the public acceptance of the fuel transition, but also derived various implications by proposing and applying a framework for analyzing the impacts of the factors on the acceptance. The results of this study could be important information for South Korean policymakers. Although the main findings of this study are unique to South Korea, the structure of this study can be extended to other countries as much as possible. The main qualitative findings of this study are not so different from those of the research that has analyzed the public acceptance of RE expansion policy [63], large-scale offshore wind power generation construction [40], and energy transition policy [4] in South Korea.

South Korea's CO_2 emissions in 2018 totaled about 728 million tons. However, the 2030 CO_2 emissions target submitted to the United Nations Framework Convention on Climate Change is 536 million tons. Ultimately, South Korea needs to reduce its CO_2 emissions drastically, and the power generation sector should play an important role in this as it is very difficult to reduce CO_2 emissions in other sectors such as transportation, industry, building, and agriculture. To that end, South Korea has set out a plan to convert 24 coal-fired power plants into NG-fired power plants by 2034. This study sought to ascertain and analyze public acceptance of this through a survey of 1000 people nationwide. The results were statistically significant and reveal a number of useful implications.

The finding that the approval rate for the fuel transition was six times the opposition rate suggests that the government-led fuel transition should be carried out continuously. However, since respondents were concerned about three challenges that could arise in the process of fuel transition, the authors have attempted to discuss them above. These were higher electricity bills due to rising power generation costs, reduced fuel supply stability due to the increased use of NG, and job losses and a consequent negative impact on the local economy. Ultimately, dealing with these three challenges effectively will determine whether or not the fuel transition succeeds.

Of course, further challenges remain to be addressed. For example, is it socially desirable to tear down a coal-fired power plant that has only been in operation for 30 years? Given that NG-fired power plants are also fossil-fuel-utilizing facilities and there is strong opposition from residents to new construction of these, how will we facilitate fuel transition? It is clear that NG is the bridge energy to a complete RE society, but for how long will it play that role? In other words, if NG-fired power plants become stranded assets in a not-too-distant future in which carbon neutrality is realized, is it reasonable to invest in constructing NG-fired power plants now? Subsequent studies should be able to answer these questions.

Author Contributions: Conceptualization, H.-S.J. and J.-H.K.; methodology, S.-H.Y.; software, J.-H.K.; validation, H.-S.J.; formal analysis, J.-H.K.; data curation, H.-S.J.; writing—original draft preparation, H.-S.J.; writing—review and editing, S.-H.Y.; visualization, J.-H.K.; supervision, S.-H.Y.; project administration, J.-H.K.; funding acquisition, S.-H.Y. All authors have read and agreed to the published version of the manuscript.

Funding: This study was supported by the Research Program funded by the SeoulTech (Seoul National University of Science and Technology) (grant number: 2021-0787).

Institutional Review Board Statement: Not applicable.

Informed Consent Statement: Not applicable.

Data Availability Statement: The data presented in this study are available on request from the corresponding author. The data are not publicly available because they were purchased for a fee.

Conflicts of Interest: The authors declare no conflict of interest.

References

1. Prehodaa, E.W.; Pearce, J.M. Potential lives saved by replacing coal with solar photovoltaic electricity production in the US. *Renew. Sustain. Energy Rev.* **2017**, *80*, 710–715. [CrossRef]
2. Gagarin, H.; Sridhar, S.; Lange, I.; Bazilian, M.D. Considering non-power generation uses of coal in the United States. *Renew. Sustain. Energy Rev.* **2020**, *124*, 109790. [CrossRef]

3.   Kerimray, A.; Suleimenov, B.; De Miglio, R.; Rojas-Solórzano, L.; Torkmahalleh, M.A.; Gallachóir, B.P. Investigating the energy transition to a coal free residential sector in Kazakhstan using a regionally disaggregated energy systems model. *J. Clean. Prod.* **2018**, *196*, 1532–1548. [CrossRef]

4.   Kim, J.H.; Park, J.H.; Yoo, S.H. Public preference toward an energy transition policy: The case of South Korea. *Environ. Sci. Pollut. Res.* **2020**, *27*, 45965–45973. [CrossRef] [PubMed]

5.   Sharpton, T.; Lawrence, T.; Hall, M. Drivers and barriers to public acceptance of future energy sources and grid expansion in the United States. *Renew. Sustain. Energy Rev.* **2020**, *126*, 109826. [CrossRef]

6.   Haggerty, J.H.; Haggerty, M.N.; Roemer, K.; Rose, J. Planning for the local impacts of coal facility closure: Emerging strategies in the US West. *Resour. Policy* **2018**, *57*, 69–80. [CrossRef]

7.   Delborne, J.A.; Hasala, D.; Wigner, A.; Kinchy, A. Dueling metaphors, fueling futures: "Bridge fuel" visions of coal and natural gas in the United States. *Energy Res. Soc. Sci.* **2020**, *61*, 101350. [CrossRef]

8.   Vögele, S.; Kunz, P.; Rübbelke, D.; Stahlke, T. Transformation pathways of phasing out coal-fired power plants in Germany. *Energy Sustain. Soc.* **2018**, *8*, 1–18. [CrossRef]

9.   Oei, P.Y.; Hermann, H.; Herpich, P.; Holtemöller, O.; Lünenbürger, B.; Schult, C. Coal phase-out in Germany—Implications and policies for affected regions. *Energy* **2020**, *196*, 117004. [CrossRef]

10.   Xin-gang, Z.; Wen-bin, Z.; Hui, W.; Ying, Z.; Ying-zhuo, Z. The influence of carbon price on fuel conversion strategy of power generation enterprises—A perspective of Guangdong province. *J. Clean. Prod.* **2021**, *305*, 126749. [CrossRef]

11.   Cherp, A.; Vinichenko, V.; Jewell, J.; Suzuki, M.; Antal, M. Comparing electricity transitions: A historical analysis of nuclear, wind and solar power in Germany and Japan. *Energy Policy* **2017**, *101*, 612–628. [CrossRef]

12.   Maamoun, N.; Kennedy, R.; Jin, X.; Urpelainen, J. Identifying coal-fired power plants for early retirement. *Renew. Sustain. Energy Rev.* **2020**, *126*, 109833. [CrossRef]

13.   *The 9th Basic Plan for Electricity Demand and Supply (2020–2034)*; Korea Ministry of Trade, Industry and Energy: Sejong, Korea, 2020.

14.   Lim, S.Y.; Kim, H.J.; Yoo, S.H. South Korean household's willingness to pay for replacing coal with natural gas? a view from CO_2 emissions reduction. *Energies* **2017**, *10*, 2031. [CrossRef]

15.   Shin, H.; Kim, T.H.; Kim, H.; Lee, S.; Kim, W. Environmental shutdown of coal-fired generators for greenhouse gas reduction: A case study of South Korea. *Appl. Energy* **2019**, *252*, 113453. [CrossRef]

16.   *Global Renewables Outlook: Energy Transformation 2050*; International Renewable Energy Agency: Abu Dhabi, United Arab Emirates, 2021.

17.   Chen, H.; Chen, W. Potential impact of shifting coal to gas and electricity for building sectors in 28 major northern cities of China. *Appl. Energy* **2019**, *236*, 1049–1061. [CrossRef]

18.   Kim, G.S.; Kim, H.J.; Yoo, S.H. Optimal share of natural gas in the electric power generation of South Korea: A note. *Sustainability* **2019**, *11*, 3705. [CrossRef]

19.   Seo, S.J.; Kim, J.H.; Yoo, S.H. Public preference for increasing natural gas generation for reducing CO_2 emissions in South Korea. *Sustainability* **2020**, *12*, 2636. [CrossRef]

20.   Ellerman, A.D. The competition between coal and natural gas the importance of sunk costs. *Resour. Policy* **1996**, *22*, 33–42. [CrossRef]

21.   Jang, J.; Lee, J.; Yoo, S.H. The public's willingness to pay for securing a reliable natural gas supply in Korea. *Energy Policy* **2014**, *69*, 3–13. [CrossRef]

22.   Söderholm, P. Fuel flexibility in the West European power sector. *Resour. Policy* **2000**, *26*, 157–170. [CrossRef]

23.   Liang, J.; He, P.; Qiu, Y. Energy transition, public expressions, and local officials' incentives: Social media evidence from the coal-to-gas transition in China. *J. Clean. Prod.* **2021**, *298*, 126771. [CrossRef]

24.   Karasmanaki, E.; Ioannou, K.; Katsaounis, K.; Tsantopoulos, G. The attitude of the local community towards investments in lignite before transitioning to the post-lignite era: The case of Western Macedonia, Greece. *Resour. Policy* **2020**, *68*, 101781. [CrossRef]

25.   Xu, S.; Ge, J. Sustainable shifting from coal to gas in North China: An analysis of resident satisfaction. *Energy Policy* **2020**, *138*, 111296. [CrossRef]

26.   *World Energy Outlook 2020*; International Energy Agency: Paris, France, 2020.

27.   De Gouw, J.A.; Parrish, D.D.; Frost, G.J.; Trainer, M. Reduced emissions of CO_2, NOx, and SO_2 from US power plants owing to switch from coal to natural gas with combined cycle technology. *Earth's Future* **2014**, *2*, 75–82. [CrossRef]

28.   Kim, H.J.; Kim, J.H.; Yoo, S.H. Do people place more value on natural gas than coal for power generation to abate particulate matter emissions? Evidence from South Korea. *Sustainability* **2018**, *10*, 1740. [CrossRef]

29.   Di Lucia, L.; Ericsson, K. Low-carbon district heating in Sweden—Examining a successful energy transition. *Energy Res. Soc. Sci.* **2014**, *4*, 10–20. [CrossRef]

30.   Li, R.; Su, M. The role of natural gas and renewable energy in curbing carbon emission: Case study of the United States. *Sustainability* **2017**, *9*, 600. [CrossRef]

31.   Kim, H.J.; Yu, J.J.; Yoo, S.H. Does combined heat and power play the role of a bridge in energy transition? Evidence from a cross-country analysis. *Sustainability* **2019**, *11*, 1035. [CrossRef]

32.   Dong, K.; Sun, R.; Li, H.; Liao, H. Impact of natural gas consumption on CO_2 emissions: Panel data evidence from China's provinces. *J. Clean. Prod.* **2017**, *162*, 400–410. [CrossRef]

33. Dong, K.; Sun, R.; Li, H.; Liao, H. Does natural gas consumption mitigate CO_2 emissions: Testing the environmental Kuznets curve hypothesis for 14 Asia-Pacific countries. *Renew. Sustain. Energy Rev.* **2018**, *94*, 419–429. [CrossRef]
34. Assefa, G.; Frostell, B. Social sustainability and social acceptance in technology assessment: A case study of energy technologies. *Technol. Soc.* **2007**, *29*, 63–78. [CrossRef]
35. Heinz, B.; Erdmann, G. Dynamic effects on the acceptance of hydrogen technologies-an international comparison. *Int. J. Hydrog. Energy* **2008**, *33*, 3004–3008. [CrossRef]
36. O'Garra, T.; Mourato, S.; Pearson, P. Investigating attitudes to hydrogen refuelling facilities and the social cost to local residents. *Energy Policy* **2008**, *36*, 2074–2085. [CrossRef]
37. Huijts, N.M.A.; Molin, E.J.E.; van Wee, B. Hydrogen fuel station acceptance: A structural equation model based on the technology acceptance framework. *J. Environ. Psychol.* **2014**, *38*, 153–166. [CrossRef]
38. Sovacool, B.K.; Ratan, P.L. Conceptualizing the acceptance of wind and solar electricity. *Renew. Sustain. Energy Rev.* **2012**, *16*, 5268–5279. [CrossRef]
39. Hall, N.; Ashworth, P.; Devine-Wright, P. Societal acceptance of wind farms: Analysis of four common themes across Australian case studies. *Energy Policy* **2013**, *58*, 200–208. [CrossRef]
40. Kim, J.H.; Nam, J.; Yoo, S.H. Public acceptance of a large-scale offshore wind power project in South Korea. *Mar. Policy* **2020**, *120*, 104141. [CrossRef]
41. Velasco-Herrejon, P.; Bauwens, T. Energy justice from the bottom up: A capability approach to community acceptance of wind energy in Mexico. *Energy Res. Soc. Sci.* **2020**, *70*, 101711. [CrossRef]
42. Dowd, A.M.; Boughen, N.; Ashworth, P.; Carr-Cornish, S. Geothermal technology in Australia: Investigating social acceptance. *Energy Policy* **2011**, *39*, 6301–6307. [CrossRef]
43. Tabi, A.; Wuestenhagen, R. Keep it local and fish-friendly: Social acceptance of hydropower projects in Switzerland. *Renew. Sustain. Energy Rev.* **2017**, *68*, 763–773. [CrossRef]
44. Batel, S.; Devine-Wright, P.; Tangeland, T. Social acceptance of low carbon energy and associated infrastructures: A critical discussion. *Energy Policy* **2013**, *58*, 1–5. [CrossRef]
45. Devine-Wright, P.; Batel, S.; Aas, O.; Sovacool, B.; LaBelle, M.C.; Ruud, A.A. A conceptual framework for understanding the social acceptance of energy infrastructure: Insights from energy storage. *Energy Policy* **2017**, *107*, 27–31. [CrossRef]
46. Mistur, E.M. Health and energy preferences: Rethinking the social acceptance of energy systems in the United States. *Energy Res. Soc. Sci.* **2017**, *34*, 184–190. [CrossRef]
47. Fischer, B.; Gutsche, G.; Wetzel, H. Who wants to get involved? Determining citizen willingness to participate in German renewable energy cooperatives. *Energy Res. Soc. Sci.* **2021**, *76*, 102013. [CrossRef]
48. Stadelmann-Steffen, I.; Dermont, C. Acceptance through inclusion? Political and economic participation and the acceptance of local renewable energy projects in Switzerland. *Energy Res. Soc. Sci.* **2021**, *71*, 101818. [CrossRef]
49. Venkatesh, V.; Morris, M.G.; Davis, G.B.; Davis, F.D. User acceptance of information technology: Toward a unified view. *MIS Q* **2003**, 425–478. [CrossRef]
50. *The 6th Basic Plan for Electricity Demand and Supply (2013–2027)*; Korea Ministry of Trade, Industry and Energy: Gwacheon, Korea, 2013.
51. *The 7th Basic Plan for Electricity Demand and Supply (2015–2029)*; Korea Ministry of Trade, Industry and Energy: Gwacheon, Korea, 2015.
52. *The 8th Basic Plan for Electricity Demand and Supply (2017–2031)*; Korea Ministry of Trade, Industry and Energy: Sejong, Korea, 2017.
53. Matlaba, V.J.; Mota, J.A.; Maneschy, M.C.; dos Santos, J.F. Social perception at the onset of a mining development in Eastern Amazonia, Brazil. *Resour. Policy* **2017**, *54*, 157–166. [CrossRef]
54. Arrow, K.; Solow, R.; Portney, P.R.; Leamer, E.E.; Radner, R.; Schuman, H. Report of the NOAA panel on contingent valuation. *Fed. Regist.* **1993**, *58*, 4601–4614.
55. *Guidelines for Applying Contingent Valuation Method to Pre-Evaluation of Feasibility*; Korea Development Institute: Sejong, Korea, 2012.
56. Ono, K.; Tsunemi, K. Identification of public acceptance factors with risk perception scales on hydrogen fueling stations in Japan. *Int. J. Hydrog. Energy* **2017**, *42*, 10697–10707. [CrossRef]
57. Saaty, T.L. *The Analytic Hierarchy Process*; McGraw-Hill: New York, NY, USA, 1980.
58. Daykin, A.R.; Moffatt, P.G. Analyzing ordered responses: A review of the ordered probit model. *Understand. Stat. Issues Psychol. Educ. Soc. Sci.* **2002**, *1*, 157–166. [CrossRef]
59. Greene, W.H. *Econometric Analysis*, 7th ed.; Pearson Education Limited: London, UK, 2012.
60. Anarkooli, A.J.; Hosseinpour, M.; Kardar, A. Investigation of factors affecting the injury severity of single-vehicle rollover crashes: A random-effects generalized ordered probit model. *Accid. Anal. Prev.* **2017**, *106*, 399–410. [CrossRef]
61. Storchmann, K. English weather and Rhine wine quality: An ordered probit model. *J. Wine Res.* **2005**, *16*, 105–120. [CrossRef]
62. Gans, H.J. The famine in mass media research. *Am. J. Sociol.* **1972**, *77*, 697–705. [CrossRef]
63. Kim, J.H.; Kim, S.Y.; Yoo, S.H. Public acceptance of the "Renewable Energy 3020 Plan": Evidence from a contingent valuation study in South Korea. *Sustainability* **2020**, *12*, 3151. [CrossRef]

 *sustainability*

Article

Measurement of Fairness Perceptions in Energy Transition Research: A Factorial Survey Approach

Ulf Liebe [1,*] and Geesche M. Dobers [2]

[1] Department of Sociology, University of Warwick, Coventry CV4 7AL, UK
[2] Institute of Landscape Architecture and Environmental Planning, Chair of Environmental and Land
     Economics, Technische Universität, 10623 Berlin, Germany; g.dobers@tu-berlin.de
*   Correspondence: ulf.liebe@warwick.ac.uk

Received: 24 August 2020; Accepted: 26 September 2020; Published: 30 September 2020

Abstract: Justice and fairness are increasingly popular concepts in energy research and comprise several justice dimensions, including distributive and procedural justice, related to energy production and consumption. In this paper, we used factorial survey experiments—a method employed in sociological justice research—for energy transition research. In a factorial survey, respondents evaluated one or more situations described by several attributes, which varied in their levels. The experimental setup of factorial surveys is one of its advantages over simple survey items, as based on this, the relative importance of each attribute for justice evaluations can be determined. We employed the method in a study on the perceived fairness of renewable energy expansion projects related to wind energy, solar energy, and biomass in Germany, and considered aspects of procedural and distributive justice. We show that the effects of these justice dimensions can be separated and the heterogeneity in justice evaluations can be explained. Compared to previous studies applying factorial survey experiments to explain the acceptance of renewable energy projects, we employed the method to directly measure justice concerns and asked respondents to evaluate the vignettes in terms of perceived fairness. This is important because acceptance and fairness as well as inequality and injustice are different phenomena.

Keywords: causal effects; justice; factorial surveys; renewable energy; vignette study

1. Introduction

While much research on energy production and consumption is concerned with the concept of justice [1–4], there is little empirical quantitative research that directly measures citizens' justice concerns and fairness perceptions. For example, in research on the acceptance of energy infrastructure, most researchers frame their work in the context of justice but empirically measure acceptance [5–8]. A direct measurement of justice perceptions is important because social inequalities related to energy production and consumption do not necessarily imply injustice: inequality and perceived injustice regarding the exposure to environmental harms and goods are two different phenomena. Inequalities in exposure to renewable energy projects, for example, an unequal distribution of power plants across geographical areas or social strata, might be accepted by citizens because they perceive such unequal distributions as unavoidable. At the same time, support or opposition do not automatically imply that renewable energy projects and policies are perceived as fair or unfair, respectively. Although support and fairness perceptions can be gathered under the same umbrella term of social acceptance [9], they refer to distinct concepts [10]. Therefore, the direct measurement of fairness perceptions is an important aspect of empirical justice research in sociology and other social sciences [11].

Research on environmental justice differentiates between distributive justice (distribution of environmental harms/goods in society), procedural justice (participation of citizens in environmental decision-making), and recognition (attention to group differences in society) [3,12,13]. With regard to justice concerns, the environmental justice movement typically strives for an equal distribution of environmental harms and goods across social groups in society. This means that all groups in society are equally affected, for example, by renewable energy production. On the other hand, it is well known that there are many different justice theories and principles, and the question that emerges is which principle is supported by whom and how this depends on the social context [14–17]. For example, not all socioeconomic groups might perceive an equal exposure to renewable power plants or equal burden of rising energy costs as equally fair. Also, citizens in different countries might evaluate an equal share of the costs of climate change mitigation across countries differently. The same can be true for aspects of procedural justice, that is, citizens' participation opportunities.

While the literature on energy production and consumption suggests that many aspects are relevant for fairness judgements [1–3,11], including distributive and procedural justice, it is empirically challenging to disentangle the effects of these aspects. For example, using standard survey items it is difficult to clarify whether distributive justice is more relevant than procedural justice, or vice versa, for the perceived fairness and local acceptance of renewable–energy projects.

In a factorial survey experiment (FSE), also called a vignette experiment, respondents evaluate a situation (i.e., vignette) which is described by experimentally manipulated attributes (i.e., actors) which vary in their levels [18]. The respondents are then asked to evaluate these situations according to criteria such as support, agreement, or perceived fairness. Given that typically more than one attribute is manipulated, FSEs belong to multifactorial methods, which allow for the identification of causal effects due to the experimental setup [18,19]. The method was introduced in Sociology by Rossi and Lazarsfeld in the 1950s [20] and, since the 1970s, has become an important tool for the study of many phenomena, including social norms and justice concerns [18,19,21–23]. The FSE employs multiple factors and respondents have to make trade-offs, and therefore it lowers socially desirable response behavior [24]. FSEs are similar to stated choice experiments, which are often employed in energy research [6,25,26]. In stated choice experiments, respondents compare alternatives that vary in multiple attributes and choose the alternative they prefer most. This method has advantages for measuring citizens' preferences and estimating welfare measures, for example, citizens' willingness to pay for renewable energy expansion, but it is less suitable for measuring attitudes, (normative) beliefs, and (fairness) perceptions. In social science research the latter are instead examined using FSEs, where respondents can express their fairness concerns on an ordinal or rating scale [18,23].

To our knowledge, there are two previous studies applying FSEs in the context of renewable energy expansion, more specifically on the social acceptance of wind energy projects [27,28]. Yet, in these applications the explanandum is acceptance and not fairness. We go beyond these previous applications of FSEs and directly measure fairness perceptions related to renewable energy projects and compare this with an acceptance measure of such projects. We uncover the causal effects of different justice dimensions, taking the heterogeneity of justice concerns into account. Moreover, we consider three renewable energy sources—wind energy, solar energy, and biomass—and compare the importance of justice dimensions and fairness perceptions across the different energy sources.

2. Factorial Survey Experimental Design and Data

2.1. Experimental Design

In designing and conducting an FSE (see [18] for state-of-the art guidelines), researchers have to decide on the number of attributes (factors or characteristics) of a situation, and attribute levels have to be assigned. In our example on renewable energy projects, we described projects to construct a renewable energy site in respondents' vicinity (10-km radius from their place of residence) and were interested in how unfair or fair the respondents perceive these projects to be. We varied four attributes across vignettes. First, the project referred with (1) a wind farm (10 turbines), (2) a photovoltaic power station, or (3) a biogas plant to different *types of renewable energy* and, second, with (1) one, (2) three, or (3) five power plants to different *magnitudes of exposure to power plants*. Third, based on the literature on environmental justice, we included the attributes *procedural justice*, that is, citizens have (1) no say in the planning process, (2) partial say in the planning process, or (3) a say at every step in the planning process, and fourth, *distributive justice*—with the planned project respondents have (1) fewer power plants, (2) the same number, or (3) more power plants in their region than in other regions in Germany.

Combining all possible attribute combinations—3 × 3 × 3 × 3—gave a the so-called full factorial of 81 vignettes and hence 81 different project descriptions. We employed the full factorial and each respondent answered one vignette which was randomly chosen from the full factorial. Using randomization and the full factorial, we were able to experimentally isolate all main effects, two-way effects, and three-way effects between attributes. If a factorial survey study comprises more attributes or attribute levels, the full factorial is often too large to consider all vignettes. Thus, an experimental design is used to reduce the number of vignettes that respondents face, but at the same time, to maintain the possibility of separating the effects of single factors. Researchers also have to choose a response scale for recording respondents' judgments (e.g., four-point, five-point, seven-point, or eleven-point response scales). While the literature suggests longer response scales [18], in this study we opted for a four-point scale because we wanted to fully label each category of the scale using the words "fair" and "unfair". Figure 1 provides an example of a vignette as used in the study.

How fair or unfair do you find the construction of <u>three</u> <u>wind farms (with ten turbines)</u> in your surroundings (radius of 10 km around your place of residence)? You, as a resident, <u>have a say at every step</u> in the planning process of the wind farms (choice of location, design, etc.). With the construction of these wind farms, your region will have <u>more turbines than in most other regions</u> in Germany.

If you live in a larger city, please think of the nearest surrounding.

I find the construction of these <u>wind farms</u> …

Very fair	Rather fair	Rather unfair	Very unfair
1	2	3	4

Figure 1. Example of a vignette used in the factorial survey. Note: Attributes and attribute levels that varied across vignettes are underlined.

2.2. Data and Variables

We embedded the FSE in an online survey on renewable energy expansion in Germany. The survey was conducted in September and October 2013 [29]. Participants were members of an access panel who were actively recruited by phone (no opt-in panel) and represented the German population that uses the internet at least once a week. We used quota sampling representing the German population regarding gender and age as close as possible. After inspection of the data, out of 3400 completed questionnaires, 3199 usable interviews remained for analyzing the factorial survey (due to missing values and implausible answers). The response rate (standard RR1, [30]) was 26%. Prior to the survey, six focus groups and two pretest surveys were conducted.

In our sample, women (45% in the sample, 51% in the population) and those living in mid-sized cities (33% in the sample, 42% in the population) were underrepresented and those with higher education, i.e., a university entrance diploma or higher, overrepresented (61% in the sample, 31% in the population). The mean values for age (43 years, SD = 14) and household net income (3048 Euro, SD = 1.519) were fairly close to the average values for the German population [31]. While the sample was clearly not representative, it contained sufficient variance on sociodemographics in order to take heterogeneity in population characteristics into account. Individuals in rural areas are more affected by renewable energy expansion compared to those in urban areas, and our data also show considerable variance along the rural-urban continuum (31% rural areas, 33% mid-sized cities, 36% large cities).

The survey also included questions on place attachment, which we considered in the regression models on heterogeneity of fairness evaluations. The corresponding variable was an additive index of answers to the following four survey items, all answered on a four-point response scale (1 = strongly disagree to 4 = strongly agree): "I like to be in the landscape next to my place of residence", "Often, I spend my free time in the landscape next to my place of residence", "The landscape around my place of residence is a part of me", "It is very important to me that the landscape around my place of residence does not change". Cronbach's alpha for the index was 0.7714; the index ranged between 4 and 16 with a mean of 13.085 and standard deviation of 2.233.

In the survey we considered three renewable energy sources: wind energy, solar energy, and biomass. At the beginning of the survey, respondents were shown pictograms and definitions of these renewables (see Table 1). It was also clarified that the survey focused on renewables in the open landscape and did not consider energy production in urban areas, for example, through solar panels on roofs. In contrast to wind and solar energy the energy source is not unboundedly available in the case of biomass. Therefore, we asked respondents to consider the cultivation of raw material and the power plant when rating the renewable energy biomass. For the most part, biomass is used for electricity generation at the place of production.

The survey also included a question regarding the general acceptance of the construction of renewable power plants in respondents' vicinity. The exact wording of this acceptance question was as follows: "How strongly would you support or oppose the construction of the following renewable power plants [solar energy, wind energy and biomass] within a 10 km radius of your place of residence?" Respondents answered this question on a four-point response scale (strongly oppose, somewhat oppose, somewhat support, strongly support).

Table 1. Definition of renewable energy sources as used in the survey.

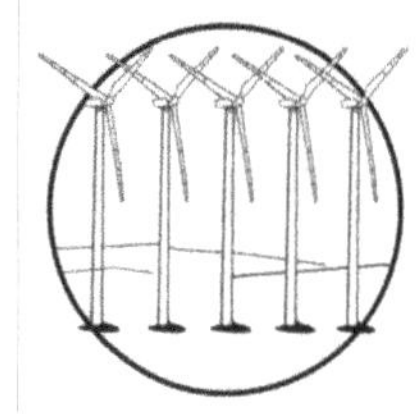 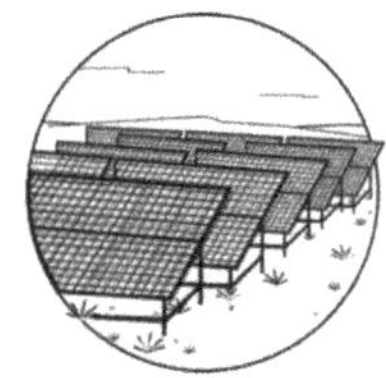 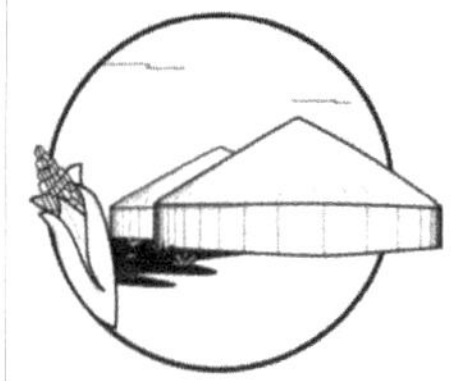

Wind Energy refers to electricity generation with single wind turbines and wind farms onshore only.	**Solar Energy** refers exclusively to the production of electricity with photovoltaic systems in the open landscape, i.e., solar fields.	**Biomass** refers to the production of biogas and its electricity and includes both the biogas plant and the cultivation of the required biomass (such as corn).

3. Results

3.1. Overall Fairness Evaluation and Acceptance Figures

Table 2 shows the fairness evaluations regarding the construction of new power plants in respondents' vicinity across all vignettes and per renewable energy type. The figures indicate that there was remarkable variance on the fairness scale. However, for each energy type the majority of respondents perceived the construction of an additional plant as rather fair or very fair. The corresponding figures were 81% for solar energy, 67% for wind energy, and 56% for biomass. We can compare these figures with those from the question on the general acceptance of the construction of additional power plants in citizens' vicinity. Both fairness perception and acceptance were measured on four-point scales. While there was a substantial positive correlation between the fairness and acceptance measure (all significant at $p < 0.001$), both were not perfectly correlated (Pearson correlations of $r = 0.529$ for wind energy, $r = 0.350$ for solar energy, and $r = 0.514$ for biomass). In other words: these measures discriminated to some extent, even if they correlated with each other. On the other hand, it needs to be kept in mind that the fairness question referred to "concrete" project descriptions presented in the vignettes, while the acceptance question referred to the construction of power plants in general; yet both questions were related to a 10 km radius of the respondents' place of residence. This could explain that, in Table 2, the mean values of the perceived fairness of the construction of "concrete" wind energy and solar energy plants are lower than the corresponding mean values of general acceptance. However, for biomass we found the opposite pattern—that is, mean fairness values for concrete projects were higher than mean general acceptance values. This can be interpreted as another indication that fairness perceptions and agreement are conceptually different. Further, the correlations between fairness and acceptance in the present study are similar to the ones presented in a vignette study on airport expansion scenarios, which included both measures at the vignette level [10], supporting our claim that fairness and acceptance are not (entirely) the same.

Table 2. Fairness evaluations and acceptance levels per type of renewable energy plant.

Plant Type	Very Unfair (Strongly Oppose) (1)	Rather Unfair (Somewhat Oppose) (2)	Rather Fair (Somewhat Support) (3)	Very Fair (Strongly Support) (4)	Mean (SD)
Wind ($n = 1051$)	7% (8%)	26% (19%)	54% (47%)	13% (26%)	2.73 (0.78) (2.91 (0.88))
Solar ($n = 1075$)	3% (2%)	16% (9%)	60% (50%)	21% (39%)	2.97 (0.71) (3.26 (0.70))
Biomass ($n = 1073$)	13% (15%)	31% (34%)	48% (41%)	8% (10%)	2.51 (0.81) (2.46 (0.87))

Note: First number in each cell refers to responses to the vignette/fairness question and the second number in parentheses to the acceptance question.

3.2. Effects of Vignette Attributes on Fairness Evaluations

In the following, we present plots for linear regression models on fairness evaluations per renewable energy type: first for models only including the vignette attributes (Figure 2) and second for models including the vignette attributes and additional variables to explain heterogeneity in fairness evaluations (Figure 3). The full regression models underlying Figures 2 and 3 can be found in Table A1 in the Appendix A. Further, Table A2 in the Appendix A contains for each renewable energy type a comparison of the results of a linear regression model, an ordered logit model and a binary logit model. In the latter, the dependent variable has value of 1 for the categories "very fair" and "rather fair", and 0 for the categories "rather unfair" and very unfair" on the four-point fairness scale. Since the results are similar across the different modeling variants, we present the results of linear regression models.

The results in Figure 2 (also Table A1) show that the number of renewable power plants does not have a significant effect on fairness evaluations regarding wind power and solar energy. There was only one negative and statistically significant effect for biomass indicating that the construction of five plants compared to one plant is associated with lower fairness perceptions. There are clear indications that procedural and distributive justice matter. With respect to all the renewable energies, having no say in the planning process was perceived as more unfair than having a partial say in the planning process. The corresponding effects were statistically significant and amounted to 0.3 points on the four-point fairness scale. Yet, there was no statistically significant difference for having a say in all steps of the planning process compared to having a partial say in the planning process. It seems that respondents valued the general possibility of participating in the planning process and not so much the extent of it. Regarding the distributive justice, respondents perceived more unfairness if the new power plants lead to overall more renewable power plants in their region as compared to other regions. The effects had a similar size to the ones for procedural justice and were all highly statistically significant. Only for solar energy did respondents perceive more unfairness also if they had fewer power plants in their region as compared to other regions. For wind energy and biomass, we found no statistically significant differences between less exposure and equal exposure to power plants across regions.

We also checked interaction effects between vignette attributes. Taking all possible two-way and three-way interaction effects into account, we only found one statistically significant two-way interaction and three-way interaction in the model for wind energy. They showed that the construction of five plants was evaluated as less unfair if respondents still had fewer renewable energy plants in their region compared to other regions. Yet, this interaction was evaluated to be less fair if residents had a say in the planning process compared to having a partial say.

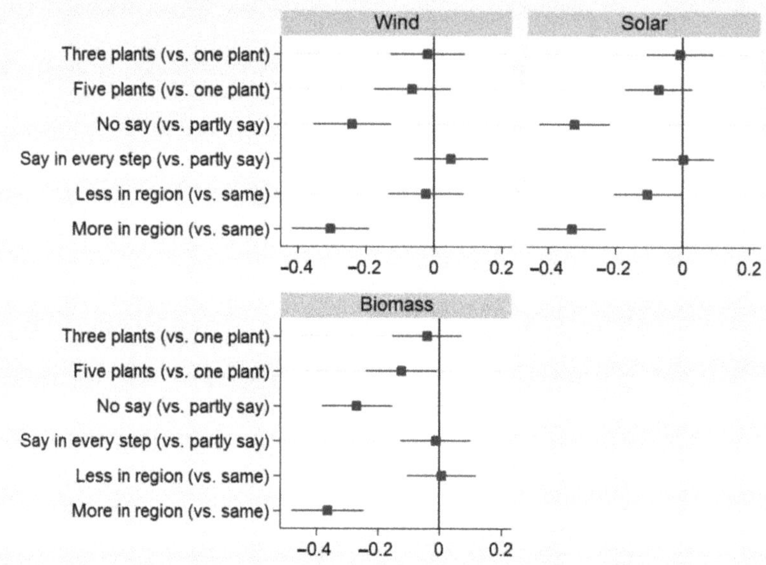

Figure 2. Regression models for fairness evaluations and vignette attributes.

Note: unstandardized coefficients and 95% confidence intervals of linear regression models with the four-point fairness scale as dependent variable and the vignette attributes as independent variables. The model characteristics are as follows: for wind energy, $F(6, 1044) = 10.85$, Prob > F = 0.000, $R^2 = 0.0596$, $n = 1051$; for solar energy, $F(6, 1068) = 15.12$, Prob > F = 0.000, $R^2 = 0.0844$, $n = 1075$; for biomass, $F(6, 1066) = 13.99$, Prob > F = 0.000, $R^2 = 0.0746$, $n = 1073$.

3.3. Effects of Respondent Characteristics on Fairness Evaluations

Figure 3 presents models that include additional variables to explain heterogeneity in fairness evaluations; the figure only depicts variables that had statistically significant effects on fairness evaluations at the 5% level (full models are presented in Table A2 in the Appendix A). The main insights are that, as already shown above, the general acceptance of new renewable power plants in respondents' vicinity did have a positive effect on the perceived fairness; yet, causation can go in both directions, i.e., acceptance can affect fairness and vice versa. The effect sizes for a unit change ranged between 0.36 (solar energy) and 0.46 (wind energy and biomass) on the four-point fairness scale. Higher education was significantly associated with higher levels of perceived fairness at the 5% level in the models on wind and solar energy.

Rural areas are more affected by renewable energy expansion than urban areas. However, we did not find remarkable differences in fairness evaluations between respondents living in medium-sized or large cities and those living in villages. Yet, there was one exception: compared to those living in villages, respondents residing in large cities perceived the construction of biomass power plants as rather fair. The effect amounted to 0.15 points on the four-point fairness scale. Place attachment did not significantly affect fairness concerns regarding solar and biomass but it had a negative and statistically significant effect on the perceived fairness of the construction of new wind energy plants. Of note, a 10-point increase on the place-attachment scale, with a minimum value of 4 and a maximum value of 16, is associated with a 0.25 decrease on the four-point fairness scale. This effect for wind energy might be due to the higher visibility of wind farms as compared with solar and biomass plants.

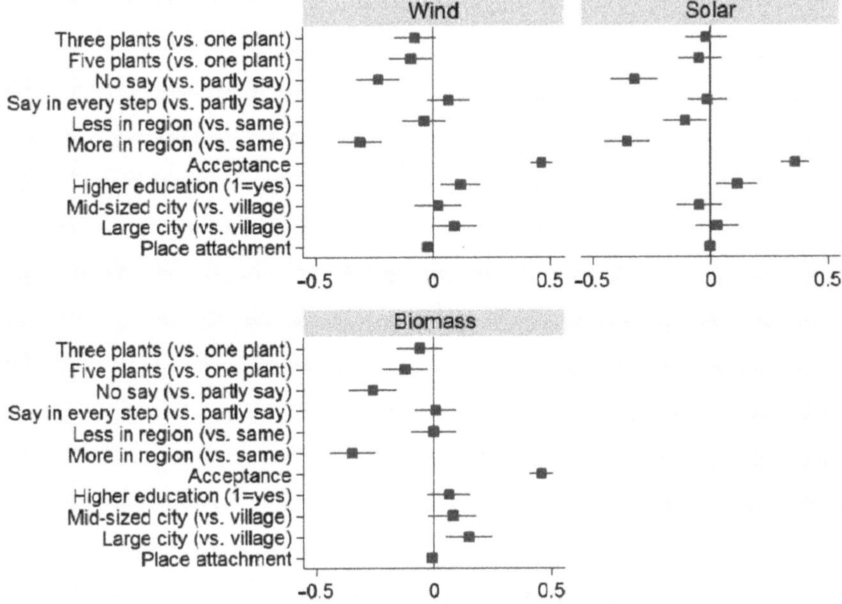

Figure 3. Regression models for fairness evaluations, vignette attributes, and respondents' characteristics.

Note: unstandardized coefficients and 95% confidence intervals of linear regression models with the four-point fairness scale as dependent variable, and the vignette attributes and respondents' characteristics as independent variables. Not all respondent characteristics are shown; the underlying models also included gender, age, income, but these characteristics had statistically insignificant effects in all three models depicted. The model characteristic are as follows: for wind energy, $F(14, 1036) = 44.12$, Prob $> F = 0.000$, $R^2 = 0.3617$, $n = 1051$; for solar energy, $F(14, 1060) = 18.18$, Prob $> F = 0.000$, $R^2 = 0.2226$, $n = 1075$; for biomass, $F(14, 1058) = 44.63$, Prob $> F = 0.000$, $R^2 = 0.3504$, $n = 1073$.

4. Discussion and Conclusion

4.1. Heterogeneity of Justice Concerns

Justice is a multi-dimensional concept and it is challenging to disentangle the importance of each of the dimensions for justice/fairness evaluations. In this paper, we focused on distributive and procedural justice related to renewable energy expansion. Both dimensions are commonly discussed in the environmental justice and energy-related literature [2,4,12,32,33]. We demonstrated how using factorial surveys can contribute to research on energy production. By directly measuring justice/fairness perceptions and varying justice-related attributes across vignettes, we examined and disentangled the relevance of different justice dimensions for energy-related projects. Our study showed, for example, that the number of renewable energy plants is less important than aspects of procedural and distributive justice. The latter justice dimensions are equally important, which is in contrast to previous FSE research on the local acceptance of wind energy plants [27], indicating that participatory justice might be more important than distributive justice. Yet, this research measured acceptance and not fairness and also included more vignette attributes. It is not clear whether the relative importance of justice dimensions depends on the outcome measure (fairness versus acceptance) and/or the information provided about renewable energy projects. For example, it could be that distributive justice related to the number of power plants across regions becomes less relevant if further information about a project is given, such as who is investing in the project and how benefits are allocated.

Our application of FSEs revealed heterogeneity regarding justice concerns. First, it is noteworthy that in terms of fairness, respondents evaluated having more power plants in their region than in other regions differently than having fewer power plants than in other regions. If outcome equality holds, respondents should also have disvalued a disproportionately lower exposure to renewable energy power plants. This was clearly not the case and only for solar energy did we find a significant effect that lower exposure levels were perceived as rather unfair compared to equal exposure levels. However, compared with equal exposure across regions, the effect for lower exposure levels was weaker than the one for higher exposure levels. The fact that there was an effect for solar energy might be associated with its large general support as compared with wind energy and especially biomass, as well as with our finding that place attachment is not a significant determinant of fairness perceptions related to photovoltaic power stations (compared to wind turbines). Such perceptions of distributive justice are likely to affect not only the acceptance of renewable energy projects at the local level but also the spatial allocation of power plants at the country level, where depending on the underlying justice principle efficient allocations can vary remarkably [34].

Second, we found a heterogeneity in justice concerns affected by education, place of residence, and place attachment as well as the type of renewable energy production. For all energy sources we found a positive effect of education on fairness perceptions related to the construction of new power plants. Education is positively related to knowledge about environmental issues, which in turn can positively affect environmental attitudes and pro-environmental behavior [35], as well as fairness concerns related to renewable energy expansion. We found that placement attachment was important for fairness perceptions related to the construction of wind turbines but not for photovoltaic power stations and biogas plants; this could be explained by the higher visibility of wind turbines as well as the fact that, at the time of the survey, respondents were more likely to be actually exposed to wind turbines compared with photovoltaic power stations and biogas plants. The place-attachment effect could be considered in decision-making processes and explicitly taken into account by addressing corresponding concerns when discussing with citizens, and in the framing of wind energy projects. As the place-attachment effect was specific for wind power it illustrates that the relevance of determinants of fairness concerns can differ across energy sources.

While our survey was carried out over five years ago and meanwhile renewable energy expansion has progressed in Germany, many of our findings are in line with more recent studies on the acceptance of renewable energy expansion in Germany. This includes the citizens' overall higher support of solar energy, followed by wind power and biomass [36], as well as the finding that the place of residence does not have strong effects regarding wind turbine acceptance [27]. Yet, in our study citizens living in large cities evaluated biogas plants more positively than those in rural areas.

4.2. The Merits of FSEs

Turning to the merits of FSEs as a methodological tool in energy research, FSEs have several advantages over standard survey items to measure justice concerns. Based on Liebig et al. [19], Table 3 provides an overview of common problems in quantitative research on energy production and consumption and refers to advantages of using FSEs to solve these problems. A standard survey item does not consider context information and this might prompt specific answers. In FSEs, respondents receive more context information, for example by combining different attributes of wind power plants and hence prompting, such as overstating the importance of one attribute (e.g., distributive justice), should be less likely. Using standard survey items, it is difficult to determine the relative importance of justice dimensions such as participatory and distributive justice related to wind power plants. The experimental design underlying FSEs makes it possible to single out the relative importance of each dimension. Responses to standard survey items might lead to biased response behavior. For example, renewable energy expansion might be perceived as socially desirable and hence respondents might tend to agree with survey items in favor of renewable energy expansion. This cannot be completely ruled out in FSEs but should be less likely because respondents have to consider and make trade-offs between several attributes. Further, in research on justice concerns related to energy production, researchers explicitly or implicitly assume causal effects of justice dimensions on outcomes related to energy production and consumption. Yet, causal effects cannot be studied based on standard survey items and cross-sectional data. They can be examined, however, in factorial surveys and other population-based survey experiments [37].

Table 3. Advantages of factorial survey experiments (FSEs) in research on energy production and consumption.

Problems of Empirical Research	Advantages of Using FSEs
Single-item measures lack context-information on different energy-related attributes and might prompt certain answers, such as overstating the importance of an attribute.	FSEs consider several energy-related attributes, e.g., regarding renewable energy power plants and hence include more context information. Respondents have to make trade-offs. This should make prompting less likely.
Uncovering the relative importance of factors relevant for justice evaluations regarding energy-related issues	Based on a multifactorial design and trade-offs between justice attributes/factor, the effect/importance of each factor for justice evaluations can be determined.
Justice as a normative concept might be prone to socially desirable response behavior, e.g., overstating support for renewable energy production	By presenting several factors at the same time, socially desirable responses are less likely. Respondents need to make trade-offs between attributes.
Causal effects, e.g., regarding renewable energy power plant attributes on fairness evaluations, cannot be identified.	By randomly varying vignette attributes causal effects can be estimated.

Note: This table is based on [19].

We believe that FSEs can complement the researcher's toolbox in energy research as a useful tool to measure justice beliefs, fairness perceptions, attitudes, and normative beliefs. In this regard they have clear advantages over single survey items or (multifactorial) stated choice experiments, which can be employed to measure preferences and to obtain welfare measures [38]. FSEs should be combined with qualitative methods such as focus groups to develop valid vignette scenarios and attributes and to obtain an impression on how respondents handle the vignettes in order to check their suitability. As any other method, FSEs are not free of methodological issues, such as the complexity of vignettes, the role of the response format (e.g., closed-ended versus open-ended question format), and order effects and fatigue, if multiple vignettes are presented per respondent [39,40]. These need to be considered when planning an FSE.

4.3. Desiderata for Future Research

In this paper, we presented a rather simple application of FSEs. As already mentioned above, in another study on the local acceptance of wind power projects in Germany and Poland, Liebe et al. [27] also included attributes on the type of investor, the use of electricity (in the region versus for export), and the tax revenue resulting from the power plant. This means more justice dimensions and context factors can and possibly should be considered in FSEs. It would be important in future research to apply such more comprehensive FSEs in a multi-country context or multi-regional context within countries to systematically explore how cultural differences, social and economic contexts affect justice evaluations. For example, it could be examined whether higher economic inequality at the country or regional level leads to differences in the relevance of distributive justice related to renewable energy projects. Also, it can be studied how the evaluation of distributive and participatory justice changes if more information is given about the renewable energy project at hand. Complementing other empirical approaches, such as case studies, FSE research can help decision-makers to better predict which contexts might result in higher or lower levels of perceived fairness of energy transition initiatives.

Previous applications of FSEs [27] measured justice/fairness concerns indirectly; they used an acceptance scale to measure respondents' evaluation of renewable energy projects. In future research it should be considered that, even if correlated, acceptance is not the same as justice, and more generally, that environmental inequality does not equal environmental injustice. Therefore, it is important to further explore differences between acceptance and justice measurement instruments, as well as under which conditions perceived unfairness results in non-acceptance of energy transition projects. In other words: similar to inequality, unfairness does not necessarily mean non-acceptance/opposition and subsequently protesting against renewable energy projects. There is a need for a better understanding of conditions and mechanisms that link justice and social acceptance, including fairness perceptions, support, and protest behavior, both at the country and regional level. Again, insights from basic research in this regard can be helpful in shaping energy transition projects with higher (local) acceptance levels.

Finally, besides energy production, FSEs can also be applied to justice concerns regarding energy poverty, involuntary resettlement, fossil fuel pollution, nuclear waste, climate change, etc. The method is also applicable in the global south [41]. The present paper paves the way for employing FSEs for direct measurement of justice concerns and for disentangling the importance of different justice dimensions and thereby complementing existing research on energy production and consumption and subsequently informing (political) decision making.

Author Contributions: Conceptualization, U.L. and G.M.D.; methodology, U.L. and G.M.D.; investigation, U.L. and G.M.D.; writing—original draft preparation, U.L. and G.M.D.; funding acquisition, U.L. All authors have read and agreed to the published version of the manuscript.

Funding: "This research was funded by the German Federal Ministry for Education and Research, grant number 01LA1110C (Research Project "Efficient and fair allocation of renewable energy production at the national level" (EnergyEFFAIR)).

Acknowledgments: We thank our colleagues from the EnergyEFFAIR project for kind collaboration.

Conflicts of Interest: The authors declare no conflict of interest. The funders had no role in the design of the study; in the collection, analyses, or interpretation of data; in the writing of the manuscript, or in the decision to publish the results.

Appendix A

Table A1. Full linear regressions models underlying Figures 2 and 3 in the main text.

	Figure 2			Figure 3		
	Wind	**Solar**	**Biomass**	**Wind**	**Solar**	**Biomass**
Three plants (vs. one plants)	−0.0177	−0.0127	−0.0457	−0.0746	−0.0233	−0.0601
	(−0.32)	(−0.25)	(−0.80)	(−1.67)	(−0.51)	(−1.23)
Five plants (vs. one plant)	−0.0660	−0.0753	−0.123 *	−0.0915	−0.0540	−0.117 *
	(−1.15)	(−1.48)	(−2.08)	(−1.88)	(−1.13)	(−2.37)
No say (vs. partial say)	−0.242 ***	−0.319 ***	−0.276 ***	−0.239 ***	−0.320 ***	−0.264 ***
	(−4.14)	(−6.00)	(−4.74)	(−4.99)	(−6.46)	(−5.21)
Say in every step (vs. partial say)	0.0526	0.000746	−0.0122	0.0636	−0.0184	0.00744
	(0.95)	(0.02)	(−0.21)	(1.41)	(−0.43)	(0.16)
Less in region (vs. same)	−0.0290	−0.105 *	0.0101	−0.0422	−0.113 *	−0.00116
	(−0.51)	(−2.06)	(0.18)	(−0.88)	(−2.36)	(−0.02)
More in region (vs. same)	−0.307 ***	−0.334	−0.360 ***	−0.316 ***	−0.360 ***	−0.341 ***
	(−5.35)	(−6.50)	(−6.09)	(−6.71)	(−7.69)	(−6.99)
Acceptance of plant in vicinity				0.459 ***	0.359 ***	0.457 ***
				(19.16)	(11.40)	(17.96)
Woman (vs. man)				−0.0356	−0.00455	−0.00282
				(−0.89)	(−0.12)	(−0.07)
Age in years				−0.000910	−0.00266	−0.00359 *
				(−0.60)	(−1.83)	(−2.32)
Higher education (vs. less education)				0.142 ***	0.130 **	0.0823
				(3.39)	(3.06)	(1.86)
Net income in Euro				0.00000384	0.0000278	0.00000583
				(0.18)	(1.34)	(0.26)
Medium-sized city (vs. small city)				0.0179	−0.0583	0.0812
				(0.35)	(−1.19)	(1.53)
Large city (vs. small city)				0.0823	0.0203	0.151 **
				(1.66)	(0.43)	(2.92)
Place attachment				−0.0251 **	−0.00297	−0.00954
				(−2.75)	(−0.31)	(−0.90)
Constant	2.931 ***	3.256 ***	2.779 ***	1.879 ***	2.132 ***	1.777 ***
	(49.33)	(58.16)	(43.85)	(11.03)	(11.92)	(9.76)
R^2	0.060	0.084	0.075	0.362	0.223	0.350
n	1051	1075	1073	1051	1075	1073

Note: t statistics in parentheses; * $p < 0.05$, ** $p < 0.01$, *** $p < 0.001$.

Table A2. Comparison of linear regression, ordered logit, and binary logit models.

	Wind	Wind	Wind	Solar	Solar	Solar	Biomass	Biomass	Biomass
	Linear	Ordered	Binary	Linear	Ordered	Binary	Linear	Ordered	Binary
Three plants (vs. one plants)	−0.0177	−0.0594	−0.149	−0.0127	−0.0137	−0.0646	−0.0457	−0.173	−0.332 *
	(−0.32)	(−0.42)	(−0.89)	(−0.25)	(−0.09)	(−0.32)	(−0.80)	(−1.24)	(−2.11)
Five plants (vs. one plant)	−0.0660	−0.169	−0.293	−0.0753	−0.213	−0.292	−0.123 *	−0.311 *	−0.321 *
	(−1.15)	(−1.16)	(−1.75)	(−1.48)	(−1.44)	(−1.50)	(−2.08)	(−2.18)	(−2.05)
No say (vs. partial say)	−0.242 ***	−0.632 ***	−0.764 ***	−0.319 ***	−0.928 ***	−1.135 ***	−0.276 ***	−0.659 ***	−0.667 ***
	(−4.14)	(−4.31)	(−4.58)	(−6.00)	(−5.90)	(−5.77)	(−4.74)	(−4.75)	(−4.30)
Say in every step (vs. partial say)	0.0526	0.0978	−0.102	0.000746	−0.0464	−0.114	−0.0122	−0.00331	0.0499
	(0.95)	(0.68)	(−0.59)	(0.02)	(−0.33)	(−0.52)	(−0.21)	(−0.02)	(0.32)
Less in region (vs. same)	−0.0290	−0.102	−0.155	−0.105 *	−0.380 *	−0.238	0.0101	0.0279	0.0149
	(−0.51)	(−0.69)	(−0.89)	(−2.06)	(−2.48)	(−1.13)	(0.18)	(0.20)	(0.09)
More in region (vs. same)	−0.307 ***	−0.849 ***	−0.965 ***	−0.334 ***	−1.046 ***	−0.924 ***	−0.360 ***	−0.873 ***	−0.943 ***
	(−5.35)	(−5.70)	(−5.76)	(−6.50)	(−6.77)	(−4.66)	(−6.09)	(−6.11)	(−6.00)
Constant	2.931 ***		1.562 ***	3.256 ***		2.454 ***	2.779 ***		0.979 ***
	(49.33)		(7.96)	(58.16)		(9.66)	(43.85)		(5.68)
Cut point 1		−3.233 ***			−4.387 ***			−2.709 ***	
		(−15.71)			(−17.72)			(−14.75)	
Cut point 2		−1.318 ***			−2.385 ***			−0.908 ***	
		(−8.17)			(−12.82)			(−5.63)	
Cut point 3		1.414 ***			0.590 ***			1.919 ***	
		(8.91)			(3.58)			(10.91)	
R^2/Pseudo R^2	0.060	0.0289	0.0513	0.084	0.0427	0.0671	0.075	0.0339	0.0566
n	1051	1051	1051	1075	1075	1075	1073	1073	1073

Note: t statistics in parentheses; * $p < 0.05$, *** $p < 0.001$.

References

1. Sovacool, B.K.; Heffron, R.J.; McCauley, D.; Goldthau, A. Energy decisions reframed as justice and ethical concerns. *Nat. Energy* **2016**, *1*, 16024. [CrossRef]
2. Wolsink, M. Wind Power wind power: Basic Challenge Concerning Social Acceptance wind power social acceptance. In *RENEWABLE Energy Systems: 3 Volumes*; Kaltschmitt, M., Ed.; Springer: New York, NY, USA, 2013; ISBN 978-1-4614-5819-7.
3. Jenkins, K.; McCauley, D.; Heffron, R.; Stephan, H.; Rehner, R. Energy justice: A conceptual review. *Energy Res. Soc. Sci.* **2016**, *11*, 174–182. [CrossRef]
4. Perlaviciute, G.; Steg, L. Contextual and psychological factors shaping evaluations and acceptability of energy alternatives: Integrated review and research agenda. *Renew. Sustain. Energy Rev.* **2014**, *35*, 361–381. [CrossRef]
5. Langer, K.; Decker, T.; Menrad, K. Public participation in wind energy projects located in Germany: Which form of participation is the key to acceptance? *Renew. Energy* **2017**, *112*, 63–73. [CrossRef]
6. Lienhoop, N. Acceptance of wind energy and the role of financial and procedural participation: An investigation with focus groups and choice experiments. *Energy Policy* **2018**, *118*, 97–105. [CrossRef]
7. Walker, C.; Baxter, J. "It's easy to throw rocks at a corporation": Wind energy development and distributive justice in Canada. *J. Environ. Policy Plan.* **2017**, *19*, 754–768. [CrossRef]
8. Walker, B.J.A.; Russel, D.; Kurz, T. Community Benefits or Community Bribes? An Experimental Analysis of Strategies for Managing Community Perceptions of Bribery Surrounding the Siting of Renewable Energy Projects. *Environ. Behav.* **2016**, *49*, 59–83. [CrossRef]
9. Wüstenhagen, R.; Wolsink, M.; Bürer, M.J. Social acceptance of renewable energy innovation: An introduction to the concept. *Energy Policy* **2007**, *35*, 2683–2691. [CrossRef]
10. Liebe, U.; Preisendörfer, P.; Bruderer Enzler, H. The social acceptance of airport expansion scenarios: A factorial survey experiment. *Transp. Res. Part D Transp. Environ.* **2020**, *84*, 102363. [CrossRef]
11. Baxter, J. Energy justice: Participation promotes acceptance. *Nat. Energy* **2017**, *2*, 17128. [CrossRef]
12. Schlosberg, D. *Defining Environmental Justice. Theories, Movements, and Nature*; Oxford University Press: Oxford, UK, 2007; ISBN 9780199286294.
13. Fuller, S.; McCauley, D. Framing energy justice: Perspectives from activism and advocacy. *Energy Res. Soc. Sci.* **2016**, *11*, 1–8. [CrossRef]
14. Liu, F. *Environmental Justice Analysis. Theories, Methods, and Practice*; CRC Press: Boca Raton, FL, USA, 2001; ISBN 1566704030.
15. Miller, D. Distributive Justice: What the People Think. *Ethics* **1992**, *102*, 555–593. [CrossRef]
16. Miller, D. *Principles of Social Justice*; Harvard Univ. Press: Cambridge, MA, USA, 1999; ISBN 0-674-70628-5.
17. Sandel, M.J. *Justice. What's the Right Thing to Do?* Penguin Books: London, UK, 2010; ISBN 9780141041339.
18. Auspurg, K.; Hinz, T. *Factorial Survey Experiments*; Sage: Los Angeles, CA, USA, 2015; ISBN 978-1-4522-7418-8.
19. Liebig, S.; Sauer, C.; Friedhoff, S. Using Factorial Surveys to Study Justice Perceptions: Five Methodological Problems of Attitudinal Justice Research. *Soc. Just. Res.* **2015**, *28*, 415–434. [CrossRef]
20. Rossi, P.H. Vignette analysis: Uncovering the normative structure of complex judgements. In *Qualitative and Quantitative Social Research: Papers in Honor of Paul F. Lazarsfeld*; Merton, R.K., Coleman, J.S., Rossi, P.H., Eds.; Free Press: New York, NY, USA, 1979; ISBN 0029209307.
21. Jasso, G.; Opp, K.-D. Probing the character of norms: A factorial survey analysis of the norms of political action. *Am. Sociol. Rev.* **1997**, *62*, 947–964. [CrossRef]
22. Jasso, G.; Rossi, P.H. Distributive Justice and Earned Income. *Am. Sociol. Rev.* **1977**, *42*, 639. [CrossRef]
23. Wallander, L. 25 years of factorial surveys in sociology: A review. *Soc. Sci. Res.* **2009**, *38*, 505–520. [CrossRef]
24. Auspurg, K.; Hinz, T.; Sauer, C.; Liebig, S. The Factorial Survey as Method for Measuring Sensitive Issues. In *Improving Survey Methods: Lessons from Recent Research*; Engel, U., Jann, B., Lynn, P., Scherpenzeel, A.C., Sturgis, P., Eds.; Routledge: New York, NY, USA, 2015; ISBN 9780415836258.
25. Dimitropoulos, A.; Kontoleon, A. Assessing the determinants of local acceptability of wind-farm investment: A choice experiment in the Greek Aegean Islands. *Energy Policy* **2009**, *37*, 1842–1854. [CrossRef]
26. García, J.H.; Cherry, T.L.; Kallbekken, S.; Torvanger, A. Willingness to accept local wind energy development: Does the compensation mechanism matter? *Energy Policy* **2016**, *99*, 165–173. [CrossRef]

27. Liebe, U.; Bartczak, A.; Meyerhoff, J. A turbine is not only a turbine: The role of social context and fairness characteristics for the local acceptance of wind power. *Energy Policy* **2017**, *107*, 300–308. [CrossRef]

28. Walter, G.; Gutscher, H. Generelle Befürwortung von Windkraftanlagen vor Ort vs. Befürwortung spezifischer Windkraftprojekte: Der Einfluss von Projekt- und Verfahrensparametern. *Umweltpsychologie* **2013**, *17*, 124–144.

29. Liebe, U.; Dobers, G.M. Decomposing public support for energy policy: What drives acceptance of and intentions to protest against renewable energy expansion in Germany? *Energy Res. Soc. Sci.* **2019**, *47*, 247–260. [CrossRef]

30. American Association for Public Opinion Research (AAPOR). Standard Definitions: Final Dispositions of Case Codes and Outcome Rates for Surveys. 2016. Available online: https://www.aapor.org/AAPOR_Main/media/publications/Standard-Definitions20169theditionfinal.pdf (accessed on 28 March 2018).

31. Federal Statistical Office of Germany. *Statistisches Jahrbuch Deutschland und Internationales, Red.-Schluss: 01. August 2015*; Federal Statistical Office of Germany: Wiesbaden, Germany, 2015; ISBN 978-3-8246-1037-2.

32. Wolsink, M. Planning of renewables schemes: Deliberative and fair decision-making on landscape issues instead of reproachful accusations of non-cooperation. *Energy Policy* **2007**, *35*, 2692–2704. [CrossRef]

33. Wolsink, M. Wind power implementation: The nature of public attitudes: Equity and fairness instead of 'backyard motives'. *Renew. Sustain. Energy Rev.* **2007**, *11*, 1188–1207. [CrossRef]

34. Drechsler, M.; Egerer, J.; Lange, M.; Masurowski, F.; Meyerhoff, J.; Oehlmann, M. Efficient and equitable spatial allocation of renewable power plants at the country scale. *Nat. Energy* **2017**, *2*, 289. [CrossRef]

35. Kollmuss, A.; Agyeman, J. Mind the Gap: Why do people act environmentally and what are the barriers to pro-environmental behavior? *Environ. Educ. Res.* **2002**, *8*, 239–260. [CrossRef]

36. Agentur für Erneuerbare Energien (AEE). Klares Bekenntnis der Deutschen Bevölkerung zu Erneuerbaren Energien. Available online: https://www.unendlich-viel-energie.de/themen/akzeptanz-erneuerbarer/akzeptanz-umfrage/klares-bekenntnis-der-deutschen-bevoelkerung-zu-erneuerbaren-energien (accessed on 21 September 2020).

37. Mutz, D.C. *Population-Based Survey Experiments*; Princeton Univ. Press: Princeton, NJ, USA, 2011; ISBN 9780691144528.

38. Louviere, J.J.; Hensher, D.A.; Swait, J.; Adamowicz, W.L. *Stated Choice Methods. Analysis and Applications*; Cambridge University Press: Cambridge, UK, 2010; ISBN 9780521782753.

39. Sauer, C.G.; Auspurg, K.; Hinz, T.; Liebig, S. The application of factorial surveys in general population samples: The effects of respondent age and education on response times and response consistency. *Surv. Res. Methods* **2011**, *5*, 89–102.

40. Auspurg, K.; Jäckle, A. First Equals Most Important? Order Effects in Vignette-Based Measurement. *Sociol. Methods Res.* **2015**, *46*, 490–539. [CrossRef]

41. Liebe, U.; Moumouni, I.M.; Bigler, C.; Ingabire, C.; Bieri, S. Using Factorial Survey Experiments to Measure Attitudes, Social Norms, and Fairness Concerns in Developing Countries. *Sociol. Methods Res.* **2017**, *43*, 1–32. [CrossRef]

MDPI
St. Alban-Anlage 66
4052 Basel
Switzerland
Tel. +41 61 683 77 34
Fax +41 61 302 89 18
www.mdpi.com

Sustainability Editorial Office
E-mail: sustainability@mdpi.com
www.mdpi.com/journal/sustainability